INTELLIGENT COMPONENTS
FOR VEHICLES
(ICV'98)

A Proceedings volume from the IFAC Workshop,
Seville, Spain, 23 - 24 March 1998

Edited by

A. OLLERO
Dpto. Ingeniería de Sistemas y Automática, Escuela Superior de Ingenieros,
Universidad de Sevilla, Sevilla, Spain

Published for the

INTERNATIONAL FEDERATION OF AUTOMATIC CONTROL

by

PERGAMON
An Imprint of Elsevier Science

UK	Elsevier Science Ltd, The Boulevard, Langford Lane, Kidlington, Oxford, OX5 1GB, UK
USA	Elsevier Science Inc., 660 White Plains Road, Tarrytown, New York 10591-5153, USA
JAPAN	Elsevier Science Japan, Tsunashima Building Annex, 3-20-12 Yushima, Bunkyo-ku, Tokyo 113, Japan

First edition 1998

Library of Congress Cataloging in Publication Data

A catalogue record for this book is available from the Library of Congress

British Library Cataloguing in Publication Data

A catalogue record for this book is available from the British Library

ISBN 0-08-043232 8

Transferred to digital print 2008

Printed and bound in Great Britain by CPI Antony Rowe, Chippenham and Eastbourne

IFAC WORKSHOP ON INTELLIGENT COMPONENTS FOR VEHICLES

Sponsored by
International Federation of Automatic Control (IFAC)
IFAC Technical Committee on Components and Instruments

Co-sponsored by
IFAC Technical Committees:
- Automotive Control
- Intelligent Autonomous Vehicles
- Aerospace
- Marine Systems

Organized by
Departamento Ingeniería de Sistemas y Automática
Escuela Superior de Ingenieros, Universidad de Sevilla

Supported by
CICYT Plan Nacional I+D
Junta de Andalucía
Universidad de Sevilla

PREFACE

This volume contains the papers presented at the IFAC Workshop on Intelligent Components for Vehicles (ICV'98) which was held in Sevilla (Spain), March 23-24, 1998. This event follows the Workshop on Intelligent Components for Autonomous and Semiautonomous Vehicles (ICASAV'95) held in Toulouse (France, October 1995). The main sponsor of both Workshops has been the IFAC Committee on Components and Instruments.

The main objective of ICV'98 was to bring together specialists on components and instruments for automotive systems, mobile robots and vehicles in general to enhance the value of their experience in both hardware and software intelligent components.

Future vehicles will deal more and more with autonomous functions to improve safety, traffic management and to reduce consumption and pollution. Numerous on-board decision systems will replace the driver in critical running phases. The problems and solutions experienced, by adopting this new technology, will bring out many common points with other transportation systems and, of course, with mobile robots. These common points have been discussed in the Workshop.

On the other hand, Research and Developments on Mobile Robotics have produced many components for perception, control and planning that can be used in vehicles for collision detection and avoidance, position estimation, guidance and maneuvering aids for drivers, advanced teleoperation, and other applications.

The topics of the Workshop are in an emerging field in which the research is being converted into industrial products very fast. Thus, the Workshop was very oriented to practical aspects. Several applications in the automotive domain, marine vehicles, agricultural and others were included in the program.

This volume has 71 papers organized in the following 18 Sections that correspond to Sessions of the ICV'98 Program: Path Tracking and Lateral Controllers, Computer Vision, Engine Control, Intelligent Components for Autonomous Navigation, Automotive Intelligent Components, Position Estimation, Obstacle Avoidance, Outdoor Navigation, Wheelchairs for the Handicapped, Mobile Manipulators, Marine Systems, Ultrasonic Sensors, Agriculture and Forestry, Perception, Suspensions, Path Planning, Control Techniques and Software Components.

In addition to the presentation of the papers, the ICV also included a Plenary talk and a round table about intelligent components for future vehicles with the participation of several industrial companies.

I would like to thank the members of the International Program Committee for their effort in the reviewing of papers and the members of the National Organizing Committee for their work and invaluable support in the organization of the Workshop and preparation of the Preprints.

We hope that the publication of these papers, which come from specialists of 17 different countries, will make a valuable contribution to the development of this important field.

March 1998

Aníbal Ollero
Editor

CONTENTS

INTELLIGENT COMPONENTS FOR AUTONOMOUS NAVIGATION

AUTOMOTIVE INTELLIGENT COMPONENTS

SAVE European Program:

POSITION ESTIMATION

OBSTACLE AVOIDANCE

OUTDOOR NAVIGATION

WHEELCHAIRS FOR HANDICAPPED

MOBILE MANIPULATORS

MARINE SYSTEMS

ULTRASONIC SENSORS

AGRICULTURE AND FORESTRY

PERCEPTION

SUSPENSIONS

PATH PLANNING

CONTROL TECHNIQUES

SOFTWARE COMPONENTS

LATERAL CONTROL OF A FOUR-WHEEL STEERED VEHICLE

D. de Bruin and P.P.J. van den Bosch

Measurement and control group, Department of Electrical Engineering
Eindhoven University of echnology
P.O. Box 513 MB Eindhoven, The Netherlands
e-mail: d.de.bruin@ele.tue.nl, Fax: +31402434582, Tel: +31402473795

Abstract: To get insight into the problems connected with the lateral control of vehicles with more than one steerable axis, we designed a lateral controller for a four-wheel steered vehicle. It is shown that under ideal circumstances, the lateral deviation of two points on the vehicle's longitudinal axis can be kept zero during cornering with a feedforward compensator only. Since, in reality, circumstances are not ideal, a feedback compensator has been added. Simulations of the total system indicate that with four-wheel steered vehicles a good performance can be achieved. *Copyright © 1998 IFAC*

Key Words: Automotive control, Control system design, Guidance systems, Mobile robots, Position control, Vehicles, Vehicle dynamics

1. INTRODUCTION

We participate in a study which has as final goal the design of a double articulated bus with four independently-controllable steering axes and six independently-controllable wheel torques. The total length of the bus is about 25 m and it has a capacity of 160 persons. The bus is equipped with four independently-controllable steering axes to give the bus a tram-like behaviour, so that the bus lane width can be kept small. The requirements are that the bus may not deviate more than 10 cm from the center of the lane during driving at velocities up to 80 km/h and that the bus has to stop within a lateral distance of less than 4 cm from the bus-stop, so that passengers can easily get on and off the bus (Bosch and Hedrikx, 1997).
For human drivers, it is impossible to steer such a vehicle satisfactorily. Therefore the bus has to be equipped with a control system.

Controlling the 3 carriage system is quite complicated. Therefore a one carriage vehicle with two steerable axes was studied first.

The aim of this study is to investigate whether it is possible to control satisfactorily the lateral deviation of two points on the longitudinal axis of the vehicle, so that as less space as possible is occupied during cornering and under the influence of wind gusts. It is assumed that the position of the sensors for measuring the lateral deviations coincide with the two points along the longitudinal axis that are controlled. This condition can be relaxed. Based on two measured positions the position of the control points can be calculated.

In this paper it will be shown that in the ideal case the lateral deviation of both these points can be kept zero during cornering with feedforward of the path curvature when four-wheel steering is applied. Furthermore, in theory the side-slip angle can be kept zero during cornering. For front-wheel steering the design of a feedforward compensator has already been illustrated (Sienel and Ackermann, 1994). However, then it is impossible to keep the side-slip angle zero.
The designed feedforward compensator is not applicable in practice since some assumptions have been made that can not be fulfilled in practice, like finite

actuator bandwidth and unknown plant parameters. To overcome this problem a feedback compensator has been added. It will be shown that with this extra feedback compensator good performance can still be achieved under more realistic conditions.

The feedback compensator has also to be used for suppressing lateral deviations due to wind gusts.

2. MODELING THE VEHICLE

The model to describe the behavior of the four-wheel steered vehicle is based on the model of Riekert and Schunk (1940). In this model both the rear wheels and both the front wheels are lumped into two wheels, that are fixed at the centerline of the vehicle at the rear and at the front respectively. The model of Riekert and Schunk is modified to bring four-wheel steering into account. Figure 1 shows a schematic representation of the model.

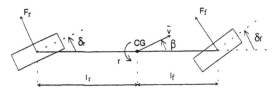

Figure 1 Single track model for four-wheel steering

In this figure:
CG= the center of gravity
δ_f = the front-wheel steering angle [rad]
δ_r = the rear-wheel steering angle [rad]
β = side slip angle at the CG [rad]
r = vehicle yaw rate [rad/s]
F_f = lateral force generated by the front tire [N]
F_r = lateral force generated by the rear tire [N]
l_f = distance from the CG to front axis [m]
l_r = distance from the CG to rear axis [m]
l = vehicle wheel base [m]
$\bar{v}$ = velocity of the vehicle at the CG [m/s]

For automatic tracking the model has to be extended, as described by Ackermann et. al. (1993). The model of Ackermann has been modified to make automatic tracking of two points at the centerline of the vehicle, instead of one point, possible. Also a modification has been made to bring disturbance forces into account.

Figure 2 shows the modified model for automatic tracking. Here, y_{CG} is the lateral deviation of the center of mass with respect to the reference line. y_f and y_r are the lateral deviations of the sensor at the front (senf) and the sensor at the rear (senr) respectively. Further φ is the vehicle yaw angle with respect to the earth-fixed coordinate frame (x_0,y_0), φ_t is the angle between x_0 and the tangent to the path and $\Delta\varphi=\varphi-\varphi_t$. (x_v,y_v) is a vehicle-fixed coordinate frame. Finally f_d is a disturbance force due to wind acting on the CG.

The lateral velocity at centre of gravity reads $y_{CG}=v\sin(\beta+\Delta\varphi)\approx v(\beta+\Delta\varphi)$, where v is the longitudinal velocity.

The lateral velocity of *senf* equals $\dot{y}_f=\dot{y}_{CG}+l_s(r-v\cdot\rho_{ref})$, where ρ_{ref} is the path curvature at CG. In the same way, the lateral velocity at *senr* equals $\dot{y}_r=\dot{y}_{CG}-l_s(r-v\cdot\rho_{ref})$.

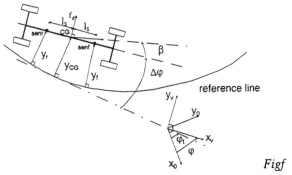

Figure 2 Model for track following

The state-space description of the vehicle then becomes:

$$\begin{bmatrix} \dot{\beta} \\ \dot{r} \\ \dot{y}_{CG} \\ \dot{\Delta\varphi} \end{bmatrix} = \begin{bmatrix} a_{11} & a_{12} & 0 & 0 \\ a_{21} & a_{22} & 0 & 0 \\ v & 0 & 0 & v \\ 0 & 1 & 0 & 0 \end{bmatrix} \begin{bmatrix} \beta \\ r \\ y_{CG} \\ \Delta\varphi \end{bmatrix} + \begin{bmatrix} b_{11} & b_{12} & 0 & \frac{1}{Mv} \\ b_{21} & b_{22} & 0 & 0 \\ 0 & 0 & 0 & 0 \\ 0 & 0 & -v & 0 \end{bmatrix} \begin{bmatrix} \delta_f \\ \delta_r \\ \rho_{ref} \\ f_d \end{bmatrix}$$

$$\begin{bmatrix} \beta \\ r \\ y_f \\ y_r \end{bmatrix} = \begin{bmatrix} 1 & 0 & 0 & 0 \\ 0 & 1 & 0 & 0 \\ 0 & l_s & 1 & 0 \\ 0 & -l_s & 1 & 0 \end{bmatrix} \begin{bmatrix} \beta \\ r \\ y_{CG} \\ \Delta\varphi \end{bmatrix} + \begin{bmatrix} 0 & 0 & 0 & 0 \\ 0 & 0 & 0 & 0 \\ 0 & 0 & -vl_s & 0 \\ 0 & 0 & vl_s & 0 \end{bmatrix} \begin{bmatrix} \delta_f \\ \delta_r \\ \rho_{ref} \\ f_d \end{bmatrix}$$

(1)

with

$a_{11}=-\mu(c_f+c_r)/Mv$ $\quad a_{12}=-1+\mu(c_rl_r-c_fl_f)/Mv^2$
$a_{21}=\mu(c_rl_r-c_fl_f)/J$ $\quad a_{22}=-\mu(c_rl_r^2+c_fl_f^2)/Jv$
$b_{11}=\mu c_f/Mv$ $\quad b_{12}=\mu c_r/Mv$
$b_{21}=\mu c_fl_f/J$ $\quad b_{22}=-\mu c_rl_r/J$

In (1), it is assumed that the longitudinal velocity v is constant. The parameters μ, c_f and c_r describe the behavior of the tires. c_f and c_r are the cornering stiffnesses of the front and rear tire respectively and μ is the road adhesion coefficient which equals 1 for dry road and 0.1 for an icy road. For the compensator design in this paper neither this parameter variation, nor the variations of other parameters will be considered.

3. FEEDFORWARD COMPENSATOR DESIGN

Ideal path tracking means $y_f=y_r=\Delta\varphi=0$. It can be shown that this can be accomplished by making y_r, y_f

and $\Delta\varphi$ equal to zero. Following (1), $\Delta\dot{\varphi}=r-v\rho_{ref}$, so for $\Delta\dot{\varphi}$ equal to zero r has to be equal to $v\rho_{ref}$. With this, $y_f=v\beta+l_s(r-v\rho_{ref})=v\beta$. In the same way $y_r=v\beta$. Thus when it is possible to steer the vehicle so that β equals zero and r equals $v\rho_{ref}$ in all circumstances, y_f and y_r equal zero in all circumstances. This in turn means that the lateral deviation stays zero at the front and rear sensor.

Now it will be shown that with four-wheel steering the requirements stated above can be met under the assumptions that $v\rho_{ref}$ is known (a condition that can be met when e.g. the discrete marker scheme discussed in (Zhang and Parson, 1990; Asaoka and Ueda, 1996) is used), that the actuators have an infinite bandwidth and that there are no parameter variations.
β and r can be written in the laplace domain as:

$$\beta(s) = \frac{a(s)}{n(s)}\delta_f(s) + \frac{b(s)}{n(s)}\delta_r(s) \qquad (2)$$

$$r(s) = \frac{c(s)}{n(s)}\delta_f(s) + \frac{d(s)}{n(s)}\delta_r(s)$$

where

$$a(s) = \frac{\mu c_f}{Mv}s + \frac{\mu c_f(\mu c_r l,l - l_f Mv^2)}{Jv^2 M}$$

$$b(s) = \frac{\mu c_r}{Mv}s + \frac{\mu c_r(\mu c_f l_f l + l_r Mv^2)}{Jv^2 M}$$

$$c(s) = \frac{\mu c_f l_f}{J}s + \frac{\mu^2 c_r c_f l}{JvM} \qquad (3)$$

$$d(s) = \frac{-\mu c_f l_f}{J}s - \frac{\mu^2 c_r c_f l}{JvM}$$

$$n(s) = s^2 - \frac{\mu(M(c_r l_r^2 + c_f l_f^2) + J(c_r + c_f))}{JMv}s +$$
$$+ \frac{\mu(c_r c_f \mu(l_r^2 + l_f^2) + Mv^2(c_r l_r - c_f l_f) + 2\mu c_r c_f l,l_f)}{JMv^2}$$

Since there are two inputs, both the required conditions $\beta=0$ and $r=0$. can be met. These conditions are met when:

$$\delta_f(s) = \frac{n(s)b(s)v}{b(s)c(s) - a(s)d(s)}\rho_{ref}(s) \qquad (4)$$

$$\delta_r(s) = \frac{-n(s)a(s)v}{b(s)c(s) - a(s)d(s)}\rho_{ref}(s)$$

So in de ideal case, when ρ_{ref} is known, both y_f, y_r and $\Delta\varphi$ can be kept zero when feedforward compensation (4) is applied. The relative degree of the feedforward compensator equals -1. The relative degree can be made zero by multiplying the feedforward compen-

sator with a high bandwidth lowpass filter, so that (4) becomes

$$\delta_f(s) = \frac{n(s)b(s)v}{b(s)c(s) - a(s)d(s)}\frac{1}{Ts+1}\rho_{ref}(s) \qquad (5)$$

$$\delta_r(s) = \frac{-n(s)a(s)v}{b(s)c(s) - a(s)d(s)}\frac{1}{Ts+1}\rho_{ref}(s)$$

where T has to be sufficiently small.

4. FEEDBACK COMPENSATOR DESIGN

By deriving the feedforward law (5) It was assumed that the actuators have infinite bandwidth. In reality the bandwidth can be as low as 5Hz (Guldner et. al., 1996; Guldner et. al., 1997), so the actuator really disturbs the ideal conditions. Furthermore it was assumed that the parameters like M, I, v and μ don't change during driving, but in reality these parameters change. Moreover there can be a timing error between the real curvature and ρ_{ref} which causes lateral deviations too.
So in practice the feedforward compensator (5) is not sufficient. An extra feedback compensator has to be added to suppress disturbances and uncertainty.

We designed a feedback compensator with H_∞ techniques (Zhou, K. et.al., 1996), since with these techniques it is easy to put constraints on output and actuator signals. Figure 3 shows the augmented plant used for the H_∞ design.

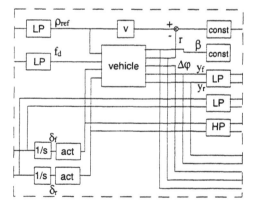

Figure 3 Augmented plant used for feedback compensator design

In this figure, LO stands for lowpass weighting filter, HP stands for highpass weighting filter and const stands for constant weighting filter.
We will not consider these filters in detail, since this is too space consuming.
The control inputs of the augmented plant are the time derivatives of the steering angles. These are chosen as inputs to make it possible to put constraints on

them. The time derivatives are integrated to get the steering angles as input for the vehicle. With the implementation of the controller these integrators are added to the feedback compensator.

It was assumed that all states where available for feedback. In practice, some states have to be estimated with an observer (Farrelly and Wellstead, 1996), but this extension was not considered here.

The requirements on the actuator inputs were that they may not exceed 0.7 [rad] and that the time derivatives of them may not exceed 0.45 [rad/s].

The final control scheme is shown in figure 4.

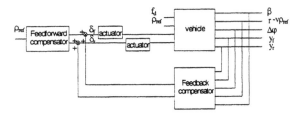

Figure 4 Final control scheme

5. SIMULATIONS

For the feedforward- and feedback compensator design a vehicle with l_f=5 [m], l_r=5 [m], c_f=c_r=300000 [N/rad], M=10000 [kg] and J=83000 [kgm^2] has been used. The two sensors were place symmetrical with respect to the CG at a distance of 2.5 m from it.

Since parameter variations where not considered in the compensator design, only the simulation results for the vehicle with nominal parameters will be discussed.

The simulations where carried out with the linear vehicle (1), with taking actuator saturation ($|\delta_i|$<0.7 [rad] and $|\dot{\delta}_i|$<0.45 [rad/s]) into account.

The next simulations where carried out:

• entering a curve of 250 [m] radius
• riding straight ahead with a disturbing lateral wind force starting at t=1 s and ending at t=4 [s].

Both simulation where carried out at v=20 [m/s].

Figure 5 shows the simulation results for entering a curve. Figure 5-a shows that y_f is max. about 0.015 [m] and both y_f and y_r equal zero in steady state. Also β equals zero in steady-state, shown in figure 5-c, which was expected from the feedforward compensator design. The deviations are not zero for t=1.5 [s] since the actuator has no infinite bandwidth. Especially the limitation on the rate of turn poses the constraint. Figure 5-b shows the steering angles occurring during cornering. Remarkable is that both the steering angles have the same sign. This is due to the fact that the lateral acceleration of the vehicle is rather high.

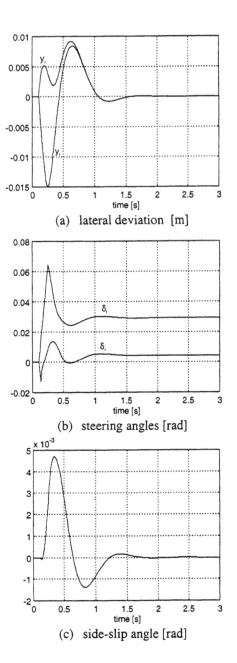

(a) lateral deviation [m]

(b) steering angles [rad]

(c) side-slip angle [rad]

Figure 5 Simulation results for entering a corner

Figure 6 shows the simulation results for riding straight ahead (ρ_{ref}=0) with a wind gust starting at t=1 s and ending at t=4 s. The disturbing force due to the wind was taken equal to 10000 N. For a vehicle with these dimensions this is equivalent with about windforce 7.

In figure 6-a it can be seen that after a peak of about 1 cm the steady state deviations of y_f and y_r are very small during the wind gust, and both are the same, which could be expected since four-wheel steering is applied. After the wind gust has disappeared, the steady state error returns to zero again. Figure 6-b shows that the steering angles are the same, which could be expected too. Finally, figure 6-c shows the side-slip angle, which has a small peak upwards and a small peak downwards during the start of the gust and the end of the gust respectively.

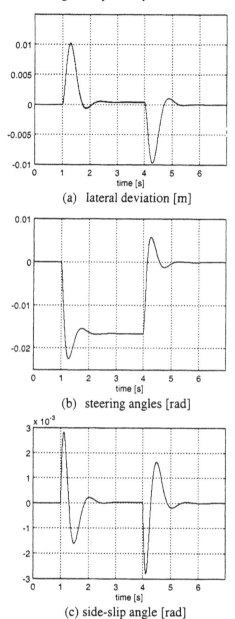

(a) lateral deviation [m]

(b) steering angles [rad]

(c) side-slip angle [rad]

Figure 6 Simulation results for wind a gust

6. CONCLUSIONS

We designed a controller for controlling the lateral deviation of two points on the longitudinal axis of a four-wheel steered vehicle so that the vehicle occupies as less road space as possible during cornering.

This controller consists of a compensator that feeds forward the curvature of the road. A feedback compensator compensates deviations that the feedforward compensator can not compensate due to non-idealities and for suppression of disturbances due to lateral wind gusts. This paper shows that it is possible to control two points on the vehicle longitudinal axis

with four-wheel steering and that with four-wheel steering a very low value for the side slip angle can be achieved during cornering.

7. FUTURE RESEARCH

In this work we didn't consider variations of vehicle mass and tire parameters. Furthermore, we didn't consider velocity changes. Future work has to bring parameter variations and changes of velocity into account.

When this work gives satisfactory results the model and controller will be extended to a fourwheel steered vehicel with a semitrailer with a steered axis.

REFERENCES

Asoaka, A. and Ueda, S.(1996). An experimental study of a magnetic sensor in an automated highway system, *Proceedings of the 1996 IEEE intelligent vehicle symposium*, pp.373-378

Ackermann, J.(1993), *Robust control, systems with uncertain physical parameters*, Springer- verlag London

Bosch, P.P.J. van den and Hendrix, W.H.A. (1997) Control of the lateral and longitudinal position of a bus *,IFAC symposium on transportation sys tems*, Chania, Greece, 1997, pp 385-390

Farrelly, J. and Wellstead, P., (1996), Estimation of vehicle lateral velocity, Proceedings of the *IEEE international conference on control appli cations*, pp.552-557

Guldner et. al. (1996), Analysis of automatic steering control for highway vehicles with look-down lat eral reference systems. *Vehicle system dynamics*, **26**, pp 243-269

Guldner et. al. (1997), Robust control design for automatic steering based on feedback of front and tail lateral displacement, *Proceedings of the 1997 European Control Conference*

Riekert, P. and Schunck, T.E., (1940) Zur fahr mechanik des gummibereiften kraftfahrzeugs ingenieur-archiv, **band 1940**, pp. 210-223

Sienel, W. and Ackermann, J (1994), Automatic steering of vehicles with reference angular veloc ity feedback, *Proceedings of the 1994 American Control Conference*

Zhang W. and Parson R.E. (1990).
An intelligent roadway reference system for vehi cle lateral guidance/control, *Proceedings of the American control conference*, pp 281-286

Zhou, K., Koyle, J.C. and Glover K., (1996), *Robust and Optimal Control*, Prentice Hall, New Jersey

MEASUREMENT OF THE LATERAL VEHICLE POSITION WITH PERMANENT MAGNETS

D. de Bruin and P.P.J. van den Bosch

Measurement and control group, Department of Electrical Engineering
Eindhoven University of Technology
P.O. Box 513 MB Eindhoven, The Netherlands
e-mail: d.de.bruin@ele.tue.nl, Fax: +31402434582, Tel: +31402473795

Abstract:. A measurement system is discussed, suitable for measuring the lateral deviation of electrical guided vehicles. It utilises permanent magnets buried in the road. It is shown that the lateral deviation of a vehicle can be determined from the longitudinal and lateral magnetic fields. This method is insensitive for changes in the fieldstrength of the magnets and changes in vertical distance between magnet and field-sensor. It is shown that part of the errors due to slant of the magnet can be eliminated by measuring with two 2-axes sensors. *Copyright © 1998 IFAC*

Keywords: Vehicles, Position estimation, Permanent magnets, Magnetic fields, Measuring range, Guidance systems, Automotive control

1. INTRODUCTION

We participate in a study which has as final goal the design of a double articulated bus with four independently-controllable steering axes and six independently-controllable wheel torques. The total length of the bus is about 25 m and it has a capacity of 160 persons. The requirements are that the bus may not deviate more than 10 cm from the centre of the bus lane during driving at velocities up to 80 km/h and that the bus has to stop within a lateral distance of less than 4 cm from the bus-stop, so that passenger can easily get on and off the bus (Bosch and Hendrix, 1997).

For human drivers, it is impossible to steer such a vehicle satisfactorily. Therefore the bus has to be steered by a control system.

One essential part of the control system is a system for measuring the lateral deviation of the bus with respect to the centre of the lane.

There are different methods to measure the lateral deviation. The most robust (robust with respect to weather circumstances ed.) methods make use of magnetic fields. These magnetic fields can be produced by electrical wires (Bosch and Hendrix, 1997), magnetic strips (Stauffer, et. al.) or permanent magnets (Chee, 1997; Choi, 1997; Johnston, et. al., 1979; Lee, et. al., 1995; Zhang and Parson, 1990; Zhang, 1991). From these methods, the method that utilises permanent magnets seems to be cheap, robust, reliable and sufficiently accurate.

One of the problems that arises by using permanent magnets is the dependency of the detected field on the detection height. Since the sensor to measure the lateral deviation has to be mounted on the vehicle's sprung mass, the lateral deviation can not be determined without special signal processing.

Most methods that utilise permanent magnets use the lateral- and vertical fields of the magnet to determine the lateral deviation. By combining these fields in a lookup-table, it is possible to eliminate the dependency of the detection height (vertical distance between magnet and sensor) (Asoaka, et. al., 1996; Chee, 1997; Choi, 1997; Johnston, et. al., 1979; Lee, et. al., 1995; Peng, et. al., 1992; Zhang and Parson, 1990; Zhang, 1991). Drawback of this method is that a complex lookup table is necessary and, to obtain a fine resolution, interpolation between the cells of the table is necessary. Furthermore, the maximal measurement range is about ± 25 cm (Peng, et.al., 1992) and the method is sensitive of changes in magnetic fieldstrength of the magnets.

Another method determines the lateral deviation by using two 3-axis sensors. This method is insensitive of changes in the magnetic fieldstrength and does not need a look-up table. A drawback of this method is that it needs two 3-axes sensors (Choi, 1997).

By measuring only the longitudinal- and the lateral fields with a (cheap) two-axes sensor, the use of look-up tables or 3-axis sensors can be omitted and the maximal measurement range can be expanded to about ± 60 cm. This depends on the strength of the magnets and on the accuracy of the measurements of the fields.

This paper describes how the deviation can be determined from the lateral and longitudinal field. Furthermore, a theoretical analysis of the influence of slant of the magnet will be discussed.

Finally, some measurement results will be discussed.

2. THEORY

The magnetic field around a circle symmetric bar permanent magnet can be described in free air by using the dipole approximation (Andrews, 1992; Zhang and Parson, 1990). This approximation yields for the fields in the x, y and z directions of the magnet:

$$B_{mx} = \frac{\mu M}{4\pi r^5} 3z_m x_m = C(x_m, y_m, z_m) x_m$$

$$B_{my} = \frac{\mu M}{4\pi r^5} 3z_m y_m = C(x_m, y_m, z_m) y_m \qquad (1)$$

$$B_{mz} = \frac{\mu M}{4\pi r^5} (2z_m{}^2 - x_m{}^2 - y_m{}^2)$$

where

$$r = \sqrt{x_m{}^2 + y_m{}^2 + z_m{}^2}, \qquad (2)$$

μ is the permeability of free space and M is the magnetic moment of the magnet. The subscript m denotes that the fields are expressed in a magnet-fixed co-ordinate frame, with axes X_m, Y_m, and Z_m.

This co-ordinate frame is fixed at the centre of the magnet. The z-axis of this frame is pointing along the axis of the magnet. The magnet is buried in the road with its z-axis pointing up. The x-axis of the magnet is pointing in the travelling direction of the road and the y direction is directed to make the right handed coordinate system complete (see Figure 2). In (1) and (2) x_m, y_m and z_m determine the position of observation of the magnetic fields, expressed in the magnet-fixed co-ordinate frame.

Figure 1 shows a plot of B_{my} versus the lateral distance y_m for different values of z_m (field strength not on scale). This figure makes clear that B_{my} depends strongly on z_m.

The lateral distance can be determined unambiguously (if z_m is known by measuring only B_{my} when the lateral deviation stays between the minimum and the maximum of B_{my}. It can be shown that the distance from the magnet to the maximum and minimum of B_{my} equals $z_m/2$. So, the maximum measurement range is equal to z_m which can be as less as 10 cm. A second drawback of this method is connected with the dependency of B_{my} on z_m. Since the sensor for measuring the magnetic field has to be fixed at the sprung mass of the vehicle, z_m can vary during driving due to bumps in the road or changing vehicle load.

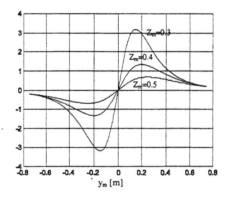

Figure 1 B_{my} (normalised) as function of the lateral distance

By measuring both B_{mx} and B_{my} these two drawbacks can be omitted. Equations (1) shows that B_{mx} equals a factor C depending on x_m, y_m and z_m times x_m and B_{my} equals the same factor times y_m. So by measuring B_{mx} at known longitudinal distance x_m ($x_m \neq 0$), this factor can be calculated by

$$C(x_m, y_m, x_m) = \frac{B_{mx}}{x_m} \qquad (3)$$

With (3) and (1) y_m can be calculated by ($B_{mx} \neq 0$)

$$y_m = \frac{B_{my}}{C(x_m, y_m, z_m)} = \frac{B_{my}}{B_{mx}} x_m \qquad (4)$$

Condition for (4) to be true is that both B_{mx}, and B_{my} are measured at the same position.

Now the problem is how to determine x_m. This can be accomplished in two ways. The first method needs only one two-axis sensor and the longitudinal velocity and acceleration of the vehicle.

By detecting the zero crossing of B_{mx}, it is known when $x_m=0$. By measuring the field again after a short time instance t, $x_m(t)$ can be determined by $x_m(t)=V*t+a*t^2$, where V is the longitudinal velocity of the vehicle and a is the longitudinal acceleration. It is even possible to measure the lateral deviation continuously after the detection of a zero crossing of B_{mx} until the signals disappear in noise. It has to be investigated how accurate the velocity and acceleration have to be measured to give satisfying results. Drawback of this method is that it is sensitive for slant of the magnet and/or vehicle, as will be shown in the next section.

The second method for determining x_m needs two sensors. These sensors are placed somewhere on the longitudinal-axis of the vehicle at a fixed distance d from each other. When the first sensor measures a zero crossing of B_{mx}, then $x_m=0$ for this sensor as equation (1) states. Since the distance between the two sensors equals d, x_m equals $-d$ for the second sensor. Now y_m can be determined with (4) where B_{mx} and B_{my} are measured by sensor 2. The same can be done when the second sensor measures $B_{mx}=0$, then $x_m=d$ for the first sensor and with this y_m can be determined from the signals measured with sensor 2. In this way, the lateral deviation can be determined two times per magnet. Figure 2 makes the measurement setup more clear. This method is also sensitive for slant of the magnet and/or vehicle. By combining the signals of both sensors in an other way part of the effects of slant will cancel out, as will be shown in the next section.

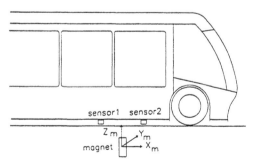

Figure 2 Measurement setup for measuring the lateral deviation

3. SLANT OF THE MAGNET

In the above analysis, it was assumed that magnetic fields were measured with a sensor that was aligned with its sensitive axes parallel to the x-axis and y-axis of the magnet. However, in reality it is impossible to place the magnet precisely perpendicular to the road and the vehicle will have a (small) roll, pitch and/or yaw angle. This can yield a wrongly measured lateral deviation.

To study this, the effect of slant of the magnet will be analysed now. Note that slant of the magnet and slant of the vehicle are different things. However the effect of slant of the vehicle can be analysed in the same way.
To study the effect of slant of the magnet a world-fixed co-ordinate frame , with axis X_w, Y_w and Z_w, and a sensor-fixed co-ordinate frame , with axis X_s, Y_s and Z_s, will be introduced. The x- and y-axis of the sensor-fixed co-ordinate coincide with the two sensitive axes of the sensor. For the analysis of the effect of slant of the magnet, it is assumed that the sensor-fixed co-ordinate frame and the world-fixed co-ordinate frame have the same orientation. Figure 3 makes things more clear.

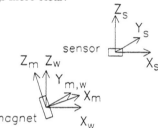

Figure 3 definition of co-ordinate frames

Only rotations of the magnet around the X_m- and Y_m-axis will be considered. Rotations around Z_m have no influence, since the magnet is assumed to be rotation symmetric.

Equation (4) yields the deviation from the track expressed in the magnet-fixed co-ordinate frame. However, the deviation has to be known in a world-fixed co-ordinate frame. To determine the deviation expressed in a world-fixed co-ordinate frame, the position of the origin of the sensor-fixed co-ordinate frame in (1) and (2) has to be expressed in the world-fixed co-ordinate frame.
This can be accomplished by the co-ordinate transformation:

$$\begin{bmatrix} x_m \\ y_m \\ z_m \end{bmatrix} = {}^m R_w \begin{bmatrix} x_w \\ y_w \\ z_w \end{bmatrix}, \qquad (5)$$

where

$$
{}^mR_w = \begin{bmatrix} \cos\beta & 0 & \sin\beta \\ -\sin\alpha\sin\beta & \cos\alpha & \sin\alpha\cos\beta \\ -\cos\alpha\sin\beta & -\sin\alpha & \cos\alpha\cos\beta \end{bmatrix} \quad (6)
$$

Here α is the rotation angle of the rotation of the magnet around X_m and β is the rotation angle of the rotation of the magnet around Y_m. Furthermore x_w, y_w, and z_w determine the position of the origin of the sensor-fixed co-ordinate frame expressed in the world-fixed co-ordinate frame.

The magnetic fields B_{mx}, B_{my}, and B_{mz} are still expressed in magnet co-ordinates. However, the fields will be measured in a world fixed co-ordinate frame. The measured fields can be expressed in B_{mx}, B_{my}, and B_{mz} with the next transformation:

$$
\begin{bmatrix} B_{wx} \\ B_{wy} \\ B_{wz} \end{bmatrix} = {}^mR_w^{-1} \begin{bmatrix} B_{mx} \\ B_{my} \\ B_{mz} \end{bmatrix} \quad (7)
$$

The lateral deviation can be determined from the measured fields by:

$$
y = \frac{B_{wy}}{B_{wy}} x_w \quad (8)
$$

Due to slant of the magnet, y will not be exactly equal to y_w. The effect of slant of the magnet in the lateral direction (a rotation of the magnet around X_m) will be described first. This means that $\beta=0$ in (5), (6) and (7). Figure 4 shows a plot of B_{my} in the ideal ($\alpha=0^0$) case and in the non-ideal case with $\alpha=5^0$, $x_w=3$ cm and the measurement height $z_w=20$ cm. The zero crossing of the lateral field has shifted somewhat due to the slant of the magnet.

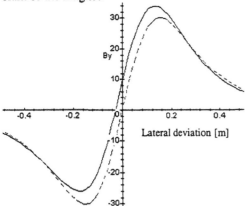

Figure 4 B_{my} without slant (dashed) and with slant (solid)

Figure 5 shows a plot of the determined lateral deviation in the ideal case and in the case when the magnet slants. Also the difference between these two has been shown in the figure. The measurement error due to 5^0 slant in the lateral direction equals 6 mm in the neighbourhood of the zero crossing. At a distance of 50 cm the error is about 2 cm. Furthermore, it can be concluded that slant in the lateral direction has almost no influence on the gain of the measurement system, since the slope of the curve with slant is almost the same as the slope of the curve without slant

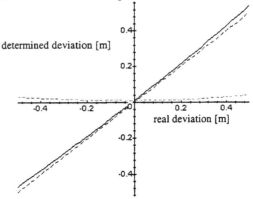

Figure 5 determined lateral displacement with slant in the lateral direction (black dashed=ideal solid=non-ideal, gray dashed=error)

Now the effect of a rotation of the magnet around Y_m (slant in the longitudinal direction) will be studied ($\alpha=0$).
Figure 6 shows a plot of the determined lateral deviation in the ideal case ($\beta=0^0$) and in the non-ideal ($\beta=5^0$) case. At 50 y_m cm the measurement error due to slant is larger than 20 cm, which is unacceptable. However, the point of zero lateral deviation can be determined without error.

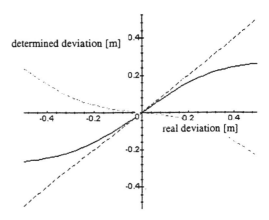

Figure 6 determined lateral displacement with slant in the longitudinal direction (black dashed=ideal solid=non-ideal, gray dashed=error)

From equations (1), (3) and (7) it can be derived that with a rotation of the magnet around Y_m

$$B_{wx} = C\frac{\frac{1}{3}\sin\beta(y_w^2 + z_w^2 - 2x_w^2) + x_w z_w \cos\beta}{r^5}$$

$$B_{wy} = C\frac{(x_w \sin\beta + z_w \cos\beta)y_w}{r^5}$$

(9)

From this equation it can be seen that the nonlinearities can be eliminated when the fields measured at $x_w=0,5d$ and $x_w=-0,5d$ are combined. Determining the lateral deviation with:

$$y = 0.5d\frac{B_{wy}(x_w = 0.5d) + B_{wy}(x_w = --0.5d)}{B_{wx}(x_w = 0.5d) - B_{wx}(x_w = -0.5d)} \quad (10)$$

yields:

$$y = 0.5d\frac{z_w y_w \cos\beta}{0.5dz_w \cos\beta} = y_w \quad 11)$$

So when there is only a rotation around Y_m, the non-linearity can be eliminated by determining y with (10). When the magnet is rotated around both X_m and Y_m, there are some cross terms that don't cancel out when the lateral deviation is determined with equation (10).
In figure 7, the effect of 5^0 slant in both the longitudinal and lateral direction is shown.

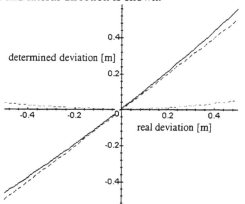

Figure 7 determined lateral displacement with slant in the lateral and longitudinal direction (black dashed=ideal, solid=non-ideal, gray dashed=error)

The measurement error due to slant at zero crossing still equals 6 mm. Drawback of determining the lateral deviation with equation (10) is that now the lateral deviation can be measured only one time per magnet with two sensors. But this doesn't need to give problems.
For determining the lateral deviation with equation (10), it has to be determined when the magnet is precisely between the sensors. This can be done by de-

tecting the zero crossing of the signal $B_{wx}(x_w=0.5d)-B_{wx}(x_w=-0.5d)$.

4. MEASUREMENT RESULTS

For the measurements an AlNiCo bar magnet was used with a length of 10 cm and a diameter of 1.5 cm. The distance d between the sensors was 30 cm and the measurement height z_w was about 35 cm. In figure 8, the measurement results of both sensors, together with the lateral deviation determined with equation (10), are shown. Although the sensors and magnet were aligned as good as possible, there is still some measurement error (possibly due to misalignment between magnet and sensors) when the lateral deviation is determined with one sensor. The figure shows that when (10) is used to determine the lateral deviation, this measurement error cancels out. Figure 9 shows the measurement error when the lateral deviation is determined with (10). As can be seen in the figure, the lateral deviation can be determined within an range of ±65 cm with sufficient accuracy. However, at a large distance, the measurement becomes less reliable. Note that these measurements where carried out under ideal conditions. Other measurements have to be carried out to investigate the influence of iron in the neighbourhood of the sensors and/or magnet.

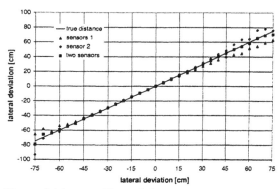

Figure 8 Measured lateral deviation

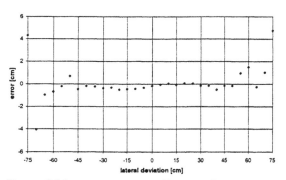

Figure 9 Measurement error (2 sensors)

5. CONCLUSIONS

We developed a method to determine the lateral deviation of a vehicle from the longitudinal and lateral field of a permanent bar magnet, placed in the road perpendicular to the road surface.

In contrary to other publications, this method doesn't need a lookup table or 3-axis sensors. The advantages of this method compared to methods that use look-up tables are:

- Larger measurement range (about two times).
- Simpler algorithm (no interpolation necessary between cells of a table).
- Zero crossing detection to determine when a magnet is passed instead of a peak crossing detection
- Insensitive of strength of the magnets.

In this paper it is shown that the method theoretically gives good result when the magnet slants.

Measurements are carried out to show that the method works well.

6. FUTURE WORK

Measurements have to be carried out to show that the method works also with slant of the magnet. Since the method has to applied to vehicles that are probably made of aluminium or steel, the influence of aluminium and/or steel in the neighbourhood of magnet and sensors has to be investigated.

Furthermore the influence of disturbance fields due to electric power cables has to be investigated.

The distance d between the sensors has to be optimised with respect to slant and signal-noise ration.

REFERENCES

Andrews, A. (1992). Theoretical and Empirical Analysys of PATH Magnetic Lane Tracking for the Intelligent Vehicle Highway System, *PATH research report, UCB-ITS-PRR-92-9*

Asoaka, A. and Ueda, S.(1996). An experimental study of a magnetic sensor in an automated highway system, *Proceedings of the 1996 IEEE intelligent vehicle symposium*, pp.373-378

Bosch, P.P.J. van den and Hendrix, W.H.A.(1997). Control of the lateral and longitudinal position of a bus *IFAC symposium on transportation systems*, Chania, Greece, pp 385-390

Chee W., (1997). *Unified approach to vehicle lateral guidance*, Ph.D. dissertation, Dept of Mech. Eng. Univ of California, Berkeley, may 1997

Choi, S.B. (1997). The design of a look-down feed back adaptive controller for the lateralc ontrol of front-wheel-steering autonomous highway vehicles, *Proceedings of the American Control Conference*, pp. 1603-1607

Johnston, A.R., Assefi T. and Lai, J.Y. (1997). Auto mated vehicle guidance using discrete reference markers *IEEE transactions on vehicular technology*, vol. **VT-28**, pp. 95-105

Lee H., Love D.W. and Tomizuka M. (1995). Lon gitudinal manoeuvring control for automated highway systems based on magnetic refer ence/sensing system, *Proceedings of the Ameri can control conference 1995*, pp 150-154

Peng, H.; Zhang, W.; Arai, A. and Lin Y. (1992). Experimental automatic lateral control system for an automobile, *PATH research report, UCB-ITS-PRR-92-11*

Stauffer D., Barett B., Demma N. and Dahlin T. Magnetic lateral guidance sensors for automated highways, *Proceedings of the spie.*, vol. **2592**, pp138-149

Zhang W. and Parson R.E. (1990). An intelligent roadway reference system for vehicle lateral guidance/control, *Proceedings of the American control conference*, pp 281-286

Zhang, W.(1991). A roadway information system for vehicle guidance/control. *Vehicle Navigation and Information system conference pro ceedings*, vol. **2**, pp. 1111-1116

STABILITY ANALYSIS OF FUZZY PATH TRACKING USING A MIMO FREQUENCY RESPONSE TECHNIQUE

G. Heredia, A. Ollero, F. Gordillo and J. Aracil

Departamento de Ingeniería de Sistemas y Automática.
Escuela Superior de Ingenieros.
Universidad de Sevilla.
Camino de los Descubrimientos, 41092 Sevilla (Spain).
Fax:+34-5-4556849. Email: {guiller,aollero,gordillo,aracil}@cartuja.us.es

Abstract: This paper presents the application of a multivariable frequency response technique to study the stability of autonomous vehicles when tracking a path. Particularly, the stability of a path tracking control loop with a fuzzy controller is considered. The technique is based on the analysis of the solutions of the harmonic balance equation. Necessary and sufficient conditions for stability are obtained by examining a family of characteristic loci parameterized in the amplitude. The method can be applied when pure delays exist in the path tracking control loop. Several simulations experiments have been done. These experiments show a Hopf bifurcation induced by a pure delay, and an unstable limit cycle around a stable origin. *Copyright © 1998 IFAC*

Keywords: Stability analysis, path tracking, autonomous vehicles, fuzzy control, nonlinear systems.

1 INTRODUCTION

Path tracking is a significant function in the control of autonomous vehicles. The objective of path tracking is to generate the control commands for the vehicle to follow a previously defined explicit path by taking into account the actual position and the constraints imposed by the vehicle and its lower level motion controllers. When autonomous vehicles are tracking a path, several disturbances arise, normally due to the vehicle-terrain interactions. These disturbances could separate the vehicle from the desired path or even can lead the vehicle to the loss of stability when the vehicle velocity and curvature are larger enough. In this case the vehicles goes out of control and a crash can occur.

Figure 1 Romeo-3R and Romeo-4R autonomous vehicles

Fuzzy logic has been used efficiently to control autonomous vehicles (Ollero *et al.*, 1994). However, stability studies are required to guarantee safe navigation conditions, particularly when the size and weight of the autonomous vehicle is significant, as occur in the ROMEO vehicles (Ollero *et al.*, 1997b), which are shown in Figure 1.

Stability of fuzzy control systems has attracted the attention of many researchers in the last ten years. The problem is difficult due to the non-linear nature of the control loop. It is well known that nonlinear control systems with unstable plants are only localy stable. Furthermore, in fuzzy path tracking only local linearization is possible and the controller is a nonlinear function with a saturation. An additional problem of the fuzzy path tracking is usually the pure delays existing in the control loop. This delay could be generated by sensing or computing functions, particularly when position estimation involving environment perception is applied (i.e. sensor fusion or image processing).

2 FUZZY PATH TRACKING

The study of the motion of a vehicle is a very complex task, when including nonlinear kinematic and dynamic motion equations, vehicle/terrain interaction, actuator dynamics, etc. However, if the vehicle is supposed to move on a plane, a simplified 2D model can be derived.

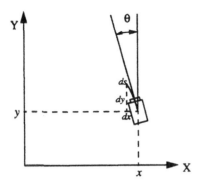

Figure 2 Vehicle kinematics

For 2D vehicle navigation, the posture or configuration of the vehicle is given by (x, y, θ), where x,y are the vehicle's global coordinates and θ is the orientation angle (see Figure 2). If ds is the distance travelled, the following expressions can be obtained:

$$dx = -\sin\theta ds \qquad (1)$$

$$dy = \cos\theta ds \qquad (2)$$

$$d\theta = \gamma ds \qquad (3)$$

assuming a constant vehicle curvature γ in the time interval.

The motion equations in world coordinates can be expressed as:

$$\dot{x} = -V\sin\theta \qquad (4)$$

$$\dot{y} = V\cos\theta \qquad (5)$$

$$\dot{\theta} = V\gamma \qquad (6)$$

where V is the longitudinal velocity, or vehicle speed, and $\dot{\theta}$ is the angular velocity. These velocities can be considered as the control variables in (4)-(6). However, in some vehicles, it has been shown that the dynamic behavior of the mechanisms to apply these velocities from the control signal is very significant for the vehicle's driving and then, should be considered for the vehicle's control (Ollero and Amidi 1991).

The dynamics of the steering actuation system can be represented by means of the following first order model:

$$\dot{\gamma} = -\frac{1}{T}(\gamma - u) \qquad (7)$$

where u is the control variable computed by the steering control algorithm and T is the time constant.

Thus, the steering control system can be represented by means of equations (4), (5), (6) and (7), where x, y, θ and γ are the state variables.

Furthermore, path tracking usually involves a pure delay in the control loop, due mainly to the mobile robot position estimation, particularly when involving environment perception, as well as to other computing and communication delays. This time delay is usually ignored in path tracking studies, but it may have a significant effect on stability.

This paper only considers path tracking at constant velocity. Then, V in equations (4)-(6) is a constant.

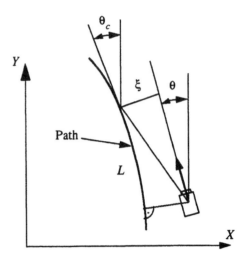

Figure 3 Path tracking with a goal point

Usually, path tracking algorithms are based on the selection of goal points in the path to track. Consider a goal point at a fixed distance from the vehicle, that is called the lookahead distance L. Using this goal point three values are obtained that are the inputs to the path tracker. These inputs are the lateral position error in vehicle coordinates ξ, the orientation error with respect to the goal point θ_e and the curvature error with respect to the goal point γ_e. The errors θ_e and γ_e are defined as:

$$\begin{aligned} \theta_e &= \theta - \theta_c \\ \gamma_e &= \gamma - \gamma_c \end{aligned} \qquad (8)$$

where θ_c and γ_c are the orientation and curvature of the path at the goal point.

2.1 Tracking of straight paths.
If the path to follow is a straight line (see Figure 4) then $\theta_c = 0$ and $\gamma_c = 0$ at all the points of the path.

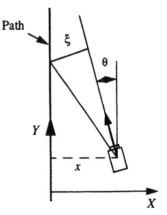

Figure 4 Straight path tracking

Therefore,

$$\theta_e = \theta$$
$$\gamma_e = \gamma \qquad (9)$$

The input u computed by the steering control algorithm can be expressed as:

$$u = f(\xi, \theta, \gamma) \qquad (10)$$

where f is a nonlinear function of the error variables. The lateral position error ξ can be expressed as a function of the state variables and the lookahead distance: $\xi = -(x\cos\theta - L\sin\theta)$

It can be observed that for straight lines the expressions of the path tracking algorithm inputs θ_e, γ_e and ξ do not depend on the y variable, and therefore equation (5) can be omitted in stability studies. Then the straight line path tracking problem can be written as:

$$\dot{x} = -V\sin\theta$$
$$\dot{\theta} = V\gamma$$
$$\dot{\gamma} = -\frac{1}{T}(\gamma - u) \qquad (11)$$
$$u = g(x, \theta, \gamma)$$

where g is a nonlinear function of the state variables. It can be noted that when the vehicle is following perfectly the path, the state variables are all zero. Therefore, the path tracking controller has to be built so that the origin of system (11) is an equilibrium point. This does not mean that there is no movement: the motion of the vehicle at the equilibrium point is given by the y coordinate.

The motion equations can also be written in dimensionless form as:

$$\dot{x}_1(t) = -\sin x_2(t)$$
$$\dot{x}_2(t) = x_3(t)$$
$$\dot{x}_3(t) = -x_3(t) + u(t - \tau) \qquad (12)$$
$$u_N = g_N(x_1, x_2, x_3)$$

where x_1, x_2 and x_3 are the dimensionless state variables x, θ and γ respectively, u_N is the non-dimensional form of the nonlinear control law and τ is the pure time delay in the control loop.

2.2 Fuzzy path tracker.

The fuzzy controller generates the control signal u from the deviations of the vehicle in position (ξ), heading (θ_e) and curvature (γ_e) with respect to the path as shown in Figure 4.

The heuristic to drive the vehicle from the deviations is very simple. Assuming that positive steering is counterclockwise, the rules are as:

R_i: if ξ is POSITIVE LARGE, θ_e is NEAR ZERO and γ_e is POSITIVE, then u_i is w_i

Some other heuristic consideration to take into account preferences to solve conflicts between rules are used (see for example García-Cerezo and others, 1996).

The above rules define a Mamdani-type controller. It is also possible to use a Sugeno-type controller for path tracking (Ollero and others, 1996), with rules like:

R_j: if ξ is POSITIVE, θ_e is NEAR ZERO and γ_e is NEGATIVE, then $\gamma_{Rj} = 0.2\,\xi - 0.1\,\theta_e$

In both cases the final resulting decision surface can be represented by the nonlinear function $u = f(\xi, \theta_e, \gamma_e)$ (García-Cerezo and others, 1996).

For straight line path tracking, the inputs to the controller are the state variables x, θ and γ. Thus, the control signal can be expressed in nondimensional form as in (12):

$$u_N = \Phi(x_1, x_2, x_3) \qquad (13)$$

In what follows it will be assumed that $\Phi(k_1 x_1 + k_2 x_2 + k_3 x_3)$ can be written $\Phi(x_1, x_2, x_3)$ where Φ is a function that saturates. This assumption is consistent with many of the usual fuzzy control laws (Yager and Filev, 1994).

Note that the k_i parameters are:

$$k_1 = \left.\frac{\partial\Phi}{\partial x_1}\right|_0 \qquad k_2 = \left.\frac{\partial\Phi}{\partial x_2}\right|_0 \qquad k_3 = \left.\frac{\partial\Phi}{\partial x_3}\right|_0 \qquad (14)$$

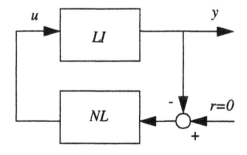

Figure 5 Structure of the system

Previous experimentation have revealed that the saturation in the fuzzy controller can generate in the above control system both stable and unstable limit cycles. To study the existence of that limit cycles a frequency response technique can be used (Khalil, 1996). This technique can also be applied to multivariable fuzzy control systems represented as shown in Figure 5, where *LI* is a linear block (in general a multivariable linear block) and *NL* is a non-linear block (Ollero *et al.*, 1997a). That can be done by using partial linearization in equations (1)-(4). Particularly, the term $\sin\theta$ can be linearized (Aracil *et al.*, 1997).

3 MIMO SYSTEM STABILITY ANALYSIS

The fuzzy path tracking problem of (12) and (15) can be formulated as in Figure 6. The system is represented by a 2x2 linear transfer matrix with the two nonlinearities in the feedback loop, which are decoupled.

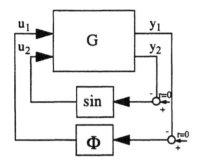

Figure 6 MIMO Equivalent system

The state space representation of the system is given by

$$A = \begin{bmatrix} 0 & 0 & 0 \\ 0 & 0 & 1 \\ 0 & 0 & -1 \end{bmatrix} \quad B = \begin{bmatrix} 0 & 1 \\ 0 & 0 \\ 1 & 0 \end{bmatrix} \quad C = \begin{bmatrix} k_1 & k_2 & k_3 \\ 0 & 1 & 0 \end{bmatrix} \quad (15)$$

and the transfer matrix is:

$$G(s) = C(sI - A)^{-1} B = \begin{bmatrix} \dfrac{k_3 s + k_2}{s(s+1)} & \dfrac{k_1(s+1)}{s(s+1)} \\ \dfrac{1}{s(s+1)} & 0 \end{bmatrix} \quad (16)$$

Defined in this way, the path tracking problem consists of a linear multi-input multi-output (MIMO) part with two inputs and two outputs conected by two nonlinear functions (sinus and Φ). The stability of this type of system can be studied using an extension of the frequency response methods to MIMO systems (Ollero *et al.*, 1997a).

The global system can be divided in two parts: a linear block (with transfer matrix $G(s)$) and a nonlinear part with a describing function matrix $N(a,\omega)$, where a is a vector with the amplitude of the inputs, and ω the common frequency of all the inputs. If the nonlinear part is memoryless, then $N(a,\omega)$ will not depend on ω and so it will be written $N(a)$. The harmonic balance equation will be used to search for limit cycles (Slotine and Li, 1991). This equation

leads to

$$G(j\omega) N(a) y = -y \quad (17)$$

where y is a vector whose components $y_i = a_i e^{j(\omega t + \theta_i)}$ are the complex representation of sinusoids, and $N(a)$ is the describing function of the nonlinear element. The equation may be rewritten as:

$$[I + G(j\omega) N(a)] a = 0 \quad (18)$$

where a is a vector of amplitudes $a = [a_i]$. For a limit cycle to exist Eq. (18) should have a nontrivial solution for the frequency ω and the amplitudes a of the limit cycle. For that it is required that

$$\det[I + G(j\omega) N(a)] = 0 \quad (19)$$

Equation (19) holds if one of the eigenvalues of $(I + G(j\omega) N(a))$ is zero, or if one of the eigenvalues of $G(j\omega) N(a)$ is $(-1,0)$ (Atherton, 1975). The eigenvalues of $G(j\omega)N(a)$ are usually called characteristic loci.

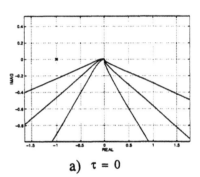

a) $\tau = 0$

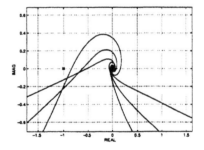

b) $\tau = 4$

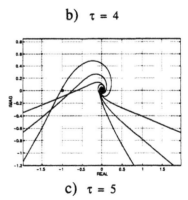

c) $\tau = 5$

Figure 7 Characteristic loci of $G(j\omega) N(a)$ for different time delays
$k_1 = -0.02$, $k_2 = 0.2$, $k_3 = 0.1$

To verify whether that happens, the characteristic loci

of $G(j\omega)N(a)$ must be drawn. Examining if the characteristic loci of $G(j\omega)N(a)$ crosses the point $(-1,0)$, the stability of the system can be analyzed. In fact there is a family of characteristic loci parameterized in a. This method provides necessary and sufficient conditions for stability.

4 SIMULATION EXPERIMENTS

To check the validity of the results, several simulation experiments have been done with the complete model of the system. A delay-induced Hopf bifurcation (Kuznetsov, 1995) and an unstable limit cycle around a stable origin are analyzed below.

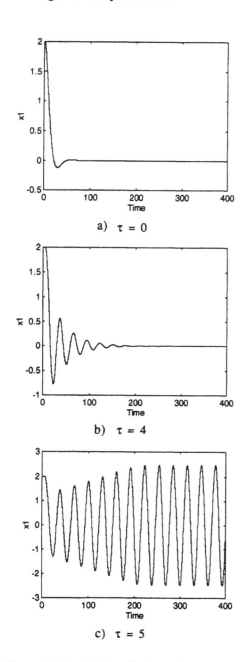

Figure 8 Simulations for $k_1 = -0.02$, $k_2 = 0.2$, $k_3 = 0.1$ and different time delays

The effect of the time delay on the input u can be analyzed by examining the characteristic loci of

$G(j\omega)N(a)$. Figure 7-a, b and c shows the plot of the characteristic loci for $a_1 \in \{0.1, 3, 10\}$ and $a_2 = 0.1$ (a_i is the amplitude cooresponding to the input u_i) for different values of the time delay ($t = 0, 4, 5$).

It can be noted that for $\tau = 5$ one of the characteristic loci encircles the critical point $(-1,0)$. Then, there exists an amplitude $0.1 < a^* < 3$ for which the characteristic locus includes the critical point. The system has a limit cycle with amplitude a^*. This result is similar to the one obtained with the SISO describing function method linearizing the $\sin\theta$ term (Aracil *et al.*, 1997), and it has been checked with computer simulations of the system. Figure 8 shows how increasing the time delay a Hopf bifurcation can be induced.

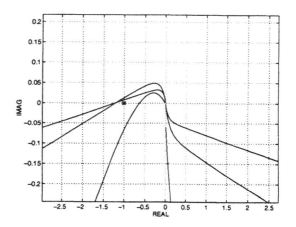

Figure 9 Characteristic loci

4.1 Unstable limit cycle around stable origin.

Consider system (12) with the following parameters: $k_1 = -0,7$, $k_2 = 0,5$, $k_3 = 0,6$. In Figure 9, the characteristic loci of $G(j\omega)N(a)$ for this system is presented. It can be observed that the characteristic locus encircles the critical point $(-1, 0)$. Therefore, a limit cycle is supposed to exist.

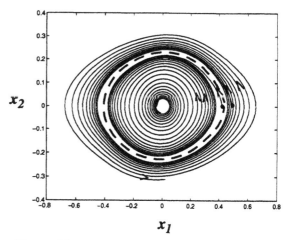

Figure 10 x_1, x_2 state space for initial conditions (0.4,0,0) and (0.5,0,0). k_1=-0.7, k_2=0.5, k_3=0.6

If the system is simulated for these parameters, the presence of an unstable limit cycle can be clearly detected. In Figure 10 the trajectories of the system for two nearby initial conditions are shown. One of them converges to the equilibrium point at the origin while the other goes away from it, showing the presence of an unstable limit cycle between them.

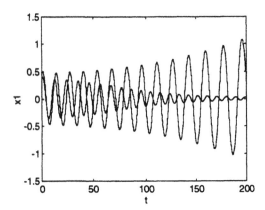

Figure 11 Simulations for k_1=-0.7, k_2=0.5, k_3=0.6 and initial conditions (0.4,0,0) and (0.5,0,0).

Figure 11 shows the time evolution of the x_1 state variable for the same nearby initial conditions than in Figure 10.

5 CONCLUSIONS

Fuzzy path tracking involves nonlinear control loops. Furthermore, pure delays could be involved due to vehicle´s position estimation, computation and communication delays. Then, the stability analysis of these control loops is a complex problem that require suitable methodologies. This paper proposes the application of multivariable frequency response techniques. These techniques can be applied to deal with pure delays and have demonstrated ability to cope with global stability phenomena.

Simulation experiments have shown how pure time delays in the fuzzy path tracking control action produce the stability loss by inducing a Hopf bifurcation. Another simulation experiment has shown how this technique can be used to detect the presence of an unstable limit cycle that limits the attraction basin of the stable origin.

Future work will involve practical experiments to validate the proposed multivariable frequency response techniques with the ROMEO-3R and ROMEO-4R autonomous vehicles.

6 ACKNOWLEDGMENTS

This work has been done in the FAMIMO Project partially funded by the European Commission (ESPRIT LTR 21911) and the CICYT projects TAP96-1184-C04-01 and TAP97-0553.

7 REFERENCES

Atherton, D.P., 1975, *Nonlinear Control Engineering*, Van Nostrand Reinhold, New York

Aracil, J., G. Heredia and A. Ollero, 1997. Global Stability Analysis of Fuzzy Path Tracking. *Proceedings of the 3rd IFAC Symposium on Intelligent Components and Instruments for Control Applications (SICICA97)*, Annecy, France.

García-Cerezo, A., A. Ollero and J. L. Martínez, 1996, Design of a robust high performance fuzzy path tracker for autonomous vehicles, *Int. Journal of Systems Science*, vol. 27, n. 8, pp. 799-806.

Khalil, H.K. 1996, *Nonlinear Systems*, Prentice Hall.

Kuznetsov, Y.A. 1995, *Elements of Applied Bifurcation Theory*, Springer-Verlag.

Mees, A.I, 1973. Describing functions, circle criteria and multiple-loop feedbck system. *Proc. IEE*, Vol. 120, No. 1, 126--130.

Ollero, A., A. García-Cerezo and J. L. Martínez, 1994. Fuzzy supervisory path tracking of autonomous vehicles. *Control Engineering Practice*, vol. 2, pp. 313-319.

Ollero, A., A. García-Cerezo and J. L. Martínez, 1996, Design of fuzzy controllers from heuristic knowledge and experimental data, *1996 IFAC World Congress*, paper 1d-01 5, Vol. A, pp. 433-438.

Ollero, A., J. Aracil and F. Gordillo, 1997a. Stability Analysis of MIMO Fuzzy Control Systems in the Frequency Domain. Submitted to *FUZZ-IEEE'98*, Anchorage, USA.

Ollero, A., B.C. Arrúe, J. Ferruz, G. Heredia, F. Cuesta, F. López-Pichaco and C. Nogales, 1997b. Intelligent Components in the Romeo Vehicles. *IFAC Workshop on Intelligent Components for Vehicles (ICV98)*, Seville, Spain.

Slotine, J-J. and W. Li, 1991, *Applied Nonlinear Control*, Prentice Hall.

Yager, R.R., and D.P. Filev, 1994, *Essentials of Fuzzy Modeling and Control*, Wiley.

EVOLUTION STRATEGIES FOR PATH TRACKING
WITH GUARANTEED STABILITY

José Cesáreo Raimúndez and Antonio Barreiro

Depto. Ingeniería de Sistemas y Automática
E-mail cesareo@centolla.aisa.uvigo.es
antonio@centolla.aisa.uvigo.es
Universidad de Vigo (Spain)

Abstract. Evolution Strategies (ES) are stochastic optimization techniques that can be used to find global optima. The current investigation focuses on finding conicity centers which guarantee global assymptotic stability for the path following control problem in a mobile robot with given nonlinear dynamics *Copyright © 1998 IFAC*

Keywords. Conicity Criterion, Evolutive Strategies, Path Following, Nonlinear Dynamics

1. INTRODUCTION

In the classical approach to optimization methods there are many procedures iterative ones and systematic others, to search for stationary points in a response surface. Those points thereafter will be qualified as local maxima or minima or even saddle. Depending on such severe hypotesis as convexity, qualification can be *global*.

In general optimization procedures require finding a set $x \in \Omega \subset \mathbf{R}^n$ such that certain quality criterion $f : \mathbf{R}^n \to \mathbf{R}$ is maximized (minimized). The solution to the global optimization (maximization) problem can be stated as: *find a vector* $x° \in \Omega$ *such that* $\forall x \in \Omega \Rightarrow f(x) \leq f(x°) = f°$. Here Ω is the set of feasible solutions, f the objective function, and $x°$ one of the solution points.

In real-world situations the objective function and the restrictions g_i which qualifie $\Omega \subset \mathbf{R}^n$ are often not analytically tractable or are even not given in closed form, when the classical procedures based mainly in the model regularity, fail to apply[Cesareo, 1997]. For solving these problems, simulated evolution is based on the collective learning processes within a population of individuals, in the quest for survival [Baeck-alii,1995]. Each individual represents a search point in the space of potential solutions to a given problem.

There are currently three main lines of research strongly related but independently developed in simulated evolution : Genetic Algorithms (GA), Evolution Strategies (ES), and Evolutionary Programming (EP). In each of these methods, the population of individuals is arbitrarily initialized and evolves towards better regions of the search space by means of a stochastic process of selection, mutation, and recombination if appropriate.

These methods differ in the specific representation, mutation operators and selection procedures. While genetic algorithms emphasize chromosomal operators based on observed genetic mechanisms (e.g., crossover and bit mutation), evolution strategies and evolutionary programming emphasize the adaptation and diversity of behavior from parent to offspring over succesive generations. The main contributions in the evolutionary computation approach are:

- Model regularity independence.

- Population search × individual search (classical).

- General meta-heuristics.

A typical field where classical optimization fails, and where Evolution Strategies could play a fruitful role is the world of nonlinear control.

In many nonlinear control problems, as those of path-following addressed here, the controller design is a difficult problem. Not only performance objectives, but even simple stability is many times difficult to guarantee under a rigorous basis.

The problem of guaranteed stability for path tracking control can be formalized in a natural way within the framework of nonlinear dynamic systems. There are several available techniques for nonlinear stability: qualitative analysis, Lyapunov methods, or passivity and conicity criteria. A complete exposition and comparison can be found in [Vidyasagar, 1993].

In this work, conicity techniques have been chosen. The advantages are that the linear dynamics can be treated by frequencial tools based on familiar Nyquist or Bode plots, and that time-delays and other infinite dimensional dynamics can be included in the models. Also, nonlinear relations have to lie inside certain conic regions, giving rise to simple geometric interpretations.

A disadvantage is that stability conditions may result too conservative. However, sufficiency is a typical problem also in other techniques (Liapunov), and can be interpreted in the sense that an extra (non-designed) robustness is achieved.

The content of the paper is as follows. In section 2, Evolutionary Meta-heuristics and Evolution Strategies are described. In section 3 a brief exposition on conic stability of nonlinear systems is given, and in section 4 examples on path-following problems are solved and discussed.

2. EVOLUTIONARY STRATEGIES

Evolution is the result of interplay between the creation of new genetic information and its evaluation and selection. A single individual of a population is affected by other individuals of the population as well as by the environment. The better an individual performs under these conditions the greater is the chance for the individual to survive for a longer while and generate offsprings, which inherit the parental genetic information.

Evolutionary algorithms mimic the process of neo-Darwinian organic evolution and involves concepts such as:

t Time or epoch.

$\sharp$ Individual.

$P(t)$ Population.

Φ Fitness.

o Variation, Selection, etc.

A simple evolutionary algorithm follows:

```
t ← 0
initialize P(t)
evaluate Φ(P(t))
while not    terminate
   P'(t) ← variation P(t)
   evaluate Φ(P'(t))
   P(t + 1) ← select P'(t) ∪ Q
   t ← t + 1
end
```

Q is a special pot of individuals that might be considered for selection purposes, e.g. $Q = \{\emptyset, P(t), \cdots\}$. An offspring population $P'(t)$ of size λ is generated by means of variation operators such as recombination and/or mutation from the population $P(t)$. The offspring individuals $\sharp_i(\pi_k) \in P'(t)$ are evaluated by

calculating their fitness represented by $\Phi(f)$. Selection of the fittest is performed to drive the process toward better individuals.

In evolution strategies the individuals consist of two types of parameters: *exogenous* π which are points in the search space ($\pi \equiv x$), and *endogenous* σ which are known too as *strategic parameters*. Then $\sharp = \sharp(\pi, \sigma)$. Variation is composed of *mutation* and *self-adaptation* performed independently on each individual. Thus

$$\sharp(\pi', \sigma') \leftarrow \texttt{mutate}(\sharp(\pi, \cdot)) \cup \texttt{adapt}(\sharp(\cdot, \sigma)) \quad (1)$$

where mutation is accomplished by

$$\pi_i' = \pi_i + \sigma \cdot N_i(0, 1) \quad (2)$$

and adaptation is accomplished by

$$\sigma_i' = \sigma_i \cdot \exp\{\tau' \cdot N(0, 1) + \tau \cdot N_i(0, 1)\} \quad (3)$$

where $\tau' \propto (\sqrt{2n})^{-1}$ and $\tau \propto (\sqrt{2\sqrt{n}})^{-1}$. $N(0, 1)$ indicates a normal density function with expectation zero and standard deviation 1.

Selection is based only on the response surface value of each individual. Among many others are specially suited:

- Proportional. Selection is done according to the individual relative fitness $p(\sharp_i) = \frac{\Phi(f(\pi_i))}{\sum_k \Phi(f(\pi_k))}$

- Rank-based. Selection is done according to indices which correspond to probability classes, associated with fitness classes.

- Tournament. Works by taking a random uniform sample of size $q > 1$ from the population, and then selecting the best as a survival, and repeating the process until the new population is filled.

- (λ, μ). Uses a deterministic selection scheme. μ parents create $\lambda > \mu$ offsprings and the best ν are deterministically selected as the next population $[Q = \emptyset]$.

- $(\lambda + \mu)$. Selects the ν survivors from the union of parents and offsprings, such that a monotonic course of evolution is guaranteed $[Q = P(t)]$

3. CONICITY CRITERION

A brief exposition on the conicity criterion will be given now, based on [Vidyasagar, 1993]. Consider two dynamical systems, G and H:

$$x_1 \xrightarrow{G} y_1, \quad x_2 \xrightarrow{H} y_2. \quad (4)$$

The (positive) feedback connection with external inputs z_1, z_2, entering additively ($x_1 = z_2 + y_2$, $x_2 =$

$z_1 + y_1$), gives rise to a feedback dynamic system, denoted by $\{G, H\}$:

$$(z_1, z_2) \xrightarrow{\{G,H\}} (x_1, x_2). \tag{5}$$

With a proper assignation to the signals of the role of reference, disturbance and error signal,this scheme represents the typical control loop. The *Small Gain* criterion says that this feedback loop $\{G, H\}$ is stable if:

$$g(G) \cdot g(H) < 1, \tag{6}$$

where $g(\cdot)$ denotes the functional or dynamical gain of a system, defined as:

$$g(G) = \sup_{\|x\| \neq 0} \frac{\|Gx\|}{\|x\|}. \tag{7}$$

Any signal norm is valid, provided that the same norm is used in both gains. The typical choice, followed here, is the L_2 norm, so that $\|x\|$ denotes the square-integral or total energy of the signal $x(t)$.

A practical problem arises here, as the computation of the gain $g(\cdot)$ of a general nonlinear system is a very hard numerical problem. For two classes of systems the gain computation is tractable. For linear systems given by a transfer function $G \equiv G(s)$, the gain is:

$$g(G) = \sup_{\omega} \overline{\sigma}(G(j\omega)), \tag{8}$$

where $\overline{\sigma}$ is the maximum singular value, so that the linear gain is the maximum modulus amplification over all frequencies. In case of nonlinear, but *static* blocks, they can be identified with the static characteristic $H \equiv y = H(x)$, so that:

$$g(H) = \sup_{|x| \neq 0} \frac{|H(x)|}{|x|}, \tag{9}$$

the simplification is due to the time-independence of the last expression. The factible computation of both gains in small-gain, requires to place an assumption, satisfied by many practical systems:

Assumption 1 (linear-static separability): The actual feedback system formed by plant P, controller K, is so that it can be equally represented, after block re-ordering, as an equivalent feedback of G and H:

$$\{P, K\} \equiv \{G, H\}, \tag{10}$$

so that G contains the *linear* dynamics, and H the nonlinear, but *static*, dynamics.

After this assumption, the small gain, can be used both as an analysis or as a design tool. However, it can be shown that it might become very conservative. For example, consider $G(s) = \frac{1}{s+1}$, so that $g(G) = 1$. Then, this implies $g(H) < 1$. This also implies for $y = H(x)$ lying in the sector $(-1, 1)$. It is easy to see that the good (non-conservative) sector is $(-1, \infty)$. A way to enlarge the conic bounds and reduce conservativeness is to use the *loop transformation* technique [Vidyasagar, 1993].

The idea is to add and substract the same block D (*center*) into the loop formed by G and H. The block D has to be linear for stability preservation, and static for practical norm computability. In other words, the center D has to be a (constant) matrix of adequate size. This loop transformation can be denoted by

$$\{G, H\} \longrightarrow \{ G(I + DG)^{-1}, H + D \}$$

the corresponding equivalent stability condition is:

$$g(G(I + DG)^{-1}) \cdot g(H + D) < 1$$

This equation is a general setting of the Conicity Criterion, from which other results, as the Circle Criterion, can easily be derived. It contains as undefined parameter the center D. The designer can use this degree of freedom for his particular problem. The question is how to find at least one center D that satisfies small-gain and then guarantees stability.

The search for an adequate center can be addressed in different ways. In [Aracil, 1993] this search is based on the condition $r(D) < d(D)$ to be held by two measures, the linear robustness $r(D)$, and the nonlinear deviation $d(D)$. In [Barreiro, 1997] an algorithm is proposed for an equivalent problem of cone maximality.

The difficult nonlinear nature of the question of cone searching, has suggested the idea in this paper of using Evolution Strategies. The advantages are the independence of the problem size (the same code is valid for SISO and for MIMO blocks), and the ability to combine performance objectives and guaranteed stability, in an easy traded-off way, into the design index.

In particular, using a C++ like pseudo language, the conic stability issue is implemented as follows:

$\mathcal{G}$ Linear dynamics (A_G, B_G, C_G, D_G)

$\mathcal{H}$ Nonlinear dynamics

$\mathcal{C}$ Conic Center $([], [], [], D)$

$\mathcal{GC}$ Auxiliar system

σ spectrum operator

$\Re$ Real component of a complex number.

$||\mathcal{H}||_\infty$ Norm

$\rightarrow$ Method or data structure

k_1, k_2 Constants $k_1 \approx 200, k_2 \approx 1000$

$$
\begin{aligned}
\mathcal{GC} &\Leftarrow \mathcal{G} \rightarrow \texttt{feedback}(\mathcal{C}); \\
\rho &\Leftarrow \max(\Re\{\sigma(\mathcal{GC} \rightarrow A)\}); \\
\rho &> 0 \text{ ? return } k_1 + \rho \cdot k_2 : \texttt{ continue}; \\
n_{\mathcal{GC}} &\Leftarrow ||\mathcal{GC}||_\infty; \\
n_{\mathcal{H}} &\Leftarrow ||\mathcal{H} + \mathcal{C}||_\infty; \\
&\text{return } (n_{\mathcal{GC}} \cdot n_{\mathcal{H}});
\end{aligned}
$$

4. CASE STUDY

Following [Cerezo-Ollero, 1996] the dynamic model of a vehicle while tracking a given path, assuming short time intervals and linear constant velocity v, the vehicle's lateral motion can be represented by the following equations.

$$
\begin{aligned}
\dot{\xi} &= -v \sin\phi \\
\dot{\phi} &= v\gamma \\
\dot{\gamma} &= -\frac{\gamma}{\tau} + \frac{u}{\tau}
\end{aligned}
\tag{11}
$$

where γ is the vehicle's curvature. This system, as many other path-following schemes, can be modelled as a nonlinear plant in the form:

$$
P : \begin{cases} \dot{x} = Ax + Bu + f(x) \\ y = Cx, \end{cases}
$$

where the observed state $y = Cx \in \mathbf{R}^2$ contains the path-following errors, A, B are matrices of adequate sizes, $C = \begin{bmatrix} 1 & 0 & 0 \\ 0 & 1 & 0 \end{bmatrix}$ and $f(\cdot)$ is an $\mathcal{O}(x^2)$ nonlinear plant contribution. The individual is thus

$$
\sharp(\pi, \cdot) = \sharp(\pi_c \cup \pi_0, \cdot)
\tag{12}
$$

representing by π_c the controller and by π_0 the conic center unknown parameters. A quite general controller structure would include both linear and nonlinear controller actions. An effective controller form could be:

$$
K : \begin{cases} u = \Phi(y, z) \\ z(s) = \mathcal{K}(s)y(s), \end{cases}
$$

Getting the idea in Assumption 1, the actual loop $\{P, K\}$ is equivalent to the linear-static loop $\{G, H\}$ given by:

$$
\begin{aligned}
G(s) &= \begin{pmatrix} I \\ \mathcal{K}(s)C \end{pmatrix} (sI - A)^{-1} \\
H(x, z) &= f(x) + B\Phi(y, z)
\end{aligned}
$$

If, for simplicity, the controller is linear, $u(s) = \mathcal{K}(s)y(s)$, then, the loop can be broken into:

$$
\begin{aligned}
G(s) &= (sI - A - B\mathcal{K}(s)C)^{-1} \\
H(x) &= f(x)
\end{aligned}
$$

In that case, after the loop transformation $\{G, H\} \rightarrow \{G_1, H_1\}$, the result is:

$$
\begin{aligned}
G_1(s) &= (sI - A - B\mathcal{K}(s)C - D)^{-1} \\
H_1(x) &= f(x) - Dx
\end{aligned}
$$

Up to now, the undefined objects are control actions $\mathcal{K}(s)$, ($\Phi(\cdot)$, if used) and auxiliary objects D (center). Using Evolution Strategies it is possible to tune them to obtain performance and stability objectives. Particularly, the local stability-performance around the origin, if $f(\cdot)$ is an $\mathcal{O}(x^2)$ nonlinearity, depends only on the local linear dynamics given by $G(s) = (sI - A - B\mathcal{K}(s)C)^{-1}$, that can be adjusted tuning $\mathcal{K}(s)$. In relation to stability, the conicity condition is:

$$
g(G_1) \cdot g(H_1) < 1.
$$

The small gain condition depends on the free objects $\mathcal{K}(\pi_c)(s)$ (controller) and $D(\pi_0)$ (center), that can be easily parametrized and coded so that Evolution Strategies could perform their search.

As an application, letting $v = 0.85[\text{m/s}]$ and $\tau = 0.1[\text{s}]$ were obtained:

```
num_individuals = 30
num_epochs      = 32
num_funct_eval  = 1167
%---------------
%   CONTROLLER
%---------------
W1 = [ 0.18354  -0.013983 0.096048  ;
      -0.073657  0.79757 -0.79186 ]';

b1 = zeros(3,1);

W2 = [ 0.48772 -0.20795 -0.33355  ;
      -0.67667 -0.8671  0.77392  ;
```

```
        0.76452 -0.83356 -0.55579 ]';

b2 = zeros(3,1);

W3 = [-2.1331  ;
      -1.073   ;
       0.027141 ]';

b3 = zeros(1,1);
g1 = 'tansig';
g2 = 'tansig';
g3 = 'purelin';
%---------------
%  CENTER
%---------------
A = [ ];
B = [ ];
C = [ ];
D = [  -3.8891  -0.33093  ;
        0.33941 -3.966  ];
```

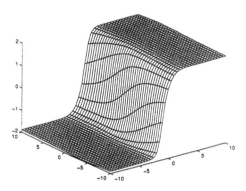

Figure 1: Actuation Surface

5. CONCLUSIONS

The conicity criterion is a suficient condition for assintotic stability under initial conditions. Good results can be obtained with ES even with the conicity condition not completely obeyed $(g(G) \cdot g(H) < 1 + \varepsilon)$. Thus, for contractive controllers, there exist a stable basin containing the origin. In abcense of performance criteria, ES gives solutions with umpredictable fashion due to its strong stochastic motivation. Depending on the seed of the randomic generators, a rich set of controllers can be obtained. The results previously presented where obtained without performance criteria and with $\varepsilon = 0.1$. Simulations shown the feedforward neural net presented as a well behaved controller.

6. ACKNOWLEDGEMENTS

This work is supported by CICYT, under project TAP96-1184-C04-03.

7. REFERENCES

[Aracil, 1993] Aracil, J., Ollero, A., Cerezo, A. and Barreiro, A. Stability of Fuzzy Control Systems, in *An Introduction to Fuzzy Control*,D. Driankov, H. Hellendoorn, M. Reinfrank (Eds.), pp. 245-292, Springer Verlag (1993).

[Baeck-alii,1995] Hans-Paul Schwefwel and Rudolph G. *Contemporary Evolution Strategies* In: Advances in Artificial Life. Third International Conference on Artificial Life, vol. 929 of Lecture Notes in Artificial Intelligence, pages 893-907. Springer, Berlin.

[Barreiro, 1997] Barreiro Blas, Antonio (1997) *On the Stabilizing Conic Sectors Obtained Using Small Gain Techniques Automatica* Vol.33, No. 6 pp 1155-1161.

[Cerezo-Ollero, 1996] García-Cerezo, A. Ollero, A. Martínez, J. L. *Design of a Robust High-Performance Fuzzy Path Tracker for Autonomous Vehicles.* In: Int. J. of Systems Science, vol 27, n° 8, pp 799-806. 1966.

[Cesareo, 1997] Cesareo Raimúndez, J. *Strategias Evolutivas y su aplicación en la Síntesis de Controladores* Doctoral Thesis Universidade de Vigo 1997.

[Schwefel-alii,1997] Thomas Bäck, Ulrich Hammel, Hans-Paul Schwefwel *Evolutionary Computation: Comments on the History and Current State* In: IEEE Trans. on Evolutionary Computation, vol. 1 n° 1, April 1997.

[Vidyasagar, 1993] Vidyasagar, M. (1993)*Nonlinear System Analysis*, Prentice-Hall.

STEERING CONTROL FOR CAR CORNERING BY MEANS OF LEARNING USING NEURAL NETWORK AND GENETIC ALGORITHM

Akihiko Shimura* and **Kazuo Yoshida****

* Graduate School of Science and Technology, Keio University
** Department of System Design Engineering, Faculty of Science and Technology,
Keio University

3-14-1, Hiyoshi, Kohoku, Yokohama, Kanagawa, Japan
Fax +81 45 560 1783, Tel +81 45 560 1289, E-mail yoshida@sd.keio.ac.jp

Abstract: Car drivers learn steering operation with exercises, but car dynamics is nonlinear at high speed situation on rough roads or low friction roads. Although skillful drivers might control cars for such nonlinear dynamics, it is difficult for ordinary drivers to control cars for such a situation. In this paper, steering operation for cornering is learned by using a neural network (NN) and a genetic algorithm (GA). The NN controller drives car autonomously with visual information and car states. The inputs to the NN controller are the direction and the curvature of the object path, and the lateral position, the yaw rate and the slip angle of the car. The output from the NN controller is the front steering angle. 4 wheel nonlinear car model with the magic formula of pure cornering is used for an analytical model. The NN controller acquires the driving operation on the curved road as a result of 30 generations iteration of the GA learning. It drives the car successfully on learned and non-learned curved roads. And, it shows the operation that is similar to the counter steering operation which is used by World Rally Championship drivers at tight curved roads. It achieves higher manoeuvrability than any other positive steering controller by using the counter steering. As a result, the availability of the NN controller learns by the GA algorithm for vehicle autonomous driving in nonlinear region is shown. *Copyright © 1998 IFAC*

Keywords: Automotive control, Genetic algorithms, Learning algorithms, Neural networks, Nonlinear control

1.INTRODUCTION

From the viewpoints of safety, comfortably and traffic efficiency, vehicle motion control is important. Cars have nonlinear dynamics, because of the characteristic of force between road and tire is saturated at large slip situation. There are many approaches and available technologies in linear region, but there are not so many in nonlinear region. Generally, combined control of steering and traction is significant in nonlinear region, because it is difficult to control cornering force at the large slip region. On the other hand, many 2WD cars match evenly with 4WD cars in World Rally Championship. And, it was demonstrated by a mathematical analysis that the counter steering is useful for cornering steering (Ono, *et al.*, 1995). Therefore, it is possible to enhance drive performance with steering control only.

Driving situation always changes, and car drivers' operation aren't so exactly. The neural network (NN) is suitable to learn the operation of drivers, since it is robust and smooth (Kageyama, *et al.*, 1994). But, it is hard to make teaching signals at various roads for the NN, and car drivers learn steering operation by trial-and-error. The Genetic Algorithms (GA) learning doesn't require teaching signals and learns by trial-and-error. Therefore, the driver's learning process is similar to the GA learning.

25

Table 1 Specifications of Car

Parameters	Value
Mass [kg]	2002
Yaw Moment of Inertia [kg m^2]	2882
Front Wheel to G.C. [m]	1.126
Rear Wheel to G.C. [m]	1.171
Tread [m]	1.4
Cornering Power in Linear Region (approx.)[N/rad]	F: 39000 R: 46000

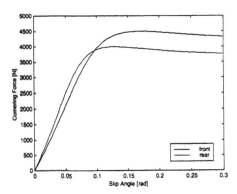

Fig. 1 Characteristic of Road-Tire Force

In this paper, nonlinear steering control strategy with NN is learned with GA. And its usefulness is examined by carrying out computer simulations.

2. DESIGN OF CONTROL SYSTEM

2.1 Car model

The vehicle model (Abe, 1992) is a 4 wheels, 2WS and 3 D.O.F. model with nonlinear road-tire force characteristics. The parameters of the vehicle model are based on the measured data of a small truck on paved road. The specifications of the model are shown in Table 1. The road-tire force model is the pure cornering magic formula (Bakker, *et al.*, 1989). The characteristic of the road-tire force model is shown in Fig. 1. The initial speed of the car is 20 m/s. There is not any traction force in the pure cornering model, so that the vehicle speed decreases gradually due to the cornering resistance.

2.2 NN controller

The NN controller drives a car like a car driver as shown in Fig. 2, and then it must sense information and operate like the car driver. The car driver senses direction and curvature of object path, the car position and motion with vision and feeling of car states. The inputs to the NN controller are two angles of object path, lateral position of the car, yaw rate and slip angle as shown in Fig. 3. These inputs include the dominant information that the car driver senses. The output from the NN controller is the steering angle of front wheel as well as

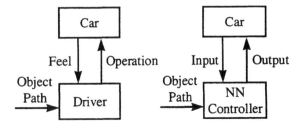

Fig. 2 Relation of Driver / NN Controller and Car

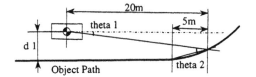

Fig. 3 Visual Information to NN Controller

Table 2 Car Driver's Feel and Operation and NN Controller's I/O

Driver's Feel	Input to NN
Car Lateral Position in Lane	d1
Direction of Object Path	theta 1
Curvature of Object Path	theta 2
Changing Rate of Car Direction	Yaw Rate
Direction of Slip	Slip Angle

Driver's Operation	Output from NN
Steering Wheel	Front Steering Angle

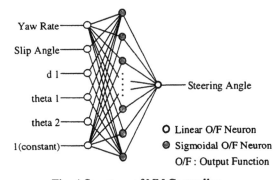

Fig. 4 Structure of NN Controller

car driver's operation. Table 2 shows the car driver's feel and operation, and the NN controller's I/O. The visual information to the NN controller include the information that is used by a conventional driver model. In addition, the usage of car states in the NN controller enables expanding controllable region in slip angle - yaw rate plane (Drive Envelope). The NN controller is a 3 layered NN as shown in Fig. 4. The hidden layer has tangential sigmoid output function with a constant input for offset. The output layer has linear output function, and doesn't have constant input, because of NN controller's symmetry. A driver's steering operation is symmetric with right and left of curvature and car motion, thus NN controller's steering operation and structure must be also symmetric. The inputs and output are normalized from -1.0 to 1.0.

2.3 GA learning strategy

The evolutionary process of the GA is shown in Fig. 5. In the initial step, 30 individuals are generated at randomly. But, individuals that don't steer straight at straight road are eliminated and regenerated. The distribution range of weight from input layer to hidden layer is -5.0 to 5.0. The distribution range of weight from hidden layer to output layer is -2.0 to 2.0. In the selection step, individuals are selected by the stochastic universal sampling method (Baker, 1987). Also the elitist individuals (Elite) are selected by the elitist preserving selection. In the crossover step, the Elite and some other individuals survive as they are, but majority individuals are replaced with children. The parents individuals of the children are selected randomly from all individuals. The children's weights of hidden and output layer neurons are generated with the 1-point crossover method (Chambers, 1995) from parents' weights. In the mutation step, some individuals mutate. Also the Elite never mutate. The mutation changes weights of a selected neuron randomly at a probability. The neuron is selected randomly in hidden layer or output layer.

The fitness index is defined as the sum of scores of numerical simulations on ten object paths. The paths are simple curved roads as shown in Fig. 6, and their radius and angle are shown in Table 3. The simulation period is 12 seconds. On all object paths, the car can pass out the curves in 12 seconds. The score of each simulation is the sum of tracing score and reaching score. The tracing score is the sum of scores of each 0.2 sec in simulation period. The score of each time period is inverse squared distance from the running trajectory to object path. If the distance is less than 1.0, the score of each time period is 1.0. Then the maximum of the tracing score is 59. The tracing score leads the running trajectories to be along the object paths. The reaching score is the sum of scores at two points shown in Fig. 6. The score of point A is 10 (15 at point B) times of inverse squared distances from the running trajectory to the point. If the distance is short enough, the score of point A is 11 (15 at point B). Then the maximum of reaching score is 26. The reaching score leads the running trajectory to reach exit of the curve. Therefore, the maximum score of each road is 85, and the maximum fitness index of each individual is 850.

3. Simulation results

After 30 generations learning, some individuals get cornering operation successfully. The learning history of the fitness is shown in Fig. 8. The running trajectory, direction and steering angle of the car with the GA learned NN controller on learned roads are shown in Fig. 9 and 10. These on non-learned roads are shown in Fig. 11. In Fig. 9 and 11, the slip angle and the steering angle are magnified 5 times. The NN controller gets 832

points fitness. It drives the car successfully at tight and loose corner on learned and non-learned roads. At the tight corner where the radius is over 40m, it can drive a car successfully, and the large slip angle and the negative steering operation appeared and look like the counter steering. Fig. 12 shows that the slip angle and the yaw rate locus at the tight corner are out of drive envelope with a positive steering operation. The drive envelope is a limit of controllable region with a positive steering operation (Inagaki, *et al.*, 1994). It shows that

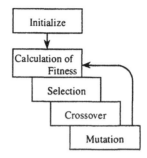

Fig. 5 GA Learning Flow

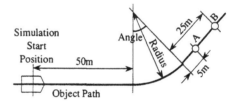

Fig. 6 Object Path Figure

Table 3 Object Path Figure

No.	Radius (m)	Angle (deg)
1,2	80	45 (right,left)
3,4	65	45 (right,left)
5,6	65	90 (right,left)
7,8	45	45 (right,left)
9,10	40	90 (right,left)

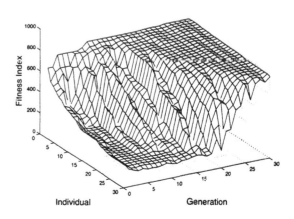

Fig. 8 Learning History of Fitness Index

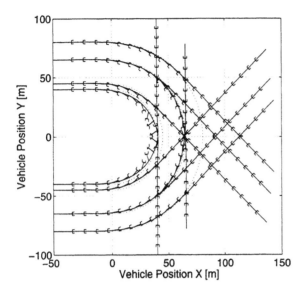

Fig. 9 Car Running Trajectory on Learned Roads

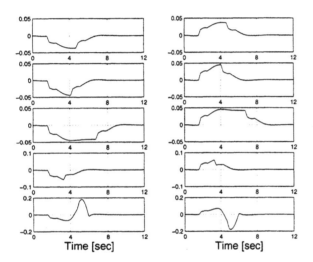

Fig. 10 Steering Angle on Learned Roads

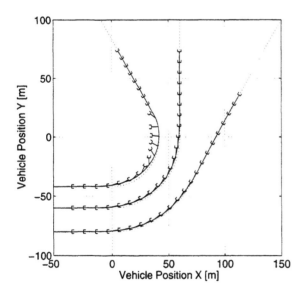

Fig. 11 Car Running Trajectory on Non-learned
Roads

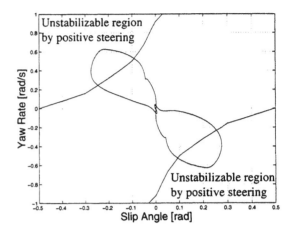

Fig. 12 Slip Angle - Yaw Rate Locus at Tight Corner
(Radius: 40m, Angle: 90deg(right,left))

the NN controller achieves higher yaw rate and more agile maneuver than any other controllers that use positive steering only. The NN controller learned and acquired the counter steering operation.

4. Conclusions

By applying the GA learning, the NN controller acquires human driver's steering operation on curved roads successfully. The inputs and output are given corresponding to human driver's senses and operation. The NN controller is learned on various simple curved roads. As a result, it is shown that the NN based on GA learning can acquire nonlinear and skillful human driver's steering operation.

References

Abe, M. (1992). *Vehicle Dynamics and Control*, Chap. 3. Sankai-do, Tokyo

Bakker, E, *et al.* (1989). A New Tire Model with an Application in Vehicle Dynamics Studies. In: *SAE paper*, No. 890087

Baker, J. E. (1987). Reducing bias and inefficiency in the selection algorithm. In: *Proc. Second International Conference on Genetic Algorithms*, pp. 14-21.

Chambers, L. (1995). *Practical Handbook of Genetic Algorithms: Applications*, Vol. 1, Chap. 4. CRC Press, Boca Raton

Doi, S., E. Ono, and S. Hosoe (1995). Anti-Spin Control by H∞ Control. In: *Proc. JSAE*, No. 954, pp. 141-144.

Inagaki, S., *et al.*(1994). Analysis on Vehicle Stability in Critical Cornering Using Phase-Plane Method. In: *Proc. of the International Symposium on Advanced Vehicle Control*, pp. 287-292

Kageyama, I., et al. (1994). Modeling of Driver-Vehicle System with Neural Network. In: *J. of JSAE*, **Vol. 48**, No. 12, pp. 5-11.

A PARAMETRIC CONTROL OF INDIVIDUAL WHEEL DRIVE VEHICLE

Pavel A. Shavrin

*Department of Electrical Engineering
Togliatti Polytechnic Institute
Belorusskaya 14, Togliatti 445667, Russia*

Abstract: The paper presents an effort to advance the problem of vehicle behavior control via "axle-axle" type individual wheel drive. The goal is to make a reasonable car such that its behavior would be similar to a conventional car in the good driving conditions but it would be able to keep its orientation and driving efficiency in the wide range of extreme situations. For this aim the vehicle parameter which represents some combination of the axle slip factors and is adjusted by the axle torque variations is used as a control input. In effect, it is the question of vehicle parametric control. The algorithm is based on the Sliding Mode Control technique. The simulations provide a comparison between the conventional and controlled cars under different simulation tests such as lane change, variation of the mass distribution, mixed braking and turn at a low speed. It is shown by theoretical analysis and computer simulations that control goal is reached and a conflict between driver habits and designer wishes is practically eliminated at the same time. *Copyright © 1998 IFAC*

Keywords: vehicle dynamics, ergonomics, parametric control, sliding mode control, computer simulation.

1 Introduction

At present there exist the stable trends to designing a control system producing not only a local regulation of the wheel forces under the different road-tire conditions but the vehicle body, speed and direction control as well (Ono, *et al.*, 1993; Amano, *et al.*, 1990; Ashley, 1995). However this problem is not only and not so much to obtain a control algorithm and some hardware. It is the question of control system ergonomics as well. Actually, the regular solution assumes that the state space variables are maintained at the desired levels or their prescribed variations. So the driver is practically imposed by someone's power. Of course, the solution of this question requires a detailed investigations

coming out of this paper limits. In the meantime, the following approach can be considered as one of the possible and reasonable solutions. It needs as much as possible to use the natural properties of the car when designing the control system making a few correction of its behavior to keep the motion stability. The driver gets used to the car such as it is and we only need to rescue him from unpleasant sensations and consequences related to vehicle unstability. In the most cases the designer uses one of the control influences such as wheel torque difference of opposite vehicle sides. Meanwhile, it is well known that road-tire adhesion characteristics also depend on tractive or braking forces though by a complex manner. Either way, an additional control helm appears for the car to adjust its behavior via road-tire adhesion variations. In effect, it is the question of vehicle parametric control.

This paper presents an effort to advance the problem of vehicle behavior control via "axle-axle" type individual wheel drive. The goal is to make a reasonable car such that its behavior would be similar to a conventional car in the good driving conditions but it would be able to keep its orientation and driving efficiency in the wide range of extreme situations.

In the first section the enough general dynamic vehicle model is substantiated. The next part is devoted to the controller design. The algorithm is based on the Sliding Mode Control technique (Utkin, 1992). The application of the sliding mode technique for control design seems to be justified by the natural property of the ABS(ACS) actuators such as discontinuity of the operational mode. The simulations provide a comparison between the conventional and controlled cars under different simulation tests such as lane change, variations of the mass distribution, mixed braking and turn at a low speed.

In conclusion some advantages of the proposed control system are outlined and the directions for further investigations are planned.

2 Dynamic Model and Problem Statement

As it is known every model should satisfy to both conflicting requirements of simplicity and adequacy. As for a vehicle the road-tire description represents the main difficulties. In particular it explains the existing variety of the vehicle models. The so called "bicycle" or single track models based on the hypothesis of linear sideslip are the most widespread (Ellis, 1975). However, their application region is limited by the small deviations of stationary mode if furthermore these modes exist. The more complex and general models described by the equations in quasicoordinates (Turaev, et al., 1987) permit to examine the vehicle behavior both in slip mode of all wheels or their part and without slip. Furthemore the vehicle motion without slip is particular case of the slip motion. But these models admit a manifold of versions caused first of all by a difference in the road-tire description. All other model occupy some intermediate position and fill by such a way the compromise field between simplicity and adequacy requirements.

Following to these rules let us consider the single mass flat model of the car with three degree of freedom in longitudinal, lateral and rotational direc-

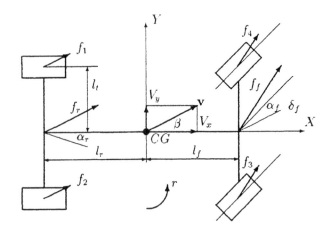

Figure 1: Model of the individual wheel drive vehicle

tions (See Fig.1) It is assumed that center of gravity occupies the stationary position and it is located on the longitudinal axle of vehicle. For description of the model the following symbols are utilized: V_x is longitudinal velocity; V_y is lateral velocity; $\mathbf{V}$ is velocity vector; r is yaw rate; β is sideslip angle; α_r, α_f are deviation angles of the rear and front axles; CG is center of gravity; l_r, l_f are distances of CG location between rear and front axles; L is wheel base; $\mathbf{f}_r, \mathbf{f}_f$ are lateral forces at the rear and front axles; $\mathbf{f}_1 \ldots \mathbf{f}_4$ are tangent reactions of road contact; δ_f is steering angle; l_t is wheel tread; m is vehicle mass; J is inertia moment with respect to the vertical axis cross with CG.

In these conditions the enough general vehicle equations have a form

$$
\begin{cases}
m\dot{\mathbf{V}} = \sum\limits_{i=1}^{4} \mathbf{f}_i; \\
J\dot{\mathbf{r}} = \sum\limits_{i=1}^{4} \mathbf{R}_i \times \mathbf{f}_i;
\end{cases}
\tag{1}
$$

Here $\mathbf{f}_i$, $\mathbf{R}_i$ denote force vectors and radius vectors of their attached points with respect to CG.

Assuming the mass being concentrated in the rear and front axles we obtain the inertia moment as

$$ J = ml_r l_f. $$

Making a discription of the road-wheel interaction we shall regard that tire properites are identical in both longitudinal and lateral directions. So we can accept as a basis the classic expression of the road-tire characteristic being a linear function of the relative slippage, expanding it on the vector case. In other words, the road surface reaction is introduced by the form

$$ \mathbf{f} = -\lambda \mathbf{s}, \tag{2} $$

where $\lambda = \lambda(\mathbf{V}, \mathbf{s}, \mathbf{R}, \mu, R_z)$ is the scalar function of vector arguments such as wheel speed $\mathbf{V}$, speed of slippage $\mathbf{s}$, wheel radius vector related to CG $\mathbf{R}$, and also adhesion factor μ and normal force R_z. In general case the slip speed vector of the i-th wheel can be written as

$$\mathbf{s}_i = \mathbf{V}_i + \mathbf{v}_i, \tag{3}$$

where $\mathbf{v}_i$ is the vector of circle wheel in the road contact. Rewrite the vehicle equations (1) taking into account (2), (3), the kinematic correlations

$$\mathbf{V}_i = \mathbf{V} + \mathbf{r} \times \mathbf{R}_i \tag{4}$$

and making notes that double vector product is

$$\mathbf{R}_i \times (\mathbf{r} \times \mathbf{R}_i) = (\mathbf{R}_i, \mathbf{R}_i)\mathbf{r} - (\mathbf{R}_i, \mathbf{r})\mathbf{R}_i = (\mathbf{R}_i, \mathbf{R}_i)\mathbf{r},$$

we obtain

$$\begin{cases} m\dot{\mathbf{V}} = -\alpha\mathbf{V} + \gamma \times \mathbf{r} - \mathbf{u}_1; \\ J\dot{\mathbf{r}} = -\beta\mathbf{r} - \gamma \times \mathbf{V} - \mathbf{u}_2, \end{cases} \tag{5}$$

where

$$\alpha = \sum \lambda_i; \ \beta = \sum \lambda_i(\mathbf{R}_i, \mathbf{R}_i); \ \gamma = \sum \lambda_i \mathbf{R}_i;$$
$$\mathbf{u}_1 = \sum \lambda_i \mathbf{v}_i; \ \mathbf{u}_2 = \sum \lambda_i \mathbf{R}_i \times \mathbf{v}_i.$$

These equations can be regarded as a most general model for the flat vehicle. The vectors $\mathbf{u}_1$, $\mathbf{u}_2$ can be examined as perturbations in this case. As you can see the motion of the imperturbed system is stable if

$$(\gamma \times \mathbf{V}, \mathbf{r}) > 0. \tag{6}$$

Indeed, for undisturbed system we have

$$(m\dot{\mathbf{V}}, \mathbf{V}) + (J\dot{\mathbf{r}}, \mathbf{r}) = -\alpha(\mathbf{V}, \mathbf{V}) + (\gamma \times \mathbf{r}, \mathbf{V}) - \beta(\mathbf{r}, \mathbf{r}) - (\gamma \times \mathbf{V}, \mathbf{r}) < 0$$

for every $\mathbf{V}, \mathbf{r} \neq 0$. Zero solution is only reached at $\mathbf{V}, \mathbf{r} = 0$.

Rewrite the equations (5) for the single track vehicle in the frame of references attached to the car so that axis OX coinsides with the longitudinal one. The kinematic expressions (4) in this frame look like

$$V_{fx} = V_x; \ V_{fy} = V_y + l_f r;$$
$$V_{rx} = V_x; \ V_{ry} = V_y - l_r r,$$

Taking into account the rules of differentiation in the moving frame we get

$$\begin{cases} m\dot{V}_x = \alpha V_x + rV_y + u_{1x}, \\ m\dot{V}_y = -\alpha V_y + \gamma r - rV_x + u_{1y}; \\ J\dot{r} = -\beta r + \gamma V_y + u_2, \end{cases} \tag{7}$$

where

$$\gamma = \lambda_r l_r - \lambda_f l_f; u_{1x} = \lambda_f v_{fx} + \lambda_r v_{rx};$$
$$u_{1y} = \lambda_f v_{fy} + \lambda_r v_{ry}; \ u_2 = \lambda_f l_f v_{fy} - \lambda_r l_r v_{ry};$$
$$v_{fx} = -v_f \cos\delta; v_{rx} = -v_r; v_{fy} = -v_f \sin\delta; v_{ry} = 0;$$
$$v_f = r_w w_f; \ v_r = r_w w_r; \ l_f = \|\mathbf{R}_f\|; \ l_r = \|\mathbf{R}_r\|;$$

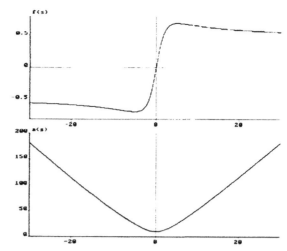

Figure 2: $f - s$ diagram and expression $a(s)$ as functions of absolute slippage.

v_f, v_r, v_{fx}, v_{fy}, v_{rx}, v_{ry} are circle wheel speeds in the road contact and their projections on the co-ordinate frame (sign "-" shows that speed vectors v_f, v_r are opposite to the direction of axis OX at zero steering angle δ); r_w, w_f, w_r are the dynamic rolling wheel radius and rotational velocities of the front and rear wheels respectively. The condition (6) in so doing is converted into

$$\gamma V_y r > 0. \tag{8}$$

Thus the solution of the control problem first of all means keeping the unequality (8) in the all driving conditions. So the examined system is the system (7) with a control input γ and a criterion (8).

3 Control design

The solution of the problem (7), (8) is considered in the class of discontinuous control systems (Utkin, 1992). With this aim in mind select a switching function

$$s_1 = \gamma - pV_y r, \quad p > 0. \tag{9}$$

Enlarge the system (7) with the equations of wheel rotation

$$\begin{cases} J_w \dot{w}_f = T_f + \lambda_f s_f r_w; \\ J_w \dot{w}_r = T_r + \lambda_r s_r r_w, \end{cases} \tag{10}$$

where

$$s_f = V_x \cos\delta + (V_y + l_f r)\sin\delta - r_w w_f; \ s_r = V_x - r_w w_r$$

are the wheel slip velocities of the front and rear axles in the road contact; T_f, T_r are torques.

Let us introduse into consideration the parameters $a_f = 1/\lambda_f$, $a_r = 1/\lambda_r$. The typical $f - s$ diagram and coefficient a are displaied in the Fig.2. As you can see everywhere

$$\begin{aligned} \frac{\partial a}{\partial s} > 0, & \quad \text{at} \quad s > 0; \\ \frac{\partial a}{\partial s} < 0, & \quad \text{at} \quad s < 0. \end{aligned} \tag{11}$$

Write (9) taking account of new symbols

$$s_1 = a_r l_f - a_f l_f - a_r a_f p V_y r = \frac{\lambda_f l_f - \lambda_r l_r - p V_y r}{\lambda_f \lambda_r},$$
(12)

and differentiate (12) along the solutions of (7), (10). So we obtain

$$\dot{s}_1 = l_f \left(\frac{\partial a_r}{\partial w_r} \dot{w}_r + \left(\frac{\partial a_r}{\partial V_x} \dot{V}_x + \frac{\partial a_r}{\partial V_y} \dot{V}_y + \frac{\partial a_r}{\partial r} \dot{r} \right) \right) -$$

$$l_r \left(\frac{\partial a_f}{\partial w_f} \dot{w}_f + \left(\frac{\partial a_f}{\partial V_x} \dot{V}_x + \frac{\partial a_f}{\partial V_y} \dot{V}_y + \frac{\partial a_f}{\partial r} \dot{r} \right) \right) -$$

$$p \dot{V}_y r - p V_y \dot{r}.$$

The enlarged system (7), (10) is a system of singularly perturbed equations because the wheel inertia moment is much less of the vehicle one. As it is known (Vasilyeva and Butuzov, 1973) this system can be decoupled if the fast motions (10) are stable at V_x, V_y, $r = const$ and they can be thrown out by $J_w = 0$. Let us assume for a while that the decomposition conditions are satisfied. Therefore, following to the standard procedure of the singular perturbed system analisys ignore the derivatives $\dot{V}_x$, $\dot{V}_y$, $\dot{r}$, just keeping the terms with $\dot{w}_f$, $\dot{w}_r$. So you can write

$$\dot{s}_1 = l_f \frac{\partial a_r}{\partial s_r} \frac{\partial s_r}{\partial w_r} \dot{w}_r - l_r \frac{\partial a_f}{\partial s_f} \frac{\partial s_f}{\partial w_f} \dot{w}_f,$$

and considering (10) together with $\frac{\partial s}{\partial w} = -r_w$, can get

$$\dot{s}_1 = \frac{r_w}{J_w} \left[-l_f \frac{\partial a_r}{\partial s_r} (T_r + r_w \lambda_r s_r) + l_r \frac{\partial a_f}{\partial s_f} (T_f + r_w \lambda_f s_f) \right].$$
(13)

Let us form the torques at the vehicle axles as

$$T_f = -u_f - r_w \lambda_f s_f + v_1; \qquad |T_f| < T_{f\,max};$$
$$T_r = u_r - r_w \lambda_r s_r + v_2; \qquad |T_r| < T_{r\,max}.$$
(14)

Here

$$u_f = u l_f \mathrm{sign}(s_f); \ u_r = u l_r \mathrm{sign}(s_r); \ u = k \mathrm{sign}(s_1);$$
$$k > 0; \quad v_1 = l_r U; \quad v_2 = l_f U;$$

u, U are the new controls; $T_{f\,max}$, $T_{r\,max}$ are the maximum wheel torques caused by the road-tire adhesion; $u_{f\,eq}$, $u_{r\,eq}$ are the equivalent controls (Utkin, 1992); k is an adjusted factor. Substituting (14) into (13) you can find

$$\dot{s}_1 = -\frac{r_w l_f l_r}{J_w} \left[\left(\frac{\partial a_r}{\partial s_r} \mathrm{sign}(s_r) + \frac{\partial a_f}{\partial s_f} \mathrm{sign}(s_f) \right) u + U \left(\frac{\partial a_f}{\partial s_f} - \frac{\partial a_r}{\partial s_r} \right) \right].$$
(15)

Following to (11)

$$\left| \frac{\partial a_r}{\partial s_r} \mathrm{sign}(s_r) + \frac{\partial a_f}{\partial s_f} \mathrm{sign}(s_f) \right| = \left| \frac{\partial a_r}{\partial s_r} \right| + \left| \frac{\partial a_f}{\partial s_f} \right|$$

and, hence,

$$\left| \frac{\partial a_f}{\partial s_f} - \frac{\partial a_r}{\partial s_r} \right| \le \left| \frac{\partial a_r}{\partial s_r} \right| + \left| \frac{\partial a_f}{\partial s_f} \right|.$$

It follows that inequalities

$$|U| < |u| \ \text{или} \ |v_1| < |u_f|, \ |v_2| < |u_r|, \quad (16)$$

being satisfied allow the condition $\dot{s}_1 s_1 < 0$ for the existence of the sliding mode on the surface $s_1 = 0$.

Let us spend the rest of control resource U for driving speed regulation. Since in this case the control accuracy do not have a particular sense, select for simplicity the average rotation wheel speed of the front and rear axles $w = w_f + w_r$ as another controlled variable. Substituting (14) into the wheel rotation equations (10) you can obtain

$$J_w \dot{w} = J_w (\dot{w}_f + \dot{w}_r) = -u_f + u_r + v_1 + v_2 =$$
$$(l_r \mathrm{sign}(s_r) - l_f \mathrm{sign}(s_f)) u + (l_f + l_r) U.$$

Let us choose this control in the form

$$U = -C \mathrm{sign}(w_f + w_r - 2w_z)$$

where $w_z \sim const$ is a desired riding speed. Making notes that $u = u_{eq}$ and

$$|l_r \mathrm{sign}(s_r) - l_f \mathrm{sign}(s_f)| < |l_f + l_r|$$

when sprung the sliding mode on the surface $s_1 = 0$ we reveal that the sliding mode will rise on the surface $s_2 = w_f + w_r - 2w_z$ if $|u_{eq}| < |U|$. In fact, given algorithm realizes the hierarchy control method (Utkin, 1992) in which the sliding mode appears on the surface $s_1 = 0$ first but then on the surface $s_2 = 0$, and the required hierarchy is set by (16). In particular the selection of the controls v_1, v_2 in the form

$$v_1 = l_r U = -C_1 \mathrm{sign}(w_f + w_r - 2w_z);$$
$$v_2 = l_f U = -C_2 \mathrm{sign}(w_f + w_r - 2w_z).$$

where

$$C_1 \le T_{f\,max} - |u_{f\,eq} + r_w \lambda_f s_f|;$$
$$C_2 \le T_{r\,max} - |u_{r\,eq} - r_w \lambda_r s_r|,$$
(17)

satisfies to (16). It means that only the rest of control is spent on the riding speed regulation, and effective speed control is possible when the car orientation is kept.

The vehicle dynamics in the closed control loop is described by the equations

$$\begin{cases} m \dot{V}_x = -\alpha V_x + r V_y + u_{1x}; \\ m \dot{V}_y = -(\alpha + p r^2) V_y - r V_x + u_{1y}; \\ J \dot{r} = -(\beta + p V_y^2) r + u_2, \end{cases}$$
(18)

where u_{1x}, u_{1y}, u_2 are limited and controlled functions.

4   Simulation results

The simulation study was carried out for the comparison of the vehicles supplied with the different control systems. As a prototype for comparison named as conventional car was taken the vehicle with typical ABS/ACS control algorithm which is realized in this case by $k = 0$ and respects to a four wheel drive car with "axle-axle" type drive. The controlled car is denoted as Car A. Car B denotes the conventional car.

The vehicle parameters: $m = 1400\,[Kg]$; $l_r = 1.2\,[m]$, $l_f = 1.2\,[m]$; $l_t = 0.7\,[m]$; $r_w = 0.28\,[m]$. The nonlinear model of the most general form (1) was used for simulation. Furthermore the wheel rotation equations were added to the vehicle system. The road-tire adhesion was approximated by an expression (Dik, 1987)

$$\mathbf{f} = -\mu \frac{R_z}{\|\mathbf{s}\|} \sin\left(k \arctan\left(n \frac{\|\mathbf{s}\|}{\|\mathbf{v}\|}\right)\right) \mathbf{s} \qquad (19)$$

Here $k = 1.5$; $\mathbf{s}$, $\mathbf{v}$ are the slip speed vectors in road contact and wheel speed vectors respectively; μ, R_z are adhesion factor and normal force; the parameter n treats tire stiffness.

Four kinds of simulation tests are considered:
a) *Lane Change.*
The driver turns the steering wheels left with 0.1 $[rad]$ at the speed of $110\,[kmh^{-1}]$ and holds them for one second. Then he sets the wheels with zero angle. After four second time interval the driver turns the steering wheels right at the same manner. The adhesion factor μ is 0.5.
b) *Variations of Mass Distribution.*
The mass distribution is varied from $l_r/l_f = 1$ to $l_r/l_f = 0.5$ at the conditions of the previous test.
c) *Mixed Braking.*
The car brakes at the speed of $110\,[kmh^{-1}]$. The adhesion factors are distributed between right and left sides in the ratio of $0.3/0.7$.
d) *Turn at a Low Speed.*
The car turns around with the steering angle of $0.5\,[rad]$ at the riding speed of $20\,[kmh^{-1}]$. and $\mu = 0.7$.
It is assumed that the driver does not react to disturbances but only fulfills the prescribed maneuvers. The simulation results are shown in the Figures 3– 6. The vehicle position is indicated with a line.

Evaluating the simulation results as a whole you can conclude in a quite certain manner that stated control goal is reached. An enough reasonable car is obtained. Indeed, when the conventional car has a good behavior so the the response of the controlled car is practically the same. It regards in particular to the test of the turn at a low speed (Fig.6). In this sense the ergonomics requirements are practically satisfied in full measure by the elimination of the conflict between driver habits and designer wishes. At the same time in the crucial cases for example at the maneuver of the lane change (Fig.3) or at the mixed braking (Fig.5) when the car gets into the extreme situation and a much experience is demanded from the driver to correct it, the controlled car keeps its orientation and the driver needs only to direct the movement do not worrying about the vehicle stability. Moreover, it seems to be in the worst conditions, for example, when shifted CG to the rear axle (See Fig.4, $l_r/l_f = 0.5$) the control resources on the contrary is increased and the behavior of the controlled car becomes more preferable. Finally, the controlled car has practically the same braking efficiency (Fig.5).

5   Conclusion

In this study an enough simple controller is proposed for the vehicle with individual wheel drive of "axle-axle" type. It is substantial that it does not use any torque difference of each wheel axle as a control input and the vehicle motion is adjusted only via the slip factor variations. In addition it does not impose any reference path, or optimal criterion, or desired transient dynamics to the car but only adjusts in some manner the natural parameter responsible for the vehicle stability. Thereby a parametric control of the vehicle is actually realized and, in fact, a conflict between driver habits and designer wishes is eliminated.

Thus, even on the given design step it becomes clear that an individual wheel drive car can be provided with practically every properties yielding nothing to the four wheel steering vehicles (Ono, et al., 1993; Whitehead, 1988) and wining at the same time in the price. Moreover, it is possible to obtain the properties inaccessible even to the FWS vehicles , for example, it is possible to force a car to creep out of an impasse or to make a turn around its vertical axis without full slippage of the tires, or even to compel the vehicle to bounce and to jump over the obstacles. In particular it seems not difficult to realize an active suspension just via an individual wheel drive not resorting to additional devices like any controllable snubbers. However, it is a direction for further investigations.

References

Amano Y., Asano K., Okada S., Mori T., Iwama N. (1990). Model following control of hibrid 4WD vehicle. In:*Proc. 11th IFAC World Congress*, **vol.8**, (Tallinn), pp.130-135.
Ashley S.(1995). Spin control of cars. *Mechanical*

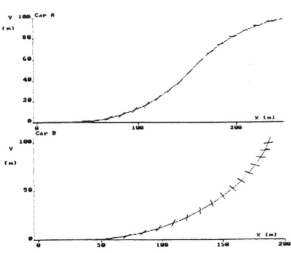

Figure 3: Maneuver of the "lane change" type.

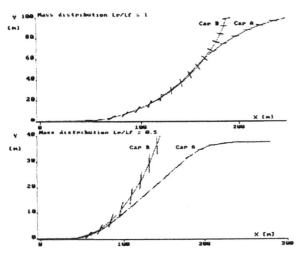

Figure 4: Reaction on the mass distribution.

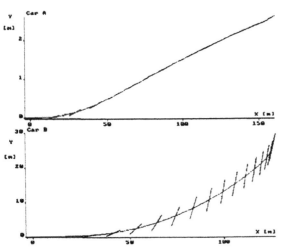

Figure 5: Mixed braking.

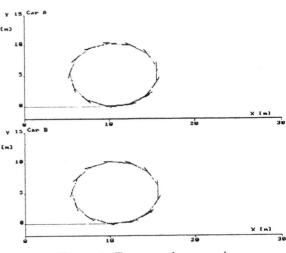

Figure 6: Turn at a low speed.

Engineering, 6.

Dik A.B.(1987). *Calculation of stationary and transient characteristics of braking wheel at sideslip motion*, Ph.D. thesis, Omsk, OmPI. (in Russian)

Ellis D.R.(1975). *Controlability of vehicle*, Mashinostroenie. Moscow. (in Russian)

Ono E., Takanami K., Iwama N., Hayashi Y., Hirano Y., Sato Y.(1993). Robust coordinated control for 4-wheel-steering and 4-wheel-drive vehicle. In: *Proc. 12th IFAC World Congress*, **vol.1**, (Sydney), pp.1-6.

Turaev H.T.. Fufaev N.A., Musarskii R.A. (1987). *Theory of rolling system's motion*, Fan, Tashkent. (in Russian)

Utkin V.I.(1992).*Sliding modes in control and optimization*, Springer-Verlag, Berlin, Germany.

Vasilyeva A.B., Butuzov V.F.(1973). *Asimptotic decomposition of the solutions of singular perturbed equations*, Nauka, Moscow. (in Russian)

Whitehead J.C.(1988). Four wheel steering: Maneuverability and high speed stabilization.*SAE paper 880642.*

HIERARCHICAL STEREO MATCHING USING CHROMATIC INFORMATION

E. de la Fuente, F.M. Trespaderne, J.R.Perán *

* *Instituto de Tecnologías Avanzadas de la Producción*
ETS Ingenieros Industriales, Universidad de Valladolid
Paseo del Cauce s/n 47011 Valladolid, Spain
E-mail: eusfue@eis.uva.es

Abstract: The key issue in making stereo vision practical is to find a combination of algorithms and hardware that led to reliable, real-time range estimation with a computer system small enough to use on robotic vehicles. In this paper, a matching algorithm to solve the central problem of binocular stereo vision systems is proposed. The algorithm utilizes two primitives of different abstraction under a hierarchical scheme: regions and linear edge segments. The hierarchical scheme greatly improves the efficiency because provides only the needed depth information with the needed degree of resolution. Chromatic information has been used. Color significantly reduces the ambiguity between potential matches and increases the accuracy of the resulting matches. *Copyright © 1998 IFAC*

Keywords: Stereo vision, Image matching, Hierarchical structure

1. INTRODUCTION

Mobile robots are desired for many tasks involving remote or hazardous operations, including aspects of planetary exploration, mining and subterranean tasks, waste clean-up, and national defense. In their operations, advanced mobile robots can utilize many different sensory modalities such as vision, force and touch but always they must include some kind of range-finder. Range-finders provide the 3D information required for automatic obstacle avoidance and navigation.

For robotic vehicle applications, there is a wide array of range-finding techniques which can be loosely classified as either active or passive. Active techniques use artificial sources of energy to illuminate the workspace while passive techniques do not require such energy sources. Time-of-flight range finders such as ultrasonic and laser range finders are examples of active methods. Time-of-flight techniques estimate the distance from the elapsed time between the transmission and the re-

ception of a ultrasonic or a laser signal. Ultrasonic range-finders are simple and cheap but present many drawbacks such as poor angular resolution, vulnerability to specular reflections and low reliability if the incident angle is higher than 15°. Laser range finders are accurate and reliable but are expensive and unsuitable in many applications due to the delicate mechanical components they need. Another example of active technique are laser light stripers. There are many variations on laser light stripers. but all of them employ a camera and a narrow light beam to illuminate the scene. By knowing the distance between the camera and the light source and by finding the light stripe in the image, triangulation can be used to find the depth of a "line" of points in the image. The simplest method sweeps a single laser line across the scene and needs to take a sequence of images to obtain a 3D representation of the scene.

One of the main disadvantages of active sensors is that they are unsuitable when two or more mobile robots are working in the same area because the sensorial systems might interfere. Furthermore, in many application areas, it may be impractical or hazardous to utilize active illumination sources, because it might obstruct human operators normally present in these environments. In these cases, passive methods are more adequate because they provide sensing of distance without the use of an artificial energy source. The passive range finding technologies are numerous: depth from defocus, shape from shading, binocular stereo vision... Depth from defocus works by taking two images of the same scene at different camera focal settings and comparing the blurriness of the two images. The approach is computationally expensive due to the Fourier transforms used to evaluate blurriness. In the shape from shading approach, surface orientation is derived from the gray-level intensity values, using information about the light source and the surface reflectivity of objects. One disadvantage of this approach is that it requires many photometric assumptions. Accuracy of shape from shading as well as depth from defocus is reduced as the range increases.

A passive technique that has been found to operate flexibly under a wide range of conditions without requiring strong photometric assumptions is binocular stereo vision. The ability to produce precise depth measurements over a wide range of distances, and the passivity of the approach, makes binocular stereo a very attractive tool for range sensing. Although the high computational cost of the algorithms has prevented it from being widely used in robotic systems, binocular stereo has been the subject of renewed attention due to the recent proliferation of highly agile stereo heads with a large number of degrees of freedom. Within the current paradigm of active vision concentrating on salient information, stereopsis can provide a fast and reliable means of extracting three dimensional information from a scene.

The key issue in making stereo vision practical is to find a combination of algorithms and hardware that led to reliable, real-time range estimation with a computer system small enough to use on robotic vehicles. In this paper a hierarchical algorithm is presented. The hierarchical structure makes possible to focus on critical areas of the image, eliminating the need for detailed examination of the entire image. The hierarchical system greatly improves the efficiency because provides only the needed depth information with the needed degree of resolution.

One mode of information that has been largely neglected in computational stereo algorithms is chromatic information. The developed algorithm demonstrates that chromatic information can be used to significantly reduce the ambiguity between potential matches, while increasing the accuracy

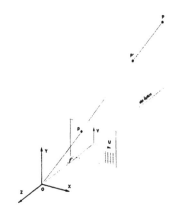

Fig. 1. The camera projection model.

of the resulting matches. Perhaps the most important advantage of having color information in addition to image irradiance information is that the color is more stable under changes in geometry that the corresponding image irradiance values (G.Healey, 1989).

2. THEORY

2.1 Perspective Projection

The imaging performed by a TV camera may be approximated by a perspective projection with a center of projection, O, and an image plane not containing O (Fig.1).

The distance between O and the image plane is the focal length f. The image, p, of a point, P, is that point where the straight line OP intersect the image plane. Let the coordinates of an object point, P, be (X, Y, Z). Then its image, p, has the coordinates $(fX/Z, fY/Z, f)$.

For any point P in the scene there exists thus a unique image point, p. The inverse is not true; P and P' in Fig.1, for instance, have the same image, p. Given only the image point, p, it is, therefore, impossible to decide whether the corresponding object point is P or P', or any other point on the straight line OP.

The idea of stereo is very simple: if one more picture of the scene is taken from another angle the position of the point P can be recovered at the point of intersection of the corresponding lines of sight. This process is called *triangulation* process. Triangulation requires to previously identify which features in the two images are projected by the same 3-D entity. This process of fusing two images taken from different viewpoints to recover depth information in the scene is known as *stereo correspondence*. Three dimensional data are inferred from the differences in the image structures provided the geometric transformation between the two cameras is determined accurately by the stereo head geometry.

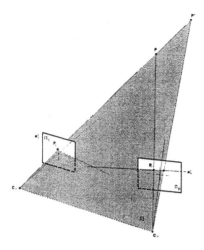

Fig. 2. Binocular stereo system geometry.

2.2 Feature Selection

The depth of world events may be computed provided their image projections are robustly segmented and successfully identified in the two images. The comparison of image brightness between the elements of the two images is a poor indication of the correspondence. Camera characteristics, scene geometry and noise make correspondent image points differ in their brightness value. Accordingly most stereo algorithms include a previous extraction feature stage. Features are more resilient to noise and permit to carry out the correspondence problem accurately. A survey of computational stereo algorithms shows a wide set of features utilized for stereo matching. Most stereo algorithms have used low-level features such as edge-based primitives. Line segments, for instance, have been widely used as features to be matched across scenes.

Although low-level features do not require a sophisticated analysis in their extraction, low-level representations makes the matching process really intricate due to the high number of features to match and to the lack of semantic content of these. Higher level image features, such as homogeneous regions, permit to share the computational burden between the feature extraction and the matching process.

In general, high-level image features have two major advantages in the matching process. First, they are usually fewer in number resulting in much faster correspondence algorithms. Second, they possess a high semantic content and, hence, high discrimination capability which can both reduce the number of match candidates, and significantly increase the matching accuracy. Although high-level features present many desirable properties, the reduction in positional accuracy is one potential disadvantage of some extensive features such as homogeneous regions. Such positional ambiguity can seriously degrade the depth estimates

computed for each match. Thus, high-level feature matching alone is unsuitable for developing an accurate stereo algorithm and an low-level feature based technique must also be included.

2.3 Hierarchical approaches

It is clear the use of a single-primitive, used in most stereo approaches, exhibit several deficiencies. It seems that these deficiencies could be overcome if features of different level would be utilized together. During the last years, hierarchical matching schemes, integrating multiple matching features of increasing complexity have been presented. The concept of hierarchy provides an efficient alternative to exhaustive search methods. Higher level features can be quickly and accurately matched while lower level feature matching is guided by the obtained results in the upper level. The major advantage is the significant computational savings resulting from the reduced search space derived from higher level matches.

One of the earliest hierarchical stereo systems was presented by (Lim and Binford, 1988). Their image hierarchy included *bodies, surfaces, junctions, curves and edgels* related by *part of* relationships. The matching begins at the highest level, bodies, and proceeds in a hierarchical way, with results at the higher levels used to constrain the matching at lower levels. (Chung, 1992) also match complex images using a hierarchy integrated by structures of line segments. Another attempt at developing a cooperative framework for multiple stereo algorithms has been presented by (Marapane and Trivedi, 1994). Their stereo approach uses regions, line segments and edgels extracted through independent monocular analysis instead of the bottom-up constructed primitives used in previous approaches. The hierarchical algorithm of (Marapane and Trivedi, 1994) embodies ideas similar to ours with respect to the use of regions and line segments in the hierarchy. However, we use a more stable similarity attributes in the high level primitive such as color and contextual information. In hierarchical systems color attributes in high features are specially valuable because lower levels depend on the accuracy of high level features matches.

A more complete discussion of hierarchical schemes can be found in (Jones, 1997).

3. CHROMATIC INFORMATION

Promising results have been attained through such hierarchical stereo algorithms. However, robotic vehicles, which must be capable of analyzing a wide array of scenes including outdoor environments, require a more robust solution. One approach to developing such a improved solution is to seek a more complete use of available image information.

One mode of information that has been largely neglected to solve the stereo correspondence problem is chromatic information. Recently, the color stereo correspondence approaches have shown that the use of color information significantly improve the discrimination and recognition capability over purely intensity-based methods (J.R.Jordan and A.C.Bovik, 1992). Perhaps the most important advantage of having color information in addition to image irradiance information is that the color of a surface, when it is adequately specified, is more stable under changes in geometry that the corresponding image irradiance values (G.Healey, 1989). There are other motivations for using chromatic information such as chromatic intensity information is an easily obtained source of information from a CCD camera and the role that color plays in human perception (Livingstone and Hubel, 1987).

3.1 *Color Representation*

The selection of a suitable color representation is one of the most vigorously discussed problems in color vision. Some of the color coordinate systems are more related to technical requirements and others are more related to human perception.

The RGB space is the most common color space. The coordinates (R, G, B) correspond to the three monochromatic primaries red, green and blue. The (R, G, B) signals are provided by almost all CCD cameras, however this representation seems not very interesting because (R, G, B) components are highly correlated. There exists a color space called HSI that represent colors using perceptual attributes described by hue, saturation and intensity. These three components can be derived from RGB coordinates:

$$I = \frac{R + G + B}{3}$$
$$H = \cos^{-1} \frac{(R - G) + (R - B)}{2[(R - G)^2 + (R - G)(G - B)]^{0.5}}$$
$$S = 1 - min(R, G, B)/I$$

The major advantage of this representation is that, under some weak assumptions (which include non-homogeneous surfaces) the color variation in a surface due to a change in the point of view makes vary only the intensity component (Fathima, 1992).

4. A HIERARCHICAL STEREO SYSTEM

Our hierarchical system uses regions and line segments extracted through independent monocular analysis. Thus, the domain of applicability is any scene whose contours can be approximated with a sequence of straight line segments (they need not lie on a plane). This includes urban, indoor and factory scenes.

Regions are extracted by modeling the image as a markov random field (MRF) (Daily, 1989) and line segments result from poligonalization of contours extracted by using a Canny - Deriche detector (Deriche, 1990).

4.1 *Structural description.*

As unary constraints such as epipolar and color are not sufficient to completely eliminate ambiguity, even in high level primitives, we will exploit the vicinity relations between them as binary constraints. The property we use to relate a couple of regions is *proximity*.

Once the regions are related in accordance with their proximity, the relative position of the two regions is defined as *above of, under of, on the left of* and *on the right of*.

This structural description is stored in a graph, where each region is represented in a node, containing individual properties such as color and area and each arc connecting two nodes contains the topological properties between regions.

The structural description for line segment is defined in a similar fashion, containing individual properties such as length and orientation.

These two structural descriptions are arranged hierarchically according to part/whole relation, so that a simple region decomposes into the several line segments which form its boundary. The new arcs introduced to couple the region graph and the line segment graph denote the relations among the two levels of the hierarchy that will be used to constraint the line segment matching.

5. MATCHING FEATURES

The feature hierarchies together with feature relationships forms the relational graphs: one representing the left image and the other, the right one. The objective is to match this two attributed graphs. The hierarchical matching start by matching at the region level and proceed to the line level. Results of stereo analysis at the region level are used to guide the analysis at the line segment level.

The constrains used to establish a set of hypothetic matches between two regions are:

Epipolar constraint: For each region in the left image, matching regions are sought in the right image within the epipolar lines spanning the region. Epipolar lines are determined by the geometry of the stereoscopic system (Fig.2).

Color constraint: Color comparison tests are applied to select the most likely matches. The stability of color attributes permit to reject unlikely matches in a reliable manner. The region R_j in the right image with color (H_j, S_j, I_j) is considered an hypothetical match of region $R_i(H_i, S_i, I_i)$ in the left image respect the color threshold δ_{color} if:

$$D_C(R_i, R_j) = \sqrt{d_I^2 + d_C^2} \leq \delta_{color} \qquad (1)$$

$d_I = |I_i - I_j|$ y $d_C = \sqrt{S_i^2 + S_j^2 - 2S_iS_j\cos\theta}$
where
$$\theta = \begin{cases} |H_i - H_j| & \text{if } |H_i - H_j| \leq \pi \\ 2\pi - |H_i - H_j| & \text{if } |H_i - H_j| > \pi \end{cases}$$

This color difference in the HSI color space has been suggested by (Tseng and Chang, 1992). Euclidean distance is not valid in this perceptual space to establish color differences.

Any area similarity constraint has been imposed because areas of homologue regions can vary dramatically due to the scene geometry and occlusions.

Let's be $\mathcal{R}^L = \{r_1^L, \ldots, r_n^L\}$ and $\mathcal{R}^R = \{r_1^R, \ldots, r_m^R\}$ the region sets of the left and right image respectively. Using the epipolar and the chromatic constraint a matching hypothesis set $\mathcal{H}_{r_i^L} = \{r_{1,i}^R, \ldots, r_{u,i}^R\}$ is obtained for each left image region r_i^L. This set of potential matched regions have been established considering only individual matches. However a pair of potential matches can be mutually incompatible. This incompatibility can be expressed using *binary constraints* derived from uniqueness and topology. The uniqueness have been implemented including a one-to-one mapping term in the region matching algorithm. The topological constraint implies that the relative positions of regions in both images remains similar.

Binary constraints between two pairs of regions are measured by a compatibility function $c(.,.;.,.)$. It provides a coefficient depending on the degree of mutual compatibility between the pairs of regions. The compatibility information is propagated modeling the matching problem as a relaxation labeling problem stated as (Hummel and Zucker, 1983): Given an initial set of labeling $\mathbf{H} = \{\mathcal{H}_{r_1^L}, \ldots, \mathcal{H}_{r_n^L}\}$ of the set of left image regions, $\mathcal{S}_L = \{r_1^L, \ldots, r_n^L\}$, find a consistent assignment with respect to the constrains over the set of nodes neighbors $\mathbf{V}_L = \{\mathcal{V}_{r_1^L}, \ldots, \mathcal{V}_{r_n^L}\}$ and $\mathbf{V}_R = \{\mathcal{V}_{r_1^R}, \ldots, \mathcal{V}_{r_m^R}\}$. The solution of this problem by a relaxation labeling procedure permit to establish the pairs of homolog regions.

Once the pairs of regions are established, the displacement in the images needed to produce the maximum overlap between homologous regions (*disparity*) is computed. This displacement is directly converted to actual depth measurements using the geometry parameters of the stereo head.

The previous process provides a 3-D map formed with frontoparallel surfaces. This map is particularly suitable for tasks like collision avoidance. However, there are others tasks such as docking that required a better accuracy. In these cases, the match information can be propagated down the hierarchy to a line-segment level. Topological constraints and *part of* relationships between regions and its boundary segments has been used to implement the hierarchical control. Hierarchical constraints can be included in a match functional as before to obtain a more formal approach. However thresholding filters have been used at this stage because hierarchical constraints and information provided by the region matching module is sufficient to establish segment matches without any ambiguity. The work of (Marapane and Trivedi, 1994) contains a complete discussion of the different types of hierarchical constraints available.

The algorithm is being experimented using indoor and outdoor scenes and promising results have being obtained. Fig.3 shows an urban scene and the matching results. The matching algorithm successfully establish the homologs of most regions even when occlusion is important. Image region segmentation and relaxation labeling are computationally expensive problems. Fortunately, they are completely parallelizable.

6. CONCLUSIONS

The proposed hierarchical stereo matching algorithm attempts to match groups of related features rather that individual units. To accomplish this, a structural description graph for each image is obtained. At the first stage the system perform the matching process at the region level using chromatic information and the vicinity relations between regions. Once a set of potential correspondences between regions are found, a compatibility function between every two couples of possible matchings is evaluated. The complete process is then modeled as a labeling problem finally solved by a relaxation process that guarantees a unique global optimum. A 3-D map can be obtained suitable for obstacle avoidance.

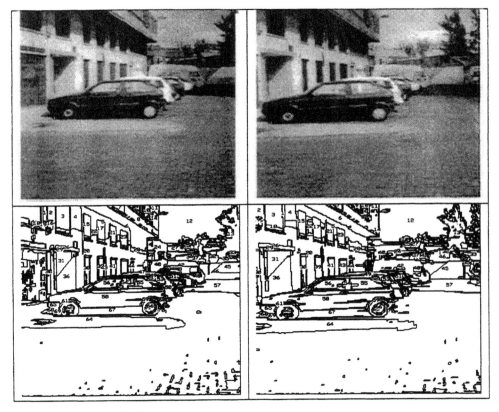

Fig. 3. a) Stereoscopic pair. b) Matching result.

If better accuracy in is required a second stage of the hierarchical system can be performed using line segments. Line segments are matched using not only individual attributes but also related properties between the line segments and regions. Hierarchical constraints provide an efficient and natural way to accomplish this task.

The robustness and flexibility of hierarchical stereo systems have been improved incorporating information about the color of object surfaces. Color information provides to region primitives a high stability and descriptive power in a natural way. Color regions are more tolerant to noise and changes in illumination increasing the accuracy of stereo matching. In hierarchical systems where the strictness of the hierarchy means that matching at lower level is heavily dependent on establishing the higher level correctly, color attributes in high level primitives are specially valuable.

7. REFERENCES

Chung, C. R. (1992). *Deriving 3-D Shape Descriptions From Stereo Using Hierarchical Features*. Univ. of Southern California. Los Angeles, CA.

Daily, M. J. (1989). Color image segmentation using markov random fields. *IEEE Conference on Vision and Pattern Recognition*.

Deriche, R. (1990). Fast algorithms for low-level vision. *IEEE Transactions on Pattern Analysis and Machine Intelligence* 12, 78–87.

Fathima, S.T. (1992). Data and model-driven selection using color regions. *Proc. Image Understanding Workshop* pp. 705–716.

G.Healey (1989). Using color for geometry-insensitive segmentation. *J. Optical Society of America* 6, 920–937.

Hummel, R.A. and S.W. Zucker (1983). On the foundations of relaxation labelling processes. *IEEE Transactions on Pattern Analysis and Machine Intelligence* 5, 267–287.

J.R.Jordan and A.C.Bovik (1992). Using chromatic information in dense stereo correspondence. *Pattern Recognition* 25, 367–383.

Lim, H.S. and T.O. Binford (1988). Structural correspondence in stereo vision. *Proc. Image Understanding Workshop* 2, 794–808.

Livingstone, M.S. and D.H. Hubel (1987). Phychophysical evidence for separate channels for the perception of form, color, movement and depth. *J. Neuroscience* 7, 3416 – 3468.

Marapane, B. and M.M. Trivedi (1994). Multiprimitive hierarchical (mph) stereo analysis. *IEEE Transactions on Pattern Analysis and Machine Intelligence* 16, 227–240.

Tseng, D.C. and C.H. Chang (1992). Color segmentation using perceptual attributes. *Proceedings 11 th IAPR International Conf. on Pattern Recognition* 3, 228 –231.

MOTION PERCEPTION VIA TWO CHANNELS CORRELATION MECHANISM: VLSI PROPOSAL FOR AN OBJECT TRACKING SYSTEM IMPLEMENTATION

Goce V. Shutinoski[1], Tomislav A. Dzhekov[1], Vanco B. Litovski[2]

[1] *"St.Cyril and Methodius" University - Faculty of Electrical Engineering*
P.O.Box 574   91000 Skopje - Republic of Macedonia
phone: +38991 363566 fax: ++364 262 E-mail: sugo@cerera.etf.ukim.edu.mk
[2] *Department of electronic engineering, University of Nis, beogradska 14, 18000 Nis, Yugoslavia*

Abstract: The paper proposes neural network architecture that can be implemented in sofisticated object tracking system. It employs neural correlation of motion detection via two channels mechanism similar to the visual system of invertebrates. The correlation based scheme with mutual lateral inhibition is adopted to implement VLSI neuromorphic cells with the dynamics characterized by first order nonlinear differential equation. Combining these cells in suitable manner in an 2-D array it is possible to provide simple mechanism to achieve direction sensitive elementary motion detectors. Both spatial and temporal adaptation responses of the VLSI model are investigated and simulation results are presented. *Copyright © 1998 IFAC*

Keywords: Neural dynamics, Motion estimation, Obstacle detection, Autonomous mobile robots

1. INTRODUCTION

Intelligent vision systems will be an inevitable component of future intelligent mobile systems. Vision chips, as a front-end part of the autonomous mobile robots, will integrate both the photosensors and. parallel processing elements at the same implementation level that precedes higher level processing modules, (Franceschini, *et al.*, 1991; Abott, *et al.*, 1994). Perception and estimation of apparent motion in real time are an integral part of vision guidance and involve computation of the parameters of 3-D motion and prediction of future positions of moving objects. They are prerequisites for achieving vision chips capable to detect the direction of arrival of the moving objects as well as obstacle detection, tracking objects or navigating through the 3-D environment. Visual motion detection is an early preatentive mechanism preceding many higher levels of information processing (Ogmen and Gagne' 1990; Koch and Mathur, 1996). It comprises fundamentals in early vision and is important step toward seeing as an intelligent act. It is not surprising

that many research efforts have been directed toward better understanding of motion perception in the visual system of invertebrates, as well as in vertebrates. Knowledge accumulated over the past decades can successfully be implemented in building proprietary vision chips as a part of the vision-guided robot controllers or position and force control in coordinated manipulators (Franceschini, *et al.*, 1991; Delbruck, 1993; Abott, *et al.*, 1994; Koch and Mathur, 1996).

While the techniques of classical motion analysis are still being improved interest in the field has been moved over the past decade to exploring some different approaches that may compensate the limitations of the traditional methods. There are several application oriented classes of motion estimation algorithms such as gradient-based techniques, pel-recursive, block-matching, frequency domain techniques and techniques based on neural networks implementation concept. It is well known that the 3-D motion interpretation and motion parameters derivation are strongly related to the estimation of optical flow. Using this approach Hutchison, *et al.*, (1988) developed neuromorphic silicon model and produced a chip that emulates vertebrate retina which can adapt locally to brightness changes, detect edges and compute motion. In the work of Delbruck, (1993) 2-D silicon retina is reported that computes a complete set of local direction-selective outputs using unidirectional delay lines as tuned filters for moving edges. The work of Sarpeshkar, *et al.*, (1996), describes an implementation of Reichardt's correlation model for motion detection with a fixed time constant and how it can be employed to compute direction of a moving subject and to estimate the time-to-contact between the sensor and a moving object. A review of the state-of-the-art retinal processing and future application of neuromorphic vision chips can be found in the paper of Koch and Mathur (1996).

2. BASIC NEURAL VISION SYSTEM ARCHITECTURE

Neuronal activity, in general, can be modeled in terms of the all-or-none response of a cell to incoming stimuli. The resulting firing patterns in networks of such neural cells arise from highly nonlinear processes related with neuron membrane activity, generation of spikes, their transmission and further higher level processing mechanisms. Certain important aspects of visual processing, which occur at retinal level, include spatial and temporal enhancement of stimuli. It has been shown that such spatio-temporal adaptation arises from inhibitory connection in the retina both between neighboring cells and via feedback loops to the cell under the consideration making a process of self-inhibition (Taylor, 1990). The most of biologically inspired models utilized in motion detection systems can be grouped in two basic classes: delay and compare schemes and feedforward shunting inhibitory systems, (Grossberg, 1980; Taylor, 1990; Sarpeshkar, *et al.*, 1996).

Whatever the biological model is, the global architecture of the motion processing system comprises photodetection elements followed by adaptive spatio-temporal filtering and elementary motion detection stages as depicted in Fig. 1. The output signals comprising motion information parameters may then be used as inputs in various sofisticated vision guided control system applications. This paper presents a plausible elementary motion detector model similar to the fly visual system and its VLSI proposal which performs a functional tasks shown in the shaded part in Fig. 1. First we start with short review of the dynamics of the simple cell and then by combining such cells in suitable manner, we present a gated dipole structure with potentiality of both temporal and spatial adaptations. In the next section we explore CMOS capabilities for circuits

implementation. More specifically, using VLSI implemented neuromorphic cells with the dynamics of

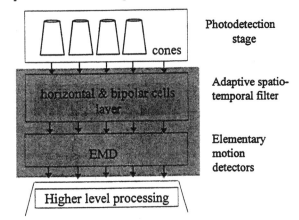

Fig. 1 Simplified adaptive motion processing system. At the input level is the photodetection layer followed by the adaptive spatio-temporal filtering layer which feeds elementary motion detection stage. Output can be used for further processing.

the transmitter characterized by first order differential equation, the bipolar cells are formed. Combining these devices in more complex structures it is possible to provide simple mechanism to achieve direction sensitive elementary motion detectors (Ogmen and Gagne', 1990; Taylor, 1990). In the last section simulation results of both temporal and spatial behavior of the device are presented and appropriate conclusions are given.

3. DYNAMIC OF A CELL IN TWO CHANNELS CORRELATION METHOD

Let us consider the neuronal cell with the dynamics characterized by the following set of equations:

$$\frac{dz(t)}{dt} = a[b - z(t)] - f[x(t), z(t)], \text{ and}$$

$$y(t) = f[x(t), z(t)] = x(t)z(t) \quad (1)$$

where *ab* is the transmitter production rate, $-az(t)$ is the feedback self-inhibition and *b* denotes maximum amount of a transmitter. The signal transmitted to the postsynaptic cell *y(t)* for simplicity is supposed to be proportional not only to the input stimuli *x(t)* but also to the available amount of transmitter *z(t)*, (Grossberg, 1980; Ogmen and Gagne', 1990). The equation (1) states that the input signal *x(t)* depletes the transmitter proportionally to its strength and to the available amount of transmitter *z(t)*. In other words, input is gated by *z(t)* to yield output *y(t)*.

Applying a step stimulus, $[x(t) = x, t>0]$, and if the transmitter is fully accumulated, $[z(0) = b]$, one can find that for time $t >0$ the response of the cell can be expressed as:

$$y(t) = \frac{abx}{a+x}\left(1 - e^{-(a+x)t}\right) + bxe^{-(a+x)t} \quad (2)$$

which means that after the overshoot [*bx*], depletion process produces a temporal adaptation and the output settles to the value: $y(\infty)=abx/(a+x)$.

It is well known that each cell's receptive field is sensitive to visual stimuli only within a spatially local region or aperture. Concerning the motion estimation the question of spatial adaptation arises which means the discrimination of spatial inputs from spatial background. So, the aperture problem arises as a consequence of uncertainty about both the position and orientation of a visual stimulus determination by a single cell, since the local direction of motion signaled by the activity of such cells is ambiguous (Taylor, 1990; Marshall, 1990). The task associated with aperture problem is to specify how such ambiguous local activation can be combined coherently to form unambiguous global motion percepts. Thus, two degrees of freedom in motion measurements must be introduced which is equivalent to two channels processing mechanism.

To show how spatial adaptation can arise from inhibition process let us consider two similar channels with lateral inhibition connections as presented in Ogmen and Gagne' (1990) and depicted in fig. 2. The interconnections at the second stage (neurons 3 and 4) mathematically are described by following (membrane or shunting) equations, (see Grossberg, 1980):

$$\frac{dy_{on}}{dt} = -Ay_{on} + (B - y_{on})[I + J_{on}]z_{on} - (D + y_{on})[Iz_{off}] \quad (3)$$
$$\frac{dy_{off}}{dt} = -Ay_{off} + (B - y_{off})[I]z_{off} - (D + y_{off})[I + J_{on}]z_{on}$$

where, A is the rate of spontaneous decay of the cell activity, B and $-D$ are upper and lower levels of saturation of the activity, $I+J_{on}$ is the net excitatory input or net inhibitory input depending on the cell under the consideration, I is the arousal signal feeding both channels, and notation y designates both cell and its activity. Applying signal $x = I+J_{on}$ to the left channel and $x = I$ to the other and taking into account slow time scale of the transmitter dynamics, one can

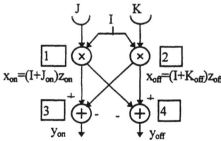

Fig. 2 A gated dipole of Ogmen and Gagne' (1990). The "x" cell exhibits temporal adaptation dynamics while the "+" cell performs shunting equation.

find the equilibrium states as:

$$y_{on} = \frac{B[I + J_{on}]z_{on} - DIz_{off}}{A + [I + J_{on}]z_{on} + Iz_{off}}, z_{on} = \frac{ab}{a + I + J_{on}} \quad (4)$$

$$y_{off} = \frac{BIz_{off} - D[I + J_{on}]z_{off}}{A + [I + J_{on}]z_{on} + Iz_{off}}, z_{off} = \frac{ab}{a + I}$$

Inspecting the equations (4) for $J_{on} = 0$, $t<0$, it is obvious that activities of both cells will equilibrate to the value

$$y_{on} = y_{off} = \frac{[B - D]Iz_{on}}{A + 2Iz_{on}}, z_{on} = z_{off} = \frac{ab}{a + I} \quad (5)$$

and if spatially homogeneous input is applied, than $J_{on}=K_{off}=I$ and the overall response of the cells is

$$y_{on} = \frac{B[I + J_{on}]z_{on} - DIz_{off}}{A + [I + J_{on}]z_{on} + Iz_{off}}, z_{on} = \frac{ab}{a + I + J_{on}} \quad (6)$$
$$y_{off} = \frac{BIz_{off} - D[I + J_{on}]z_{off}}{A + [I + J_{on}]z_{on} + Iz_{off}}, z_{off} = \frac{ab}{a + I}$$

That is, because of lateral interactions of the cells spatial adaptation is generated. Extending this structure in a 2-D lattice and incorporating low-pass filtering as shown in the Fig 3., it is possible to achieve device that is capable to detect the direction of visual motion. The directional selectivity is signaled by output cells activity above certain threshold level. Further details can be found in Shutinoski and Dzhekov, (1997).

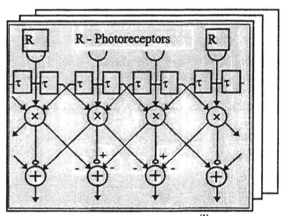

a) Detector output signals $y_{ij}^{(1)}$

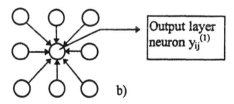

Fig. 3. 2-D neural network model for motion detection. (a) Signals from the photoreceptors are first delayed and than passed through the temporal and spatial adaptation layers. The low pass filtering is performed with fixed time constant τ. (b) Lateral connections of cell $y_{ij}^{(1)}$.

4. ON MODELING A CMOS NEURAL CELL

Monolithic integration becomes a necessity for applications such as artificial neural networks where the large scale of components has to be implemented. According to the continuous-time properties of proposed system, modeled by the ordinary differential equations we decide to implement monolithic CMOS realization of elements in 2 μm n-well technology. The useful range of component values for this technology is determined by sizes and shapes of corresponding physical components. Although the absolute accuracy of integrated components is very poor and depends on temperature variations, process of aging and component spreading, taking into account the good tolerance of ratios between similar components (as low as 0.1%), encountering the influence of temperature, aging or even nonlinearities (cancellation of nonlinearities in a current mirror or differential pair), benefits the use of this technology. In standard CMOS technologies both passive quasi-linear resistors (with area occupancy $\sim 10^3$ $\mu m^2/K\Omega$) and active elements resistive components based on MOS ohmic region properties can be fabricated. Also thick oxide capacitors can be formed at the price of area occupation, or by exploiting capacitive effects between gate and channel of an MOS transistor, Wu and Lan (1996).

Let us now consider the structure of fig. 4. The capacitor potential V can be found by solving the first order differential equation of the form:

$$\frac{dV(t)}{dt} = \frac{1}{CR}\left[I_0 R - V\right] - \frac{1}{C}I_s \qquad (7)$$

where I_0 represents constant current source and I_S represents a voltage controlled current source. By a simple change of variables $a = 1/R$, $b = I_0/(RC)$ and $I_S = Cx(t)z(t)$, this is the same relation as (1). Thus, our goal is to design proprietary CMOS building blocks to implement the proposed cell dynamics.

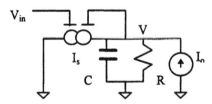

Fig. 4 Circuit that emulates simple cell dynamics

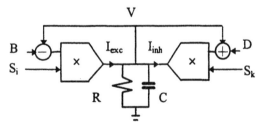

Fig .5 Circuit that emulates membrane equation dependence of the cells activities in the output layer where $S_i=J+I$ and $S_k=K+I$.

Conducting the same approach as above to the potential V, fig. 5, one finds the equivalence between proposed circuit functioning and membrane equation (3).

5. A CMOS NEURAL NETWORK PROPOSAL

A current-mode approach is adopted for the implementation of the voltage controlled current source I_S. The actual realization comprises a high-output impedance transconductance element used as a controllable current source linear in the driving voltage V_{in} over a wide range and the double-stack transistor configuration of the differential pair (Cauwenberghs, 1996; Rodriguez-Vasquez and Delgado-Restituto, 1993). Figs. 6a) and 6b) show the schematic of the transconductance element and its application respectively. The device utilizes MOS transistor M1 biased in the triode region and connected to a cascode transistor M2 via high-gain feedback circuit. Therefore, the triode transistor drain voltage is forced to a constant level largely independent of the input voltage. The supplied output current is proportional to the driving voltage V_{in} while invariant to the output voltage of the device,

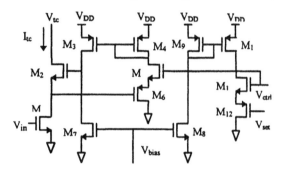

Fig 6a). Transconductance element with high output impedance

($I_{tc} = G_m V_{in}$). The bias circuit that generates the control voltage for transconductance element is specified by an externally supplied voltage, allowing approximately linear control of the G_m value of the element, keeping the output voltage as low as possible (~0.5V). The output current of the transconductance element feeds into a differential

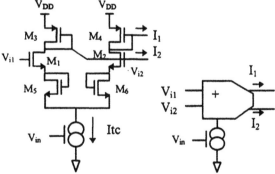

Fig. 6b). Implementation of a transconductance element as controllable current source.

pair injecting the current $g_m(V_{i1}-V_{i2})I_{tc}$ into the diode-connected transistors M3, M4 Fig. 7shows the dependence of the generated output current of the device on the input voltage V_{i1} and the control voltage V_{in}. In the fig. 8 is presented active components implemented resistor utilizing ohmic region of depletion type transistor M1 while the other MOS transistors are biased in saturation. Providing that the applied resistor voltage is grater than $\max\{V_{T1},V_{T2}\}$, the nonlinear term of the implemented

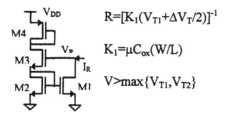

Fig. 7. Dependence of the generated output current of the device in fig.6a) on the input voltage V_{in} and the control voltage V_{ctrl}.

element is canceled which produces satisfying linear V/I dependence of the element. Although, there have been many proposed improved mirror structures (Wu

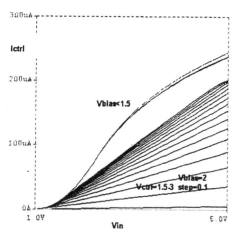

$R=[K_1(V_{T1}+\Delta V_T/2)]^{-1}$

$K_1=\mu C_{ox}(W/L)$

$V>\max\{V_{T1},V_{T2}\}$

Fig. 8. An active components resistor

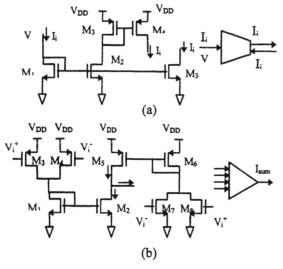

(a)

(b)

Fig. 9. A current duplicator device (a) and (b) a current summing element.

and Lan, 1996; Rodriguez -Vasquez and Delgado-Restituto, 1993), a simple four-transistor current mirrors are chosen since the device functionality and simulation validity were the most important concern in this application. Implemented current duplicating devices so as current summing structures, (fig. 9), have a wide range of linearity on the voltage at the sink node providing that this voltage is kept in the range of the values $(V_{DD}-V_T)/2$ and $(V_{DD}+3V_T)/2$. Fig. 10 shows realization of the cell with membrane equation dynamics.

6. SIMULATIONS AND RESULTS

Extended structure of fig. 10 was generated by a subcircuit model definition and simulations on Spice program were performed. We experimented with various input signal levels that represent sudden

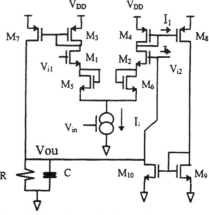

Fig. 10 Realization of the cell with membrane equation dynamics

changes of environmental signal. Fig. 11 shows temporal adaptation process of a simulated neuromorphic circuit over various levels of excitation in both directions. Notice that the temporal adaptation strongly depends on the accumulated transmitter and applied stimulus. The change of the stimulus strength produces the change of the overshoot value which is followed by a settling plateau. Fig. 12 depicts the spatial adaptation process. Applying the same excitation input for the on-cell as above while holding the off-cell at a level of arousal signal results in overshoot response followed by a plateau for on-cell but the activity of the off-cell is suppressed. When the excitation to the on-cell becomes low, it responses undershoot and the off-cell will rebound. Upon a same stimulus to both of the cells they respond equally reaching approximately same activity level.

7. CONCLUSION

In this paper the analog current-mode design technique is used to realize the neuromorphic cell capable to respond on temporal and spatial variations

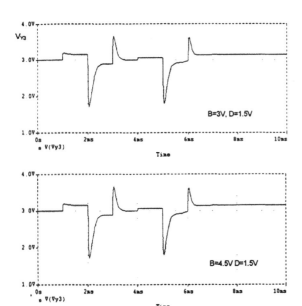

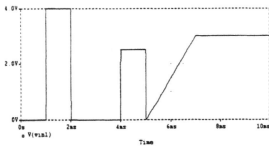

Fig. 11 Responses of the second layer neuron 3,(V_{Y3}), depending on the constant B; D=1.5 V.

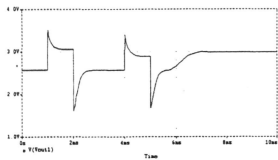

Fig.12 a) Input signal to the neuron 1, b) output signal from the neuron 1

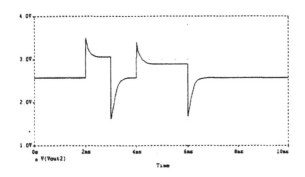

Fig.12 c) Response of the neuron 2 of the first layer.

of stimuli. SPICE simulation of 2mm CMOS level 3 components implemented circuits gives results that are in good accordance with the mathematical background. Concerning the actual motion direction and estimation of motion parameters, more realistic results should be expected using real 2-D network of such cells.

REFERENCES

Abott, D., A. Moini, A. Yakovleff, X.T. Nguyen, A. Blanksby, G. Kim, A. Bouzerdoum, R.E. Bugner and K. Eshraghian, (1994). A new VLSI smart sensor for collision avoidance inspired by insect vision. *Intelligent vehicle highway systems, SPIE*, **2344**, pp. 105-115.

Cauwenberghs, G. (1996). An analog VLSI recurrent neural network learning a continuous-time trajectory. *IEEE Tran. on NN*, **7**, No.2, pp. 346-361.

Delbruck, T. (1993). Silicon retina with correlation-based, velocity-tuned pixels. *IEEE Trans.NN* , **4**, No.3, pp. 529-541.

Franceschini, N., J. Pichon and C. Blanes (1991). Real-time visuomotor control: from flies to robots. *In IEEE Fifth Int. Conf. Advanced Robotics*, Pisa, Italy, pp. 91-95.

Grossberg, S. (1980). How does a brain build a cognitive code. *Psychological Review* **87**, pp. 1-57

Hutchison, J, C. Koch, J. Luo, and C. Mead (1988). Computing motion using analog and binary resistive networks. *IEEE Computer Mag.*, **21** No.3, pp. 52-63.

Koch, C. and B. Mathur, (1996). Neuromorphic vision chips. *IEEE Spectrum*, **33**, p.p. 38-46.

Marshall, J. (1990). Self-organizing neural networks for perception of visual motion. *Neural Networks*, **3**, pp. 45-74.

Ogmen, H. and S. Gagne' (1990). Neural network architectures for motion perception and elementary motion detection in the fly visual system. *Neural Networks*, **3**, pp. 487-505.

Rodriguez-Vasquez, A. and M. Delgado-Restituto (1993). CMOS design of chaotic oscillators using state variables:a monolithic Chua's circuit. *IEEE Tran. on CAS-II*, **40**, No.10, pp. 596-613,.

Sarpeshkar, R., J.Kramer, G.Indivieri and C.Koch, (1996). Analog VLSI architectures for motion processing; from fundamental limits to system applications. *Proc. IEEE*, **84**, pp. 969-987.

Shutinoski,V.G. and T.A.Dzhekov, (1997).VLSI design of neural motion detector. *Proc. of the 4th Seminar on Neural Network Applications, NEUREL '97*, Belgrade, Yugoslavia, pp.99-105.

Taylor, J.G. (1990). A silicon model of vertebrate retinal processing. *Neural Networks*, **3**, pp. 171-178.

Wu, C.-Y. and J.-F.Lan (1996). CMOS current-mode neural associative memory design with on-chip learning. *IEEE Tran .on NN*, **7**, No.1, pp. 167-181.

OPTICAL FLOW CALCULATION USING LOOK-UP TABLES

José Otero, Jose A. Cancelas, Rafael C. González

Universidad de Oviedo ETSIII de Gijón
Manuel Llaneza, 75, 33208 Gijón Spain
Tel. 34-8-5182067   Fax 34-8-5182156
jotero@lsi.uniovi.es

Abstract: This work presents a new approach to optical flow calculation from an image sequence. Previously published methods were characterised by their high computational cost and complexity, limiting their applicability. This approach is characterised by avoiding floating point calculations or its emulation using integer arithmetic. Instead of this, starts with a previous off-line calculation of some LUT containing the resulting values of all the necessary operations for flow calculation. In the on-line phase of the method, the LUT is directly accessed using the gray-level of the point neighbourhood in one frame and the corresponding search space in the next frame. The vector formed by the gray levels of both neighbourhoods is interpreted as a number in a numerical base equal to image depth. Those numbers are the indexes to access the LUT and obtain the correct displacement value directly. In order to have some preliminary results, the test of the algorithm was performed with Matlab 5.0, and the results were compared with traditional correlation based methods. This implementation performed nine times faster than the simple correlation based algorithm. Despite of the reduction in the number of gray levels of the image, achieved precision is equal to the one from correlation. According to these results, near real-time optical flow computation can be achieved, with no special hardware requirements. *Copyright © 1998 IFAC*

Keywords: Robot vision, image analysis, optical flow, inverse problem, hardware, RAM.

1. INTRODUCTION

From all the possible tasks to be performed by an artificial vision system, there is a group focused in movement detection from an image sequence. In the lasts years, the industrial use of robots has became very common. In general, the robot is used in repetitive tasks, where every action and movement to be taken is perfectly known. During task execution, it is supposed that no object may interfere with the robot, but this assumption can not be completely guaranteed.

In the development of this work, different approaches

On the other hand, the "motion is basic" paradigm

Table 1

Methods based in:	Advantages.	Disadvantages.
Matching (static is basic).	• Precision. • Easy mathematical formulations.	• Features definition. • Features actualization. • Sparse information. • Difficult to implement.
Time-space derivatives (motion is basic).	• Dense information. • Easy to implement. • Interesting qualitative properties.	• Derivative calculation. • Sensitive to illumination changes. • Complex mathematical formulation.

to the problem of range information extraction from visual information were investigated. Several solutions to this problem have been found, but all of them have a difficult application in practice, because of its computational cost. A practical solution would allow the introduction of this techniques for autonomous robot operation in a non controlled environment (Uhlin and Johanson 1996; Giachetti, *et al.* 1993).

One fact to be taken into account is that not always is necessary to calculate the exact distance of every point of the scene. In fact, knowledge of which objects are near or far is enough, just changing the trajectory in order to avoid nearer obstacles. In this case, a reduction in precision can be acceptable in order to speed up the process, and to obtain a more robust calculation.

According to this ideas, depth and speed from an intensity image are calculated, without an a priori knowledge of the scene of hypotheses about physical properties of the objects. Of course calculation should be done in real time.

2. DIFFERENT APPROACHES TO THE PROBLEM

There are different approaches to this problem, that can be classified according to different criteria (Murray and Buxton 1990), although the different obtained classes are usually the same. As an example, Schalkoff (Schalkoff 1989) proposes to paradigms to analyse movement, "static is basic" and "motion is basic".

Following the "static is basic" paradigm, each image in the sequence must analysed independently to determine which characteristics identify every entity in the scene (Feddema and Lee 1990; Feddma, *et al.* 1991). Once those characteristics have been defined, a frame to frame matching is performed to calculate the path of the object through the sequence.

avoids the analysis of dynamic image properties. Approaches following this base principle use time-space derivatives and relationships between them to calculate speed projection into the image plane (Papanikolopoulos, *et al.* 1993; Papanikolopoulos and Khosla 1993; Luo *et al.* 88). From this basic information, it is necessary to obtain the biggest possible knowledge of the scene properties. Table 1 summarizes the main advantages and disadvantages of each approach.

3. OPTICAL FLOW: DEFINITION AND CALCULATION

There are different definitions for optical flow, all of them are very similar but with different shades. As an example, Horn (Horn 1986) states that the optical flow is the apparent movement in the brightness patterns of an object, when the camera moves with respect to the object. Verri, Poggio and Torre (Verri, *et al.* 1991) define the optical flow as the apparent speed field in a time varying image. Sonka, Hlavac and Boyle (Sonka, *et al.* 1993) state that optical flow reflects image changes due to movement and therefore, the optical flow is the 2D speed field that represent the 3D movement of objects in a 2D image. Basically, there are three approaches for optical flow calculation,: techniques based in time-space derivatives, techniques based in correlation and phase based approaches. The third approach are more accurate than the other two (Barron, et al. 1994), but the computation time is so high that are not useful to real time applications (Nesi 1994). In the next sections the other techniques are summarized, see (Otero 97a) for details.

3.1. Time-Space derivatives methods.

All these methods are based in the assumption that pixel intensity does not change between frames (apart from movement). From this statement equation (1) is obtained, best known as optical flow constrain.

$$-f_t = f_x \frac{dx}{dt} + f_y \frac{dy}{dt} - f_t =$$
$$f_x u + f_y v = \mathrm{grad}(f) \cdot \mathbf{C}$$

(1)

This equation relates image temporal derivatives to spatial gradient and pixel speed. Several authors has developed this equation following their own methods (Nagel 1983; Shalkoff 1989; Horn and Schunk 1980). Table 3.1.1 summarize the most significant methods and their main characteristics.

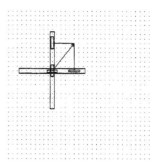

Fig. 3

Table 2

Method.	Basis.	Characteristics.
Direct formulation.	Constant brightness equation.	An incomplete estimation of optical flow is obtained.
Horn & Schunk algorithm.	Regularization of the problem through maximization of a smoothness norm.	Convergence is not ensured in the edges.
Neighborhood algorithm.	Least Squares estimation in a neighborhood	Erroneous calculation in edges.
Second order derivatives.	Second order derivatives are not neglected.	Bad estimation of second order derivatives.

The main problem of this methods is the necessity of derivative calculation, reducing precision and increasing noise sensitivity.

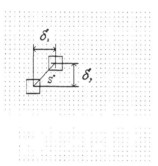

Fig. 1

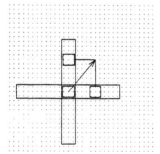

Fig. 2

3.2. Correlation based algorithms.

This algorithms are based on effecting the correlation between a 2D mask from a frame and a 2D search space of the next frame, by this reason is called the 2D2D algorithm. The base idea of the 2D2D algorithm consist on measure for each pixel the displacement (x,y) that maximize the correlation between a neighborhood centred in the pixel in the first frame and other in the second frame belonging a certain search space. The correlation between frame regions in t and $t+\delta t$ is defined as:

$$\Phi(\delta x, \delta y; t) \equiv I^w \otimes I =$$
$$\int I^w(\xi,\eta;t) I(\delta x+\xi, \delta y+\eta; t+\delta t) d\xi d\eta$$

(2)

Where $I^w(\xi,\eta;t)$ is a matrix equal to the frame in a window w and zero out of it. The distance L_2 is defined as the square root of the addition of the squares of the differences between the corresponding pixels of the regions of each frame. This distance has properties similar to the correlation, verifying that when the correlation is maximum L_2 is minimum. The commented method is schematically shown in the Fig 1. Due to the enormous quantity of floating point operations necessary to perform the algorithm, Ancona and Poggio in (Ancona and Poggio 1993) propose do a simplification in the algorithm, this consists on separately search in x and y directions the displacements that maximize the correlation. In this way the number of floating point operations needed is reduced, because the search space is unidimensional instead of bidimensional. Due to the algorithm is known as 1D2D (see Fig 2). In addition, if the row and columns average is calculated, it is possible to perform the unidimensional correlation in regions that are also unidimensional, decreasing more the amount of floating point operations needed. This algorithm is graphically shown in Fig 3.

In this paper a new approach for obtaining the optical flow without floating point operations is shown.

3.3 Optical flow from LUT tables.

Camus in (Camus 1994) introduces the possibility of calculate L_2 distance using LUT tables. This magnitude is used in the correlation based algorithm of Ancona and Poggio (Ancona and Poggio 1993) instead of the correlation.

In this paper the method founded in Otero research report (Otero 1997 b) is shown, where the original idea of an algorithm using LUT tables was proposed. This innovation gets the optical flow based techniques closer to real-time applications, because no floating point operations are necessary in the on-line phase. The complexity of the algorithm is decreased because there are only two nested loops for searching the correlation values. In the next

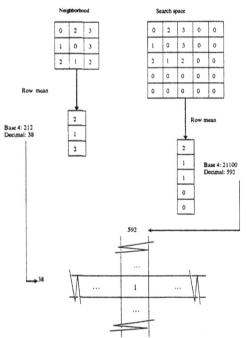

Fig. 4

paragraph the algorithm is explained with an example.

Given a 3x3 neighborhood, a 5x5 search space and four gray levels, the displacements in x and y directions are separately found, according to the 1D1D algorithm of Ancona and Poggio (Ancona and Poggio 1993). The key of the algorithm is that directly founds the correct displacement of the pixels in the LUT, instead of use it only just for obtaining the L2 distance as Camus did in (Camus 1994). Using directly the gray levels of the image, the displacements of the pixels are given without any operations, they are pre-calculated in the LUT table. This LUT is calculated in a of line stage, every possible pairs of row (and column) means of neighborhood and search space are generated and the shift that maximizes the correlation is calculated. Note that the same LUT works for row and column means. For each pair, the corresponding element of the LUT is filled with the value of the calculated shift. The position of the element in the LUT is determined using the vectors of row (or column) means as row and column indices expressed in a certain numerical base, 4 in this example.

In order to give the LUT table in a convenient size it was necessary to decrease the amount of gray levels used in the representation of the image. If the lengths are 3x1 and 5x1 vectors the limit is 6 gray levels (the size of the LUT will be $6^3 x 6^5 = 1679616$ elements). If the size of the vectors are 3x1 and 7x1 the limit is four gray levels ($4^3 x 4^7 = 1048576$). This limit of the gray levels are valid in a PC with Matlab and 32 Mbytes. The number of gray levels and the size of the vectors can easily increased just with more memory in the system.

In figure (4) the calculation of the row average is shown. Columns average is performed in a similar way. This process allows to search separately the displacements in x and y that maximize the correlation. In the same figure is show also how the LUT table is accessed to obtain the correct displacement. The displacement in y direction is obtained in a similar way.

In order to convert the numbers from base 4 to base 10, it will be necessary to do some arithmetic operations. For example, the number in base 4 "012" is $2x4^0 + 1x4^1 + 0x4^2 = 6$ in base 10. Instead of this a faster conversion algorithm is used, based in a tree with 4 child per node and as many levels as digits has the number to be converted. This tree is expressed as a bidimensional table in Matlab.

In Fig. 5 a sequence of synthetic images is shown. The optical flow obtained is shown in Fig. 6. The images were 50x50 and the algorithm performs nine times faster than the simple correlation method.

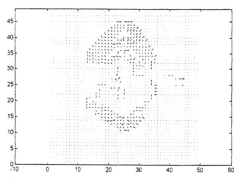

Fig. 5

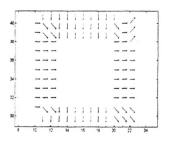

Fig. 8

This problem has been partially solved with a technique that, once well-known, it seems simple. It is to decompose each frame of the sequence of images in different ranges of gray levels. Each range is represented later on with four gray levels, the optic flow is calculated for each part of the image and finally the different optical flows obtained are added. For example, if one works with 16 gray levels, each image of the sequence breaks down in 4. In each one of the images will be only the pixels whose values are in the appropriate range. That is to say, in the first one they will be the pixels of values between 0 and 3, in the second those that have levels between 4 and 7 until arriving to the image where the pixels of intensities are between 12 and 15. In the Fig 9 a decomposition of an image of 16 gray levels in four images is shown, each one of them is represented with four gray levels.

Fig. 6

The optical flow obtained has enough quality to obtain useful information from it via divergence analysis, for example a time to contact estimation as in (Otero 1997 a).

3.4 Optic Flow From L.U.T. Applied To Regions.

In the previous paragraph it was shown that, to obtain a LUT of viable size, it is necessary to reduce the number of gray levels of the image. This circumstance reduces the quantity of present information in the image and it will cause that areas that originally present some type of variation become uniform. In the areas that are uniform the optical flow algorithms return a null value for the speed of the pixels, since just by local information it is impossible to determine if a pixel, similar to other many, has moved or not. Under these conditions non null values of the speed will only be obtained in the contour of the objects with gray levels different from those of the bottom.. In Fig. 7 an example of this fact is shown. The rectangle that moves a pixel to the right and another down only produces optic flow in its contour.

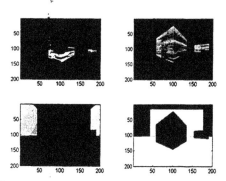

Fig. 9

A sequence of images of this type can talk to the algorithm of optic flow based on LUT with much bigger success that if you uses binary images, like the one that was shown in the Fig 7. Unfortunately, the preliminary results of this technique are not good, because the effect of the movement of the places were there are pixels that are not in a given range. The absence of a pixel is represented with a zero gray value, and the movement of the place were the pixel lays is detected in the same way as a pixel that is in the range of gray values being treated. It's necessary to improve this method in order to increase the performance of the results.

Fig. 7

The optic flow is appreciated with more clarity in the Fig 8. This figure shows the presence of the optic flow in the contour of the rectangle.

51

4 CONCLUSIONS

The algorithm here presented is expensive in terms of the amount of memory needed, but today, memory is almost the cheapest element of a conventional computer. So is better the need of more memory than a faster CPU or a DSP board. In addition, the memory of a computer can be easily increased, but not the CPU power. In the Computer Vision area, is not frequent to find an algorithm that doesn't need CPU power or a DSP capable of millions of flops. In other words, the algorithm is close to the real time with zero cost, because no additional and specialized hardware is needed. The number of gray levels and the size of the vectors can be increased easily if more memory is supplied. This will result in a better quality of the optical flow obtained. The algorithm works in a very low level stage of image processing, some post-processing technique is needed to increase the accuracy of the results. Finally, note that the algorithm here presented is in a very early stage of development, a better structure of the LUT table and the elimination of redundant elements will decrease the amount of memory needed or increase the number of gray levels used in the images, the major lack of this new approach.

REFERENCES

Aggarwal J. K., N. Nandhakumar (1988). On the Computation of Motion from Sequences of Images. *Proceedings of the IEEE*, **vol 76, No 8.**

Ancona N. and T. Poggio (1993). Optical Flow from 1D Correlation: Aplication to a simple Time-To-Crash Detector. Massachusetts Institute of Technology, Artificial Intelligence Laboratory and Center for Biological and Computational Learning.

Barron J. L., S. S. Beauchemin, D. J. Fleet (1994). On Optical Flow. *AIICSR Proceedings*, Bratislava, Slovakia, **pp3-14.**

Camus T. (1994). Real-Time Optical Flow. Department of Computer Science. Brown University. Providence, USA. September 1994.

Feddema J. T., G. Lee (1990). Adaptative Image Feature Prediction and Control for Visual Tracking with a Hand-Eye Coordinated Camera. *IEEE Transactions on Systems, Man, and Cybernetics*, **vol 20, No 5.**

Feddema J. T., G. Lee, O. R. Mitchell (1991). Weighted Selection of Image Features for Resolved Rate Visual Feedback Control. *IEEE transactions on Robotics and Automation*, **vol 7, No 1.**

Giachetti A., M. Campani, V. Torre (1993). The use of Optical Flow for the Autonomous Navigation. Dipartamento di Fisica Università di Genova.

Horn B. K. P, B. G. Schunk (1980). Determining Optical Flow. *A.I. Memo N°572*. Massachusetts Institute of Technology, Artificial Intelligence Laboratory.

Horn B. K. P (1986). *Robot vision*. Mc Graw Hill.

Luo R. C., R. E. Mullen Jr, D. E. Wessell (1988). An Adaptative Robotic Tracking System Using Optical Flow. Robotics and Intelligent Systems Laboratory, Departament of Electrical and Computer Engineering, North Carolina State University.

Murray D. W., B. F. Buxton (1990). *Experiments in the Machine Interpretation of Visual Motion*. Massachusetts Institute of Technology.

Nagel H. H. (1983). Displacement Vectors Derived from Second Order Intensity Variations. *Computer Vision, Graphics, and Image Processing* **21.**

Nesi P. (1994). Real-Time Motion Estimation. Department of Systems and Informatics, Faculty of Engineering, University of Florence.

Otero J.a (1997). Detección y Localización de Obstáculos en Entornos no Estructurados Mediante Técnicas de Flujo Óptico. Master Thesis. Department of Automation and Systems Engineering University of Oviedo.

Otero J.b (1997). Optimización en el cálculo del Flujo Óptico. Research Report. Department of Automation and Systems Engineering University of Oviedo.

Papanikolopoulos N. P. (1993), P. K. Khosla, T. Kanade. Visual Tracking of a Moving Target by a Camera Mounted on a Robot: A Combination of Control and Vision. *IEEE Transactions on Robotics and Automation*, **vol 9, No 1.**

Papanikolopoulos N. P. (1993), P. K. Khosla. Adaptative Robotic Visual Tracking: Theory and Experiments. *IEEE Transactions on Automatic Control*, **vol 38, No 3.**

Schalkoff R. J. (1989). *Digital Image and Computer Vision*. Departament of Electrical and Computer Enginerring, Clemson University.

Sonka M. (1993), Hlavac, Boyle. *Image Processing, Analysis and Machine Vision*. Chapman & Hall.

Uhlin T. (1996), K. Johanson. Autonomus Mobile Systems: A Study of Current Research. The Royal Institute of Technology, Estocolm.

FAST INDOOR MOBILE ROBOT LOCALIZATION
USING VISION AND GEOMETRIC MAPS

Enrique Paz Domonte, Ricardo Marín Martin

Dpto. Ingeniería de Sistemas,   E.T.S. Ingenieros Industriales
Universidad de Vigo (SPAIN)
email: epaz@uvigo.es

Abstract. A vision and CAD map based architecture for mobile robot localization that uses naturally occurring landmarks of indoor environments is presented. It employs an original iconic map matching technique. This technique limits image processing to a minimum: it does not require extracting geometric features from image and therefore does not include additional errors. Map matching is performed on the image space: instead of extracting features from image, the CAD model is projected over the image. It implicitly handles the matching or correspondence determination problem. Therefore it is efficient and fast and achieves accurate pose estimation. With this architecture our indoor mobile robot is able to navigate at speeds as high as 0.5m/s while maintaining its absolute pose with an accuracy better than five centimeters. *Copyright © 1998 IFAC*

Keywords: mobile robot navigation, absolute positioning, computer vision, uncertainty.

1. INTRODUCTION

Mobile robot navigation usually requires accurate knowledge of robot's position and orientation (pose) relative to some static reference. This is especially true for indoor mobile robots evolving on a restricted environment.

A number of methods for computing and/or controlling the pose of a mobile system with respect to some metric (Cox, 1991; Kosaka and Kak, 1992; Atiya and Hager, 1993) or topological map (Meng and Kak, 1993; Thrun and Bücken, 1996) has been recently proposed.

The topological approach is robust and well suited to cope with noisy sensors, but its accuracy does not suffice for many indoor applications. The metric approach, using either geometric or iconic map matching, grants good pose accuracy but its computing requirements limits the maximum velocity allowable for the robot. Geometric map matching is –in general– faster and more robust than the iconic matching in the sense that it does not

require a good initial estimation of pose (Shaffer, *et al.*, 1992). However, the image processing and feature extraction phase makes it slower in vision-based applications, and more inaccurate than the iconic matching.

The geometric map along with the knowledge of robot's pose and its uncertainty allows to predict the expected location of landmarks and their uncertainty region (validation window) on the image. If the uncertainty regions are narrow, feature extraction and correspondence finding with expected landmarks is simplified. If validation regions of expected features has null intersection and if in each validation window there are at most one image feature, the solution of the correspondence problem is trivial. Both conditions hold if uncertainty is small and if the selected landmarks are isolated and easily distinguishable from background.

To restrain the growing of pose uncertainty requires a high rate of absolute pose measurements (2 to 5 updates per second) to ensure that cumulated odometric errors are small.

In this paper it will be assumed that pose uncertainty can be kept under ±10cm and ±1°. It allows to use an original and efficient measurement of image distance as a simplified iconic map matching technique.

2. MODEL OF ENVIRONMENT

Indoor scenes like offices, rooms and corridors, usually have a rectangular structure with three prominent orientations for 3-D line segments: one vertical and two horizontal orientations perpendicular to each other. There are not many lines in any other orientations than these three predominant (Shakunaga, 1992; Lebegue and Aggarwal, 1993). It is also assumed that floors are horizontal allowing to represent the pose of robot by three coordinates, the x and y location and the rotation θ around a vertical axis.

In those conditions, a geometric map of segments grouped by faces can be used to represent the mayor features of environment (figure 1): selected vertical and horizontal lines will be used as visual landmarks.

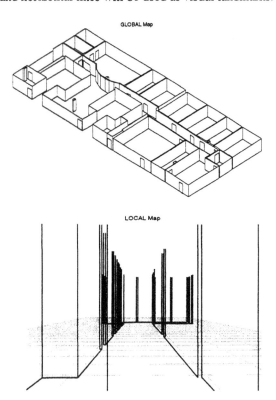

Fig. 1. Global (top) and local map (bottom)

After the clipping and hiding processes required to discard occluded elements, a number between five to ten meaningful segments are selected according to their length, orientation, distance to other segments, and expected visibility on the image (Paz, 1998).

The resulting local 3D map projected over the image is the expected map.

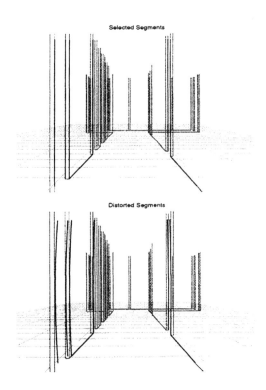

Fig. 2. Top: expected map (selected segments)
Bottom: expected image (distorted segments).

Before matching the image and map, the expected map must be distorted in the same way that the lens distorts the image (Nomura, *et al.,* 1992). It is computationally much more efficient distorting the map –using polygonal approximation of curved contours– than recovering the image distortion. The distorted projection on the image space of the selected segments using the currently estimated pose, will be denoted the expected image (figure 2).

3. MAP MATCHING

Some methods to calculate image distance or dissimilarity have been proposed (Brown, 1994; Huttenlocher, *et al.,* 1991). A common measure of image distance is de sum of squared differences. Let A an B two gray scale images of size (mxn)

$$D(A, B) = \sum_{i=1}^{m} \sum_{j=1}^{n} \left[A(i,j) - B(i,j) \right]^2 \quad (1)$$

When several images A_h must be compared with another image B with the purpose of determining which of them are most similar to image B, it can be shown that if the mean squared gray level of images A_h is constant, finding the minimum of $D(A_h, B)$ is equivalent to find the maximum of

$$S(A_h, B) = \sum_{i=1}^{m} \sum_{j=1}^{n} A_h(i,j) * B(i,j) \quad (2)$$

being $S(A_h, B)$ a measure of image similarity.

This measure of similarity is used to compare the image viewed by the robot with the expected and distorted map, and can be interpreted as a map matching measurement.

In this case, the image B is the magnitude gradient of the image currently viewed by the robot (*Grad*) and image A_h is the selected map projected and distorted form a valid robot pose X (Map_X) (it will be denoted expected image). Then the measure of similarity is

$$S(Map_X, Grad) = \sum_{MapImg_X(i,j) \neq 0} \sum Map_X(i,j) \cdot Grad(i,j) \quad (3)$$

where the sum includes only pixels of gradient image that are no null on the expected image (usually less than the 5% of the total number of pixels in the image):

$Map(i,j)=0$ if pixel (i,j) does not pertains to any segment

$Map(i,j)=a_r$ being a_r the weighing factor for segment S_r which lies over pixel (i,j)

Note that the expected image is composed by a one pixel width polygonal approximation to curved segments. Then the computation of equation (3) can be efficiently implemented using the Bresenham's fast algorithm for drawing straight lines.

The measure of similarity presented on equation (3) does not offer comparable results when the view point and/or lighting conditions changes. Some kind of normalization is required to make de measurement independent of the number and length of segments, and independent of image brightness:

$$S_N(Map_X, Grad) = \frac{\frac{1}{\sum a_r} \sum_{r=1}^{nsegs} \left(\frac{a_r}{length(s_r)} \sum_{(i,j) \in s_r} Grad(i,j) \right) - mean(Grad)}{max(Grad)}$$

$$(4)$$

The normalized similarity is strictly less than unity and greater than zero if the expected image matches with the actual image. It is assumed that this function presents an absolute maximum on the actual pose.

At each time step the expected map is constant, then the normalized similarity can be interpreted as a map matching function: it returns the matching degree as a function of pose.

$$M(X) = S_N(Map_X, Grad) \quad (5)$$

Figure 3 shows two examples of matching. The image captured with robot in an estimated pose (0,0,0) is compared with two expected images corresponding to location (0,0,0) and location (015,0.015,1.25). Fig. 3.a presents worse normalized similarity than figure 3.b which confirms the visual

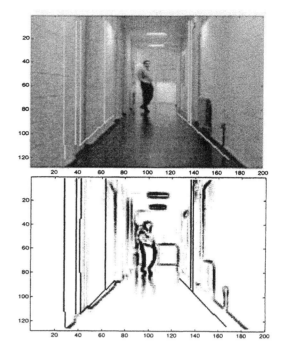

Figure 3.a Pose=(0,0,0), match=−0.013

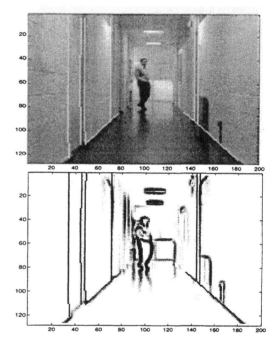

Fig. 3.b. Pose=(0.15,0.015,1.25), match=0.126

fact that actual image matches better with the expected image using the second pose than using the first one. This suggests that the actual pose of robot must be much closer to (015,0.015,1.25) than to the initially estimated pose.

4. POSE ESTIMATION

The measurement supplied by the vision system must be the pose which maximizes the matching, that is, the pose which produces the expected image with best similarity with the actual image. Unfortunately, the matching function is not differentiable and presents

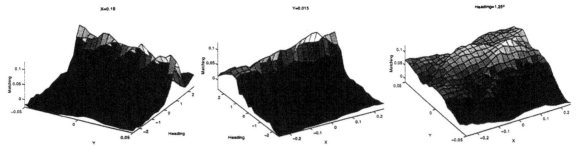

Figure 4. Three sections of the matching hyper-surface

many local maxima that make it difficult to find the absolute maximum. Three sections of the matching hyper-surface are presented on figure 4. This hyper surface is the result of matching function evaluated on a uniform 3-D grid around the estimated pose.

Consider 27 locations around the estimated pose by shifting the estimated pose a constant amount on each component.

$$(X+dx, Y+dy, \theta+d\theta) \rightarrow \begin{cases} dx = -\Delta x, 0, \Delta x \\ dy = -\Delta y, 0, \Delta y \\ d\theta = -\Delta\theta, 0, \Delta\theta \end{cases} \quad (6)$$

Each shifted location is a candidate for the actual pose. However a location with small matching is less likely to represent the true pose than a point with a high matching. Thus, the matching or response distribution could be interpreted as a frequency distribution in the space of pose: the matching at a location depicting the frequency of occurrence –or likelihood– of being the actual pose. This interpretation due to Singh (1991) in the optic flow framework, lets us use a variety of estimation-theoretic techniques to compute the robot pose and associate a notion of confidence with it. An estimate of the actual pose can be computed using a weighed least-squares approach.

$$x_c = \frac{\displaystyle\sum_{dx=-\Delta x,0,\Delta x;} \sum_{dy=-\Delta y,0,\Delta y;} \sum_{d\theta=-\Delta\theta,0,\Delta\theta;} M(x+dx,y+dy,\theta+d\theta)\cdot dx}{\displaystyle\sum_{dx=-\Delta x,0,\Delta x;} \sum_{dy=-\Delta y,0,\Delta y;} \sum_{d\theta=-\Delta\theta,0,\Delta\theta;} M(x+dx,y+dy,\theta+d\theta)}$$

$$y_c = \frac{\displaystyle\sum_{dx=-\Delta x,0,\Delta x;} \sum_{dy=-\Delta y,0,\Delta y;} \sum_{d\theta=-\Delta\theta,0,\Delta\theta;} M(x+dx,y+dy,\theta+d\theta)\cdot dy}{\displaystyle\sum_{dx=-\Delta x,0,\Delta x;} \sum_{dy=-\Delta y,0,\Delta y;} \sum_{d\theta=-\Delta\theta,0,\Delta\theta;} M(x+dx,y+dy,\theta+d\theta)}$$

$$\theta_c = \frac{\displaystyle\sum_{dx=-\Delta x,0,\Delta x;} \sum_{dy=-\Delta y,0,\Delta y;} \sum_{d\theta=-\Delta\theta,0,\Delta\theta;} M(x+dx,y+dy,\theta+d\theta)\cdot d\theta}{\displaystyle\sum_{dx=-\Delta x,0,\Delta x;} \sum_{dy=-\Delta y,0,\Delta y;} \sum_{d\theta=-\Delta\theta,0,\Delta\theta;} M(x+dx,y+dy,\theta+d\theta)}$$

(7)

The uncertainty of the estimate given above can be represented by a covariance matrix

$$C_c = \begin{bmatrix} Cxx & Cxy & Cx\theta \\ Cyx & Cyy & Cy\theta \\ C\theta y & C\theta y & C\theta\theta \end{bmatrix} \quad (8)$$

being the variances

$$Cxx = \frac{\displaystyle\sum_{dx}\sum_{dy}\sum_{d\theta} M(x+dx,y+dy,\theta+d\theta)\cdot(x+dx-x_c)^2}{\displaystyle\sum_{dx}\sum_{dy}\sum_{d\theta} M(x+dx,y+dy,\theta+d\theta)}$$

$$Cyy = \frac{\displaystyle\sum_{dx}\sum_{dy}\sum_{d\theta} M(x+dx,y+dy,\theta+d\theta)\cdot(y+dy-y_c)^2}{\displaystyle\sum_{dx}\sum_{dy}\sum_{d\theta} M(x+dx,y+dy,\theta+d\theta)}$$

$$C\theta\theta = \frac{\displaystyle\sum_{dx}\sum_{dy}\sum_{d\theta} M(x+dx,y+dy,\theta+d\theta)\cdot(\theta+d\theta-\theta_c)^2}{\displaystyle\sum_{dx}\sum_{dy}\sum_{d\theta} M(x+dx,y+dy,\theta+d\theta)}$$

(9)

and the cross covariances

$$Cxy=Cyx=$$
$$\frac{\displaystyle\sum_{dx}\sum_{dy}\sum_{d\theta} M(x+dx,y+dy,\theta+d\theta)\cdot(x+dx-x_c)\cdot(y+dy-y_c)}{\displaystyle\sum_{dx}\sum_{dy}\sum_{d\theta} M(x+dx,y+dy,\theta+d\theta)}$$

$$Cx\theta=C\theta x=$$
$$\frac{\displaystyle\sum_{dx}\sum_{dy}\sum_{d\theta} M(x+dx,y+dy,\theta+d\theta)\cdot(x+dx-x_c)\cdot(\theta+d\theta-\theta_c)}{\displaystyle\sum_{dx}\sum_{dy}\sum_{d\theta} M(x+dx,y+dy,\theta+d\theta)}$$

$$Cy\theta=C\theta y=$$
$$\frac{\displaystyle\sum_{dx}\sum_{dy}\sum_{d\theta} M(x+dx,y+dy,\theta+d\theta)\cdot(y+dy-y_c)\cdot(\theta+d\theta-\theta_c)}{\displaystyle\sum_{dx}\sum_{dy}\sum_{d\theta} M(x+dx,y+dy,\theta+d\theta)}$$

(10)

4.1 Additional degrees of freedom

It was assumed that the floors are horizontal and that the camera location with respect to the robot is perfectly known. Neither are really true: there are floor undulations and errors mainly on the orientation of camera. It is shown in (Paz, 1997) that the main effect of these errors depicts as an error on the *tilt* angle of camera. It was necessary to consider at less one additional degree of freedom to cope with that kind of unmodeled errors: the camera *tilt* angle. This angle is included as a non-constant parameter of the matching function that is modified on execution-time to achieve maximum matching.

5. IMPLEMENTATION AND RESULTS

The system proposed has been implemented on a network including four transputer modules (TRAMs) (INMOS, 1991) as shown on figure 5. A TRAM includes in a single board the μP along with some amount of memory, timing and control, and communication interface.

A general purpose IMS-B411 TRAM, is responsible of interfacing the network with the computer that commands the robot and provides a basic interface with the user. It also coordinates the data flow on the network, executes the preprocessing of maps and, using a Kalman filter, integrates the odometric pose and the sensed pose. A second IMS-B411, projects the map over the image and evaluates the matching function around the estimated pose. The result is a corrected measure of pose and its covariance. An IMS-B429 Video Image Processing TRAM captures and processes the image. Processing includes filtering, decimation and magnitude gradient extraction. A fourth TRAM, an IMS B437-16 Compact Display TRAM, displays the image along with some graphic and numeric information for debugging purposes.

Using low resolution images (128x200x8bit) this implementation achieves –depending on the complexity of the map and on the number of segments being effectively used– an updating rate about one to three iterations per second.

In the tests presented below, the robot is located by hand at the beginning of a corridor, approximately at pose (0,0,0). An uncertainty of (10cm x 5cm x 2°), is assigned to that pose. Then the robot is moved to the other end of corridor and then return back to the starting point; meanwhile, the proposed vision system is periodically updating the estimated pose.

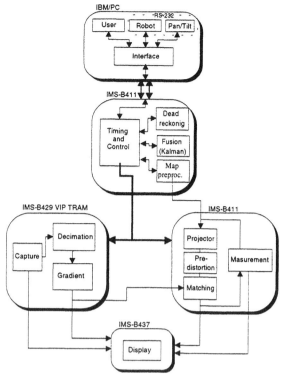

Figure 5. Hardware/Software architecture

To avoid wheel slippage, the acceleration is under $0.5m/s^2$, and the cruise speed is limited to 0.5m/s which is a convenient navigational speed.

The estimated pose represented with dark line, and the odometric pose in gray. The single path experiment consists of 31 meters and two 180° turns. The double path consists of 70m approximately, including four 180° turns and several people passing in front of the robot at various instants. After the courses, the actual or true location of robot was measured, and was verified that the error in the estimated pose was within the initial tolerance while the odometric pose cumulated larger errors (80cm and 2°).

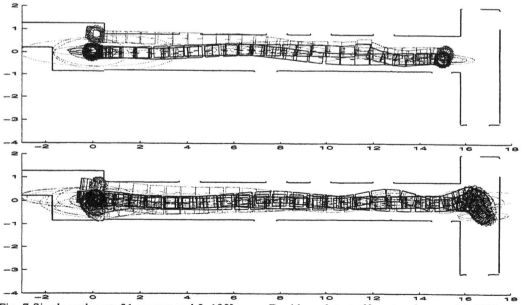

Fig. 7 Single path test: 31 meters and 2x180° turns. Double path test: 68 meters long and 4x180° turns

Figures 7 show that at both ends of corridor the uncertainty of the sensed pose is poor: the uncertainty ellipses are larger than in the rest of the path. This fact is due to the lack of enough visible features to locate the robot with confidence. In such locations the uncertainty ellipses grow until new features come into the image. These situations –for example when the robot is located in front of a flat wall, or when an unexpected obstacle completely occludes the viewing angle– can be detected by tracking the uncertainty of the sensed pose. In that case an active vision system to select alternative targets (Paz,1997) can be a good solution.

Furthermore, it has been verified that this technique is sensitive to non systematic catastrophic odometric errors (Borenstein and Feng, 1994). If not corrected, this kind of errors can cause the robot to get lost. They are consequence of bad floor conditions or excessive acceleration that cause one or both wheels to slip. Catastrophic odometric errors can be easily detected and partially corrected by means of a cheap solid state gyro (Borenstein and Feng, 1996). After detecting one of such errors, the robot must perform a repositioning exercise using a technique not so susceptible to initial uncertainty, for example by means of a geometric matching (Kosaka and Kak, 1992; Paz, 1997, 1998).

6. CONCLUSIONS

It has been presented an original technique for mobile robot absolute positioning that allows our (indoor) mobile robot to navigate at 0.5m/s without modifying the environment. Two important points in the system presented are :

- no geometric features are extracted from image
- matching the image viewed by the robot (the sensed map) and the expected map is based on a very efficient technique for measuring image distance.

Results have confirmed that the system presented is fast, accurate, and robust against variations in lighting conditions and partial occlusions.

7. ACKNOWLEDGMENTS

This work was done in the framework of programs CICYT TAP95-0968-E and Spanish/German WTZ.

REFERENCES

Atiya S., Hager G. D. (1993) Real-Time Vision-Based Robot Localization, *IEEE Transactions on Robotics and Automation*, **Vol. 9, No. 6**

Borenstein J., Feng L., (1994) *UMBmark, a Method for Measuring, Comparing, and Correcting Dead-reckoning Errors in Mobile Robots*, UM-MEAM-94-22, Mobile Robotics Laboratory, University of Michigan.

Borenstein J., Feng L., (1996) Gyrodometry: A New Method for Combining Data from Gyros and Odometry in Mobile Robots. *IEEE Int. Conf. on Robotics and Automation* (Mineapolis)

Brown R. L.(1994) The Fringe Distance Measure: An Easily Calculated Image Distance Measure with Recognition Results Comparable to Gaussian Blurring, *IEEE Trans. on Systems, Man, and Cybernetics*, **Vol. 24, No. 1**

Cox I. J. (1991) Blanche: An Experiment in Guidance and Navigation of an Autonomous Mobile Robot. *IEEE Trans. on Robotics and Automation*

Huttenlocher D. P., Klanderman G.A., Rucklidge W.J., (1991) *Comparing Images Using the Hausdorff Distance*, Cornell University, Tech. Report CUCS TR 91-1211

INMOS,(1991). *The Transputer Development and iq systems Databook.* SGS-Thomson Microelectronics.

Kosaka A., A.C. Kak. (1992) Fast Vision-Guided Mobile Robot Navigation Using Model-Based Reasoning and Prediction of Uncertainties *CVGIP: Image Understanding*, **Vol 56, No. 3**.

Lebègue, X., J.K. Aggarwal (1993). Significant Line Segments for an Indoor Mobile Robot. *IEEE Transactions on Robotics and Automation*, **Vol. 9, No. 6**

Meng M., Kak A., (1993) Mobile Robot Navigation Using Neural Networks and Nonmetrical Environment Models *IEEE Control Systems*

Nomura Y., Sagara M., Naruse H., Ide A., (1992). Simple Calibration Algorithm for High Distortion Lens Camera. *IEEE Trans. on Patt. Analysis and Machine Intellig.*, **Vol 14, No 11**

Paz E. (1997). *Sistema de posicionamiento basado en visión y planos CAD para navegación de robots móviles de interiores.* Dpto. de Ingeniería de Sistemas y Automática, Universidad de Vigo.

Paz, E. (1998). Robot localization from meaningful segments. Accepted for presentation on the *IFAC Symposium on Intelligent Autonomous Vehicles*, IAV'98, Madrid.

Shaffer G., González J., Stentz A., (1992) Comparison of Two Range-based Pose Estimators for a Mobile Robot. Proceedings of the 1992 S PIE Conf. on Mobile Robots, Boston

Shakunaga T. (1992). 3-D Corridor Scene Modeling from a Single View under Natural Lighting Conditions. *IEEE Trans. on Pattern Analysis and Machine Intelligence*, **Vol. 14, No.2**

Singh A., (1991) *Optic Flow Computation. A Unified Perspective* IEEE Computer Society Press. ISBN:0-8186-2602-X

Thrun S., Bücken A., (1996) *Learning Maps for Indoor Mobile Robot Navigation*, CMU. Tech. Report CMU-CS-96-121

A NEW LASER TRIANGULATION PROCESSOR FOR MOBILE ROBOT APPLICATIONS: PRELIMINARY RESULTS

H. Lamela[1], E. García[1], A. de la Escalera[2] & M.A. Salichs[2]

Grupo de Optoelectrónica y Tecnología Láser, Área de Tecnologia Electrónica (1),
Área de Ingeniería de Sistemas y Automática (2)
Departamento de Ingeniería Eléctrica, Electrónica y Automática,
Universidad Carlos III de Madrid

Abstract: Triangulation laser rangefinders using a laser sheet of light and a video camera are one of the most popular active optical vision systems for short range (indoor) autonomous robot applications. Conventional implementations of this system are based on digitising the video signal which is output from the camera, and processing the digitised images in order to obtain the points of incidence of the reflection of the laser sheet of light on the objects in front of the robot on the camera, from which the range information can be directly acquired. In this paper we propose a system is based on direct hardware detection of the pixels that receive maximum illumination (i.e. those corresponding to the intersection points between the laser beam and the objects in the scene). In this way, only the co-ordinates of these pixels are fed into the vision computer, and range information is obtained in a straightforward manner, without any need to digitise full images. Experimental results and performance tests of a vision system prototype based on this idea, are presented in the paper. *Copyright © 1998 IFAC*

Keywords: Range data, Range images, Vision, Sensor systems, Obstacle detection.

1. INTRODUCTION

Ultrasound sensors are still the most widely sensor used for mobile robotics research. A ultrasound ring is the base of almost every mobile robot. As every sensor they have some drawbacks. The centimeter resolution is an obstacle to localize the robot with high precision although good works have been carried out with them; in the ultrasound ring only opposite sensors can be fired simultaneously to avoid interference, and because of that the rate of the overall sensorial system is low (4-5Hz); finally ultrasound detection depends very much on the kind

of material and the relative angle between the sensor and the obstacle.

These are the reasons why a lot of research is done in other sensors, and mainly in computer vision (Jarvis, 1983; Strand, 1985). Within this kind of sensor the analysis of the reflection of a laser plane on a camera (structured light, see fig. 1) has been studied for some years (Besl, 1988; Yuta, 1991). The classical approach is to use an image processing card to analyze the image and find the brightest pixels in it.

This approach has some inconveniences. First there was the price of the image processing boards that could process the images in real-time. With improvements in computers, such as the PCI bus and the Pentium processor, these inconveniences have disappeared, because a frame-grabber can acquire and transfer the images in real-time. Nevertheless two inconveniences remain: the necessity to have a computer dedicated to analyzed the images with the associated space and energy problems and secondly the limit due to the standard video signal, CCIR in Europe and RS-170 in USA, that put a limit to the speed of the system: 25 images for the CCIR and 30 for the RS-170 because the video signal is interlaced.

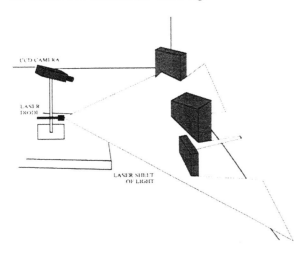

Figure 1: *Camera-laser structure of conventional sheet-of-light triangulation laser rangefinder.*

Our system has two boards. One is a video triangulation hardware processor and the second is a ISA interface for a PC. In this way, the computer is free to do another tasks and access the range measurements trough memory positions when needed. Besides, the system analyses the even and odd fields independently and thus 50 rangemaps per second are obtained instead of 25, as is the case with standard frame grabber configurations.

The sensor described in this article will be located on the two B-21 mobile robots. The ultrasounds sensor give the general information of the obstacles around the robot. The laser will provide a much more detailed information of the obstacles present in the path of the robot. Thus, the navigation through narrow places such corridors or the crossing through doors will be made easier.

2. THE TRIANGULATION PROCESSING SYSTEM

If we take a second look at fig. 1, we can observe that, due to the geometrical configuration of the system, we should expect a single pixel of maximum brightness (corresponding to the planar laser beam intersection with scene objects), per vertical line

scanned in the CCD image. This leads to a first step in the simplification of image processing, which takes advantage of the serial nature of the analog video signal provided by the CCD camera, in which the scene is scanned in horizontal lines. By rotating the camera 90° on its axis, this can be changed, so that the scene is scanned in vertical lines, and we can expect a single maximum brightness pixel per image line. As we shall see, this is of the utmost importance in the conception of our system.

The main idea of our system, is to extract the synchronism signals (pixel clock, line sync. and frame sync.) from the video signal, and then pass the analog (illuminance) video signal through a peak follower circuit, which can detect the point of maximum brightness in each line, and generate a trigger signal (which, from now on, shall be called "MAX" signal), that stores the corresponding value of the pixel counter for each line in the system memory.

With this idea in mind, a first prototype of the system was designed, using a video processor based on a high speed analog peak follower. This system demonstrated good performance (Lamela, 1996), and was very important for establishing the practical possibilities of the idea described so far. However, there were two main drawbacks caused by the analog design of the video processor: on the one hand, the system speed was limited by the analog peak follower and, on the other hand, the analog video processor had to be matched to a single camera model, requiring specific modifications in order to adapt it to different video cameras.

This led us to the design of a second prototype (Lamela, 1996) in which the design specifications were to overcome both of the problems shown by the first one. To achieve this, a fully digital video processor was designed to replace the analog one.

This second prototype was designed to work with any CCD camera following the CCIR standard, and it is based on a digital peak follower which doesn't set a limit to the speed of the system.

The architecture of the second prototype is displayed in fig. 2, in which both the digital video processor and the digital circuit block can be seen (Lamela, 1996).

The video signal coming from the CCD camera enters into the digital video processor, where it is digitised in a video ADC, and fed into a video sync. separator, to obtain the necessary video sync. signals.

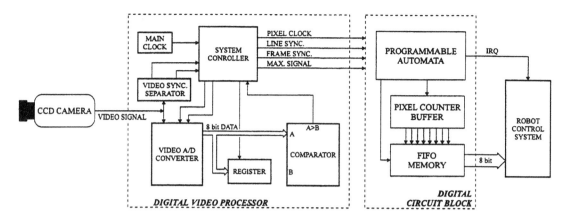

Figure 2: *Architecture of the video signal processing prototype with a digital video processor.*

The 8 bit digitised video signal is fed into one of the inputs of a digital comparator. The second input of this comparator is connected to the output of an 8 bit register, which holds the intensity value corresponding to the last pixel with maximum brightness in the line. When the light intensity in the pixel which is currently being scanned is higher than the value stored in the register, the "A>B" output of the comparator is activated, which causes the system controller to trigger the "LOAD" input of the register (see fig. 2), so that the intensity value of the new maximum is stored in the register. Apart from this, the system controller also sets the "MAX" signal "high" each time the A>B signal is activated. As we shall see, this signal is used in the digital circuit block to mark the pixel that showed maximum brightness in each line.

The "MAX" signal generated by the digital peak follower and the analog video signal coming out of the video camera are shown in fig. 3, in which we can see how the pixel corresponding to maximum brightness in each line (i.e. the point corresponding to the image of the laser beam in the scene), is marked by the last time the "MAX" signal goes high during that line. In fig. 3a, we can see the video signal (top trace) and "MAX" signal output during a full line scan (between the two line sync. pulses). In fig. 3b we can see a "zoom in" version of fig. 3a, in which the intensity maximum corresponding to the laser sheet of light and the individual "MAX" signal pulses can be clearly seen.

The last function performed by the system controller in the digital video processor, is to generate, from the signals supplied by the video sync. separator, the necessary synchronism signals for the digital circuit block, i.e. the "PIXEL CLOCK", "LINE SYNC." and "FRAME SYNC." signals. From these three signals and the previously mentioned "MAX" signal, the digital circuit block obtains the pixel with maximum brightness per line, and feeds this data into the robot control computer.

From these three signals and the previously mentioned "MAX" signal, the digital circuit block obtains the pixel with maximum brightness per line, and feeds this data into the robot control computer.

The "PIXEL CLOCK" signal is used to clock the pixel counter in the digital circuit block, so that, each time that the "MAX" signal is activated, the value of this counter is stored in the pixel counter buffer (see fig. 2). When a line synchronism pulse is detected through the "LINE SYNC." signal, the value contained in the pixel clock buffer is stored in the FIFO memory, and the pixel counter and the peak follower (in the digital video processor) are reset. In this way, each word in the FIFO memory contains the position of the pixel with maximum brightness in each vertical line (i.e. the vertical position of the pixel in the CCD camera), and the address of each word in the FIFO represents the line number (i.e. the horizontal position of the pixel in the CCD camera).

This information of pixel co-ordinates in the image plane, can be directly converted to corresponding values of the range-maps using a L.U.T. (Look-Up Table) allocated in the control computer memory. The L.U.T. could be easily implemented as part of our system, but placing it in the computer memory allows for very simple software calibration of the system, without imposing any load on the robot control computer.

In conclusion, it must be noted that a novel approach to range-map acquisition by direct triangulation laser/camera systems has been developed, which can be directly applied to any existing hardware configuration (as long as a CCIR video camera is used), and that effectively eliminates the speed problems associated with the digitised video signal processing presented by standard systems.

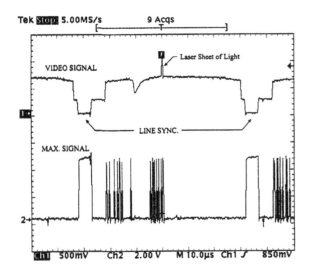

Fig. 3a.: *Oscilloscope display of video signal (top trace) and "MAX" signal (bottom trace).*

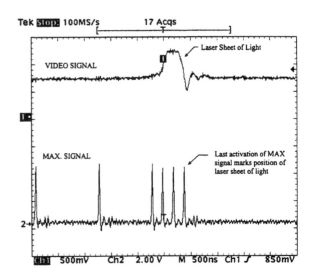

Fig. 3b.: *"Zoom in" of fig. 3.a, showing relationship between last "MAX" signal pulse and position of laser sheet of light in the video signal.*

Figure 3. *Oscilloscope displays of "MAX" signal*

3. EXPERIMENTAL RESULTS

3.1. Static Calibration

The system that we have described, consists of two printed circuit boards, one for the digital video processor, and the other for the digital circuit block. This second board has been designed as a direct plug-in for the host computer, so that it fits a standard ISA slot.

The system was installed into a personal computer, for calibration and performance tests so that range-maps could be acquired and displayed on the computer monitor.

Once installed, the first thing that we did was to calibrate the measurement system in the laboratory, in order to establish its accuracy in the measurement of distance. The measurement error relative to the measured distance was below 5% for all experimental data points, and below 3% for most of them. Absolute measurement error (i.e. the difference between the ideal calibration line and the experimental points measured with our system), increases with measured distance. This is to be expected in triangulation systems, due to the hyperbolic relationship between real distance in the scene, and distance in camera image (Jarvis, 1983). In our system this is augmented by the fact that greater distances in the real scene, correspond to the lowest values of the pixel counter, which are more

affected by the ±1/2 pixel spatial resolution of the camera. Such behaviour of the measurement error is an advantage if the system is to be used as a navigation guidance in autonomous robots, since the dimensions and position of those obstacles closest to the robot will be measured with the highest accuracy.

After these preliminary calibration tests, the performance of the vision system in the application it has been developed for (i.e. range-map acquisition for autonomous robot navigation) had to be evaluated. For this purpose, the system was taken out of the laboratory and different object patterns were placed in front of the camera, so that the obtained range-maps could be compared to the real scene. We shall now present some of the experimental results obtained.

In the first example (see fig. 4), the ability of the system to follow different profiles (Lamela, 1997) is shown, by using a real scene consisting of a waste-paper basket and a plastic board behind it (see left side of fig. 4). In the range-map acquired by our system, which is shown to the right in fig. 4, we can see how the experimental measured points fit the profile of the real scene, following the slowly varying waste-paper basket profile, and the abrupt change from the borders of this object and to the plastic board behind it. When examining fig. 4, it must be noted that, in order to avoid distorting the circular section of the waste-paper basket, both axial and transverse axis cover the same range (of 120 centimetres). This makes it impossible to show the part of the wall behind the object pattern falling within the field of view of our system.

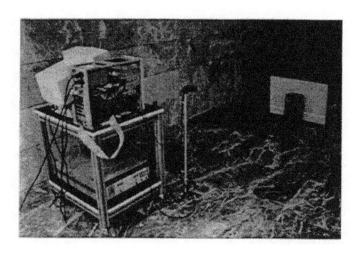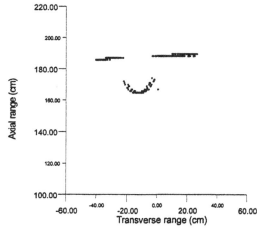

Figure 4. *Vision system ooperating on a scene consisting of a waste-paper basket and a white plastic board (left), and acquired rangemap of this scene (right).*

A second object pattern, consisting of two white plastic boards, and two square-section aluminium bars, 4 centimetres wide, is presented in fig 5. Again on the left hand side, a photograph of the real scene is displayed, showing both boards, and the aluminium bars, one in front of the boards and the other behind them. To the right we can see the corresponding range-map acquired by our system is displayed, showing how even the small-section aluminium bars are detected. In this second example the range axis covers the field of view of our system, so that we can see the wall at the top of the rangemap, at a distance of about 6 metres from the CCD camera.

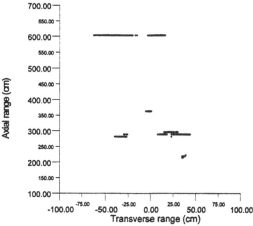

Figure 5. *Vision system ooperating on a scene consisting of two white plastic boards and two aluminium profiles (left), and acquired rangemap of this scene (right).*

3.2. Installation on the mobile robot

The triangulation vision system is now installed for performance tests on a B-21 mobile robot (see fig. 6). This system acts as a complement to the ultrasound sensor ring installed around the robot, covering about the same front vision angle as a group of sonar sensors (see fig. 7).

The robot control computer will be programmed to ignore the data of the ultrasound sensors if they are contradictory to the ones provided by the triangulation vision system, since this is more reliable than the one provided by the sonar sensors. This will be very important when navigating through narrow passages, such as doors, where ultrasound sensors are liable to detect false obstacles.

Figure 6. Triangulation laser vision system installed on B-21 mobile robot.

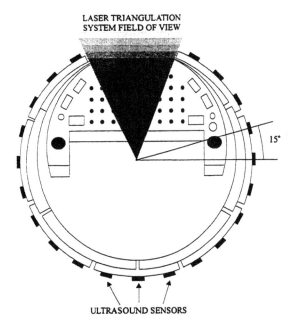

LASER TRIANGULATION
SYSTEM FIELD OF VIEW

15°

ULTRASOUND SENSORS

Figure 7. Detail of installation of triangulation laser vision system on top of B-21 mobile robot.

4. CONCLUSIONS AND FUTURE WORK

In this work, we have presented a novel system for high-speed video signal processing in laser triangulation rangefinders, well suited for indoor robotics vision. It is based on hardware processing of the signal coming from the CCD video camera, which directly detects the pixels receiving maximum illumination. By handling the serial video information provided by the camera in this way, we avoid using a dedicated vision computer for processing the huge data amounts that are obtained when digitising full images with a frame grabber. In fact, the speed limitations of the presented system arise from the CCD camera itself (50 frames per second), not from the triangulation hardware processor.

Our system was first installed in a host computer for performance tests, showing its ability to detect the profiles corresponding to different object patterns that were placed in front of the system.

This sensor is now installed on a B-21 mobile robot which is being programmed to test its performance as a vision system for navigation purposes (passage through narrow places, such as corridors and doors), shall be evaluated.

REFERENCES

Jarvis, R. A. "A Perspective on Range Finding Techniques for Computer Vision", *IEEE Transactions on Pattern Analysis and Machine Intelligence*, **Vol. PAMI-5(2)** pp122-139 (1983).

Strand, T. C. "Optical Three-Dimensional Sensing for Machine Vision", *Optical Engineering*, **Vol. 24(1)**, pp. 33-40 (January/February 1985).

Besl, P. J. "Active, Optical Range Imaging Sensors", *Machine Vision and Applications*, **Vol. 1**, pp. 127-152 (1988).

Yuta, S. S. Suzuki, Y. Saito, S. Iida, "Implementation of an Active Optical Range Sensor Using Laser Slit for In-Door Intelligent Mobile Robot", *IEEE/RSJ International Workshop on Intelligent Robots and Systems, IROS' 91*, pp 415-420 (1991).

Lamela, H., S. Gómez, G. Carpintero, P. Acedo, L. Hernández & E. Olías, "High Speed Range Map Obtaining in a Triangulation Range Finder for Mobile Robots", *IEEE IECON' 96*, Proceedings, Vol. 2, pp. 1566-1571, (1996).

Lamela, H., S. Gómez, G. Carpintero & E. Olías, "Triangulation Video Signal Pocessing in Mobile Robots Laser Range Finders", *DCIS' 96*, Proceedings, pp. 399-404, (1996).

Lamela, H., S. Gómez, A. Varo y E. García, "Utilización de Técnicas de Procesamiento Hardware en Sistemas de Visión Basados en Triangulación con Láser de Semiconductor para Robots Móviles", *URSI' 97*, Actas, Vol II, pp. II-469 a II-472, (1997).

ON THE USE OF QUANTIZED λ-SENSOR MEASUREMENTS IN RECURSIVE IDENTIFICATION OF AIR-FUEL MIXING DYNAMICS

Torbjörn Wigren and Bengt Carlsson

Systems and Control Group, Uppsala University, P.O. Box 27, SE-751 03 Uppsala, Sweden

Abstract: Current legislation has put strong pressure on air-fuel ratio control research for compliance with emission limits for sparc ignition engines. This paper focuses on the highly nonlinear λ-sensor that is used in many systems. The main idea is to include the relay-type sensor characteristics in a nonlinear input-output model for joint state and parameter estimation. An estimation algorithm is derived that can be used for adaptive trim table buildup and internal model control schemes. *Copyright © 1998 IFAC*

Keywords: Automotive control, extended Kalman filter, engine modelling, recursive estimation, nonlinear system

1. INTRODUCTION

To maintain low emissions from todays sparc-ignition engines, tight air-fuel (A/F) ratio control is necessary. The catalysts used for reduction of exhaust emissions typically require engine operation very close to the stoichiometric ratio of 14.64. As stated in (Heywood, 1988) and (Chang *et al.*, 1995), control within 0.05 % is preferable. To fulfill this requirement, feedback control from the exhaust gas oxygen (EGO) sensors can be applied, see for example, (Chang *et al.*, 1995), (Grizzle *et al.*, 1991) and (Åström and Wittenmark, 1989). Two common sensors are the universal air-fuel ratio heated exhaust gas oxygen (UEGO) sensor and the λ-sensor. The UEGO sensor has the advantage of being linear while the λ-sensor is highly nonlinear with a relay-type measurement characteristics indicating lean/rich conditions. However, the cost of the UEGO sensor is high which makes λ-sensor feedback control commercially interesting. The sensor nonlinearity typically results in a limit cycle oscillation of the controlled A/F-ratio.

The engine dynamics from the throttle and the fuel injector is normally nonlinear. At least for close to steady state operation it can be convenient to use trim tables in combination with gain-scheduled controllers (Åström and Wittenmark, 1989). One of the possible application of the algorithm developed in this paper is on-line adaptation of such trim tables. This procedure could reduce the need for factory tuning but also increase the control efficiency. The adaptive observer that is derived here could also be used in combination with an internal model control (IMC) scheme. This has the potential advantage of more accurate control of the A/F-ratio. A further discussion of the implications of the suggested adaptive observer is given in the technical report (Wigren and Carlsson, 1996), which may be consulted for further details. The work presented in (Chang *et al.*, 1995) and (Chang, 1993) indicates experimentally that a λ-sensor feedback controller does not necessarily lead to limit cycle oscillations in the A/F-ratio. This results was obtained when feedback was applied from observed states. The observer and the controller were designed assuming linear UEGO measurements and known dynamics from off-line adaptation.

The contributions of this paper are based on the idea to include the nonlinear λ-sensor characteristics in the dynamic model of the engine from the throttle and the fuel injectors to the measured EGO output. Recent results of recursive system identification, (Wigren, 1994), (Wigren, 1995) and (Wigren, 1996) show that arbitrary linear dynamics can be recursively identified using only binary (possible disturbed) output measurements. This clearly relaxes previous constraints on linearity that leads to a need for UEGO sensors for identification purposes, see (Jones *et al.*, 1995). Based on a joint input-output model, a nonlinear and adaptive observer is constructed that simultaneously estimates the states and the parameters of an assumed engine model. The estimator can be interpreted as a modified extended Kalman filter (EKF). The main difference as compared to previous work is the treatment of the λ-sensor as an integrated part of the model. *Since the main purpose is to illustrate the possibilities of this idea*, several simplifying assumptions are made. First, a linear engine model is used. Secondly, the λ-sensor is treated as a cascade of a linear filter and a static nonlinearity with relay characteristics. More advanced nonlinear models and live data trials are topics left for further research.

2. NONLINEAR STATE AND PARAMETER ESTIMATION

2.1 *General models*

Both the engine and the λ-sensor are nonlinear dynamic systems. Although relatively simple models will be used for illustrative purposes, it is useful to review the more general case since the principal development of algorithms is independent of the model structure. A fairly general description of the engine dynamics is given by the following state space model

$$x_e(t + T) = f_e(x_e(t), u(t), \theta_e(t), w_e(t))$$

$$y_e(t) = h_e(x_e(t), \theta_e(t)) \tag{1}$$

where T is the (possibly time varying) sampling period. The input vector $u(t)$ is typically fuel injection or throttle commands. The state vector $x_e(t)$ is used in order to model dynamics and transport delays of the engine in terms of the unknown, possibly time varying, parameter vector $\theta_e(t)$. Process disturbances are denoted by $w_e(t)$. The output $y_e(t)$ is the actual quantity that is measured by the λ-sensor to produce air-fuel ratio measurements. Detailed development of engine models is, for example, discussed in (Heywood, 1988), (Powell, 1979) and the references therein.

The λ-sensor is a nonlinear device, the most pronounced effect is the relay-type characteristics (Åström and Wittenmark, 1989). There are also inherent dynamic phenomena in the sensor that need to be accounted for. This dynamics enters before the relay characteristics and most often dominates over the dynamics of the engine. The response time of the sensor is usually different on rising and falling edges, although linear models with symmetric responses are sometimes used (Grizzle *et al.*, 1991) and (Cook *et al.*, 1983).

A general λ-sensor model can be described as follows

$$x_\lambda(t + T) = f_\lambda(x_\lambda(t), y_e(t), \theta_\lambda(t), w_\lambda(t))$$

$$\tag{2}$$

$$y_\lambda(t) = h_\lambda(x_\lambda(t), \theta_\lambda(t), e_\lambda(t)).$$

Note that $h_\lambda(\cdot, \cdot, \cdot)$ may not be formally differentiable because of an assumed relay-type sensor characteristic in the model. In (2), $e_\lambda(t)$ denotes measurement disturbances, the remaining notation parallels that of (1).

It is straightforward to transform (1) and (2) to a form allowing joint state and parameter estimation using, for example, the extended Kalman filter, c.f. (Ljung and Söderström, 1983). Simply introduce the parameters as (extended) states and write (1) as

$$
\begin{aligned}
x_e(t + T) &= \begin{pmatrix} x_e(t + T) \\ \theta_e(t + T) \end{pmatrix} \\
&= \begin{pmatrix} f_e(x_e(t), u(t), \theta_e(t), w_e(t)) \\ \theta_e(t) + w_{\theta_e}(t) \end{pmatrix}
\end{aligned}
\tag{3}
$$

$$y_e(t) = h_e(x_e(t), \theta_e(t))$$

and similarly for (2).

To illustrate the procedure, a low order example will be treated in the next subsection that focuses on the λ-sensor characteristics. It is stressed that the development of algorithms in the full nonlinear case is analogous. Hence, the treatment of the model below is not a severe practical restriction. The simplifications are introduced in order to highlight the effect of the λ-sensor.

2.2 *System assumption*

For the purpose of illustration, a one cylinder engine, with approximations similar to those of (Grizzle *et al.*, 1991) will be used for generation of the data. Thus, the engine dynamics (the system) is composed of a gain factor K_o and a time delay T_o as

$$y_e(t) = K_o u(t - T_o) \tag{4}$$

where the input is assumed to be fuel injection commands. The λ-sensor is given by a unity gain one pole filter in cascade with an ideal relay i.e.

$$x_\lambda(t + T) = a_o x_\lambda(t) + (1 - a_o) y_e(t) + w_{\lambda,o}(t) \tag{5}$$

$$y_\lambda(t) = \text{sign}(x_\lambda(t) - b_\lambda) + e_{\lambda,o}(t)$$

Here, the parameter b_λ is the sensor bias of 14.64. The unity gain of the sensor dynamics corresponds to an assumption that the sensor is well calibrated.

2.3 Engine and sensor model

Considering (4)-(5), it will be practically important to be able to capture fast variations of the time delay T_o, resulting from varying engine speeds (Grizzle et al., 1991). These variations can in practice be expected to be quicker than the variation of, for example, a_o. Hence, it will be important to allow for different tracking bandwidths for different parameters in the corresponding estimation algorithm to be derived below. This is one reason that the extended Kalman filter is advantageous as compared to the schemes of (Wigren, 1995) and (Wigren, 1996) that do not allow for individual tuning of the tracking bandwidth. Typically, assumed levels of system noise and measurements noise are exploited in this tuning procedure. Note that it is common practice to use engine speed dependent sampling rates to reduce the variation of the dynamics as much as possible (Grizzle et al., 1991).

The following time-varying tapped delay line engine model will be used in the estimation algorithm

$$y_e(t, \theta_e(t)) = \sum_{j=0}^{nb} b_j u(t - jT) \tag{6}$$

$$\theta_e(t) = (b_o(t) \quad b_1(t) \ldots b_{nb}(t))^T \tag{7}$$

$$\theta_e(t + T) = \theta_e(t) + w_e(t) \tag{8}$$

The noise vector $w_e(t) = (w_e^o(t) \quad w_e^1(t) \ldots w_e^{nb}(t))^T$ is assumed to be white and Gaussian with covariance R_e and it is used in order to describe time variations of the model parameters. Note that the gain K_o is naturally absorbed in (6). Provided that zero order hold sampling is used, the model (6) is capable of modeling fractional lags since a fractional lag just results in an additional tap in the discrete time model (6), see (Åström and Wittenmark, 1989).

The λ-sensor is modeled by

$$x_\lambda(t+T, \theta_\lambda(t)) = a(t) x_\lambda(t, \theta_\lambda(t)) + y_e(t, \theta_e(t)) \tag{9}$$

$$y_\lambda(t, \theta_\lambda(t)) = \text{sign}(x_\lambda(t, \theta_\lambda(t)) - b_\lambda) + e_\lambda(t) \tag{10}$$

$$\theta_\lambda(t) = a(t) \tag{11}$$

$$\theta_\lambda(t + T) = \theta_\lambda(t) + w_\lambda(t) \tag{12}$$

The noises $w_\lambda(t)$ and $e_\lambda(t)$ are assumed mutually uncorrelated and Gaussian with variances r_λ and σ^2, respectively. The noise variance $w_\lambda(t)$ is used in order to describe the time variation of the sensor dynamics while $e_\lambda(t)$ describes the level of the measurement noise. The quantity $b_\lambda = 14.64$ is assumed to be known. Equation (9) does not model the unity gain of the sensor filter. The engine gain thus has to be factored out after filtering in order to obtain the estimate of $y_e(t)$, which is the quantity to use for internal model control feedback.

2.4 The estimation algorithm

As stated in the previous subsection, one important capability of the algorithm for air-fuel ratio dynamics estimation, is tracking of varying transport delays in the engine. The second capability, which is the main contribution of the paper, is to perform the estimation using only inputs $u(t)$ and quantized outputs $y_\lambda(t)$ as measurements. The latter property will be obtained from an approximation in the extended Kalman filter.

To derive the algorithm from (6)-(12), these equations are first written in state space form

$$x(t+T) = \begin{pmatrix} x_\lambda(t+T, \theta_\lambda) \\ x_1^D(t+T) \\ x_2^D(t+T) \\ \vdots \\ x_{nb}^D(t+T) \\ a(t+T) \\ b_o(t+T) \\ \vdots \\ b_{nb}(t+T) \end{pmatrix} =$$

$$\begin{pmatrix} a(t)x_\lambda(t) + b_o(t)u(t) + \ldots b_{nb}(t)x_{nb}^D(t) \\ u(t) \\ x_1^D(t) \\ \vdots \\ x_{nb-1}^D(t) \\ a(t) + w_\lambda(t) \\ b_o(t) + w_e^o(t) \\ \vdots \\ b_{nb}(t) + w_e^{nb}(t) \end{pmatrix}$$

$$= f(x(t), u(t), w_\lambda(t), w_e(t)) \tag{13}$$

$$y_\lambda(t) = h(x(t), e(t))$$
$$= \text{sign}(x_\lambda(t, \theta_\lambda(t)) - b_\lambda) + e_\lambda(t) \tag{14}$$

The states $x_i^D(t) = u(t - i)$ are used in order to delay the samples of the input signal to obtain the time delay of the engine model. The variance of the state noise becomes

$$R_1 = E \begin{pmatrix} 0 \\ \vdots \\ 0 \\ w_\lambda(t) \\ w_e(t) \end{pmatrix} (0 \ldots 0 \; w_\lambda(t) \; w_e(t))$$

$$= \begin{pmatrix} 0 & \cdots & & & 0 \\ & \ddots & & & \\ \vdots & & 0 & & \vdots \\ & & & r_\lambda & \\ 0 & & & & R_e \end{pmatrix} \quad (15)$$

The variance of the measurement noise is

$$R_2 = E e_\lambda^2(t) = \sigma^2 \quad (16)$$

In order to derive the extended Kalman filter, (13) needs to be linearized around the latest estimates and (14) needs to be linearized around the latest one step prediction. Linearization of (13) gives

$$F(t) = \frac{\partial f}{\partial x} \mid_{x=\hat{x}(t|t)}$$

$$= \begin{pmatrix} A & B \\ C & D \end{pmatrix} \quad (17)$$

where

$$A = \begin{pmatrix} a(t) & b_1(t) & \ldots & b_{nb-1}(t) & b_{nb}(t) \\ 0 & 0 & \ldots & & 0 \\ 0 & 1 & 0 & \ldots & 0 \\ & & \ddots & & \\ & & & 1 & 0 \end{pmatrix},$$

$$B = \begin{pmatrix} x_\lambda(t) & u(t) & x_1^D(t) & \ldots & x_{nb}^D(t) \\ 0 & & \ldots & & 0 \\ \vdots & & & & \vdots \\ 0 & & \ldots & & 0 \end{pmatrix},$$

C is a $nb + 2|nb + 1$ zero matrix, and D is a $nb + 2|nb + 2$ identity matrix.

There is no need to linearize with respect to the noises, since these enter linearly in f. The main problem is now that the measurement equation (14) is not differentiable. Furthermore, if the formal Dirac-pulse derivative would be used, the estimation algorithm would never update since the derivative would be zero almost everywhere. To circumvent this, $h(x(t), e(t))$ is replaced by a smooth approximation $k(x(t), e(t))$ which is used

only in order to compute an approximative gradient. The original equation (14) is retained for prediction error computations. Hence,

$$\bar{H}(t) = \frac{\partial k(x(t), e(t))}{\partial x} \mid_{x=\hat{x}(t+T|t)} \quad (18)$$

is used as a replacement for $H(t)$ in the extended Kalman filter. This gives the following algorithm, cf (Söderström, 1994)

$$\hat{x}(0 \mid 0) = x_o$$
$$P(0 \mid 0) = P_o$$
$$\hat{x}(t+T \mid t) = f(\hat{x}(t \mid t), u(t))$$
$$F(t) = \frac{\partial f}{\partial x} \mid_{x=\hat{x}(t|t)}$$
$$P(t+T \mid t) = F(t)P(t \mid t)F^T(t) + R_1(t) \quad (19)$$
$$\bar{H}(t+T) = \frac{\partial k(x(t), e(t))}{\partial x} \mid_{x=\hat{x}(t+T|t)}$$
$$K(t+T) = P(t+T \mid t)\bar{H}^T(t+T)$$
$$\times [\bar{H}(t+T)P(t+T \mid t)\bar{H}^T(t+T) + R_2]^{-1}$$
$$\hat{x}(t+T \mid t+T) = \hat{x}(t+T|t) + K(t+T)$$
$$\times [y_\lambda(t+T) - h(\hat{x}(t+T \mid t))]$$
$$P(t+T|t+T) = P(t+T \mid t) - P(t+T \mid t)\bar{H}^T(t+T)$$
$$\times [\bar{H}(t+T)P(t+T \mid t)\bar{H}^T(t+T) + R_2]^{-1}$$
$$\times \bar{H}(t+T)P(t+T \mid t)$$

2.5 Theoretical Background

What guarantees are there that the approximation (18) will result in a working algorithm? It is not intuitively obvious that identification of the unknown parameters is possible using only quantized output measurements. Local convergence properties of related stochastic approximation algorithms were, however, recently analyzed in (Wigren, 1995). There it was shown that the algorithm can be expected to be locally convergent to the true parameters when the following heuristically formulated list of conditions hold

A1) The system is contained in the model set.
A2) The linear part of the system is controllable and observable.
A3) The model is not overparametrized.
A4) The quantizer and the approximation are either both increasing or both decreasing.
A5) The input signal is persistently exciting of sufficiently high order.
A6) There is signal energy (the pdf is nonzero) in the quantization step. The step does not occur for zero input signal to the quantizer.

Conditions A1)-A3) and A5) also occur in linear system identification and do not pose any particular problems here. Condition A4) holds provided that the approximation (18) is chosen in a correct manner. Because of the bias $b_\lambda = 14.64$, the step does occur for a nonzero input to the quantizer.

Since the control objective is to keep the output at exactly 14.64, condition A6) may be hard to fulfill during closed loop control. For open-loop experiments A6) can easily be satisfied. The conclusion is that the proposed algorithm can be expected to work well in the particular application.

Further, global convergence to a zero output error parameter setting can be proved in the FIR model case see (Wigren, 1996). The algorithm suggested in this paper is based on an IIR-model and hence only local convergence results apply.

3. NUMERICAL EXAMPLES

The simple engine model (4)-(5) was used in order to generate data for illustration of the estimation performance. The parameters in (4) were initially taken as $K_o = 2$ and $T_o = 0$. The parameters of the λ-sensor (5) were $a_o = 0.75$, $b_\lambda = 14.64$.

The system was excited with a normally distributed random sequence with a standard deviation of 1.0 and a mean value of 7.0. White noise with a standard deviation of 0.1 was added to the output from the sensor. The system was simulated for $N = 2000$ data points. The data used in the estimation experiment is shown in Figure 1.

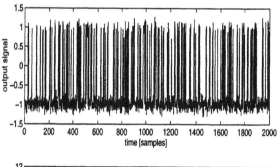

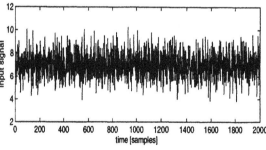

Fig. 1. Data used in the estimation experiment.

In the estimation algorithm, a fourth order tapped delay line model was used, see (6). This gives the possibility to handle an unknown time delay in the system. The used λ sensor model is given in (9)-(12) where the correct value of b_λ was used. The parameter vector to be estimated was thus

$$\theta = (a_o \ b_o \ b_1 \ b_2 \ b_3)^T. \qquad (20)$$

The true system corresponds to the following parameter vector

$$\theta_o = (0.75 \ 0.5 \ 0 \ 0 \ 0)^T. \qquad (21)$$

The following covariance matrix was used

$$R_1 = \begin{pmatrix} 0 & & \cdots & & 0 \\ & \ddots & & & \\ \vdots & & 0 & & \vdots \\ & & & \alpha & \\ 0 & \cdots & & & \alpha I \end{pmatrix} \qquad (22)$$

with $\alpha = 4 \ 10^{-8}$. The measurement noise level was $R_2 = 0.1$.

The gradient of the λ sensor nonlinearity was approximated with (c.f. (18)).

$$\bar{H}(t) = \frac{\partial}{\partial \hat{x}_\lambda} \frac{2}{\pi} \tan^{-1}(2(\hat{x}_\lambda - 14.64))$$

$$= \frac{4}{\pi} \frac{1}{1 + 4 * (\hat{x}_\lambda - 14.64)^2}$$

The estimation algorithm was initialized as follows

$$\hat{x}(0 \mid 0) = (0 \ 0 \ 0 \ 0 \ 0.1 \ 0.1 \ 0.1 \ 0.1 \ 0.1)^T$$
$$P(0 \mid 0) = 0.1I$$

The estimation results are shown in Figure 2 and Figure 3. The estimated parameter vector obtained at the end of the data run was

$$\hat{\theta}(2000) = (0.72 \ 0.53 \ -0.01 \ 0.01 \ 0.03)^T (23)$$

It can be seen that the parameter estimates quite quickly come close to the true system parameters despite the binary and noisy sensor output.

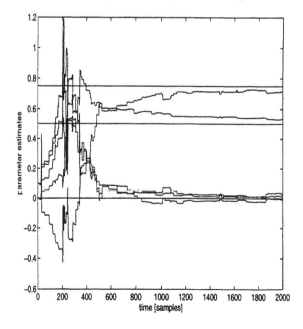

Fig. 2. Parameter estimates.

As a comparison, the estimation results are presented when the value of the time delay $T_o = 3$.

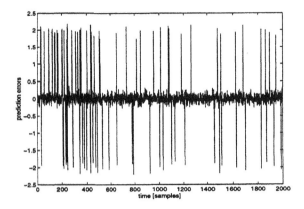

Fig. 3. Prediction errors.

The same set-up as in the previous example was used. The estimated parameter vector obtained at the end of the data run was

$$\hat{\theta}(2000) = (0.74 \quad 0.00 \quad -0.01 \quad 0.02 \quad 0.51)^T \quad (24)$$

Also, in this case, the parameter estimate is close to the true system parameters and the time delay is easily detectable.

In summary it can be concluded that the estimation results are quite promising. The sensor and engine dynamics could be accurately estimated from noisy binary data. There is, however, a number of topics to be studied further (which, however, is outside the scope of this paper). First, the algorithm seems to be sensitive with respect to the covariance matrix R_1. In the examples above a parameter α was introduced, see (22). The value of α must be chosen small otherwise the parameter estimates become biased. On the other hand, to track parameter variations α should not be chosen too small. It is hence relevant to study what the limiting performance is in terms of tracking and bias. Further, it would be of interest to evaluate how the input signal affects the estimation results. Note that from the theoretical analysis it is required that the mean value of the input to the sensor should be different from the switching point.

4. CONCLUSION

The problem to find an engine model from data obtained from a nonlinear λ-sensor was treated. A recursive estimation algorithm was derived for this purpose. In a simulation study, the performance of the estimation algorithm was illustrated and an accurate engine and sensor model was obtained from noisy binary outputs. An engine model estimated on-line has a number of interesting applications such as adaptive trim table buildup and internal model control. First, however, a closer study of the estimation performance

should be conducted. This and the control applications are important topics for further research.

5. REFERENCES

Chang, C.-F. (1993). Air-fuel ratio control in an IC engine using an event based observer. PhD thesis. Stanford Univ. Stanford, CA.

Chang, C.-F., N. P. Fetete, A. Amstoutz and J. D. Powell (1995). Air-fuel ratio control in sparc-ignition engines using estimation theory. *IEEE Trans. Contr. Syst. Technol.* **3**, 22–31.

Cook, J. A., D. R. Hamburg, W. J. Kaiser and E. M. Logothetis (1983). Engine dynamometer study of the transient response of zro$_2$ and tio$_2$ exhaust gas oxygen sensors. *SAE Tech. Paper Series, no 830985.*

Grizzle, J. W., K. L. Dobbins and J. A. Cook (1991). Individual cylinder air-fuel ratio control with a single ego sensor. *IEEE Trans. Veh. Technol.* **40**, 280–286.

Heywood, J. B. (1988). *Internal Combustion Engine Fundamentals.* McGraw-Hill. New York.

Jones, V. K., B. A. Ault, G. F. Franklin and J. D Powell (1995). Identification and air-fuel ratio control of a sparc ignition engine. *IEEE Trans. Contr. Syst. Technol.* **3**, 14–21.

Ljung, L. and T. Söderström (1983). *Theory and Practice of Recursive Identification.* MIT, Press. Cambridge, MA.

Powell, B. K. (1979). A dynamic model for automatic engine control analysis. In: *Proc. IEEE Conf. Decision, Contr.* Vol. 66. pp. 120–126. Boston, MA.

Söderström, T. (1994). *Discrete-Time Stochastic Systems - Estimation and Control.* Prentice Hall. Hemel Hempstead, UK.

Åström, K. J. and B. Wittenmark (1989). *Adaptive Control.* Addison–Wesley. MA.

Wigren, T. (1994). Convergence analysis of recursive identification algorithms based on the nonlinear wiener model. *IEEE Trans. Automat. Contr.* **39**, 2191–2206.

Wigren, T. (1995). Approximate gradients, convergence and positive realness in recursive identification of a class on non-linear systems. *Int. J. Adaptive Contr., Signal Processing* **9**, 325–354.

Wigren, T. (1996). Adaptive filtering using quantized output measurements. *Submitted.*

Wigren, T. and B. Carlsson (1996). On-line estimation of air-fuel mixing dynamics using nonlinear λ-sensor measurements. Technical Report UPTEC 96121R. Systems and Control Group, Uppsala University. Uppsala, Sweden.

FUZZY MODELING AND CONTROL OF AN ENGINE AIR INLET WITH EXHAUST GAS RECIRCULATION.

P. Bortolet[a,c], E.Merlet[b], S. Boverie[c]

[a] *Laboratoire L.A.A.S./C.N.R.S. - 7, av. du Colonel Roche - 31077 TOULOUSE Cedex France.*
[b] *I.N.S.A. - Complexe Scientifique de Rangueil - 31077 TOULOUSE Cedex France.*
[c] *SIEMENS Automotive SA - B.P.1149 av. du Mirail - 31036 TOULOUSE Cedex France.*

Abstract:

The purpose of this document is to present an original fuzzy modeling method applied to a physical highly non linear system: an engine air inlet with exhaust gas recirculation. This system is modeled with fuzzy logic rules of Takagi-Sugeno type. The rule base switches between local linear models defined in the whole state space. The control objective is to preserve a linear behavior of the closed loop system for all operating conditions. To reach this objective, the linear automatic tools are applied to each local linear model. The fuzzy model rule base structure is then used to switch between local controllers. *Copyright © 1998 IFAC*

Keywords: Fuzzy System, Fuzzy Modeling, Fuzzy Control, Automotive Engine.

1. Introduction.

In order to limit automotive fuel consumption and pollutant emission, a precise control of fresh air quantity introduced in the cylinder has to be developed. The manifold pressure is a relevant image of this fresh air quantity. But the engine manifold is a strongly non linear system, difficult to control with conventional means. The objective of the study was to obtain the same closed loop linearized behavior of the manifold pressure for all operating conditions. To reach this objective, a piece wise linear fuzzy model of the process to be controlled is identified. On the basis of this fuzzy partitioning, a set of linear controllers is then designed. The physical model of the engine manifold process is presented in section 2. This model will be next used to test our methodology and will be considered as a reference benchmark.

In section 3, The fuzzy modeling method is described. The method is then applied for the benchmark identification. The fuzzy model is compared to the benchmark. Results are exposed in

section 4.The control strategy and results are developed in section 5.

2. The engine air inlet with exhaust gas recirculation.

The benchmark proposed in this section describes the behavior of an engine air inlet. (Bidan, P. (1989); Chaumerliac, V. (1995)).

Referring to Figure 1, the system inputs are:

- Throttle command for the fresh air valve (ϕ_{CmdF}). Fresh air is taken directly from atmosphere through an air filter.
- Valve position for the exhaust gas recirculation (ϕ_{EGR}). In modern cars, a part of the exhaust gas are taken from the exhaust manifold and reinjected in the inlet manifold. EGR are used in order to reduce pollutant emissions.
- Engine speed (N).

The system outputs are the manifold pressure (P_{Mani}) and its derivative ($\dot{P}_{Mani}$).

Referring to Figure 1, QThr is the gas flow through the fresh air valve. QThr is issued from a look-up table with fresh air valve position and manifold pressure (P_{Mani}) - atmospheric pressure (P_{atmo}) ratio as inputs. The expression of the throttle valve position for fresh air ϕ_{Fresh} is given by:

$$\phi_{Fresh} = \frac{1}{1 + \tau \cdot s} \cdot \phi_{CmdF} \qquad (1)$$

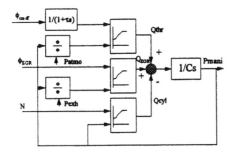

Figure 1: Engine air intake benchmark.

QEGR is the gas flow through the exhaust gas valve. QEGR is issued from a non-linear look-up table with exhaust gas valve position and manifold pressure (P_{Mani}) - exhaust gas pressure ($P_{Exhaust}$) ratio as inputs. The exhaust gas pressure ($P_{Exhaust}$) is issued from a non-linear look-up table with manifold pressure (P_{Mani}) and engine speed (N) as inputs. QCyl is the gas flow injected in the cylinders. QCyl is issued from a non-linear look-up table with manifold pressure (P_{Mani}) and engine speed (N) as inputs. The manifold pressure (P_{Mani}) is given by:

$$\frac{P_{Mani}}{dt} = \frac{1}{C}(Q_{thr} + Q_{EGR} - Q_{cyl}) \qquad (2)$$

C is the manifold capacity.
Both tmospheric pressure, exhaust gas pressure and manifold capacity are supposed constant during simulations.
Note: all the coefficients (tables and data) are extracted from measurements on real engine.

3. The fuzzy modeling method.

The objective of this procedure is to approximate by a fuzzy Sugeno's model the second member of equation (3) dealing with a non-linear process:

$$\dot{x} = f(x, u) \qquad (3)$$

where x is the state vector, $\dot{x}$ is the state vector derivative, u is the control vector, $f(x, u)$ is a non linear multi-variable function.
The rules of the fuzzy model will be expressed in a conventional way by the expression (4).

$$if \left(\left(x_1 \; is \; A_1^j\right) \; and \; \left(x_2 \; is \; A_2^j\right) ... \; and \; \left(x_i \; is \; A_i^j\right)\right) \qquad (4$$

$$then \; s = \alpha_1 \cdot x_1 + \alpha_2 \cdot x_2 ... + \alpha_i \cdot x_i + \alpha_{i+1}$$

where $\left(\alpha_1, \alpha_2, ... \alpha_i, \alpha_{i+1}\right)$ are constant and $\left(x_1, x_2, ... x_i\right)$ are model inputs. The identification procedure can be decomposed in four sequential operations (Figure 2) (Bortolet, P. *et al*, (1997)).

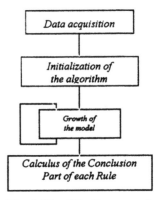

Fig. 2: Identification procedure.

The "initialization of the algorithm" and "growth of the model" have been inspired by the work exposed in (Nakoula Y.*et al* (1996)); Babuska, R. (1995)). The last step "Calculation of the Conclusion Part of each Rule" provides a linear piece-wise model on a fuzzy partition of the input state space.

3.1 Data acquisition

A test signal u is applied in order to obtain testing trajectories in whole state space. For each time step, a data is then constituted by all the inputs and all the outputs of the system.

3.2 Algorithm initialization

In the initialization step, a first approximate reference model on the whole state space is defined. The first membership functions of the rule base are initialized considering the extreme values of each inputs in the data set. The membership functions are centered on the data. A Bezdek partition is respected. For example, for a Double Inputs - Single Output (DISO) model, the initialization is done with four membership functions for each input (see Figure 3a).
The center of each membership function for the first input x_1 are defined by:

- x_1 minimum (Msf1),
- x_1 maximum (Msf2),
- x_1 when x_2 is minimum (Msf3'),
- x_1 when x_2 is maximum (Msf4').

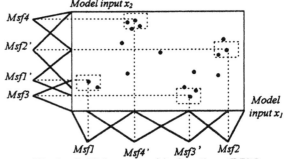

Fig. 3a: Initial membership functions (DISO case).

All combinations of the initial membership functions are implemented in the initial fuzzy model rule base. In the case of a DISO system (see Figure 3b), 16 rules are implemented in the initial fuzzy model rule base. This process defines an initial fuzzy partition of the input state space.

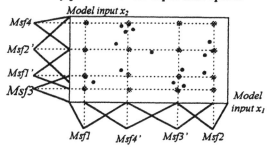

Fig. 3b: Initial fuzzy model rule base (DISO case).

In a first step, the conclusion part of each rule is crisp: it corresponds to the average value of the data contained in a predefined zone. A zone is defined by the intersection of two membership functions, one for each input. Such an allocation minimize the white noise influence on the data. If there is no data in the zone defined for the intersection of two membership functions, the closest data is considered.

3.3 Growth of the model.

From the first approximate reference model, the rule base is progressively expanded. For a better understanding, the expansion principle is explained in the case of double inputs - single output system (this methodology could be then easily extended to MIMO systems):

i) Evaluation of the maximum error area within the input space partitioning:

The objective is to evaluate the region of the fuzzy partition where the error between the model and the real process output is maximum. In the following sections, i is an index corresponding to the first model input, j is an index corresponding to the second model input. For each region (i,j) (see Figure 4) the error $\varepsilon_{i,j}$ is given by (5):

$$\varepsilon_{i,j} = \frac{\sum_{n=0..N_{i,j}}(y_n - \bar{y}_n)}{N_{i,j}} \cdot \frac{l_i}{L_1} \cdot \frac{l_j}{L_2} \qquad (5)$$

for a given set of inputs, y_n is the real process output, $\bar{y}_n$ is the model output, $N_{i,j}$ is the number of data in the given region (i,j), l_i (respectively l_j) is the size of the region (i,j) according to the input 1 (respectively the input 2), L_1 (respectively L_2) is the range (difference between the maximum and the minimum value of the input 1) (respectively the model input 2). A comparison between the errors

calculated for each region gives the maximum error area: (i,j) for the example (see Figure 4).

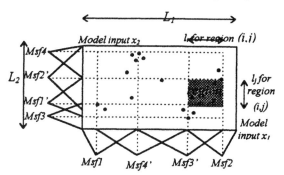

Fig. 4: Determination of the region where ε_i is maximum (DISO case).

ii) Selection of the most critical input:

The objective of this second step is to select the input which has the most important influence on the model error in the maximum error area.

The errors between real output data and model output are calculated considering the lanes which are orthogonal in the maximum error area zone (i,j), with regards to each variables:

$$\varepsilon_i = \frac{\sum_{m_i=0..M_i}(y_{m_i} - \bar{y}_{m_i})}{M_i} \cdot \frac{l_i}{L_1} \qquad \varepsilon_j = \frac{\sum_{m_j=0..M_j}(y_{m_j} - \bar{y}_{m_j})}{M_j} \cdot \frac{l_j}{L_2}$$

$$(6)$$

Where y_{mi} (respectively y_{mj}) is the real process output, $\bar{y}_{mi}$ (respectively $\bar{y}_{mj}$) is the model output, M_i (respectively M_j) is the number of data in the lane defined by the region where ε is maximum and its neighbors orthogonal to the model input i (respectively j) (see Figure 5), l_i (l_j) is the length of the region according to the model input i (j), L_1 (L_2) is the range (difference between the maximum and the minimum value range of the input i (j)).

The errors with regard to each variable are compared. The highest values of error gives the most critical input (x_1 for the example). See Fig. 5 & 6.

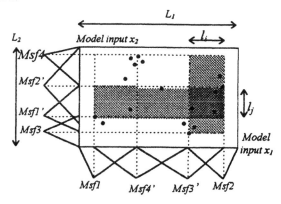

Fig. 5: Determination of the region where ε_k is maximum (DISO case).

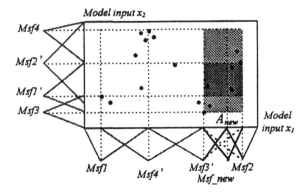

Fig. 6: Addition of a new membership function Msf_new (DISO case).

iii) Calculation of the center of gravity:
In this third step, we calculate the center of gravity of the data inputs with regard to the most critical input (see equation (7)).

$$A_{new} = \frac{\sum\limits_{m_i=0..M_i}\left(x_{m_i} \cdot \varepsilon_{m_i}\right)}{\sum\limits_{m_i=0..M_i}\left(\varepsilon_{m_i}\right)} \qquad (7)$$

where x_{m_i} is the data input, , M_i is the number of data in the lane defined by the shaded region where ε_i is maximum, ε_{mi} is the difference between real data output and corresponding model data output (see Fig6).

iv) Addition of a new input membership function:
On the universe of discourse of the most critical variable x_k a new membership function centered on A_{new} is added (see Fig. 6).
v) Addition of new output singletons:
a set of new output singletons, related with the different new combinations of the input membership functions is then added.
The centers of these new output membership functions are the center of gravity of the corresponding output data. The weight affected to each corresponding data output is characteristic of the distance between this data and the membership functions intersection. If there is no data in the zone defined by the intersection of two membership functions, the closest output numerical data is considered.

vi) The procedure is repeated until the maximum model error is lower than a predetermined reference.
The maximum model error is calculated by (8):

$$\varepsilon_D = \frac{\sum\limits_{n=0..D}\left(y_n - \bar{y}_n\right)}{D} \leq \varepsilon_{MAXIMUM} \qquad (8)$$

for all available data, y_n is the real process output,

$\bar{y}_n$ is the model output, D is the number of data in the whole state space, $\varepsilon_{MAXIMUM}$ is the predeterminated reference error.

Remark 1: the number of the input membership functions growths independently from each other.

Remark 2: in order to reduce the rule number, as for the initialization procedure, an aggregation of the closest membership functions can be done.

3.4 Calculation of the conclusion Part of each Rule.

The fuzzy model obtained in step vi of the modeling process has singletons as conclusion of the rules.
These rules can be expressed by (9):

$$if\ x_1\ is\ A_i\ and\ x_2\ is\ B_j\ then\ s = a_{i,j} \qquad (9)$$

where A_i and B_j are fuzzy sets characterized by triangular membership function. These membership functions are centered on the data and a Bezdek partition is respected.
To express a fuzzy Sugeno dynamic model the conclusion part of each rule (singleton) is replaced by the equation of the plane including the singleton and defined by the vectors $a_{i,j-1}a_{i,j+1}$ and $a_{i-1,j}a_{i+1,j}$, where $a_{i,j-1}$, $a_{i,j+1}$, $a_{i-1,j}$, $a_{i+1,j}$ are the conclusions of the adjacent rules.
Then the rules (9) can be expressed as (10).

$$if\ x_1\ is\ A_i\ and\ x_2\ is\ B_j\ then\ s = \alpha_1 \cdot x_1 + \alpha_2 \cdot x_2 + \alpha_0 \qquad (10)$$

Where $s = \alpha_1 \cdot x_1 + \alpha_2 \cdot x_2 + \alpha_0$ is the second member of a linear differential equation.

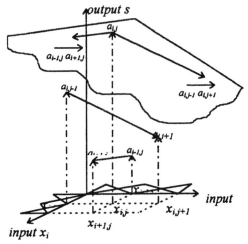

Fig. 7: Addition of a new membership function Msf_new (DISO case).

The aggregation of the different rule outputs on the fuzzy partition generates a linear piece-wise differential equation which is an approximation of the function f defined in the equation (3).Then α_i

coefficients depend on the fuzzy partition of the input state space (for each rule, there is a set of α_i coefficients). Referring to this linear piece-wise differential equation, the conventional linear theory can be directly applied for each fuzzy partition to generate a linear piece-wise controller.

Remark: in the case of N inputs - M outputs process, the rule conclusions can be defined by (11).

$$y_i = \sum_{j=1..N}\left(a_j \cdot x_j\right) + \sum_{\substack{j=1..M \\ j \neq i}}\left(b_j \cdot y_j\right) + c_i \qquad (11)$$

where y_i is the ith output, x_j is the jth input, a_j, b_j and c_j are constants determined by the hyper surface linear approximation.

4. Identification of air inlet process.

4.1 Purpose.

The method developed in section 3 is applied to the benchmark exposed in section 2. Finally, the fuzzy model and the benchmark are compared.

Referring to Figure 1 and equation (3), fuzzy model inputs are:

- Throttle position for the fresh air valve (ϕ_{Fresh}).
- Throttle position for the exhaust gas (EGR) valve (ϕ_{EGR}).
- Engine speed (N).
- Manifold pressure (P_{Mani}).

Fuzzy model output is manifold pressure derivative ($\dot{P}_{Mani}$). Equation (3) becomes:

$$\dot{x} = \dot{P}_{Mani} = f\left(\phi_{Fresh}, \phi_{EGR}, N, P_{Mani}\right) \qquad (12)$$

500 000 data vectors are generated by running the benchmark. The fuzzy modeling method is applied to this data base. The resulting fuzzy model has 12 membership functions for the model input ϕ_{Fresh}, 13 for P_{Mani}, 15 for ϕ_{EGR}, and 10 for N. The fuzzy model rule base is constituted with 23 400 rules, witch represents a big rule base. But the modeling absolute middle error is equal to 243.1 hPa/s for all the data issued from the database. The relative absolute middle error is equal to 1.04% for the same data. This is a good result. It concerns the single model output $\dot{P}_{Mani}$.

The identified rules can be expressed as:

if $\left(\phi_{Fresh} \text{ is} A_i\right)$ and $\left(\phi_{EGR} \text{ is} B_i\right)$

and $\left(N \text{ is} C_i\right)$ and $\left(P_{Mani} \text{ is} D_i\right)$ (13)

then $\dot{P}_{Mani} = a_{1,i} \cdot \phi_{Fresh} + a_{2,i} \cdot N$

$+ a_{3,i} \cdot P_{Mani} + a_{4,i} \cdot \phi_{EGR} + a_{5,i}$

Rule format of the benchmark fuzzy model.

Referring to equation (13), A_i, B_i, C_i and D_i are the input variable membership functions of the considered model. $a_{1,i}$, $a_{2,i}$, $a_{3,i}$, $a_{4,i}$ and $a_{5,i}$ are

constants defined through the identification procedure.

4.2 Model cross-validation.

The benchmark and the fuzzy model are supplied by the same inputs. The inputs are shown in Figures 8. 9 10. They are different from those which have been used for the identification process.

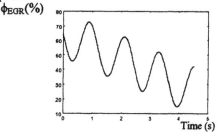

ϕ_{EGR}(%)

Figure 8: Exhaust gas valve evolution.

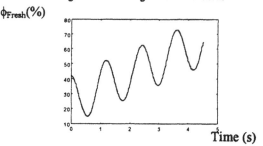

ϕ_{Fresh}(%)

Figure 9: Fresh air throttle command evolution.

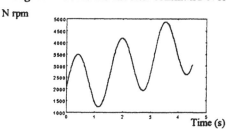

N rpm

Figure 10: Engine speed evolution.

Figure 11 represents both the benchmark and the fuzzy model manifold pressure evolutions. The absolute middle error between the two processes is equal to 3.1674 hPa for the simulation horizon. The absolute relative middle error is equal to 0.38% for the simulation horizon.

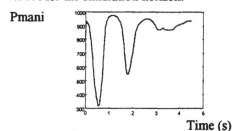

Pmani

Time (s)

Figure 11: Manifold pressure evolutions (benchmark and model).

5. The control results.

5.1 The control strategy.

The closed loop process behavior for all operating conditions should be:

$$\ddot{P}_{Mani} = -\alpha_1 \cdot \dot{P}_{Mani} - \alpha_2 \cdot P_{Mani}$$
$$+\alpha_3 \cdot \int \left(P_{Desired} - P_{Mani}\right) \cdot dt + \alpha_4 \cdot P_{Desired} \qquad (14)$$

Referring to equation (14) P_{Mani} is the manifold pressure, $P_{Desired}$ is the desired manifold pressure, and α_1, α_2, α_3, α_4 are constants determined by the user. The desired manifold pressure derivative is not present in (14), in order to avoid eventual steps of the signal.

The controller is described by equation (15).

$$\phi_{CmdF} = K_{P,i} \cdot \left(b \cdot P_{Desired} - P_{Mani}\right)$$
$$+K_{I,i} \cdot \int \left(P_{Desired} - P_{Mani}\right) \cdot dt - K_{D,i} \cdot \dot{P}_{Mani}$$
$$-K_{1,i} \cdot N - K_{2,i} \cdot \dot{N} - K_{3,i} \cdot \phi_{EGR} - K_{4,i} \cdot \dot{\phi}_{EGR} - K_{5,i}$$

(15)

Considering the desired behavior for each rule i, it is possible to express the $K_{i,j}$ parameters from equation (15). By introducing (13) and (15) in (1) equation (16) is obtained.

$$\ddot{P}_{Mani} - a_{2,i} \cdot \dot{N} - a_{3,i} \cdot \dot{P}_{Mani} - a_{4,i} \cdot \dot{\phi}_{EGR}$$
$$+\frac{1}{\tau} \cdot \dot{P}_{Mani} - \frac{a_{2,i}}{\tau} \cdot N - \frac{a_{3,i}}{\tau} \cdot P_{Mani} - \frac{a_{4,i}}{\tau} \cdot \phi_{EGR} - \frac{a_{5,i}}{\tau}$$
$$= \frac{K_{P,i} \cdot a_{1,i} \cdot b}{\tau} \cdot P_{Desired} - \frac{K_{P,i} \cdot a_{1,i}}{\tau} \cdot P_{Mani} \qquad (16)$$
$$+\frac{K_{I,i} \cdot a_{1,i}}{\tau} \cdot \int \left(P_{Desired} - P_{Mani}\right) \cdot dt - \frac{K_{D,i} \cdot a_{1,i}}{\tau} \cdot \dot{P}_{Mani}$$
$$-\frac{K_{1,i} \cdot a_{1,i}}{\tau} \cdot N - \frac{K_{2,i} \cdot a_{1,i}}{\tau} \cdot \dot{N}$$
$$-\frac{K_{3,i} \cdot a_{1,i}}{\tau} \cdot \phi_{EGR} - \frac{K_{4,i} \cdot a_{1,i}}{\tau} \cdot \dot{\phi}_{EGR} - \frac{K_{5,i} \cdot a_{1,i}}{\tau}$$

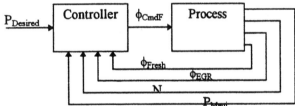

Figure 12: Air inlet closed loop description.

The K parameters are useful for the control. They are identified by similarity between (14) and (16)
The throttle position for the exhaust gas (EGR) valve (ϕ_{EGR}) and the engine speed (N) are considered as disturbances for the air inlet system. Their effect must be rejected by the closed loop system behavior.
This allows the identification of K parameters:

$$-a_{3,i} + \frac{1}{\tau} + \frac{K_{D,i} \cdot a_{1,i}}{\tau} = \alpha_1 \qquad -\frac{a_{3,i}}{\tau} + \frac{K_{P,i} \cdot a_{1,i}}{\tau} = \alpha_2$$
$$\frac{K_{I,i} \cdot a_{1,i}}{\tau} = \alpha_3 \qquad \frac{K_{P,i} \cdot a_{1,i} \cdot b}{\tau} = \alpha_4 \qquad (17)$$
$$-\frac{a_{2,i}}{\tau} + \frac{K_{1,i} \cdot a_{1,i}}{\tau} = 0 \qquad -a_{2,i} + \frac{K_{2,i} \cdot a_{1,i}}{\tau} = 0$$
$$\frac{a_{4,i}}{\tau} + \frac{K_{3,i} \cdot a_{1,i}}{\tau} = 0 \qquad -a_{4,i} + \frac{K_{4,i} \cdot a_{1,i}}{\tau} = 0$$
$$-\frac{a_{5,i}}{\tau} + \frac{K_{5,i} \cdot a_{1,i}}{\tau} = 0$$

5.2 Application to the benchmark.

To control the benchmark, the parameters to be set in (17) are given by:
$$\alpha_1 = 145 \quad \alpha_2 = 7000 \quad \alpha_3 = 112500 \quad \alpha_4 = 2250$$
Then, the K parameters from (15) are identified by (17). The system can evolve under control.
ϕ_{EGR} and N are considered as disturbances for the air inlet system. Their behavior are presented on Figure 13 &14

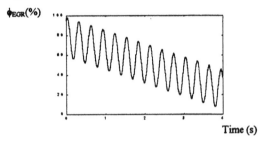

Figure 13: Exhaust gas valve evolution.

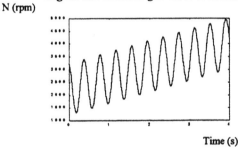

Figure 14: Engine speed evolution.

5.3 Single P.I.D. controller.

The benchmark is approximated by only one linear model on the whole state space. A single hyperplane approximation is calculated from the data. Then, the controller is determined and integrated as exposed above. Figure 15 shows the throttle command for the fresh air valve (ϕ_{CmdF}) in dashed line, and the throttle valve position for fresh air ϕ_{Fresh} (solid line).

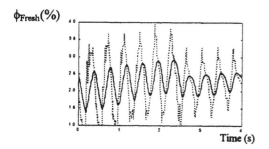

$\phi_{Fresh}(\%)$

Figure 15: Fresh air throttle command (dashed line) and valve position (solid line).

Figure 16 shows the manifold pressure evolution (hPa) during the simulation. The desired manifold pressure evolution is a step. The manifold pressure oscillates around the desired value. The middle value of absolute error between desired and simulated manifold pressures is equal to 5.8909 hPa for the simulation horizon. This is due to the modeling error. A single linear model for the whole state space is insufficient for this benchmark.

$P_{Desired}$ & P_{Mani} (hPa)

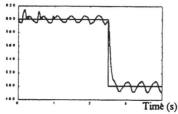

Figure 16: Manifold pressure (desired and from the benchmark).

5.4 Piecewise linear controller.

Piecewise linear PID controller is issued from the model obtained and cross-validated in section 4. 23 400 rules give 23 400 local controllers. They reject the EGR valve and engine speed influence as shown in Figure 17. Figure 18 shows the manifold pressure evolution (hPa) during the simulation. The desired manifold pressure evolution is a step. The manifold pressure oscillates around the desired value but is quickly stable. The middle value of absolute error between desired and simulated manifold pressures is equal to 2.8643 hPa for the simulation horizon.

$\phi_{Fresh}(\%)$

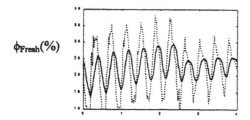

Figure 17: Fresh air throttle command (dashed line) and valve position (solid line). Time (s)

$P_{Desired}$ & P_{Mani} (hPa)

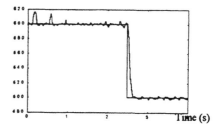

Figure 18: Manifold pressure (desired and from the benchmark).

6. Conclusion.

The fuzzy modeling and control method exposed in this paper gives good results when applied to an engine air inlet benchmark. Nevertheless, due to a need of very good model accuracy, too many rules have been automatiquely generated. At each step of the identification procedure, expert knowledges should be taken into account in the fuzzy model. By adding manually some membership functions or rules, the algorithm should converge faster to a precise model. In this case, the rule base of the model (and of course of the controller) would be smaller. On the other hand the rule number can be drastically reduced if the model accuracy is reduced.

Acknowledgments.
This paper presents the first results obtained by SIEMENS Automotive in the context of the European project FAMIMO.
All these works have been performed thanks to ADEME (French Agency for the Environment and Energy Management)

References.

Babuska, R. 1995.Fuzzy Modeling. $4^{Th.}$ *IEEE International Conference on Fuzzy Systems*, Yokohama, Japan, March 95, vol. 4, pp. 1897-1902.

Bidan P. 1989. Modélisation et Commande d'un Moteur à Allumage Commandé d'Automobile. *Thesis from L.A.A.S./C.N.R.S.* defended at "Université Paul SABATIER de Toulouse"

Chaumerliac ,V. 1995.Commande Multivariable d'un Moteur à Allumage Commandé: Amélioration du Rendement sous Différentes Contraintes. *Thèse du L.A.A.S ./C.N.R.S.* defended at "Université Paul SABATIER de Toulouse"

Nakoula, Y, Galichet, S., Foulloy, L. 1996 Simultaneous Learning of Rules and Linguistic Terms. $5^{Th.}$ *IEEE International Conference on Fuzzy Systems*, New-Orleans, USA, Sept. 96, Vol. 3, pp. 1743-1749.

Bortolet, P., Boverie, S., Titli A. 1997. Engine Control-Oriented Fuzzy Modeling. $6^{Th.}$ *IEEE International Conference on Fuzzy Systems*, Barcelona, Spain, June 97, Vol. 1, pp. 409-414.

FAULT DIAGNOSIS OF EMISSION CONTROL SYSTEM OF AUTOMOTIVE ENGINES VIA FUZZY ARTMAP NEURAL NETWORK.

Luis J. de Miguel [*,1] José R. Perán [**]

Department of Systems Engineering and Control, ETSII
University of Valladolid
Paseo del Cauce, s/n, Valladolid, 47011, Spain
Phone: +34 83 423355
Fax: +34 83 423358
E-mail: luimig@eis.uva.es
***I.T.A.P. University of Valladolid*
E-mail: peran@eis.uva.es

Abstract:
The aim of this work is to show the performance of a neural network architecture, Fuzzy ARPMAP, when it is used for fault diagnosis of the emission control system of an automotive engine. The data set has been taken from the on-line (on-board) application. The experiments has focus the attention in the identification of the MAP variable. *Copyright © 1998 IFAC*

Keywords: Fault Diagnosis, Emission Control, Automotive Engine, Neural Network.

1. INTRODUCTION

Model-based fault detection and isolation techniques have been developed in the last decades (Basseville and Nikiforov, 1993; Frank, 1990; Isermann, 1984; Gertler, 1988) and have been applied successfully to real systems (Patton *et al.*, 1989). However some problems may appear on special cases. The emission control system of automotive engines is a hard non-linear system with significant load perturbance. These conditions are difficult to manage with the common model-based F.D.I. methodologies and has been a research topic of several papers. (Rizzoni and Min, 1991; Gertler and Costin, 1994) Robustness and fault sensitivity have been the aim of several research lines by fitting identification methods to fault isolation purposes. Neural networks have

been also applied to fault diagnosis of different systems.(J.Y. Fan and White, 1993; Miguel *et al.*, 1992; Sorsa and Koivo, 1991) This work tries to measure the performance of a Fuzzy ARPMAP neural network in the automotive engine case, using some practical index criteria. The paper is based on the previous work and real data from (Gertler and Costin, 1994).

2. PROBLEM DESCRIPTION

The importance of the reliability of the engine's emission control system has grow due to the new European Community regulation about the emission of pollutant. The critical component of this system is the catalytic converter. Its function is to remove from the exhaust gas the three major pollutant: carbon monoxide, unburnt hydrocarbons and nitrous oxides. In order to guarantee a reasonable performance of the catalytic converter the air-to-fuel ratio must be kept close to the

---

[1] This work was partially developed in the Department of ECE of the George Mason University, with the useful comments of Prof. J. Gertler.

stoichiometric value. It will be the main aim of the engine's fuel control system. A complete description of that system is presented in (Gertler and Costin, 1994).

A short description of the main variables of the system is given:

- *MAP*: Intake Manifold Pressure.
- *thr*: Throttle position.
- *egr*: Exhaust gas recirculation.
- *fuel*: fuel injection.
- *rpm*: Revolutions per minute.
- *bat*: Battery charge.
- *iac*: Idle air control.
- *clt*: coolant temperature.
- *mat*: manifold air temperature.

From the set of equations to isolate faults that were used on (Gertler and Costin, 1994) two of them have been selected.

The first equation which has been identified is:

$$MAP(t) = \mathbf{f}(iac, thr, egr, fuel, rpm, clt) \quad (1)$$

The second equation which has been identified is:

$$MAP(t) = \mathbf{f}(iac, thr, egr, fuel, rpm, mat, clt, bat) \quad (2)$$

2.1 *Data sample*

The data has been taken from the real on-board application. Each data file has 10.000 samples, with a sample time of 25 msec. For the identification of the model were used 7 dada files. These data files were taken under city normal driving conditions and highway driving conditions. It implies a different load perturbance and different operation conditions. The goodness of the model was checked with the two different data files, for validation purposes.

3. FUZZY ARTMAP MODEL

3.1 *Introduction*

The Fuzzy ARTMAP (Carpenter *et al.*, 1991) system is a neural network architecture based on the Fuzzy Sets Theory and the ART architectures. The complete Fuzzy ARTMAP architecture includes two ART modules called ART_a and ART_b. Each of them creates stable recognition categories in response to arbitrary sequences of input patterns. During the supervised learning period, the ART_a module receives a vector of input patterns $\mathbf{U}^r$, while ART_b receives the vector of input patterns $\mathbf{V}^s$, which is the known prediction given $\mathbf{U}^r$. These modules are linked by an associative learning network and an internal controller, which is designed to create the minimal number of recognition categories, at F_2 field, needed to get accuracy criteria. It is done by using a Minimax Learning Rule that enables the system to learn quickly, looking for minimizing the predictive error and maximizing the predictive generalization.

The main parameters of the system are:

- Vigilance parameters ρ_a and ρ_b are dimensionless parameters used to establish a matching criterion. Vigilance parameter ρ_a of ART_a is increased by the minimal amount needed to correct a predictive error at ART_b. However, lower values of ρ_a lead to form larger categories, broader generalization and higher code compression.
- Choice parameter, α, which tunes the choice function at the category field. Small values of α tend to minimize recoding during learning.
- Learning rate parameter, β, enable to get fast learning, with $\beta = 1$, and slow recoding learning, with $\beta < 1$.

The convergence of the learning algorithm is ensured because all adaptive weights, in the Fuzzy ART systems, are monotone increasing. This leads to a problem of categories proliferation, which is solved by a pre-processing scheme called complement coding. It consists of representing each feature by two input variables, one to encode the presence of a feature and the other to encode its absence.

3.2 *System identification*

The input vectors of the neural-net are:

$$\mathbf{U}^i = \{u_1^i, u_2^i, ..., u_j^i, ..., u_r^i\}$$

and

$$\mathbf{V}_b^i = \{v_1^i, v_2^i, ..., v_j^i, ..., v_s^i\}$$

where,

$$i \in (1, N)$$

being r the number of input variables, s the number of output variables and N the number of learning samples. Therefor, u_j^i will be the input variable jth of the equation (1), and v^i, with $s = 1$, the output variable.

In order to identify the equation (1) some design criteria must be chosen. The vigilance parameters ρ_a and ρ_b fix the precision degree at the category level. It means that a high value of ρ_b will generate many different categories. The maximum number of categories will be the number of different vectors $\mathbf{V}^i$, that is the number of different values of the output to be predicted in the equation.

However, for this application (note that $\mathbf{V}^i = v_1^i = v^i$) the tolerance for the prediction error ($\delta map = 10$) is quite enough to choose only one category, (e.g. $F_{2b}^{v^i}$) for each interval (e.g. $(v^i - \delta, v^i + \delta)$) of output values. This point is important to reduce the number of weights and consequently the computational cost. The initial value for ρ_a is chosen small and it will be increased gradually and automatically by the neural-net if there is not match at the category map. Therefor, the minimal value of ρ_a will be used to fit with the category chosen at ARTb, and it will imply a minimal number of categories at ARTa. Another important issue is the selection of the learning data set. As much wider the set is chosen as higher accuracy criteria will be get, but also the number of weights will be meanly increased. The computational cost must be balanced with the required accuracy.

The parameters which have been used in equation (1) are:

- $\rho_a = 0.7$
- $\rho_b = 0.97$
- $\alpha = 0.001$
- $\beta = 1$

The choice of $\rho_b = 0.97$ means that there will be only 39 different categories at $ARTB$. It implies that the model can be insensitive to changes in the output variable under $\Delta map = 6$. The choice of $\rho_a = 0.7$, is only the initial value. That means it will be increased automatically as much as it is necessary to match with the adequate category at $ARTB$. The choice of $\alpha = 0.001$ will minimize recoding during learning.

and in the equation (2):

- $\rho_a = 0.65$
- $\rho_b = 0.95$
- $\alpha = 0.001$
- $\beta = 1$

The choice of $\rho_b = 0.95$ has been made in order to reduce the number of weights, which is already very high. It increases the insensitivity interval to $\Delta map = 10$, and obviously it also decreases the performance indexes. As can be see in the tables the accuracy has decreased meaningfully due to the decreased of the vigilance parameter to $\rho_b = 0.95$. For this choice the number of weights is approximately 42000 (single precision float), which implies a low speed performance for an on-line application. On the other hand, the possibility to increase the accuracy under the lack of speed performance makes the model useful for of-line application. Furthermore, the ability to learn new inputs allows to improve the model easily.

3.3 *Output prediction.*

It has been made using a $\rho_a = 0.7$ for the equation (1) and $\rho_a = 0.8$ for the equation (2) Higher values of ρ_a would improve weakly the accuracy but not enough to compensate the increase of computational time.

4. PERFORMANCE INDEXES.

In order to reach a high robustness, a general fault detection algorithm uses at least two thresholds. The first the check the residual as a function of its standard deviation and the second as a counter of samples which are over the first one.

The performance indexes for identification of the unfaulty model may be:

- MR : Mean of the residuals.
- SR : Standard deviation of the residuals.
- ME : Maximum value of the residual.
- $T(5)$: Number of residuals over 5 (absolute value).
- $T(10)$: Number of residuals over 10 (absolute value).
- $T(15)$: Number of residuals over 15 (absolute value).

The performance indexes for fault detection, testing a fault case may be:

- MSF : Mean of the residuals.
- SRF : Standard deviation of the residuals.
- MEF : Maximum value of the residual (absolute value).
- $TF(5)$: Number of residuals over 5 (absolute value).
- $TF(10)$: Number of residuals over 10 (absolute value).
- $TF(15)$: Number of residuals over 15 (absolute value).

5. EXPERIMENTAL RESULTS.

The experimental results are shown in the tables.

Comments. The computational cost is an important difference between both methods. For the equation (1), the election of $\rho_b = 0.97$ implies 39 categories at ART_b and 1497 at ART_a. The number of weights (parameters) associated with those categories and the 12-dimension input vector is 18042! Furthermore, for the equation (2), the election of $\rho_b = 0.75$ means a lack of precision with only 22 categories at ART_b, but now the input vector at ART_a is 16-dimensional, and it implies an additional difficulty to classify. In other words, the number of categories at ART_a is now 2592, which means 41516 weights!! Obviously, the identification of such a high number of parameters takes a several hours, even days in a Sparc Work Station as S-4000. The fault detection process, by checking the model, takes around 30 minutes for a 10000-samples file. However, for a multiple output system the increase of parameters would be very small because a high degree categorization of the input vectors has been already made for any number of outputs. In a large scale MIMO system it may be an important issue.

6. TABLES.

6.1 *Equation 1*

Id. Ind.	MR	SR	ME	T(5)	T(10)
I	-0.34	2.4	6	538	0
T	0.26	6.09	19	2865	978

Table 1. IDENTIFICATION INDEXES FOR CITY DRIVING CONDITIONS

Id. Ind.	MR	SR	ME	T(5)	T(10)
I	-0.22	2.57	7	709	0
T	-1.32	4.14	18	2268	309

Table 2. IDENTIFICATION INDEXES FOR HIGHWAY DRIVING CONDITIONS

where the meaning of the capital letters is:

- I : Data were taken from the same identification data set.
- T : Data were taken from a new data set.
- NN : Fuzzyartmap neural-net model.
- Id. Ind. :Identification Indexes.

FILES	MRF	SRF	MEF	TF(5)	TF(10)
File 1	-2.37	5.38	30	3128	900
File 2	-1.86	5.64	18	2334	1146
File 3	0.27	5.17	29	2253	612
File 4	21.87	346.56	inf	474	488
File 5	11.81	56.76	193	530	493

Table 3. NEURAL NET INDEXES FOR DIFFERENT FAULTY CASES

6.2 *Equation 2*

FILES	MRF	SRF	MEF	TF(5)	TF(10)
Non-faulty case 1	-0.57	4.22	19.5	2544	121
Non-faulty case 2	0.5	3.77	14.5	1947	134
Non-faulty case 3	11.75	62.5	6894	4091	
Non-faulty case 4	-3.76	8.09	39.5	4482	2964
Faulty case 1	-9.02	15.35	58.5	7804	6442
Faulty case 2	-1.84	11.15	36.5	6514	3960
Faulty case 3	-3.29	15.45	61.5	6642	4639
Faulty case 4	-1.55	14.41	64.5	7908	4828
Faulty case 5	-6.32	15.14	81.5	7805	5908

Table 4. NEURAL NET INDEXES FOR DIFFERENT CASES

7. REFERENCES

Basseville, Michélle and Igor V. Nikiforov (1993). *Detection of Abrupt Changes. Theory and Application.* Prentice-Hall, Inc.

Carpenter, G.A., S. Grossberg, N. Markuzon, J.H. Reynolds and D.B. Rosen (1991). Fuzzy artmap: A neural network architecture for incremental supervised learning of analog multidimensional maps. Technical Report CAS/CNS-TR-91-016. Boston University. Boston.

Frank, P. M. (1990). Fault diagnosis in dynamic systems using analytical and knowledge-based redundancy: A survey and some new results. *Automatica* **26**(3), 459–474.

Gertler, Janos (1988). Survey of model-based failure detection and isolation in complex plants. *IEEE Control Systems Magazine* pp. 3–11.

Gertler, Janos and Mark Costin (1994). Model-based diagnosis of automotive engines. In: *Proceedings of the IFAC Symposium on Fault Detection , Supervision and Safety for Technical Processes.* Espoo , Finland. pp. 421–430.

Isermann, Rolf (1984). Process fault detection based on modeling and estimation methods— a survey. *Automatica* **20**, 387–404.

J.Y. Fan, M. NIkolau and R.E. White (1993). An approach to fault diagnosis of chemical processes via neural networks. *AIChe Journal* **39**(1), 82–88.

Miguel, L. J., E. Baeyens and Y.A. Dimitriadis (1992). Fault diagnosis in dynamic systems using an art3 neural network architecture. In: *Proceedings of the Research Conf.:Neural Networks for Learning, Recognition, and Control.* Boston, USA.

Patton, R., P. Frank and R. Clark (Eds.) (1989). *Fault Diagnosis in Dynamic Systems. Theory and Applications.* Prentice Hall Int.

Rizzoni, G. and P. S. Min (1991). Detection of sensor failures in automotive engines. *IEEE Transactions on Vehicular Technology* **40**, 487–500.

Sorsa, Timo and Heikki N. Koivo (1991). Neural netwoks in process fault diagnosis. *IEEE Transactions on Systems , Man , and Cybernetics* **21**, 815–825.

ON-BOARD COMPONENTS DETECTION OF DAMAGE
FOR CONTROL SYSTEM OF DIESEL ENGINE - EXAMPLE

Rafal Klaus

Institute of Computing Science
Poznan University of Technology
ul. Piotrowo 3a, 60-965 Pozna , Poland
tel. +48 61 87-82-574, fax. +48 61 87-71-525
E-mail: klaus@.put.poznan.pl

Abstract: The Diesel engine without load and seed controller is a non-linear astatic object. Hence, the Diesel engine is equipped with a speed governor. The main task of the governor is counteract exceeding the speed limit. We did research on the engine with on injection pump but where the conventional centrifugal governor replace the microprocessorial controller. Microcontroller because of using a large number of electronic elements is smaller reliability in comparison with mechanical governors. In order to protect the engine from results of the controller's defect we invented a safety system. The system guarantees the controlled stoppage of the engine's work in dangerous states.

Informations in this article are presented concerning protection and self-diagnostics of SW 400 engine's controlling system that makes use of DP 535 controller with Siemens 80C535 microcontroller. The being presented controlling system is successfully tested in an excavator. *Copyright © 1998 IFAC*

Keywords: safety, on-line diagnostic, microcontroller application, Diesel engine control

1. INTRODUCTION

The compression-ignition engines as astatic objects require the controllers to be applied. Hence, improving quality of engines' running and avoidance of damage is a strategic purpose of controllers application.

The conventional controllers applied in the diesel engines make work possible only in limited range of parameters. This is the reason for many defects of the engines, such as the increased smoke level, noise, bad dynamics *(Ochocki et al.1993)*.

At the present stage of internal-combustion engines development mechanical governors of rotation speed are more and more frequently replaced by controllers which make possible numerical control of engines. The controllers which don't have the mechanical

governors' defects appeared together with progress in application of microprocessors.

The digital controller's task is to work out appropriate setting of control rod of the injection pump and injection advance angle which depends on the present engine's working point.

In microprocessor controllers it is very easy to change the control structure, dynamic characteristic and working points in relation to the required criterion.

Usually the following criteria of control:
- maximisation of rated the power,
- minimisation of the smoke level,
- minimisation of the fuel consumption.

2. CONTROL SYSTEM STRUCTURE

The task of engine controller is to maintain appropriate settings of the control rod of the injection pump.

The controller performs following functions to enable correct work of the engine *(Klaus, 1996)*:

the measurement of the rotation speed;

the measurement of the injection advance angle;

the regulation of the injection advance angle as a function of the current rotation speed-minimisation of the toxic combustion gas;

the control of the rod position by means of a step motor;

the diagnostic of sensors;

the service of the A/D converter for reading of a potentiometer setting a required rotation speed;

the service of the display indicating an assigned rotation speed of the engine;

the service of the RS232 interface for communication with the master computer;

checking whether the maximum permissible rotation speed of the engine is not exceeded;

Combining mechanics, electronics and software increases adaptability of Diesel engine. Module construction and functions partition realised by hardware and software increase flexibility and reliability in Diesel engine.

The modular construction of the whole system has been applied here with the separation of the injection pump subsystem and the digital control subsystem and injection advance angel.

The advantage of such a construction is the possibility of applying it in the new engines as well as in the already exploited ones. In the exploited machines it is very easy to replace the old control system with the new one. It requires only dismantling the injection pump with the centrifugal governor and then installing the new one together with an equipment.

Moreover, the modular construction guarantees system's flexibility and make the service time much shorter.

Every module can be developed improved or replaced independently, obviously with regard to the rules of co-operation.

The control system of the engine make also possible the co-operation with the master microcomputer of the whole machine, such as excavator.

3. SAFETY ALGORITHMS

Usually, the electronic controllers characterises the low level of reliability in comparison of mechanical governors. The most dangerous situation is the overspeed of the engine. Based on the construction analysis the most important elements are selected. There are used two approach for safety analysis: Failure Mode Effect Analysis (FMEA) and Fault Tree Analysis FTA *(Redmill et al.,1993)*. The FMEA method allows to determining of the functional relations between particular elements which proper running is critical for system working. The FTA method identifies the sets of events which can to provide to breakdown of the system. The error tree is created for each hazardous situation. There are represented the hardware, software events and operators actions. The both method allow to analyse the whole system. The events are described using natural language thus the formal semantic description are introduced. Based on this analysis the hierarchical safety system was implemented.

The safety improvement can be obtained by the redundancy technique, too.

Self-diagnosis algorithms are suitable to use in real time the specific redundancy elements, in this case - mechatronics elements which prevent overspending of the engine and its defects. Mechatronics is a way for effectively combining of mechanics, electronics and software into solutions that increase flexibility, adaptability, safety and reliability machines and systems.

Reliability improvement of the systems concerns first of all safe stoppage of the engine in different break down states, and thus the exclusion of over speeding of the engine and its damage. In the system there was applied the redundancy technique characterised by the usage of additional components duplicating the work function.

Diagnosis software monitors the sensors and actuators condition and top speed protection electronic devices. In case of extensive damage of all systems the machine operator can pass on manual control.

On the basis of Faults' Tree Analysis there were selected the most crucial points of the controlling system and they were served determine the safety system. The presented system is equipped with the protection means as follows *(Klaus et al.,1994)*:

- check controller's feeding
- checking of the rotational speed's sensor and of the injection advance angle's sensor
- counteract exceeding the speed limit
- check of the software
- check of the hardware
- supervisor working of the master computer.

3.1. Formal description

The formalisms description for FMEA and FTA methods utilize the natural language. However, words in the natural language have more then one meaning usually. The utilization of formal semantic gives the higher credibility then natural language. Described event is defined as a logic expression with attributes defined previously in the system. The next needed elements are time dependencies between events. Below, in the logic expression the attributes of the system are utilized as a variables.

$$A1(Time,t) \equiv \exists e \in \phi(PA1) \cdot duration(e) \rangle t_B \wedge start(e) \langle t \langle end(e) \quad (1)$$

$$PA1(Time,t) \equiv \left(\left(MO = less \right) \wedge \left(BPD = vdanger \right) \right) / (Time,t) \quad (2)$$

The presented segment describes event A1 from the Fig 1, where the *Time* is a function defining the system state in the time. The set of all events PA1 in the whole range of *Time* is denoted as $\phi(PA1)$. Thus the event A1 is the same as a event PA1 which exist during the time t. The functions *start(e)* and *end(e)* return respectively start time and end time of event e but *duration(e)* expresses the duration time of event e, i.e., *duration(e)=end(e)-start(e)*.

The gates of the fault tree describe the causing and timing dependencies between events (gates AND and OR). For example gate G1 (Fig.1) describes that engine can go to destruction without properly load and during time t_B the velocity will growing up to maximal velocity n_{gr}.

$$M\left(G1(B1,B2,A1) \right) \equiv$$
$$\forall a1 \in \phi(A1) \cdot occur(a1) \Rightarrow$$
$$\exists b1 \in \phi(B1), b2 \in \phi(B2) \cdot \quad (3)$$
$$occur(b1) \wedge occur(b2) \wedge duration(b1 \wedge b2) \rangle t_B \wedge$$
$$start(a1) = max\left(start(b1), start(b2) \right) + t_B$$

The expression presented above is interpreted in sense of possibility state of analyzed system, i.e. it is connected with general quantification with variable *Time*. The condition *occur(e)* assumes that event *(e)* occurs in the range *Time*.

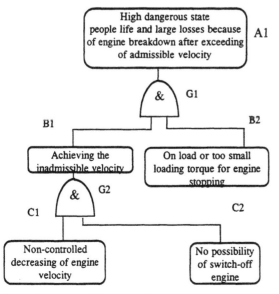

Fig.1. Example FTA metod

4. SELF-DIAGNOSIS AND REDUNDANCY

4.1. Check controller's feeding

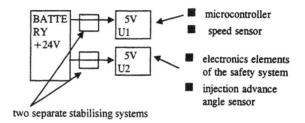

two separate stabilising systems

Fig.2. Check controller's feeding

The controller is feed from 24V electrical installation of the engine. Because actuator needs at least 13V from its correct working, the valve of the voltage is monitored. The electronic controller needs two independent 5V voltages. The first voltage feed the microcontroller and speed sensor. It is monitored by the hardware. The second voltage feed the electronics elements of the safety system and the injection advance angle sensor. It is monitored by the software.

Lack of any of them causes stoppage of the engine.

4.2. Checking of the speed sensor and of the injection advance angle sensor

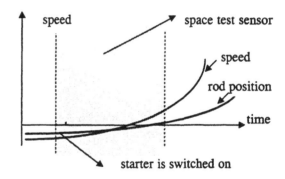

Fig.3. Checking of the speed sensor

These sensors are installed on the injection pump's shaft. The sensors' impulses have to come in respective sequence, i.e. after one impulse sent by detector of the injection advance angle it is due to be followed by six impulses coming from speed sensor. The sensors are tested during the engine's working and before it starts (before fuel rod is pulled out). The appearance of the respective impulses' sequence from the sensors is absolutely necessary to start the engine, when the starter is already switched on. Lack of any of impulse (when fuel rod is pulled out) makes the start of the engine impossible.

4.3. Counteract exceeding the speed limit

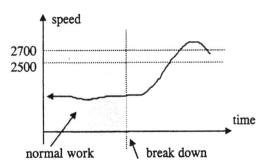

Fig.4. Counteract exceeding the speed limit

The overspeeding protection system has two levels: software level, hardware level. On the lowest software level the safety system checks exceeding the 2500 rpm seed. (The normal working speed is 500 - 2200 rpm.) Stoppage of the engine is result of drawing in fuel rod to its resting position. The software safety is due to act as the first one. If the microprocessorial controller or actuator is defective then the hardware protection system operate. The hardware protection system works independently from microprocessorial module. When engine's speed is exceeding the 2700 rpm, than safety system immediately shuts off the fuel.

4.4. Supervisor working of the master computer

The microprocessorial controller of the Diesel engine can work autonomically or subordinately. In the latter mode the controller co-operates with the master computer. The master computer has information that come from many points unavailable for the controller and has capability of making decision to stop the engine - protecting in this way the machine from destruction.

5. CONCLUSIONS

The safety system has been tested in the laboratorial conditions. Finally the safety system has been tested in excavator. The tests results do not indicate coming up of any dangerous situation that could put human life at risk or destroy the equipment *(Dabrowski et al., 1997)*. This construction of the controlling system gives the following advantages: easy and fast assembly, possibility of modernisation and easy service, east adaptability to different devices.

REFERENCES

D browski D, J. Szlagowski (1997). Test of the excavator with working equipment localization system, *Conference on Engineering Machines Problems*, Zakopane, vol.2, pp.77-79

Klaus R., A. Urbaniak (1994). Self-diagnosis and redundancy for electronic control of Diesel engine, *2nd IFAC Symposium SICICA*, Budapeszt, pp.167-171

Klaus R. (1996). Microcomputer Control System for Diesel Engines, *12th International Confrerence on Process Control and Simulation, ASRTP'96*, Kosice, Vol. 1, pp.218-222

Ochocki W., R. Klaus, P. Rybarczyk (1993). Selected results of numerically and conventinally controlled Diesel engine research. *Symposium Engines Control*, Stawiska, pp.68-75

Redmill F., T.Anderson (1993). Directions in safety-critical systems, *First Safety-Critical Systems Symposium*, Springer-Verlag, Bristol

THERMAL SIMULATION OF REFRIGERATION CONTROL SYSTEM IN COMBUSTION ENGINE TEST-BENCH

F.V. Tinaut Fluixá*, A. Melgar Bachiller*, L.J. de Miguel González** and
J. Rivera Rodríguez***.

(*) Departamento de Ingeniería Energética y Fluidomecánica (IEF),
E.T.S.I. Industriales de Valladolid, Pº del Cauce s/n, 47011 Valladolid, SPAIN.
Tf.: 34-83-423367,   Fax: 34-83-423363
(**) Departamento de Ingeniería de Sistemas y Automática,
E.T.S.I. Industriales de Valladolid, Pº del Cauce s/n, 47011 Valladolid, SPAIN.
Tf.: 34-83-423355,   Fax: 34-83-423358
(***) Centro de Investigación y Desarrollo en Automoción (CIDAUT),
Parque Tecnológico de Boecillo, Parc. 209, 47151 Boecillo (Valladolid), SPAIN.
Tf.: 34-83-548035,   Fax. 34-83-548062

Abstract: A thermal simulation model for transient process of the coolant temperature regulation system in a combustion engine test bench is presented. The simulation model has been implemented in a MATLAB environment with SIMULINK tools. The model includes a reciprocating combustion engine, a heat exchanger and a regulation system. The regulation system consists of a PID temperature regulator that actuates on a flow regulation valve. An experimental test has been performed, the simulated results are compared with the experimental data. Finally, a study of the influence of different parameters of the system have been performed. *Copyright © 1998 IFAC*

Keywords: Heat exchangers, Engine modelling, PID control, Valves, Modelling, Temperature control.

1. INTRODUCTION

When combustion engines are tested in engine test-bench, some elements of the vehicle are replaced by other systems that have to keep the same functions. This is the case of the radiator of the refrigeration system, that uses the air flow produced by the movement of the vehicle to transfer heat from the coolant to the ambient air, working as an air/water heat exchanger. But when the engine is working in an engine test bench (static system), it is necesary to replace the radiator by other element. Habitually, a liquid/liquid exchanger is used since it occupies less space and generally has a quicker response.

In order to obtain a good agreement in the results of different engine tests, it is necessary to precisely control the coolant temperature. This control is obtained by adjusting the heat transfer from the engine coolant to the secondary liquid in the heat exchanger.

Sometimes, tests require that the temperature of the coolant evolve in a certain way to simulate given conditions. For all that, the temperature regulation system must have a quick and precise response, existing different configurations to get these objectives.

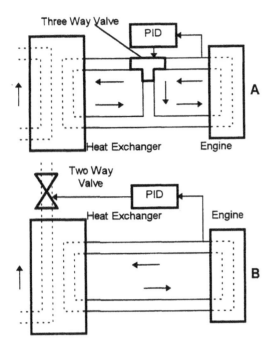

Fig. 1. Scheme of two typical cooling systems for a combustion engines test bench.

Figure 1 shows two possible configurations for the cooling system. The configuration of figure 1-A has the advantage of a fast response, but the flow across the engine varies very much with the valve position because the coolant circuit is modified. In the second configuration (figure 1-B) the coolant circuit is not modified by the regulation, but the response is slower. The purpose of the present work is to simulate configuration 1-B.

2. MODEL APPROACH

The model is based on the simulation of different elements by means of linear equations.

2.1 Heat exchanger.

The heat exchanger is characterized by the effective area of exchange and the global transmission coefficient. This global coefficient depends little on the fluid speeds, whenever these speeds do not differ much from the nominal values.

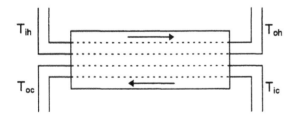

Fig. 2. Scheme of the circulation of the fluids in the heat exchanger.

In the figure 2 is represented the scheme of a counterflow heat exchanger. Considering the global transmission coefficient constant, the rate of heat exchange is controlled by the mass flows of the working liquids that pass through the heat exchanger. When the flow of one of the fluids is increased, its temperature variation across the heat exchanger is less, and therefore the difference of temperatures between the fluids is bigger, and the heat flux increases, as it can be deduced from equation (1) when heat exchanged Q can be expressed through the logarithmic mean temperature difference.

$$Q = U A \frac{\Delta T_1 - \Delta T_2}{\ln \frac{\Delta T_1}{\Delta T_2}} \qquad (1)$$

where:

T_{ih}	Input temperature of hot fluid
T_{oh}	Output temperature of hot fluid
T_{ic}	Input temperature of cold fluid
T_{oc}	Output temperature of cold fluid
ΔT_1	High terminal diference of temperature: T_{ih}-T_{oc}
ΔT_2	Low terminal diference of temperature: T_{oh}-T_{ic}
U	Global transmission coefficient
A	Exchange area

Under transient conditions, the energy conservation equation has to consider heat accumulation in the masses of both heat exchanger and fluids. This can be written for each fluid as:

$$G_c c_p \left(T_{ic} - T_{oc}\right) + Q = \left(m_c c_p + \frac{m_{exc} c_{sp}}{2}\right) \frac{dT_{mc}}{dt} \qquad (2)$$

$$G_h c_p \left(T_{ih} - T_{oh}\right) - Q = \left(m_h c_p + \frac{m_{exc} c_{sp}}{2}\right) \frac{dT_{mh}}{dt} \qquad (3)$$

where:

T_{mh}	Mean temperature of hot fluid: $(T_{ih}+T_{oh})/2$
T_{mc}	Mean temperature of cold fluid: $(T_{ic}+T_{oc})/2$
c_p	Specific heat of fluid
c_{sp}	Specific heat of exchanger material
G_c	Mass flow rate of cold fluid
G_h	Mass flow rate of hot fluid
m_c	Mass of cold fluid in heat exchanger
m_h	Mass of hot fluid in heat exchanger

where the mass of the heat exchanger has been split in two parts, one for each circuit. This leads to linear equations.

2.2 Combustion engine.

About one third of the fuel energy content is transferred by the engine to the coolant, the exact amount depending on the operating conditions. In order to consider this, the empirical expression of

Taylor and Toong has been used. This expression predicts the mean heat flow transferred from the engine chamber to the coolant fluid inside the engine block. This equation was developed for steady conditions. For transient processs, it is necessary to separate the heat tranfer in two steps: Heat flow from combustion gases to engine block (Q_g), and heat flow from engine block to coolant (Q_r). Each step can be characterized by a particular heat-transfer coefficient, while the total heat-transfer coefficient is given by:

$$h = \frac{1}{\frac{1}{h_g} + \frac{1}{h_r}} \qquad (4)$$

The heat flows are:

$$Q_g = h_g (T_g - T_{eng}) \qquad (5)$$

$$Q_r = h_r \left(T_{eng} - T_m\right) \qquad (6)$$

with

$$h = 10.4 \, k_g \left(\frac{c_m \rho}{\mu}\right)^{0.75} D^{-0.25} \qquad (7)$$

The energy conservation equation for the engine block must consider the heat accumulation in the metal and coolant masses:

$$Q_r + G \, c_p (T_i - T_o) = m_{ref} c_p \frac{d}{dt}\left(\frac{T_i + T_o}{2}\right) \qquad (8)$$

$$Q_g - Q_r = m_{eng} c_{sp} \frac{dT_{eng}}{dt} \qquad (9)$$

where:

Q_g	Heat flux from combustion gases to engine
Q_r	Heat flux from engine block to coolant fluid
h	Total heat-transfer coefficient
h_g	Gas-to-block heat-transfer coefficient
h_r	Block-to-coolant heat-transfer coefficient
k_g	Thermal conductivity of combustion gases
c_m	Piston mean linear speed
D	Piston diameter
ρ	Mean density of combustion gases
μ	Gas viscosity
G	Mass flow of coolant fluid through engine
m_{ref}	Mass of coolant fluid in engine
m_{eng}	Engine mass
c_p	Specific heat of coolant fluid
c_{sp}	Specific heat of engine material
T_g	Mean temperature of gases
T_m	Mean temperature of coolant fluid
T_{eng}	Engine block temperature
T_i	Inlet temperature of coolant fluid
T_o	Outlet temperature of coolant fluid

2.3 Control Actuation.

The control system consists, essentially, in a temperature sensor which measures the engine coolant outlet temperature a PID regulator system and a valve that controls the liquid flow in the secondary circuit of the heat exchanger. The temperature measured value is compared with the set temperature in the PID to calculate the error e(t) and the output of regulator system y(t) is calculated whit the equation (10) and is applied to operate the valve, consequently the flow rate that passes through the secondary is a function of the signal provided by the regulator system.

$$y(t) = K_p \left(e(t) + \frac{1}{T_i} \int e(t) dt + T_d \frac{de(t)}{dt} \right) \qquad (10)$$

where:

K_p	Proportional parameter
T_i	Integral time
T_d	Derivative time

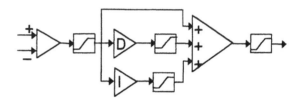

Fig. 3. Scheme of the structure of PID.

Figure (3) show the implementation of equation (6) en SIMULINK, due to physical limitations of the circuits is necessary to limit the outputs of each module, Also it is necessary take into account the delays that are produced in the temperature sensor as well as in the valve.

The Ziegler-Nichols rules provide a first approximation to adjustment the constants of PID. The final values of parameters are obtained by comparison with experiments.

3. EXPERIMENTAL APPLICATION AND RESULTS

3.1 Experimental results.

To check the model validity, an experiment in a engine test bench with refrigeration system shown in figure 1-B has been performed. The test has consisted of making to operate the motor without load and modifying the engine speed and the set temperature in the regulation system according to the table 1.

Table 1 Test conditions

Time interval (s)	Engine speed (rpm)	Set temperature (°C)
0-5100	800	80
5100-6900	800	65
6900-8160	1300	65
8160-9120	2200	65
9120-9720	2200	80
9720-10500	1300	80
10500-11420	800	80

The figure 4 front sample to the time the evolution of the input and output temperatures of the coolant fluid in the engine (these are respectively those of output and input in the primary of the heat exchanger) also is represented the engine speed and the set temperature, each one of the regions of the test are indicated in the graph. It can be seen in the figure as upon increasing the engine speed the oscillation frequency of the system increases, this is due to the fact that increases the flow of the coolant fluid and the processes of heat transmission occur more quickly.

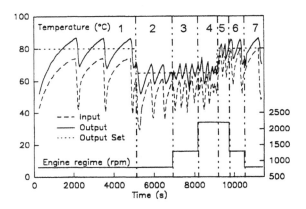

Fig 4. Evolution of the variables during the test.

3.2 Adjustment of the model.

To identify in one way but simple each one of the systems and power to adjust more exactly the model, firstly it is simulated in isolation the system heat exchanger/engine, for this it is used as entry the position of the valve that had been measured during the test and it is used to calculated the evolution of the temperature of coolant fluid at the input and output of engine. In the figure 5 is represented a comparison of experimental and calculated output engine temperature, also is represented the valve position during the test, this is the same that the value used for the simulation. The results of output engine temperature are adjusted well to the experimental results.

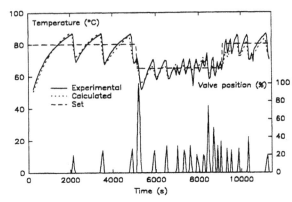

Fig 5. Comparison of the experimental and calculated temperature values using as entry the position of the valve.

The figures 6 and 7 present the results that are obtained when is introduced in the model the system of control of temperature, in the figure 6 is used a PI control and in the figure 7 a PID control. It is observed as the simulated controller procures a better regulation of the output engine temperature. The PI system is behaved just as the experimental in the region 1, losing this behavior when is decreased the set temperature. In the last region when is returned to the initial conditions returns to improve his behavior.

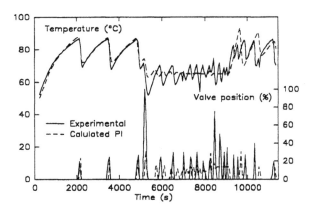

Fig 6. Comparison of the experimental and calculated temperature values using a PI control.

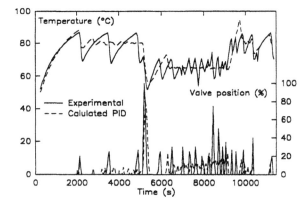

Fig 7. Comparison of the experimental and calculated temperature values using a PID control.

3.3 Parametric study.

To analyze the different parameters influence of design of the installation, it been used the model to simulate different conditions, two from those which result more interesting are the engine speed and the size of the heat exchanger. in the figure 8 can be observed the influence that has the engine speed on the stability the system, how much greater is the engine speed more quickly is stabilized the temperaure and as it had been commented previously the oscillation frequency of the system is greater. In the case of modifying the size of the heat exchanger can be observed that exists an optimum size with the one which is arrived before to a stationary situation, possibly the heat exchanger smallest is adapted better to the simulated situation but one must take into account than must serve for a wide range as operation conditions and it is important than in no case the heat exchanger will be small, since this would suppose an overheating of the motor.

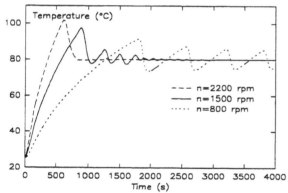

Fig 8. Output engine temperature comparison for diferent engine speeds.

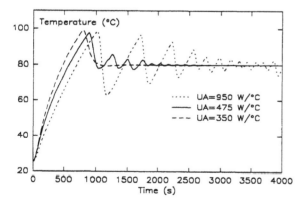

Fig 9. Output engine temperature comparison for diferent heat exchanger sizes.

4. CONCLUSIONS

It has been accomplished a model based on MATLAB to simulate the regulation system of the coolant temperature in an engine test bench. The model permits to calculate the coolant temperature at the characteristic points of the system, as well as modify any design or operation parameter of the system. In order to validate the model an experimental test facility has been used, in which the operation conditions of engine speed and set temperature have been modified. The results obtained with the model show a good agreement with the experimental results when the regulation valve position is used as input variable to the model. When the temperature PID regulation system is used as input in the model, the simulated system results more stable than the real, though the results maintain the same trend. It can be concluded therefore that the simulation of the thermal part is better adjusted than the control part, possibly due to the fact that the PID regulator used is not adapted to the assumed ideal behaviour.

The accomplished model is easily adaptable to any other configuration of regulation systems for fluids temperature, being a useful tool in the design of this type of systems.

REFERENCES

Andrés y Rodríguez-Pomatta, J.A. de (1987). Calor y Frío Industrial I, Vol. 1, Chap. 11-12, pp. 383-448. U.N.E.D.

Benajes, J., Galindo, J., Reyes, E. de los, Serrano, J.R (1997). Estudio teórico de las temperaturas de pared en el cilindro de un M.C.I.A. durante un transitorio de carga. *Anales de Ingeniería Mecánica*, Year 11, Vol. 2, pp. 287-294.

Muñoz, M., Payri, F.(1989). Motores de Combustión Interna Alternativos, Chap. 4, pp. 61-82. Sección de Publicaciones de la E.T.S.I.I. Fundación General-U.P.M.

Ogata, K. (1993). Ingeniería de Control Moderna. Prentice-Hall Hispanoamericana, S.A.

Özisik, M. Necati (1985). Heat transfer: A Basic Approach, Vol. II, Chap. 11, pp. 524 592. McGraw-Hill Book Company.

Simulink. Dynamic System Simulation Software. *User's Guide*.

Weeks, R. (1996). Control de un motor de automóvil utilizando Matlab y Simulink, DYNA, Year LXXI-2, pp. 22-23.

S.I. ENGINE IDLE ASSISTANCE USING A SYNCHRONOUS MACHINE : EXPERIMENTAL RESULTS

P. Bidan*, L.K. Kouadio*, M. Valentin* and S. Boverie**

* L.A.A.S./C.N.R.S., 7 Avenue du Colonel Roche, F-31077 Toulouse Cedex O4 France (bidan@laas.fr)
** Siemens Automotive SA, Avenue du mirail, BP 1149, F-31036 Toulouse Cedex, France

Abstract: This paper presents the experimental validation of an original way to improve idle of a spark-ignition engine. It is well known that inappropriate idle-speed control increases fuel consumption, pollutant emissions and also compromises stability. The solution proposed here has the distinctive feature of combining the traditional airflow rate control, simultaneously with the usual automobile alternator operating as a synchronous motor to provide a fast supplementary torque. This strategy is compared with the standard one in terms of stability, consumption and pollution. *Copyright © 1998 IFAC*

Keywords: Spark-ignition engine, idle speed control, synchronous machine, airflow control, multi-input system

1. INTRODUCTION

All the politics of energy savings and environment protection initiated since the last twenty years require improvements in the spark-ignition engine operation particularly during urban traffic. While steady state operations are considered to be more or less clean, transients generates several difficulties. Namely maintaining optimal fuel/air ratio is more difficult due to the intake subsystems non-linearities and dynamics (Aquino, 1981; Bidan , et al, 1993; Turin and Geering, 1994; Ault, et al, 1994; Hasegawa, 1994; Bidan , et al, 1995). As a consequence, fuel consumption and pollutant emission increase. Additionally, engine idling operation occurs frequently in city traffic and is critical in terms of energy savings, reduction of pollutant emissions and also in terms of idling quality. A constant and low average idle-speed and a minimum oscillation are necessary. The crucial question is how to reduce engine idle-speed and at same time improve its stability and its robustness with respect to disturbances (Nishimura and Ishii, 1986). Air-flow ratio and spark-advance control have been more thoroughly studied (Francis and Fruechte, 1983; Powers, et al, 1983; William and Citron, 1984) to perform engine idle-speed control. Hitoshi and Shoichi (1992) propose a control system based on compensating the variation of the

alternator's current seen as an electrical load disturbance. This control induces throttle opening and thereby anticipates idle-speed drop. The solution experimented in this paper combines an air-flow ratio control (via an electric auxiliary throttle) simultaneously with the traditional automobile alternator operating as a synchronous motor, to provide a supplementary torque. This control approach leads to a multivariable system with, on the one hand, two input variables defined by the throttle opening and the synchronous machine R.M.S. current and, on the other hand, one output represented by the engine speed. Fig. 1 shows the architecture of the system.

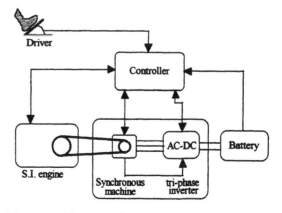

Fig. 1. Architecture of the system

Another purpose of this configuration is to take advantage of the S.M. faster torque response to impose a smooth gradient of the throttle opening for better fuel/air ratio control.

2. ELECTRICAL ISSUES

As explained previously, the principle of the proposed idle-speed control system is based on the capability of the automobile's alternator to operate as a motor. Several experiments and tests have been made on an usual three phase Wound-Rotor Synchronous Machine (see characteristics in appendix A.1). The back e.m.f. and the static torque are almost sinusoidal. Thus, using sinusoidal phase current, the theoretical torque is given by the simplified expression:

$$\Gamma_{MS} = 3 \cdot p \cdot \Phi_e \cdot I \cos\psi , \quad (1)$$

where p is the number of pole pairs, Φ_e the flux created by the rotor, I the R.M.S. value of the stator currents and ψ the angle between the current's stator and the back e.m.f.

The main modification of the traditional electrical on-board network concerns the static converter between the alternator and the battery. The rectifier bridge that is neither controllable nor power reversible have been replaced by a three phase inverter especially designed for this application. A complete self-control system of the synchronous machine has been developed (Kouadio, 1996) using an hysteresis current controller (Lajoie-Mazenc, et al, 1985). Through preliminary simulation tests, a 24V battery was chosen to produce a significant torque in order to effectively prevent stalls of the S.I. engine. Obviously this voltage level seems to be much higher than the classical 12V in the automobile. However, due to the growth of electrical on-board components in new generation of cars, manufacturers are considering a voltage level increase. Another possibility is to have an electrical on-board network using two voltage levels, as electric vehicles (Kouadio, 1996).

The current I and the angle ψ (see equation 1) are selected for each speed to keep sinusoidal currents and maximum torque. For idle speed control, the synchronous motor is then considered as a torque generator. To get a linear control on the torque, a map of optimal tuning has been generated. From the desired torque and the current speed of the electrical machine, the pair $\left(I, \psi \right)$ is selected.

Fig. 2 shows the maximum torque versus speed measured for the synchronous motor without exceed a stator R.M.S. current of 60A. The speed range corresponds to the range of idle operation.

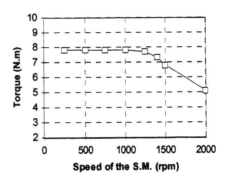

Fig. 2. Torque of the synchronous machine (S.M.)

3. IDLE SPEED CONTROL

3.1 S.I. engine model

Various models of S.I. engine have been presented in the past twenty years (Dobner, 1980; Heywood, 1988; Thompson and Duan, 1991; Moskwa and Hedricks, 1992; Chaumerliac, et al, 1994). To achieve a control-oriented model, the system was evaluated using the mean values of the state variables and divided in three basic subsystems (see Fig. 3).

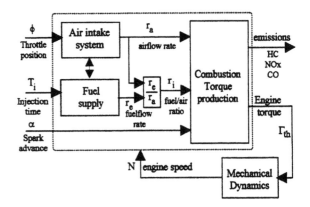

Fig. 3. Engine model description

The intake manifold portion of the model has two highly non-linear fundamental dynamics. The first one is the air filling effect due to the intake manifold volume and the second one depends on the wall wetting by the fuel. In this paper it is assumed that the electronic injection device compensates the fuel film dynamic to maintain the Fuel/Air ratio (r_i) to its nominal point during transients. Thus, the throttle opening (ϕ) and the spark advance (α) are the only input engine parameters available for idle-speed control. A local linear engine model can be obtained by a first order approximation of the non-linear model around the nominal point of idling (Francis and Freuchte, 1983 ; Powers, et al, 1983 ; Kouadio, et al, 1996). The model parameters have been identified from step responses (Kouadio, 1996) on the engine.

3.2 Conventional idle-speed control

The Fig. 4 shows the structure of the conventional idle-speed control. It exhibits two feedback loops. The first one of Proportional-Integral type, acts slowly using the throttle opening for the speed regulation of the nominal point. The second is faster, using spark advance and occurs especially during great perturbations; the corresponding adjustment of engine torque is very quick because no time delay occurs. However, in the aim of ensuring a sufficient action area, the spark advance corresponding to the nominal point must be tuned to a lower value than the optimal one regarding consumption and pollution.

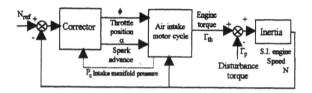

Fig. 4. Idle-speed control using throttle opening and spark advance

3.3 Idle-speed control using electrical assistance

In this approach, the spark advance loop is replaced by a synchronous machine torque control (see Fig. 5). The spark advance is tuned to the value yielding the maximum S.I. engine torque. A Proportional controller is used for the electrical assistance and a Proportional Integral one for the air flow control. Then, the electrical assistance especially acts when a great disturbance occurs, while the air flow loop provides the torque for permanent operation. The basic choice of corrector parameters is made by pole assignment using roots locus (Kouadio 1996), by considering the electrical assistance as a fast internal loop and the air flow control as a slow external one.

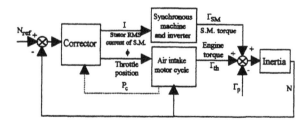

Fig. 5. Idle-speed control using throttle opening and electrical assistance

4. EXPERIMENTAL VALIDATION AND RESULTS

Fig. 6 shows the configuration of the test bench. To simulate a mechanical load on the engine crankshaft, the standard alternator has been preserved with variable electrical load.

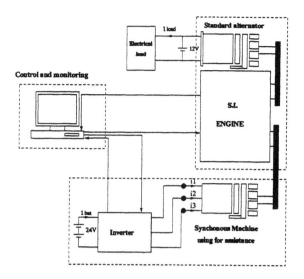

Fig. 6. Test bench

Two types of experiments have been realised : the first one in steady state and the second one in transients. In each case, the above principles of regulation have been tested.

Steady state operation.
Fig. 7 allows to compare the results concerning the nominal idle-speed. The stability is better with electrical assistance. The mixture preparation is easier because the manifold pressure is more regular. The important consequence is the reduction of the injection time which induces a consumption reduction about 20%.

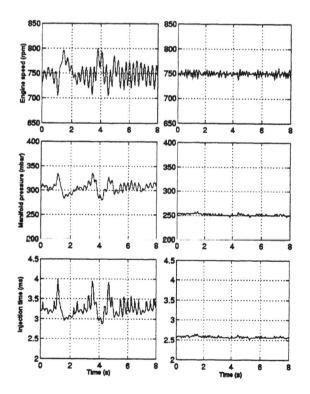

Fig. 7. Steady state operation of regulations (conventional on left, electrical assistance on right)

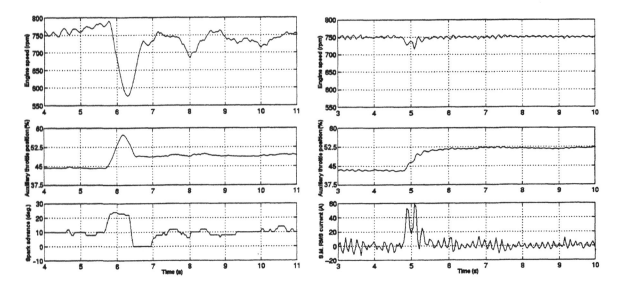

Fig. 8. Idle-speed control during torque disturbance (advance and throttle on left, electrical assistance on right)

Compensation of disturbance torque.
Fig. 8 shows the reaction of the two control loops in the presence of a 10 N.m perturbing torque step on the crankshaft The speed fall is smaller with electrical assistance. At the same time, the throttle opening is smaller and slower.

Reduction of idle-speed reference.
Fig. 9 shows the good reaction of the regulation loop with electrical assistance in the case of 500 rpm reference and the same perturbing torque as above. In the same condition, the engine stalls with classical control. Another consequence is the reduction of consumption of fuel of about 40% in steady state operation, in comparison with idle-speed control at 750 rpm using classical regulation.

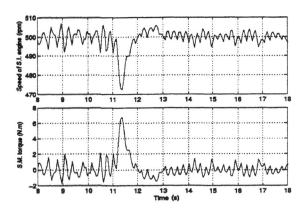

Fig. 9. Idle-speed control at 500 rpm (Electrical assistance)

Conclusion. Electrical assistance improves robustness of idle with respect to disturbances, without inducing fast opening of throttle and without requiring the use of spark advance. This permits a significant consumption reduction. In the same time pollution should decrease because of better idle stability and "optimal" advance adjustment.

5. CONCLUSION

The proposed electrical assistance for idle-speed control improves engine stability and reduce the risk of stalling in presence of external torque disturbance. The preparing of mixture is then easier, the spark advance may be set to its optimal value and the idle-speed reference can be reduced. The major consequence is the reduction of consumption of fuel and pollution. This assistance could be extended to the transient phases, to join or leave idle. Futhermore, the speed measurement is at present available only at each TDC (Top Dead Center); but with instantaneous speed measurement, the electrical assistance could be used as a dynamic flywheel to regularise the engine rotation.

In term of automotive product, the study of mecatronic integration should be undertaken for the inverter and the S.M, and the structure of the electrical on-board network still remains to be defined. Finally, such a system will permit to improve energy management with regard to the growth of electrical servitudes on vehicles.

ACKNOWLEDGEMENTS

The authors would like to thank J.P. Berry and G. Montseny (LAAS) for their help in realising the inverter and J.M. Lequellec (Advanced Development Department of SIEMENS Automotive) for his support in experimentation and computerised tools. This work have been subsidised by "Région Midi-Pyrénées" (France).

APPENDICES

A.1 Alternator Technical Specification in standard configuration

Type	Wound-rotor Three phase
pairs of poles	6
Nominal voltage	14 V
Max. rectified current	110 A
Max. speed	15000 rpm
Armature inductance	192 μH
Armature resistance (20°C)	30 mΩ

A.2 S.I. Engine Technical Specification

Type	Four-cycle Atmospheric aspiration
Cylinders	4 in-line
Swept volume	1324 cm^3
Bore × stroke	71×83.6 mm
Compression ratio	9.7
Intake manifold volume	1.28 dm^3
Injection system	multi-point sequential
Ignition	electronic

REFERENCES

Aquino C.F. (1981). Transient A/F Control of characteristics of the 5 liter central fuel injection engine. SAE Papers n°810494.

Ault A., V.K. Jones, J.D. Powell and G.F. Franklin (1994). Adaptive Air-Fuel ratio control of a S.I. engine. SAE paper 940373.

Bidan, P., S. Boverie and V. Chaumerliac (1995). Non linear Control of a Spark-Ignition Engine. IEEE Transactions on control systems technology, Special Issue on Automotive Control Systems, Vol.3, N°1, 4-13.

Bidan, P., S. Boverie and J.C. Marpinard (1993). State feedback linearizing control: Application to an engine car. 12th IFAC World Congress, Sydney, Australia, vol. 1, 373-376.

Chaumerliac V., P. Bidan and S. Boverie (1994). Control-oriented spark engine model. Control Engineering Practice, vol.2, n°3, 381-387.

Dobner D.J. (1980). A mathematical Engine model for development of dynamic engine control. SAE paper n°800054

Francis E.C. and R.D. Fruechte (1983). Dynamic engine models for control development- Part.2: Application to idle-speed control. International Journal of Vehicle Design. Special publication SP4, 75-88.

Hasegawa Y. (1994). Individual cylinder Air-Fuel ratio feedback using an observer. SAE paper 940376.

Heywood J.B. (1988). Internal combustion engine fundamentals, Mc Graw-Hill.

Hires S.D. and M.T. Overington (1981). Transient mixture strength excursions - an investigation of their causes and the development of a constant mixture strength fuelling strategy. SAE paper 810495.

Hitoshi I.. and W. Shoichi (1992). A performance improvement in idle-speed control system with feed-forward compensation for the alternator current. IEEE International Conference on Power Electronics in transportation, Dearborn, USA.

Koudio L.K. (1996). Assistance d'un moteur thermique d'automobile par une machine synchrone autopilotée. Thesis of the Paul Sabatier University, France.

Kouadio L.K., P. Bidan, M. Valentin and J.P. Berry (1996). S.I. engine idle control improvement by using automobile reversible "alternator". 13th World IFAC Congress (IFAC'96), Vol.Q, pp.93-98, San Francisco (USA).

Lajoie-Mazenc M., C. Villanueva and J. Hector (1985). Study and Implementation of Hysteresis Controlled Inverter on a Permanent Magnet Synchronous Machine. IEEE transactions on industry applications, vol.IA-21, n°2.

Moskwa J.J. and J.K.Hedricks (1992). Modelling and validation of automotive engines for control algorithm development. ASME Journal of Dynamic Systems, Measurement and control, vol.114, 278-285.

Nishimura Y. and K. Ishii (1986). Engine idle stability analysis and control. SAE paper n°860412.

Powers W.F., B.K. Powell and G.P. Lawson (1983). Applications of optimal control and Kalman filtering to automotive systems. International. Journal of Vehicle Design. Special publication SP4, 75-88.

Thompson S. and S.Y. Duan (1991). Modelling, parameter selection and simulation of a single-cylinder four-cycle engine. Proc. Instn Mech. Engrs, vol. 205 n° I1, 49-57.

Turin R.C. and H.P. Geering (1994). Model-based fuel control in an S.I. engine. SAE paper 940374.

William P.M. and S.J. Citron (1984). An adaptive idle mode control system. SAE paper n°840443

A new driving concept for a mobile robot

R. Graf, M. Rieder, R. Dillmann *

*Institute for Process Control and Robotics (IPR), University of Karlsruhe, Kaiserstrasse 12, 76128
Karlsruhe, GERMANY, Fax: ++49 721 608 7141, EMail: {graf|rieder|dillmann}@ira.uka.de*

Abstract. *There are several driving concepts for mobile robots. The most propagated ones are differential and synchro drives. Both of them have their specific advantages and disadvantages. The new concept described in this paper deals with a new drive which combines the advantages of both systems without keeping the disadvantages. The system offers a nearly omni-directional movability with less effort than other solutions. Copyright © 1998 IFAC*

Key Words. Mobile robots, drives

1 Introduction

The basis for a mobile robot's capabilities is the drive system. The configuration of the drive determines the sensor architecture and the navigation strategy as well.

Generally two drive types are used in robotics, either a synchro or a differential drive. Both of them have their specific advantages, therefore the choice of the drive is application specific.

The synchro drive enables a robot to move in any direction, but its orientation cannot be changed. This requirement restricts the use of the synchro drive system to robots with a symmetrical shape like circles or octagons, since the space needed for the movement is independent of the direction.

However, the differential drive offers the possibility to turn the robot to any orientation, but the movement is limited to forwards and backwards. Therefore sidewards docking activities are not practicable.

A desirable drive system should combine the mentioned features. This has been realized in the Mecanum-wheel-drive. Unfortunately not every ground material permits the use of these wheels. Additionally this drive system is complex to handle and very expensive. The possibilities given with the Mecanum-wheel-drive are oversized for common applications resigning on the capability of changing independently orientation and position. The possibility of using both movements one after another would be sufficient for almost any imaginable application.

In this paper we present a new solution to combine the advantages of the synchro and the differential drive. It is implemented in the robot VIPER[1] (see figure 1). VIPER is a prototype for a driver-less transport system in industrial environment.

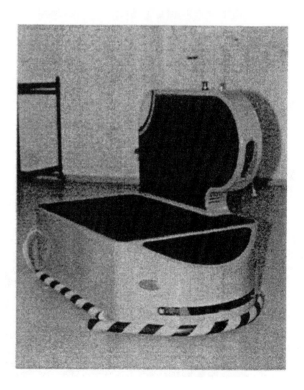

Figure 1   VIPER

[1] Vierrad des Institutes für Prozessrechentechnik und Robotik

2   The mechanical structure

VIPER is equipped with a new, nearly omni-directional driving concept firstly developed by "Maschinenbaubetrieb Gronau". The driving concept is based on a combination of a differential and a synchro drive(J. Bohrenstein, 1996). VIPER has a rectangular shape sized $650mm$ by $900mm$. Additionally, a half-circular shape is added at the robot's front.

Figure 2   VIPERs internal mechanical structure

Figure 2 shows the internal mechanical structure of VIPER. It is build of aluminum to reduce its weight. At the front there is a planar laserscanner for collision avoidance and at the side are the two batteries. The hood of VIPER can be opened very easy.

all computers will be installed in the case on the top. The window at the front is for a camera to observe the environment. At the backside of the top case there is a joystick for controlling the robot by hand.

3   The kinematic system

The drive system consists of four single wheels (see figure 3. The front and rear wheels are connected by an axis and motor driven. The side wheels act

passively and have a flexible wheel suspension to ensure that the active wheels have contact to the ground.

All wheels are rotatable and connected by a chain, which can be rotated arbitrarily changing the orientation of all wheels simultaneously.

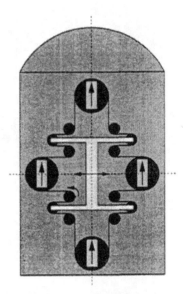

Figure 3   Drive configuration

The overall wheel orientation is measured by an incremental encoder mounted on the rear wheel. In addition to that four wheels for the chain are mounted on a H-shaped plate in the middle of the robot. This plate can be moved sidewards.

3.1   Lateral mode

Driving the front and rear wheels does not change the wheel orientation. Rotating the chain the vehicle can be aimed in any direction without changing the robots orientation. This procedure is used for docking activities, but also for collision avoidance in narrow passages, where the robot is unable to turn.

There is no limit for this rotation. So it is possible to move on a full circle without changing the robot's orientation.

The configuration of the drive during this movement is shown in figure 4. This driving mode with four wheel is similar to a synchro drive(Holland, 1983).

3.2   Differential mode

Since the robot does not have symmetrical shape, it needs the capability to change its orientation.

If all wheels are oriented in straight forward direction, the drive system can be switched into the second type of movement, the differential drive.

To turn the front and rear wheel in opposite direction (but with the same angle) the H-shaped plate in the robot's drive is moved sidewards (see left picture in figure 5). The side wheels remain in straight forward orientation, because they are blocked by the motor controlling the chain. As result of this operation the vehicle is able to move on a circular arc.

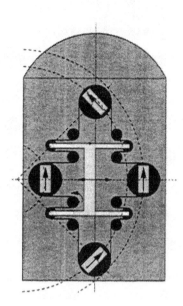

Figure 5   Drive configuration for differential movement

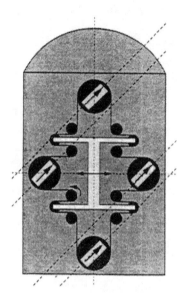

Figure 4   Drive configuration for lateral movement

The maximum inclination of the wheel is ±90°. In this case the front and rear wheel are turned perpendicular to the side wheels and the robot is able to turn on the spot.

However, the simultaneous use of differential and synchro drive mode is impossible. The wheels would block each other. Therefore, the controller ensures, that the driving mode can only be changed, if all wheels point in the straight forward direction. On the other hand the robot does not have to stop in order to change the driving mode.

In comparison to the Mecanum-drive the effort for this powerful movability is low, since only three motors are requiered.

4   Mathematical model

The position of the robot can be described with three degrees of freedom x, y and α. The lateral mode only affects x and y, whereas the differential mode affects all three DOF.

The next paragraphs give a description of the kinematic. The center of the robot is set to the intersection point of the lines connecting both wheel pairs. The parameters of the kinematic are the same as in (J. L. Crowley, 1992) and some more.

- the number of pulses of the encoder of the drive motor N,

- the gear ratio of the reduction gear between motor[2] and wheel n,

- the encoder resolution C in pulses per revolution,

- the diameter of the wheels D in m,

- the angle of the inclination in the lateral mode ϕ ($0° \cdots 360°$),

- the angle of the inclination in the differential mode ρ ($-90° \cdots +90°$) and

- the distance between the two driven wheels l in m.

4.1   Lateral mode

Let all wheels aim to the angle ϕ and the mode is set to lateral. If the drive motor rotates by ΔN pulses, the distance traveled by the wheels is

$$\Delta U = \frac{\pi D}{nC} \Delta N$$

---

[2] The encoder is mounted on the motor

The change in position is

$$\Delta\alpha \;=\; 0 \tag{1}$$

$$\Delta x \;=\; \Delta U \cos\phi \tag{2}$$

$$\Delta y \;=\; \Delta U \sin\phi \tag{3}$$

The inverse kinematics are also quite simple. The robot should move from point $A(x_1, y_1)$ to $B(x_2, y_2)$. Then

$$\Delta\alpha \;=\; 0 \tag{4}$$

$$\Delta x \;=\; x_2 - x_1 \tag{5}$$

$$\Delta y \;=\; y_2 - y_1 \tag{6}$$

The parameter for the drive are

$$\Delta U \;=\; \sqrt{\Delta x^2 + \Delta y^2} \tag{7}$$

$$\phi \;=\; \arctan\phi \tag{8}$$

4.2 Differential mode

In the differential mode the passive wheels aim in front and the other wheels are inclined in opposite direction by the angle ρ (see figure 6). Figure 6 equals figure 5, but is rotated by 90 degrees. The point C is the middle of the robot as defined above. The point M is the midpoint of the circular arc the robot is moving on.

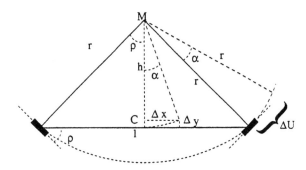

Figure 6   Configuration in differential mode

The drive motor rotates by ΔN pulses again, the distance ΔU remains the same. The robot moves on a circular arc with the radius

$$r = \frac{l}{2\sin\rho}$$

To determine the change in position the distance $h = \overline{MC}$ can be computed by

$$h \;=\; r\cos\rho$$
$$\;=\; \frac{l}{2}\cot\rho$$

Assumed $\rho \neq 0$, the change in position and orientation can now be described by

$$\Delta\alpha \;=\; \frac{2\Delta U \sin\rho}{l} \tag{9}$$

$$\Delta x \;=\; h\sin\alpha \tag{10}$$

$$\Delta y \;=\; h(1 - \cos\alpha) \tag{11}$$

If $\rho \to 0$, the values can be determined by calculating the limit of the equations 9 to 11. Equation 9 is easy

$$\lim_{\rho\to 0}\Delta\alpha = 0 \tag{12}$$

The limit for Δx is calculated from equation 10

$$\begin{aligned}
\lim_{\rho\to 0}\Delta x &= h\sin\alpha \\
&= \frac{l}{2}\cot\rho\sin\frac{2\Delta U\sin\rho}{l} \\
&= \frac{l}{2}\frac{\cos\rho}{\sin\rho}\frac{2\Delta U\sin\rho}{l} \\
&= \Delta U\cos\rho \\
&= \Delta U \tag{13}
\end{aligned}$$

The limit for Δy is calculated from equation 11

$$\begin{aligned}
\lim_{\rho\to 0}\Delta y &= h\left(1 - \cos\alpha\right) \\
&= \frac{l}{2}\cot\rho\left(1 - \cos\frac{2\Delta U\sin\rho}{l}\right) \\
&= \frac{l}{2}\frac{\cos\rho}{\sin\rho}\left(1 - \left(1 - \left(\frac{2\Delta U\sin\rho}{l}\right)^2\right)\right) \\
&= \frac{l}{2}\frac{\cos\rho}{\sin\rho}\left(\frac{2\Delta U\sin\rho}{l}\right)^2 \\
&= \frac{2\Delta U^2\cos\rho\sin\rho}{l} \\
&= 0 \tag{14}
\end{aligned}$$

In summary the movements for $\rho \to 0$ are the following, which are obvious

$$\lim_{\rho\to 0}\Delta\alpha \;=\; 0 \tag{15}$$

$$\lim_{\rho\to 0}\Delta x \;=\; \Delta U \tag{16}$$

$$\lim_{\rho\to 0}\Delta y \;=\; 0 \tag{17}$$

Since equations 1 to 3 and equations 15 to 17 are the same, if $\rho = 0$ and $\phi = 0$, it is save to change the driving mode in this state without affecting the calculation of the robot position.

The inverse kinematics are similar to a normal differential drive and a little bit more complex. To move from A to B, either the robot has to turn in

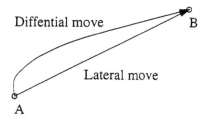

Figure 7　Driving towards a goal in both modes

the direction pointing from A to B and then move straight forward or it moves on a circular arc from A to B.

The first case is simple. The values for phi and ΔU are the ones from equations 7 and 8. ρ is set to $90°$ and the robot turns by phi. After that ρ is set to $0°$ and the robot drives to point B. The trajectory equals the trajectory of the lateral move except the robot orientation. At the goal VIPER turns to the requested angle α.

The second way is to start rotating, so the angle to the goal decreases. If the angle is under $45°$ the robot starts moving and the rotation is reduced, until the robot aims exactly to the goal. If the goal is reached, the robot turn on the spot to the requested angle as described above. Figure 7 shows both trajectories.

4.3　Numerical example

The parameters as defined above have the following values for VIPER:

- the gear ratio $n = 30$,

- the encoder resolution $C = 4000$,

- the diameter of the wheels $D = 0.1225\,m$,

- the distance $l = 0.7\,m$

Let say, the motor is at $R = 3000\,rpm$. The velocity v of VIPER is

$$
\begin{aligned}
v &= \frac{\pi D}{nC}\,\Delta N \\
&= \frac{\pi D}{nC}\,RC \\
&= \frac{\pi D\,R}{n} \\
&= \frac{\pi * 0.1225m * 3000}{30 * 60s} \\
&= 0.64\frac{m}{s}
\end{aligned}
$$

The maximum numbers of revolutions is $R_{max} = 5000\,rpm$, the maximum velocity $v_{max} = 1\frac{m}{s}$.

5　Controlling architecture

The whole controlling concept and sensor system are used in the robot MORTIMER(R. Graf, 1998). The controlling unit of VIPER consists of the three parts navigation, sensor data acquisition and drive control.

Each part is represented by a single computer on board. All computers are running under $VxWorks$, a real-time-operating system with a micro-kernel structure. Every CPU has a specific kernel depending on the required functions. The data-flow between all levels is described in figure 8.

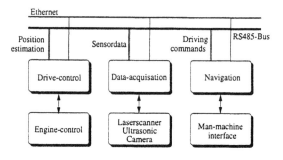

Figure 8　Controlling architecture of VIPER

The drive control consists of two modules. The low level is the electronics which controls the motors. These electronics have been developed at our institute(Woersching, 1997). The system consists of a main board with a C167 microprocessor, a logic board for signal processing and the power electronics. A detailed description can be found in (R. Graf, 1998).

There is one problem with the controllers that control the position of the chain and the H-shape. The encoder to determine the angle of the chain is mounted at the rear wheel to have a direct dependence without knowing the gear ratio of the chain motor. If the robot is moving in the differential mode, the H-shape affects the position of the rear wheel as well. Therefore the controller of the chain has to be disabled, since the two controllers would block each other.

The high level is one of the mentioned computers. The software is called CAMPCO[3] and consists of a position controller and a subordinated speed controller. The communication to the C167 is performed by a RS232 serial interface. Every

---

[3] CAscaded Multi Purpose COntroller

20 ms the actual values of the encoder are sent to the computer to calculate the movement.

The position controller runs at 50 Hz, the speed controller at 100 Hz, which is enough for smooth movements. The interface to the navigation computer is realized both by Ethernet and by a special real-time serial bus based on RS485.

6 Conclusion

A new driving concept has been presented, which combines the advantages of the most used drives, the synchro and the differential drive. This provides a large movability with less effort than other concepts like Mecanum drive or other.

The kinematics of this system are easy to handle, so the controlling unit does not require much cpu power.

The controlling and sensor architecture is the same as the one of the MORTIMER system.

7 Acknowledgment

This work was performed at the Institute for Real-Time Systems and Robotics (IPR), Prof. Dr.-Ing. U. Rembold, Prof. Dr.-Ing. H. Wörn, Prof. Dr.-Ing. R. Dillmann, Faculty for Information Science, University of Karlsruhe. The authors thank Mr. Gronau for his great idea and his support.

8 REFERENCES

Holland, J. M. (1983). *Basic Robotics Concepts.* Howard W. Sams, Macmillan, Inc.

J. Bohrenstein, H. R. Everett, L. Feng (1996). *Navigating mobile Robots.* A K Peters.

J. L. Crowley, P. Reignier (1992). Asynchronous control of rotation and translation for a robot vehicle. In: *Robotics and Autonomous Systems.* Vol. 10. pp. 243–251.

R. Graf, P. Weckesser (1998). Roomservice in a hotel. In: *Intellegnt Autonomous Vehicles.* Madrid, ES.

Woersching, U. (1997). Modulare Motorsteuerung. Master's thesis. Universität Karlsruhe, Institut für Prozeŝrechentechnik und Robotik.

INTELLIGENT COMPONENTS IN THE ROMEO VEHICLES

**A. Ollero, B.C. Arrue, J. Ferruz, G. Heredia, F. Cuesta, F. López-Pichaco
and C. Nogales**

*Departamento de Ingeniería de Sistemas y Automática.
Escuela Superior de Ingenieros.
Camino de los Descubrimientos, 41092 Sevilla (Spain).
E-mail: aollero@cartuja.us.es*

Abstract: This paper presents the most significant components in the control architecture of the ROMEO outdoor autonomous vehicles. A short description of the general characteristics of the ROMEO-3R (tricycle) and ROMEO-4R (four wheels) vehicles is included. These autonomous vehicles are the result of the adaptation of conventional electric vehicles and they are currently used for experimentation in autonomous navigation and teleoperation in outdoor environments. The paper presents the general characteristics of the vehicles and the control architecture. Then describes the position estimation and the tracking components. The former components include: position estimation using dead reckoning sensors and orientation sensors, global position estimation using GPS, and position estimation with respect to the environment. On the other hand, tracking components include: tracking of explicit paths, moving objects, and environment features. *Copyright © 1998 IFAC*

Keywords: Autonomous vehicles, position estimation, visual tracking, reactive navigation, control architecture.

1. INTRODUCTION

The traditional mobile robot control architecture is based on planning using environment models. In the last years reactive architectures have gained popularity. These architectures have been proposed to deal with complex, uncertain and dynamic environments where models are not available nor reliable. The concept of behaviour has played an important role in these architectures.

In some cases the mobile robotic research is only oriented to demonstrate Artificial Intelligence and computational capabilities. Then, simulation is applied. However, it has been noticed that computer simulation usually deals with an idealized world where many of the real characteristic of physical components are not taken into account. Some techniques have been also illustrated by using small robots, and special purpose omnidirectional vehicles have been also designed and built to illustrate these robotic capabilities.

The emphasis of the above mentioned research is in autonomous processing capabilities including adaptation and learning. However, the control strategies are usually designed without taking into account the kinematic and dynamic models. Thus, when trying to apply this mobile robotic research to drive real vehicles performing productive tasks many difficulties arise. In these cases attention should be devoted to mechanical and power components which impose significant constraints in sensors and actuators. Furthermore, the kinematic and dynamic characteristics play an important role.

This paper deals with the implementation of the mobile robot reactive control architectures to control real outdoor vehicles. Many different experimental outdoor autonomous vehicles have been developed. Between them it can be mentioned the Navlab family in Carnegie Mellon University (Thorpe, 1990;

Pomerleau, 1993; Hebert et al., 1997), and other many vehicles resulting from the adaptation of conventional ones. The ROMEO vehicles fall in this category.

This paper describe some significant components used in the ROMEO vehicles developed in the School of Engineers at the University of Seville, and how these components can be used in a reactive control architecture of the vehicle. The direct predecessors of these vehicles are the RAM (Ollero, et al., 1993) and the Aurora (Mandow, et al., 1996).

2. THE ROMEO VEHICLES

Figures 1-3 show the two ROMEO vehicles currently developed. Both are electrically powered. The vehicles can be driven automatically and from a teleoperation station. A driver in the vehicle can also takes the control if occasion demands.

Fig. 1: ROMEO-3R vehicle.

ROMEO-3R is a tricycle (steering and driving in the front wheel) with a clutch to switch between manual and automatic drive (see Figure 2). In the automatic mode the computer control system generates the steering command to the steering motor and the speed command to the driving motor. These commands are applied to the motors using PWM electronic regulators. The power and control electrical systems are independent.

Fig. 2: Steering gears and clutch (ROMEO-3R).

ROMEO-4R is a four-wheeled vehicle with Ackerman steering (see Figure 3). Six conventional batteries provide the power to the driving motor. There is a separate battery for the automatic control system. Automatic steering is implemented by using a DC motor which is connected to the steering wheel axle through a reduction gear.

Fig. 3: ROMEO-4R vehicle.

The control system is composed of two computers connected by an ethernet link. The low level control is carried out by an industrial PC with a motion controller board. The reactive components generating vehicle commands from the proximity sensors are also implemented in this computer. Real-time image processing capabilities and path planning have been implemented in the second computer. The ROMEO vehicles carry a variety of sensors to implement intelligent functions: optical encoders for motion control and dead-reckoning, gyroscopes and compass for the orientation, inclinometers, 2D scanner laser and sonars as proximity sensors and active environment perception, video cameras for passive environment perception, and a Differential Global Positioning System. Furthermore, the vehicles carry a controlled pan and tilt unit with a camera for the teleoperation.

The ROMEO vehicles are linked to the teleoperation station by means of a radio modem and a video link.

3. THE ROMEO CONTROL ARCHITECTURE

The architecture of ROMEO is based on the combination of components at several levels. At the higher levels there are two components: autonomous navigation and teleoperation. Both components cooperate in the vehicle control. There are several possible combinations from the basic teleoperation, in which the operator generates directly the steering and motion commands to the fully autonomous navigation, in which the operator only supervises and acts in case of emergency.

The position estimation is a very important function of the ROMEO vehicles. Several position estimation components have been implemented. These components can be used both for autonomous navigation and teleoperation.

Tracking components are basic elements of the ROMEO autonomous navigation. These components include both tracking of explicit paths, tracking of environment features, and tracking of mobile objects. These tracking components are implemented using behaviours. Another basic behaviour is the obstacle avoidance. Each behaviour is designed independently from the others. The autonomous navigation is based on the combination of behaviours to reach an objective or to execute a plan. These objectives or plans can be provided by the teleoperator or by a planning program.

There are two different basic strategies in the ROMEO architecture to combine the behaviours:

1. Define a sequence of behaviours to execute a plan like: "follow a corridor and turn left, after that follow a right wall, and then turn right".

2. Multiple simultaneous behaviours working in a cooperative scheme toward a given goal. In this strategy each behaviour working in parallel produces its own control command, named c_i.

In the strategy 1., the implementation of the sequence involves the detection of the transition between behaviours.

In the strategy 2., the output of each behaviour is weighted by its behaviour weight (BW_i):

$$c = \frac{\sum_i (BW_i * c_i)}{\sum_i BW_i}, \quad i = flw, frw, oa, trn, ... \quad (1)$$

The behaviour weights are calculated dynamically taking into account the situation the mobile robot is in, and determining the applicability of each behaviour according to the context. It is possible to define priorities between behaviours, taking into account the weight of one behaviour to compute the weight of the other ones. For example, if the behaviour weight for left wall following BW_{flw} and for obstacle avoidance BW_{oa} get a high value at the same time, the priority of the left wall following behaviour will decrease using the following equation:

$$BW_{flw} = BW_{flw} * (1 - BW_{oa}) \quad (2)$$

4. POSITION ESTIMATION AND PERCEPTION COMPONENTS

4.1 Position and orientation estimation.

Many sensors exist than can be used for position estimation, but there is no single sensor that provides a good solution in every situation. Therefore, a combination of sensors has to be used, and the readings of all these sensors need to be combined in some way to obtain an estimation of the actual position of the robot (what is usually called "sensor fusion"). In the ROMEO vehicles, sensors of several types are available. Depending on the application, some or all of them are used for the estimation of the position and orientation of the vehicle.

Two types of dead reckoning sensors have been installed in the ROMEO vehicles: odometry sensors (incremental and absolute optical encoders) and attitude and heading sensors (magnetic compass, inclinometers and gyroscopes).

Furthermore, the above position and orientation estimation can be integrated with two different higher level estimators: position estimation with respect to the environment, and absolute position estimation. The former technique is described in next section while the second is being implemented by using differential Global Positioning System (DGPS). The system installed in the ROMEO vehicles uses carrier phase and double differencing techniques to achieve accuracies of 20 cm. or less. The main problem with GPS is the need of the receivers to be in direct sight of the satellites, and thus periodic signal blockage occurs due to buildings, foliage and hilly terrain. Furthermore, differential GPS also has insufficient position accuracy for primary (stand-alone) position estimation systems.

All of these sensors have their own characteristics of accuracy, sensor noise and sampling frequency. To obtain an estimation of the position, an Extended Kalman Filter (Gelb, 1986; Kelly, 1994) is used. The

Extended Kalman Filter provides an estimation of the position, and this estimation is updated as new observations became available.

4.2 Position estimation with respect to the environment.

Involves the environment perception during navigation. That includes passive techniques using a video camera and active techniques by means of proximity sensors.

Position estimation by using a sequence of video images will be considered in 4.2.1. Active techniques have been implemented using a 2D scanner laser and sonars. Procedures to build maps using a 2D scanner laser and to estimate position with respect these maps (Gonzalez, Stenz and Ollero, 1995) were previously developed. A new technique to perceive the environment using sonars and taking into account the sensorial limitations and the ROMEO-3R kinematic constraints has been implemented. This technique is summarized in 4.2.2.

4.2.1 Position estimation through image processing.

The position estimation method is based on a robust matching algorithm, which allows to the tracking of automatically selected points over long sequences of monocular images, without any need of environment structuring. Generic, block-based features, defined as fixed-size windows, are automatically selected and tracked over a stream of images. In figure 4, a 7x7 pixel window is shown, along with its matching window in the next image. Image intensity is displayed as a 3-D surface and a density plot, as a function of pixel coordinates.

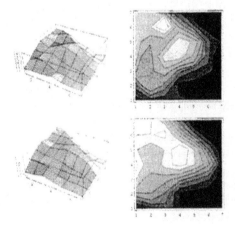

Fig. 4: Feature 51 is tracked from the first image (up) to the second (down).

Matching pairs are validated with two different constraints: Best square error fitting and approximate similarity of shape between clusters of features (Ferruz, 1997; Tomasi, 1991). In figure 5 the results

of matching for two images are shown. Some matching pairs are highlighted; window 51 is the same as in figure 4.

The position estimation method is based on the hypothesis of planar motion. Motion parameter estimation is decoupled from the computation of feature coordinates, which are initially unknown. A single-variable non-linear minimization process is used to determine the rotation angle, while the translation vector can be computed in closed-form. The translation modulus is computed from the relative motion of floor points, by using the camera height as a reference. Once the motion parameters are known, the estimated absolute robot position can be updated and sent to a path-tracking task. This method is described in Ferruz and Ollero (1997).

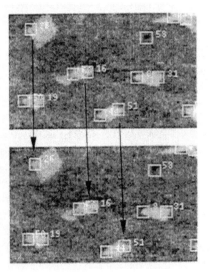

Fig. 5: Detail of feature matching results between the first (up) and second (down) images.

High speed processing is achieved with the help of a multiprocessing system based on a network of 320C40 DSP processors. A parallel implementation of the most time-consuming stages of the matching process significantly reduces the total cycle time to about 240 ms.

4.2.2 Virtual perception from proximity sensors.

The distance measurements (d_s) of an ultrasonic sensor (u) (see Figure 4) can be transformed into a virtual perception p by means of the perception function defined as: $p = f(d, v)$. At the same time, it allows to assign a perception value to a distance d taking into account the angle between the direction of attention a_1 and p. Furthermore, data from different types of ultrasonic sensors can easily be combined into one perception. This is fully explained by the authors in Arrue, et al., (1997a, 1997b).

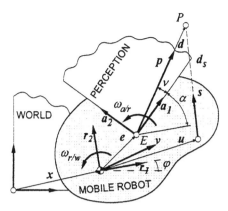

Fig. 6: Virtual perception p of a mobile robot.

The perception is updated as the robot moves on taking into account the odometry and the actions that compute the changes in the angle (v) and the length (p) of the perception vector:

$$\dot{v} = \frac{1}{g}\left[(\dot{x} + \omega_{r/w} \times \mathbf{e})(\mathbf{r}_1 sin(\alpha + v) - \mathbf{r}_2 cos(\alpha + v))\right] - \omega_{r/w} - \omega_{a/r}$$

$$\dot{p} = -\frac{1}{\frac{\partial g}{\partial p}}\left[(\dot{x} + \omega_{r/w} \times \mathbf{e})(\mathbf{r}_1 cos(\alpha + v) + \mathbf{r}_2 sin(\alpha + v)) + \frac{\partial g}{\partial v}\dot{v}\right]$$

(3)

where $g(p,v)$ is the inverse perception function, $\omega_{a/r}$ is the angular velocity of the virtual perception coordinate system relative to the robot, $\omega_{r/w}$ is the angular velocity of the mobile robot relative to the world coordinate system, and $\dot{x}$ is the linear velocity of the mobile robot.

5. TRACKING COMPONENTS

Includes tracking of explicit paths and tracking of environment features (Ollero, et al., 1997):

5.1 Tracking of explicit paths.

The explicit paths can be previously recorded using the vehicle itself or computed from environment maps.

Then, the execution of the behaviour consist of the generation of steering commands to track the path taking into account the actual position-orientation and the constraints imposed by the vehicle and its motion controllers.

Several path tracking strategies have been implemented for evaluation and comparison:

- Pure pursuit. It is based on simple geometric considerations: the path is tracked adjusting at each point circular arcs from the actual position of the robot to goal points on the path, usually chosen at a fixed distance called the lookahead distance.

- Generalized predictive control path tracking (GPPT). In this method, the goal is to minimize a cost index with the errors between future desired outputs and predicted outputs, and the future increments in control. The method is based on a receding horizon approach similar to that used by the human operator when driving. The parameters of this path tracker are the time horizon (time interval in which future positions of the robot are predicted), the length of the control sequence and the weighting of the control variable in the index.

- Direct fuzzy control (Garcia-Cerezo, et al., 1996). It uses a fuzzy controller that has as inputs the lateral displacement of the vehicle from the goal point on the desired path, the deviation angle from the goal point and deviation in curvature. The output of the fuzzy controller is the curvature commanded to the vehicle.

Furthermore, the above path tracking methods involve a substantial amount of heuristic knowledge in the selection of the parameters for each different situation. A fuzzy supervisory controller can be used to consider this heuristic knowledge in the vehicle's real time controller (Ollero, et al., 1994). Thus, this fuzzy supervisor is responsible of the automatic tuning in real time of the control parameters.

5.2 Tracking of environment features.

Two different tracking with proximity sensors behaviours have been described: wall following and corridor following. In these behaviours the lateral proximity sensors are used to maintain the vehicle parallel to the wall at a certain distance, or to pass by the center of a corridor. In these behaviours the position estimation with respect to the environment is given by the virtual perception vector (Arrue, et al., 1997a, 1997b).

5.3 Visual tracking of moving objects.

The image matching method described in Ferruz (1997) is used as source of data for mobile object detection and tracking. After detection, the object is represented by a set of low-level features (fixed-size windows) and a reference point, which is to be kept in the same position relative to the object features.

No model is needed for the tracking activity, which can be used with general-shaped objects. The set of features associated to the object is validated at each tracking cycle, after the next frame has been processed. In the validation process a subset of object windows is selected under the constraint of approximate rigid motion of its projection, after allowing for a scale factor. The worst-fitting windows in the object set are discarded, while new candidates may be accepted.

Using the DSP-based image processing system, the processing time per frame can be reduced to less than 200 ms.

6. CONCLUSIONS

This paper has presented the most significant intelligent components of the ROMEO vehicles. This architecture is a behaviour-based one, but maintain some characteristics of the traditional hierarchical architectures based on planning to improve the performance in the execution of planned tasks.

It should be noted that, in addition of the implementation of autonomous navigation strategies, the components presented in the paper can also be used to implement functions in conventional vehicles such as obstacle detection and avoidance, automatic tracking of features, guidance of vehicles for handicapped people, and other navigation aids.

Acknowledgment: This work has been partially supported by the CYCIT TAP96-1184-C04-01.

REFERENCES

Arrue B.C., F. Cuesta, R. Braunstingl and A. Ollero (1997a). *Application of Virtual Perception Memory to Control a Non-Holonomic Mobile Robot.* 3rd IFAC Symposium on Intelligent Components for Control Applications. SICICA'97, Annecy, France.

Arrue B.C., F. Cuesta, R. Braunstingl and A. Ollero (1997b) *Fuzzy behaviours Combination to Control a Non-Holonomic Mobile Robot Using Virtual Perception Memory.* Proceedings of the 6th IEEE International Conf. on Fuzzy Systems, Barcelona, Spain.

Ferruz J. (1997), *Sistema para establecimiento de correspondencias en secuencias de imágenes. Aplicaciones en robótica móvil,* Universidad de Sevilla.

Ferruz J. and A. Ollero (1997). *Autonomous Mobile Robot Motion Control in Non-Structured Environments Based on Real-Time Video Processing.* IEEE/RSJ International Conference IROS'97, Vol. 2, pp 725-731. Grenoble, France.

García Cerezo A. , A. Ollero and J.L. Martínez (1996). *Design of a robust high-performance fuzzy path tracker for autonomous vehicles.* "International Journal of Systems Science,* Vol. 27, No. 8, pp 799-806.

Gelb A (1986), *Applied Optimal Estimation,* The MIT Press.

González J., A. Stenz and A. Ollero (1995). *A Mobile Robot Iconic Position Estimator using a Radial Laser Scanner,* Journal of Intelligent and Robotic Systems, Vol. 13, pp 161-179. Kluwer Academic Publishers.

Hebert, M., Pomerleau, D., Stentz, A., Thorpe, C., (1997). *A behaviour-Based Approach to Autonomous Navigation Systems: The CMU UGV Project,* to appear in IEEE Expert.

Kelly A. (1994), *A 3D State Space Formulation of a Navigation Kalman Filter for Autonomous Vehicles,* Technical Report CMU-RI-TR-94-19, Robotics Institute, CMU.

Mandow, A., J. Gomez de Gabriel, J.L. Martinez, V.F. Muñoz, A. Ollero and A. García Cerezo (1996). *The Autonomous Mobile Robot AURORA for Greenhouse Operation.* IEEE Robotics and Automation Magazine, Vol. 3, No. 4, December 1996.

Ollero A., A. Simón and F. García (1993). *Mechanical Configuration and Kinematic Desing of a new Automous Mobile Robot.* In "Intelligent Components and Instruments for Control Aplications", A. Ollero and E. F. Camacho Editors, pp 461-466. Pergamon Press.

Ollero A., A. García-Cerezo and J. Martinez (1994). *Fuzzy Supervisory Path Tracking of Mobile Robots.* Control Engineering Practice, Vol.2, No. 2, pp 313-319, Pergamon Press.

Ollero A., A. García-Cerezo, and J.L. Martínez (1997). *Fuzzy Tracking Methods for Mobile Robots.* Chapter 25. of book"Applications of Fuzzy Logic". Vol 7. Prentice Hall.

Pomerleau, D. (1993). *Neural Network Perception for Mobile Robot Guidance.* Kluwer Academic Publishing.

Tomasi C. (1991). *Shape and Motion From Image Streams: A Factorization Method.* Ph.D. Thesis, Carnegie Mellon University.

Thorpe C. (1990). *Vision and Navigation: CMU NavLab.* Kluwer Academic.

GUIDANCE OF AUTONOMOUS VEHICLES BY
MEANS OF STRUCTURED LIGHT

**José Luis Lázaro Galilea, Alfredo Gardel Vicente, Manuel Mazo Quintas,
César Mataix Gómez, Juan Carlos García García,**

*Departamento de Electrónica. Universidad de Alcalá. Campus Universitario s/n. 28871
Alcalá de H. (Madrid). Tel.:34 1 885 48 10-13. Fax.: 34-1-885 48 04.
E-mail:{lazaro, alfredo, mazo, mataix}@depeca.alcala.es*

Abstract: This paper describes a system comprising a CCD sensor coupled with an infra-red emitter so that the emission of structured light then captured in the sensor CCD (vision angle 90°) gives the 3-D co-ordinates of the light impact points. Working from the co-ordinate matrix supplied, surrounding obstacles and vacant areas can be detected. The environment through which the robot may move is generated considering its dimensions and orientation. A check is made in the latest environment update of whether any obstacles balk the objective. If so, the path is varied so that the obstacle is avoided and the path optimum. *Copyright © 1998 IFAC*

Keywords: Obstacle detection, Robot navigation, Telemetry, Trajectory planning.

1. INTRODUCTION

One technique for modelling an unknown environment in which a mobile robot is to be guided, involves obtaining 3D co-ordinates by emitting structured light and capturing it in a CCD camera (Jarvis 83). Beforehand, the whole system (camera and emitted light analyser) has to be jointly calibrated with reference to the same co-ordinates origin.

If the structured light emitted consists of light planes, a co-ordinate can be deduced from each pixel resulting from the impact of said planes in the environment. The obtaining of so many co-ordinates allows a precise recognition of the presence of objects or limits of the physical environment. A deduction can therefore be made of the position of objects balking the robot's movement and the vacant spaces, so that the robot may move through the space without collisions. Related jobs making use of techniques to determine the object position and orientation can be found in (Blais et al., 1988; Sato and Otsuki 1993; Kemmotsu and Kanade 1995). These papers have been developed with small depths and homogeneous background images. Motyl et

al. (1993) and Khadroui et al. (1995) have made use of these techniques for positioning a robot arm in an environment with an a-priori knowledge of objects such as polygonal and spherical ones. Evans et al., (1990) and King (1990) have pointed out the use of structured light plane to detect the obstacles encountered by a mobile (MOB).

The distance maps obtained may be used to deduce vacant, occupied and safety zones, generating paths that are updated according to the characteristics of the captured field of view. From the many path options, that should be chosen which gives some guarantee of safety while also observing the path to be followed. This paper achieves paths that are based on cubic-spline curves. Latombe (1993) have developed a navigation algorithm in order to obtain the space configuration in the surrounding environment. Koch (1985), (Oommen (1887) and Krogh and Feng (1989) shows some path planning algorithms to reach out the goal point. Payton (1986); Nitao and Parodi (1986); Brooks (1986) works with dynamic trajectories recalculated only when the surrounding environment changes.

In each moment the absolute position of the robot must be known to steer it towards its final goal, but it will also be necessary to update the relative position of the objects that crop up in its path. Finally, a movement and turning control of the robot will have to be carried out, so that it can follow the defined path in each iteration of the general system control.

This application involves the need of recognise environments with a large field of view in short distances. The use of a wide-angle lens is justified by this fact. This type of optics hasn't a linear response so traditional camera calibration methods can't be used. To adjust it, the lens response has been modelled. In (Theodoracatos and Calkins 1993) the lens distortion with a large focal length is fixed up in order to take measurements. In that case the determination problem of the system parameters such as the center of image formation is not present. Stein (1993) has used the object geometry (spheres) to find out the aspect relation. And he has also used parallel lines to solve the following parameters: lens distortion, main point and focal length. This process limits the image center to a square of 20x20 pixels.

Tests carried out prove the system to be capable of following side walls, crossing doors, skirting obstacles, redefining its course and coming to a stop at a set distance from objects, with no difficulty and with absolute precision. The aim of this work is to install the system on an autonomous wheelchair which will be capable of avoiding head-on crashes and reach out the final objective without collisions.

2. ROBOT GUIDANCE

The following sequence of tasks needs to be effected for guiding the robot: firstly modelling wide-angle optic and coupling up the emitter-receiver system , then obtaining the co-ordinate matrix of the various points captured, the definition of the virtual movement space to avoid collisions, and the planning of the path to follow and control thereof.

2.1 Modelling wide-angle optics.

Two alternatives are proposed for calculating the parameters of the optical system model, both based on a knowledge of the image formation centre. The first assumes that the lens has a symmetrical revolution response, where only the distortion effect and not the assembly mechanics is influential. The second considers all the effects, so there is no revolution symmetry in this case, and it will therefore be necessary to model it in all directions.

Search for sensor centre. The algorithm consists in capturing the image of parallel straight lines in the scene. The Hough transformation is applied to each one and the curvature is studied by analysis of the accumulation of points in the transformed space, plus the adjacence of maximum accumulations. The above procedure is repeated to capture the straight lines in different directions. With the two least curved lines of each capture, the intersection pixels are obtained from the zones between these two lines, thereby limiting the image formation centre. The Hough transformation is then newly applied to straight lines crossing each of the points of the limited zone in different directions.

Radial correction. A relation is set up between a real co-ordinate and a co-ordinate of the CCD camera. This relation is obtained by means of a polynomial approximate to the values (figure 1 dashed line). The polynomial is obtained by regressive methods using least squares. The steps taken are the following: the derivative of the polynomial in point (0,0) is calculated, thereby obtaining the hypothetical ideal linear response (figure 1 straight line); tables are then drawn up with the error produced in the pixels, making a translation of the distance in the scene to the pixel captured on camera and the pixel that should be captured (independence from the depth of field is thereby achieved); the error at different points is obtained.

Working from the error table obtained, the coefficients (c_k) of a regression polynomial of error (ϵ) are calculated (eq. 1). Then, depending on the distance of the impact from the image centre $(u^2+v^2)^{1/2}$, this polynomial gives the real point (u',v') where it should have impacted (eq. 2). In equation (2) α are *atan(v/u)*

$$\epsilon = c_1 \, r + c_2 \, r^2 + c_3 \, r^3 + c_4 \, r^4 + c_5 \, r^5 \ldots\ldots \, (1)$$

$$\begin{aligned} u' &= (r + \epsilon) \cos \alpha \\ v' &= (r + \epsilon) \sin \alpha \end{aligned} \qquad (2)$$

Working from the original position of the pixel (x,y), a calculation is made of the vector of the point with respect to the optical centre of the image, modulus (r) and angle (α) and the parameters to be included in the calibration matrix, where $c_{i,j}$ are the coefficients of camera calibration (eq.3).

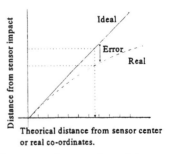

Fig.1. Ideal curve and extraction of the error.

$$\begin{pmatrix} C_{11} & C_{12} & C_{13} \\ C_{21} & C_{22} & C_{23} \end{pmatrix} \begin{pmatrix} x \\ y \\ z \end{pmatrix} = \begin{pmatrix} U \\ V \end{pmatrix} + \left(\begin{bmatrix} \cos\alpha \\ \mathrm{sen}\,\alpha \end{bmatrix} * \left(\begin{bmatrix} r & r^2 & . & r^n \end{bmatrix} * \begin{bmatrix} c_1 \\ c_2 \\ . \\ c_n \end{bmatrix} \right) \right) \quad (3)$$

Bidimensional correction. The lens will be considered to produce two errors on the image captured of each point; one error corresponds to *phase error* and the other to *modulus*, obtaining modulus and phase corrections that will depend not only on the distance from the image formation centre but also on the position of the points to be analysed in the image (figure 2).

The angle α is that with which the projection of a real point is received, and β corresponds to the angle in real co-ordinates. The data obtained from multiple points chosen randomly are then used to calculate the correction polynomials, in equations (4) and (5), by least squares method.

$$err\_phase = a\,u + b\,v + c\,uv + d\,u^2 + e\,v^2 + f\,uv^2 + \\ + g\,u^2v + h\,u^3 + j\,v^3 + k\,u^2v^2 + l\,uv^3 + m\,u^3v + n\,u^4 \quad (4)$$

$$err\_modulus = a'u + b'v + c'uv + d'u^2 + e'v^2 + \\ + f'uv^2 + g'u^2v + h'u^3 + j'v^3 + k'u^2v^2 + l'uv^3 \quad (5)$$

2.2 CCD-laser coupling. Obtaining 3-D co-ordinates.

The general method for obtaining the 3D co-ordinates involves solving the system of equations obtained from the camera model (6) for each point registered that belongs to those emitted by the laser in plane form (6), and the equation (7) of this plane. This gives three equations that in turn give the x,y,z co-ordinates.

$$a_1 x + b_1 y + c_1 z + d_1 = 0 \quad (6)$$
$$a_2 x + b_2 y + c_2 z + d_2 = 0$$

$$A x + B y + C z + 1 = 0 \quad (7)$$

The aim of the emitter-camera coupling is the modelling and calibration of the system, with respect to the same reference and with the minimum error. Once the optimal position has been analysed, the emitter is coupled to global system and this is calibrated according to a known reference system (figure 3).

Before obtaining the plane equation it is necessary to eliminate from the image everything that does not belong to the light emitted by the laser and reduce the thickness of same to a pixel. Firstly an optical filter is used to eliminate all possible light except the wave length emitted as structured light. An active search is made for very bright points belonging to the laser and a "kernelling" of the image is carried out, stressing points belonging to horizontal or near-horizontal lines and the threshold is then set. Then all those points not belonging to narrow beams are eliminated.

Thereafter, using the plane equation (which is the same at all times) and the equations afforded by the camera model for each point, the system of equations is solved, deducing the co-ordinates of the points captured. The depth from each dot is the z co-ordinate.

2.3 Virtual space definition.

With the information about the environment provided by the perception system (co-ordinate matrix) the environment is analysed with a sufficient depth (safety distance) to be able to react in the face of any obstacles. Due to the large number of points analysed, they are previously filtered to unify values of adjacent points (figure 4), thereby deducing vacant and obstructed zones in a first sectorisation.

From a practical point of view, a check is made of whether there is sufficient vacant space for the robot to pass between the obstacles detected. If not, they are unified as a single obstacle. The problem of obstruction between obstacles is also studied (figure 5), analysing the distance matrix.

After these steps, shown in more detail in figure 6, the detected outline of the obstacles is generated, modifying it if it presents concavities, since this would guide the robot into a cul de sac or cause it to make unnecessary movements.

Now the problem is tackled of determining where the robot may be set up in terms of its orientation and dimensions, and how to move it from one position to

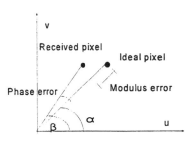

Fig. 2. Representation of modulus and phase error.

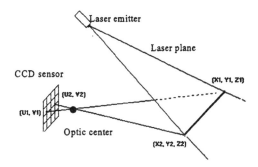

Fig. 3. System scheme and obtaining plane equation.

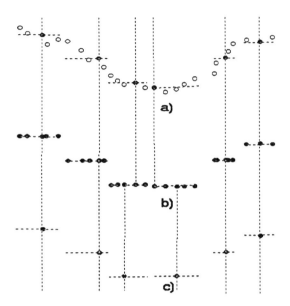

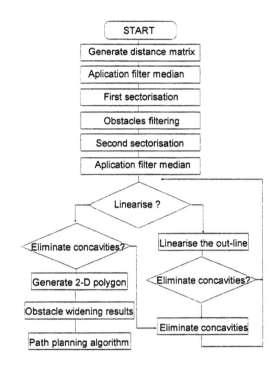

Fig. 4. Distance filtering in adjacent points. a) Obtained points from the perception system. b) Median filter. c) Selected points-obstacle vertexes.

Fig. 6. Environment analysis sequence

another without producing collisions. The solution adopted is to widen the intermediate objects in each iteration, in accordance with the size of the robot and the direction it is moving in, considering it as a one point (figure 7).

2.4 Path planning.

The next step is to decide which path should be followed to reach the goal without provoking collisions. This is done by considering the virtual vertices obtained from the widening of the obstacles in the previous passage. An algorithm is developed capable of selecting the intermediate points and temporary directions. that would guide the robot along an obstacle-free path. These intermediate objectives are selected sequentially while the robot is moving, using the most recent information suplied by the perception system.

If the obstacle is located between the robot and the goal an intermediate objective is generated and the path is deviated. A vertex of the virtual objects is chosen as the objective, laying down a path that passes by same,

so that the objects are skirted according to the criteria of safe movement and the shortest path.

The spline cubic curve has been used for obtaining the path. Beginning with n points (3 in our study: the original, final and the point to avoid the obstacle), these points have to be linked with (n-1) cubic polynoms. In order to determine the 4 coefficients of each polynom, 4(n-1) equations are needed. Every polynom has to pass the 2 points that it links and this impose to fulfil 2(n-1) equations. Besides the union of the cubic polynoms, continuity in slope and in curvature in the n-2 middle points is also imposed. This gives 2(n-2) equa-

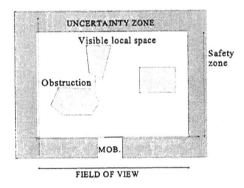

Fig. 5. Obstacles obstruction.

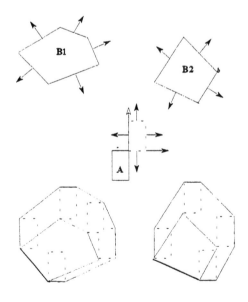

Fig. 7. Consideration of the robot as a point and obstacles widening.

tions more. And the MOB direction (slope) in the extreme points gives the last two equations that allow to obtain all the coefficients.

2.5 Execution and control of movement.

Once the path to be followed has been defined and until this is redefined by the system, the linear and angle speed of the robot must be controlled, to generate the suitable movements for following the desired path. A PID control is therefore made of the turning speed of the motors in charge of the movement to ensure that the commands sent to them are obeyed.

In this work a system (dead reckoning) has been developed for finding out what position the robot occupies at all times, in order to modify the commands to be obeyed by the motors at the time required, to correct possible errors of location and follow the desired path. The location with respect to the origin of absolute coordinates is important in terms of finding out the position, as is the relative position of the robot vis-à-vis the obstacles, so as to be able to make translations of said relative positions (e.g., when the robot is moving while environment data are being updated).

3. EXPERIMENTAL RESULTS

The equipment outlined below was used in the tests, making up a system with the structure represented in figure 8. This equipment was: a CCD camera previously corrected to present a transfer function coupled to a laser emitter generating a light plane with an openning angle of 80°, a Matrox Comet image digitizing card, a PCLTA card, Neuron Chip motor controller, cards for controlling the robot's position based on the Neuron Chip and platforms with two drive wheels and two idle wheels to support the system.

Figure 9 shows the image of the moving devices used in the tests. They are an experimental platform and a wheelchair. Firstly the emitter and CCD are set up integrally with each other and jointly calibrated. Once the system has thus been set up and calibrated, calculation is made of the maximum error due to the quantification of the sensor. Said error for scene depth is

Fig. 9 Test platforms.

shown in figure 10. A static measurement test is then carried out to evaluate errors produced. Table 1 shows the error readings obtained in a scene with a maximum depth of 300 cm., with 8 different object zones.

A host computer sees to all processing and communicates with motor and position control modules via a PCLTA card over a LonWorks network.

The next step is the modelling of the environment to obtain the virtual obstacles and vacant areas for movement, supplying the intermediate path objectives. With a precision at all times similar or superior to those outlined in table 1.

Others trials involved detection tests of the mobile set-up against obstacles (at a fixed distance from them), obtaining errors that never exceeded one centimetre for distances less than one metre. Tests were made with the robot placed directly up against a surface, having previously moved from oblique positions with respect to the same surface.

Figure 11 shows the real detection of two obstacles with one over the vehicle trajectory. In this case, the whole system was programmed to avoid obstacles in a range of two meters in front of the vehicle.

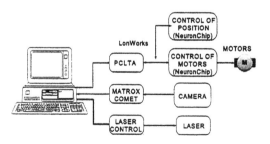

Fig. 8. Structure of system used.

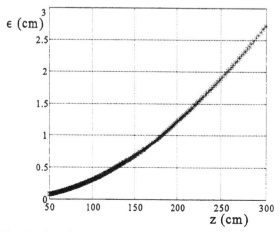

Fig. 10. Maximum error in Z co-ordinate.

117

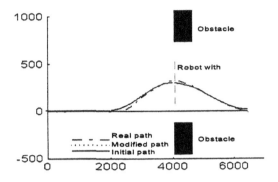

Fig. 11. Trajectories with surrounding obstacles.

Table 1. Maximum errors detected.

Zone	Maximum errors (mm.)	
	x	z
1	12	-20
2	20	9
3	19	-
4	-12	6
5	7	15
6	-6	12
7	-12	12
8	-6	13
9	-	14

Figure 12 shows the original vehicle trajectory, the theoretical corrected trajectory to avoid the detected obstacles and the real one that the system follows in order to avoid them (crossing centered between them - example of a door-).

A test was also made of its ability to follow walls, and to skirt obstacles on the way towards a predetermined position. In all cases highly satisfactory results were obtained, proving the success of this option for the modelling of environments that call for the utmost precision in the guidance of mobile robots.

4. CONCLUSIONS

The infra-red method of robot guidance has been proven to provide a high precision in the determination of environments it moves through, and therefore a control over same. Measurement error will depend on the positioning relative to emitter and detector, and the quantification of the CCD sensor. As it has been shown, this error is always very small over short distances, so the position and size of obstacles can be detected with a high precision.

Once known the position of the obstacles preventing the robot from reaching its goal, and with its absolute position, it is relatively straightforward to devise new paths to reach the desired point.

ACKNOWLEDGEMENTS

This work was supported by the CICYT (Comisión Interministerial de Ciencia Y Tecnología) under the research project n° TTER96-1957-C0-01.

REFERENCES

Blais, F., M. Rioux and J.A. Beraldin (1988). Practical considerations for a design of a high precisión 3-D laser scanner system. *Opt. and Elec. Des. of Ind. Syst.* **SPIE 958**, 225-246.

Brooks, R. (1986) A robust layered control system for mobile robot. *IEEE Jour. Rob. Aut.,* **RA-2**, 14-23.

Evans, J.M., S.J. King and C.F.R. Weiman (1990). Visual navigation and Obstacle avoidance structured light system. *U.S. patent 4 954 962.*

Jarvis, R. A. (1983). A perspective on range finding techniques for computer vision. *IEEE Trans. on Patt. Anal. and Mach Intell,.***PAMI5 (2)**, 122-139.

Kemmotsu, K. And T. Kanade (1995). Uncertainty in objetc pose determination with three light-stripe range measurements. *IEEE Trans on Rob. and Aut,,* **11 (5)**, 741-747.

Khadroui, D., G. Motyl, P. Martinet, J. Gallice and F. Chaumet (1996). Visual servoing in robotics scheme using a camera laser stripe sensor. *IEEE Trans. on Rob. and Aut.,***12 (5)**, 743-750.

King, S.J. and C.F.R. Weiman (1990). HelpMate autonomuos mobile robot navigation system. *Mob. Rob.* **SPIE 1388**, 190-198.

Koch, E. (1985) Simulation of path planning for a system with vision and map updating. *Proc. IEEE Int. Conf. on Rob. Aut..* 465-475.

Krogh, B.H. and D. Feng (1989). Dynamic generation of subgoals for autonomous mobile robots using local feedback information. *IEEE Trans. on Aut. Cont.* **34 (5)**, 483-497.

Latombe, J.C. (1993). *Robot Motion Planning.* Kluwer Academics, Norwell, MS.

Motyl, G., F. Chaumet and J. Gallice (1994). Coupling a camera and laser stripe in sensor based control. *Proc. of Intell Sym. on Mea and Cont.***ISMCR-92.** 685-692.

Nitao, J.J. and A.M. Parodi (1986). A real-time reflexive pilot for an autonomous land vehicle. *IEEE Const. Sys. Mag.,* **6** 10-14.

Oomman, B.J. (1987). Robots navigation in unknown terrains using learned visibility graphs. *IEEE Jour. Rob. and Aut.,***RA-3**, 672-681.

Payton, D.W. (1986). An architecture for reflexive autonomous vehicle control. *Proc. IEEE Int. Conf. Rob. Aut.,* 1838-1845.

Sato, Y. and M. Otsuki (1993), Three-dimensional shape reconstruccion by active rangefinder. *Proc. of IEEE I063-6919,* 142-147.

Stein, G.P. (1993)*Internal camera calibration using rotation and geometric shapes.* Massachusetts institute of technology. MS.

Theodoracatos, V.E. and D Calkis (1993). A 3-D system model for automatics objects surface sensing. *Int. Jour. of Vis.,***11 (1)**, 75-99.

PRELIMINARY DESIGN STUDY OF A GUIDANCE SYSTEM BY PHOTOSENSORS FOR INTELLIGENT AUTONOMOUS VEHICLES

Ferreiro Garcia, R., Pardo Martinez, X. C. & Vidal Paz, J.

Dept. Electrónica e Sistemas. Universidade da Coruña
Facultade de Informática. Campus de Elviña, s/n. 15071. A Coruña. Spain.
E-mail: {ferreiro, pardo}@des.fi.udc.es

Abstract: In this paper a preliminary design study of a novel guidance system to be applied on IAV path control is presented. The fundamentals of this work are based in two optional sensorial system implementations: (a) A orientation detection system which detects the left-right orientation of the sensor base with respect to the path, based in photovoltaic geometrical surfaces. (b) A deviation angle sensor system capable for measuring the deviation angle of an IAV from its reference path by means of a photovoltaic servo-sensor which find the angular direction of a reflected beam from a source light in one degree of freedom. The path to be followed consist optionally in a light band or in a narrow band of reflectant material on the ceiling and/or at the opposite surface, that is the floor.
Copyright © 1998 IFAC

Keywords: guidance control, intelligent autonomous vehicles, path scheduling, photovoltaic sensors, sequential function chart.

1. INTRODUCTION

The need for a manipulator or robotic support system in order to solve the transportation problem under any controlled paths is growing. This growth is driven by several factors such as safety assurance, risk reduction, or simply its own working needs. The typical autonomous vehicle associated to a robotic manipulator must be able to perform surveillance, quantitative inspection, repairs upgrading eventual dismantling for decommissioning tasks or simply transport spare parts to the repair place. In all this tasks, a reliable guidance system is strongly demanded.

The starting point has been generally remote operations controlled directly by the operator. Telerobot, telemanipulator, teleoperator and teleoperated manipulator are all terms used to describe a remote controlled system. Such systems are required to perform some type of positioning or mechanical manipulation under real time direction of a human operator. All mentioned tasks are performed nowadays in absence or without the interaction of human operator (Ghoshray and Yen, 1994; Vazquez

et al., 1994; Wane, 1994). Such is an autonomous vehicle in which the guidance task play an important roll. The typical guidance systems are based in artificial pattern recognition (vision), acoustic sensorial system (sonars) and acoustic and photovoltaic detectors mainly in obstacle avoidance tasks. The concept of orientation based in photovoltaic geometric surface sensors is introduced to be applied as the guidance sensorial system

2. PHOTOVOLTAIC SENSOR ARCHITECTURE

Optical position sensors utilize optical reflection properties of materials (Roberts, 1965). In some cases the light source and light detector is housed in the same enclosure in a certain geometry so that the beam emitted by the light source gets reflected from the target material and sensed by the light detector. The geometry is adjusted in such a way that whenever the object is at certain distance the presence or absence of target object is sensed for the case of light detection. In the researched case our problem is to detect the centre of gravity of a light beam orientation.

There are three basic approaches available to implement light sensors which could be useful to be applied in IAV guidance control

(a) Reflectant photovoltaic orientation sensor

A reflectant surface based light sensor, shown at figure 1 in which a light beam is projected on a reflectant surface located in opposition to the original source. The orientation of reflected light is being captured by the photovoltaic light sensor located in same symmetry plane as light projector. The orientation or deviation angle from a reference centre-line can't be determined.

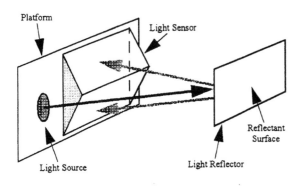

Fig. 1. The effect Source-Reflector-Sensor for one DOF.

(b) Photovoltaic orientation sensor

A direct based light sensor shown at figure 2, in which a light source beam orientation is captured but deviation angle is not quantified.

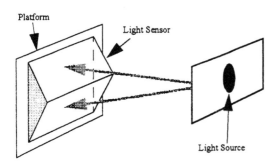

Fig. 2. The effect Source-Sensor for one DOF.

(c) Photovoltaic Servo-sensor

The proposed sensor is based in a combination of one or two pairs of flat photovoltaic cells disposed as shown in figure 3. A servo-system measure the angle between the light source and initial position of light sensor.

The servo-sensor search for a position where the potential is zero which occurs when the light source is oriented toward the sensor, due to the equal potential produced by every cell. The angle of rotation necessary to get zero differential potential is the orientation of sensor platform with respect to the light-sensor axis of symmetry. This servoing method is based on the principle that the spatial relationship between a vehicle and a target object (the light source) can be derived by using active detection to compute the position and orientation of the vehicle respect to the fixed target or alternatively with respect to another vehicle in which a light source is placed as target object.

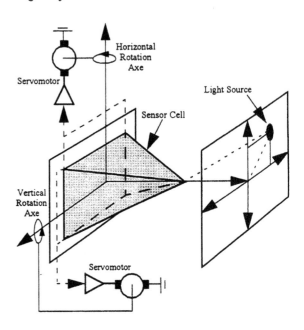

Fig. 3. The photovoltaic angle servo-sensor for two DOF.

This servoing sensorial method needs only approximate calibration, it can deal with unpredicted dynamic object motion. In figure 4 it is shown the block diagram of the servo-sensor to get the orientation of a light source with respect to the rotation in both, the horizontal and the vertical planes, that is in two degrees of freedom.

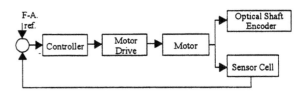

Fig. 4. Servo-Sensor block diagram for one DOF.

3. GUIDANCE SCHEDULING

Path planning is a task which depends strongly on the strategies taken into account when designing the combination of both the vehicle and its applicable environment (Ghoshray and Yen, 1994; Vazquez et al., 1994; Wane, 1994). The proper path to go from an initial point (A) to a target point (H) is selected from the complete network, shown at figure 5(b). The links between nodes A-H are scheduled so that an optimization criterion (minimum time) is applied by means of Bellman's algorithm (Bellman, 1957) which does not exclude another optimization algorithm (Bertsekas, 1987; Kouvatsos and Othman, 1989).

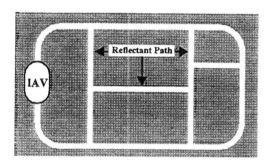

(a)

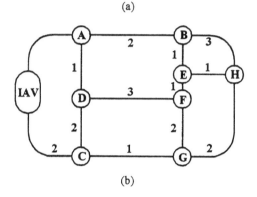

(b)

Fig. 5. (a) The path delimited by a narrow reflectant band. (b) The traffic network.

Two strategies can be applied about path planning, depending on the proposed type of sensor used:

Case 1: Path planning using a reflectant surface based light sensor

In this case the sensor shown at figure 1, is installed in a IAV that must travel on the network described by figure 5(a), the network is delimited by a reflectant band, so that the IAV can follow every link of such network. The guidance scheduling task is carried out by solving an optimization algorithm (the Bellman's algorithm in this case) and consequent decision - making strategy based in the application of an SFC (sequential function chart) (IEC, 848). The algorithm is described as follows:

Given a call from an external agent that request to travel from starting point (node A) to target place (node H), the following steps must be processed:

(a) determining the optimal path

Find the optimal path by applying an optimization algorithm (the Bellman's algorithm is appropriated)

under a performance criterion based in minimum time. As consequence of optimization procedure, a series of links between initial and end nodes are achieved. For this example the optimal path is designed by AB, BE, EH

(b) Decision making strategy through the nodes

Enter to a data-base to retrieve the nodes description, that is the information that tell us if, when crossing a node, it is necessary to turn on the right, on the left or not to turn at all in order to follow optimal path. The data-base structure supports the nodes descriptions from its antecedents and consequents In tables 1 and 2 it is shown the nodes description form for two arbitrary nodes. The database will holds the global network information.

Table 1: Description of node A

Precedence	Destination	Orientation
CA	AB	
	AD	R
DA	AC	L
	AB	R
BA	AC	
	AD	L

Table 2: Description of node B

Precedence	Destination	Orientation
AB	BH	
	BE	R
EB	BA	L
	BH	R
HB	BE	L
	BA	

For the case of the proposed example, data-base information about optimal path will be scheduled according table I and figure 5(b) such that turning orientation is for link AB no deviation, link BE right side and link EH left side.

(c) Implementation of path scheduling algorithm

SFC is the tool which fits well to implement the decision making task under optimum path scheduling. Figure 6 shows the SFC for above example.

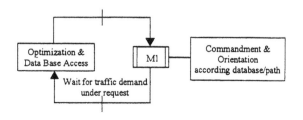

Fig. 6. Scheduling the guidance algorithm.

Case 2: Path planning using a direct incidence based light sensor

For the second case, the network is delimited by a light shining band, so that the IAV can follow every link of such network under the condition that the path to be followed is shining. The guidance scheduling task is carried out by selecting and switching on the lights corresponding to the links that conform the optimal path into the network. The proper algorithm is described as follows:

The sequence of operations to be carried out under an existing traffic motion request is,

(a) determining the optimal path

Find the optimal path by applying an optimization algorithm (the Bellman's algorithm is appropriated) under a performance criterion based in minimum time. As consequence of optimization procedure, a series of links between initial and end nodes are achieved. For this example the optimal path is designed by AB, BE, EH

(b) decision making strategy

Switch on the links between nodes achieved by applying the optimization algorithm as delimiters of the optimal path

(c) start the motion of the IAV till the end of light path appears, stopping the IAV motion.

4. RESULTS AND CONCLUSIONS

Proposed sensorial system architecture has been proved experimentally with satisfactory results in environments free of light disturbances. Light based sensing systems, when used in contaminated environments (diffusive emitted light and /or shining disturbant lights) needs more sophisticated measuring equipment, at least a combination of several measuring principles. Future work is needed to validate experimentally the performance of a complete set (sensors/actuators) because of the possibility of many different driving systems and dynamic behaviour of supported vehicles.

REFERENCES

Bellman, R. E. (1957). *Dynamic Programming*. Priceton University Press, New Jork, USA.

Bertsekas, D. P. (1987). *Dynamic programming: Deterministic and Stochastic Models*. Prentice-Hall, Englewood Cliffs, USA.

Ghoshray, S., Yen, K. K. (1994) Collision free path planning for robots with prismatic joints. In: *1994 IEEE International Conference on Systems, Man and Cybernetics*. Vol II, pp 1639-1644.

IEC publication 848, (1988). *Preparation of function charts for control systems*. (Atar S.A. (first ed.)) Geneva, Switzerland.

Kouvatsos, D. D. and Othman, A. T. (1989). Optimal Flow Control of a General End-to-End Packet Switched Network. In: *Proc. of the IMACS 1989*. (Ed. Tzafestas). Elsevier Science Publishers B.V., North Holland.

Roberts, L.G. (1965). *Machine perception of three dimensional solids. Optical and electro-optical information processing*. MIT Press Cambridge, MA, pp 159-197.

Vazquez, V., Sossa-Azuela, G. and Diaz de Leon Santiago, J. L. (1994). Auto Guided Vehicle Control Using Expanded Time B-Splines. In: *1994 IEEE International Conference on Systems, Man and Cybernetics*. Vol. III, pp 2786-2791.

Wane, S. (1994). Navigation System for a Mobile Robot in an Unstructured Environment. In: *1994 IEEE International Conference on Systems, Man and Cybernetics*. Vol. III, pp. 2797-2802

FIELD THEORY BASED NAVIGATION FOR AUTONOMOUS MOBILE MACHINES

E. E. Kadar[*] and G. S. Virk[†]

*Department of Psychology, † Department of Electrical and Electronic Engineering,
University of Portsmouth, Portsmouth, Hampshire PO1 3DJ, UK.*

Abstract: The paper discusses the role of field theory in the navigation of autonomous agents. Although the current forms of field theory based implementations have serious limitations the method is one of the most promising for providing a generic and robust methodology. The aim here is to remedy the shortcomings and provide a case study in which a biased random walk strategy is compared with a standard chemotaxis method in a chemical gradient field. Computer simulations are presented which demonstrate that the seemingly inefficient biased random walk is the better strategy under naturally unstable conditions. *Copyright © 1998 IFAC*

Keywords: Field theory, autonomous agents, navigation, mobile robots, biased random walking.

1. INTRODUCTION

Despite all the progress in Artificial Intelligence, the remarkable plasticity of human intelligence, a characteristic feature of human cognitive processes and sensory-motor performances, remains one of the most puzzling and challenging problems. Plasticity can appear in many different forms but insensitivity to initial conditions and the remarkable tolerance to perturbation in goal directed behaviour is probably the most fundamental. Plasticity is characteristic in humans as well as other biological systems. Biological creatures are goal directed systems "designed" by the survival-of-the-fittest rules of evolution and they are able to cope with the intrinsic variability and uncertainty in natural settings. In contrast, researchers in AI typically fail to achieve similar flexibility in their scientifically engineered solutions and, in trying to mimic biological systems, scientists continue to seek optimal solutions usually at the expense of flexibility.

The present paper addresses the problems of optimality and flexibility of autonomous navigation in a diffusion field. A generic field theoretical approach to navigation is outlined and related to the remarkable plasticity of animal locomotion in a chemical diffusion field (Kadar, 1996). It is believed that animals use chemotaxis in this elementary form of goal directed behaviour (Koshland, 1980; Berg, & Brown, 1972; Beer, 1990). Chemotaxis is a mechanism to generate a movement trajectory ideally connecting an initial position to the source of a diffusion field. This strategy can, in principle, be optimal if the diffusion field is stationary and free of environmental perturbations. However, if natural instability of a chemical diffusion field is considered, the seemingly primitive and inefficient strategy of bacteria, the biased random walk, is found to be superior.

To clarify these points, first a generic field theory based method for navigation is presented, then a navigational strategy for animals in diffusion fields is discussed, and, lastly, a case study with computer simulations demonstrates the superiority of biased random walking to chemotaxis.

2. THE CONJUGATE FIELD APPROACH

In psychology and AI, field theory has been used to describe navigation by several researchers. In psychology, Lewin (1936) described the motion of an actor in an environment by using the field of actual paths. In a similar vein, Gibson and Crooks (1938), postulated a so-called "Field of Safe Travel" in car driving, as all the possible "safe paths" a car driver can take in a particular traffic situation. These approaches, however, did not accommodate the possible use of perceptual information in path selection. Gibson's (1986) optic flow field was the first field theoretic approach to visual perception but the method was not linked directly to the field of paths.

In robot navigation, several researchers have already used field theoretical methods (see for example Krogh, 1984; Payton, 1990). Most of these attempts are based on a representation of the environmental layout that is used to design an optimal movement trajectory from an initial position to a target location. Although representational approaches are useful in finding global optimal paths in specific scenarios, their use in practise is limited due to unrealistic assumptions. For instance, they require that the environmental layout be either fed into the robot in advance or a cognitive map be learned by an exploratory navigation process. Furthermore, representational approaches are very sensitive to changes in the layout forcing robot navigators to include on-line instructors or built-in heuristics to continuously revise and update their inner models.

Humans and animals probably use other strategies because under natural circumstances they need to navigate mostly in an unknown or continuously changing environment. Different animals use different perceptual modalities. Moreover, many species including humans have more than one perceptual modalities which are used either simultaneously or selectively. Kadar (1996) has shown that navigation in a cluttered environment can be described by a core of fundamental field equations for all perceptual modalities. This core field can be construed from the information field, viewed as a gradient field, and the commonly used "taxis" behaviour of animals. It is well known, that animals have developed various mechanisms (phototaxis, geotaxis, etc.) by which they can move toward a target using some form of external stimulus (a light source, centre of gravity, etc.). It is also believed that navigation towards a source of chemical substance is possibly guided by chemotaxis.

The various forms of taxis behaviour are basically embodiments of the gradient technique commonly used in mathematics and applied to physical problems. Perhaps the approach can be best illustrated by the motion of a ball as it rolls down the slope of a hill. In this simple process the potential energy of the ball is converted into kinetic energy, whereas in animal navigation, the metabolic on-board energy potential is converted into energy used for the locomotion. A common feature in these techniques is that the trajectories of motion are orthogonal (conjugate) to specific fields if the momentum is ignored. In the case of the rolling ball example, the field represented by the height of the surface points given by the isocontours is well known in geography and can be the information field (see Figure 1(a)).

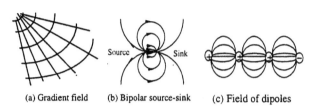

(a) Gradient field   (b) Bipolar source-sink   (c) Field of dipoles

Figure 1: Field theory based approaches

In the navigation of an animal the concentration levels of the diffusion field (of a chemical substance) can play the role of the information field. These information fields generate orthogonal conjugate fields of all possible motion trajectories and the pairs of fields can be described by a common Laplacian differential equation with the proper boundary conditions (e.g., $Lv = 0$). If the sources of an information field are included in the domain of motion then Poisson's equations (for example $Lv = S$) are needed. Kadar (1996) revised these equations in the context of animal navigation by reinterpreting the information sources as navigational sinks (the goals to be reached) and introducing the initial positions of the animals as navigational sources. In this way a simple ballistic motion can be described as a bipolar field of a source-sink pair as shown in Figure 1(b). These dipoles pairs can be concatenated as shown in Figure 1(c) to provide a basis to further generalise this approach to non-ballistic motions. Although animals often use ballistic movement (e.g. jump, dive, glide, etc.) their motion is typically periodic in carrying out the navigation. Animals propel themselves by converting metabolic energy in quanta to overcome the resistance of the medium (e.g., in air they flap their wings, in water they use fins or rotate cilia, for terrestrial locomotion limbs are used for hopping, running, walking, etc.) The field theoretic model of these rhythmic movements can be built up from bipolar fields by concatenating them in such a way that sinks and sources are replaced by sink-source dipole fields. The dipole singularities can be smoothed out and Poisson's equation has to be replaced by the Helmholtzian one, such as $Lv + v = 0$.

These technical details are seemingly light years from the simple bacteria navigation considered in section 3 but it will be shown that the very same formal tools can be used for understanding the navigational skill of bacteria, and, indeed, can be extended to autonomous machines.

3. NAVIGATION IN A DIFFUSION FIELD

It is well known that parabolic differential equations (such as $\partial u/\partial t = Lu$) with linear time derivatives can be used to describe heat and chemical diffusion processes. It is the chemical diffusion processes which are the most commonly encountered by animals and the lowest level in the hierarchy of species, bacteria use chemical substances and associated gradient fields as aids to navigate. Recognition of this has led to researchers investigating bacteria's perceptual and control processes in response to chemical gradients. An overwhelming amount of data has been accumulated that reveals the basic mechanism underlying bacteria's navigational skills (Koshland, 1980; Lackie, 1986) and there are numerous experimental results on the perception of chemicals by higher order species when looking at their actions (Kleerekoper, 1969; Papi, 1992).

From the perspective of conjugate field theory, the form of the field equation must be investigated with regard to perception and action. The diffusion equation has a first order temporal partial derivative. Since this is the only time-dependent component in the diffusion equation, it provides a deviation from the stationary core which is dependent on spatial co-ordinates only (Laplacian, or Helmholtzian equations). One can apply the Fourier method which is often used in solving partial differential equations by seeking a solution in the form of a product, such as $u(x, y, t) = v(x, y) e^{-t}$. It is possible to substitute this form of solution into the diffusion equation and, by taking the time derivative, the equation $e^{-t} Lv + v e^{-t} = 0$ is obtained. Since $e^{-t} \neq 0$ it follows that $Lv + v = 0$.

Thus, a Helmholtz equation is revealed as the core of the parabolic equation. It follows that the time dependent term is an exponential function and is "parasitic" on the Helmholtz equation. If the conjugate field approach is to be valid for navigation in a chemical gradient field that is construed as an information field, then there must be a corresponding process in the control field of action to cope with this temporal exponential term. What can bacteria (and other species with olfactory mechanisms) do to match this exponential time-dependent term in the diffusion equation?

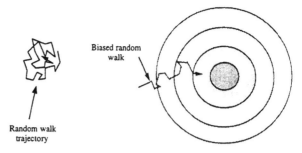

(a) RW in a homogenous field   (b) RW biased in a gradient field

Figure 2: Bacteria navigation in a chemical field

In fact, bacteria provide a simple case which we consider here (see Figure 2); in navigating toward a nutritious substance, bacteria use a biased random walk strategy (Koshland, 1980; Lackie, 1986). Here *Escherichia coli* (E. coli) bacteria provides the prototype of our analysis. Although there are other types of bacteria with different motility (e.g., *Spirochaeta aurantia* can run forward and backward and can flex (Fosnaugh & Greenberg, 1988]), functionally their movement control is the same. For *Escherichia coli* the movement trajectory is shaped by two different modes of motion, running straight ahead (translation) or tumbling (rotation). Bacteria use these two fundamental modes of the motion in an EITHER-OR strategy that creates a zigzag shaped trajectory. Without bias, the pattern would be an aimless random walk. If the length of straight movement segments (the steps or runs) depends on the chemical content of the medium, direction can be biased by the chemical gradient, thereby allowing the bacteria to move toward the source of a nutritious substance. Although this technique may not be the most efficient from an external perspective since there are lots of short detours, the strategy does have major advantages in that it can support navigation in weak gradient fields and is insensitive to external perturbations.

From the perspective of the conjugate-field approach, the most important aspect of the bacterium's strategy is that movement control is a biased random walk. A random walk process can be viewed as a diffusive control process. Navigation of the bacterium in the presence of a chemical gradient can be viewed as two coupled diffusive processes resulting in a biased random walk. There are good reasons for using the biased random walk strategy for navigation. First of all the intrinsic instability of the diffusion field can make the chemotaxis useless. Chemical gradients are usually weak and easily disturbed by other processes (e.g., wind, fluid current) and so to use such a gradient for orientation towards the source, animals must be able to obtain a valid solution in the presence of perturbations. Also, for chemotaxis two distinct but simultaneous measurements are needed and in a weak field, the two measurements (sensors) have to be sufficiently far from each other, so that a meaningful decision can be made as to which direction to move.

The proposed theoretical framework suggests that the temporal derivative of the diffusion equation is complemented by a time-dependent random factor in the control of the movement. Because detection of these diffusion processes is influenced by intensity changes in diffusion gradients and internal adaptation processes, stationarity (i.e., temporal invariance) can be achieved by balancing these two processes. Consequently, the fundamental feature of the navigational field of the bacteria (and also postulated for other species) is the conjugate portion of the diffusion equation.

Perception of chemicals and orientation via olfaction play important roles in higher order species as well (e.g., insects, birds, fish, etc.). Previous research projects on the use of chemical gradients for navigation do not provide clear-cut evidence to support a common mechanism for perceptual control because higher order species have multiple perceptual modalities that can interfere with olfactory navigation. Nevertheless, from the arguments presented it is plausible to believe that higher order species also use the same types of strategies. For example, dogs move their heads around when they are sniffing to find food and humans do the same if they need to find something smelly in a room. These obvious observations can be rigorously investigated in the future, but the fundamental tenet of this paper is to demonstrate that the theoretical arguments can lead to applications in robotics. The following case study tests and validates the proposed theory by computer simulations and will be tested on actual robotic platforms in the near future.

4. SIMULATION STUDIES

To demonstrate the key points of the proposed field theory computer simulations of navigation were used with four different conditions, namely chemotaxis and biased random walking in stable and noisy diffusion fields. Specifically, the four conditions of navigation are as follows:

(a)   chemotaxis in a stable diffusion field;

(b)   chemotaxis in a noisy (unstable) diffusion field;

(c)   biased random walking in a stable diffusion field; and

(d)   biased random walking in a noisy (unstable) diffusion field.

In each condition five trials have been carried out with the same initial position and target (where the chemical attractant is located). To be more exact

the navigator has to find its way from position (4,3) to position (0,0) using the field strength of the diffusion process at any given point along the movement trajectory. The trials of all the cases are depicted on Figures 3-6. In order to carry out the simulations, the following assumptions were used:

(i)   The navigator is a 0.4 unit wide creature and it has three chemical sensors. One is in the centre and two others are at its two ends. The central sensor is used during biased random walking and the end two are used for chemotaxis.

(ii)   The field strength is assumed to be inversely proportional to the square of the distance from the source.

(iii)   The movement is quantified under all conditions. The basic step size is 0.5 unit. During chemotaxis this step size is fixed, but for biased random walking there are three concentric regions around the chemical source where the step sizes are defined differently. There are several methods of selecting the step size, but for convenience we assume that beyond the distance of 10 units the step size is *0.5* and when the navigator is closer then 0.1 unit to the target the step size is the inverse of the field strength (*1/f*). In between these two regions, the step length is defined by a simple formula: *0.5 + f/2*, where *f* is the measured field strength.

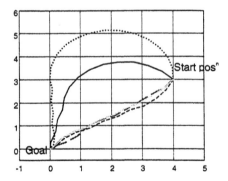

Figure 3: Navigation by chemotaxis in a stable field

During chemotaxis the rotation towards the target in each step is either 0 or 0.25 radians. It is small relative to the step size of 0.5 and this slow steering explains the curvilinear trajectories shown in Figure 3. A uniform random number generator R(-0.5,0.5) was used for both the direction selection in the random walking and in the noise creation to perturb the diffusion field.

The graphs of the four conditions in Figures 3-6 demonstrate the hypothesis proposed in this paper; chemotaxis seems optimal under stable conditions with rapid convergence but slow steering can generate long detours.

The number of iterations needed to reach the target using chemotaxis was only 14, 11, 20, 11 and 11 for the five runs in stable field conditions. However, Figure 4 shows that the chance of finding the source in a noisy diffusion field is rather slim and none of the five trials were successful - that is, using 1000 steps per trial were not sufficient to find the target. In contrast, the biased random walking strategy was successful under both stable and noisy field conditions although on average more iterations were needed (the exact number needed appear to be insensitive to noise level). The number of steps needed to reach the target using a biased random walking strategy was found to be 126, 130, 128, 126 and 140 under stable fields conditions and 135, 139, 127, 129 and 116 in unstable diffusion fields.

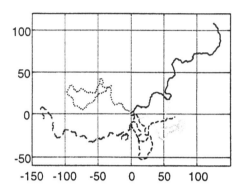

Figure 4: Navigation by chemotaxis in a noisy field

Furthermore, one can observe that the motion trajectories illustrate the basic characteristic of the biased random walking strategy as implemented in this study. In the neighbourhood of the target the bias is getting more pronounced showing that increasing concentration changes biased random walking in such a way that it becomes approximately a chemotaxis.

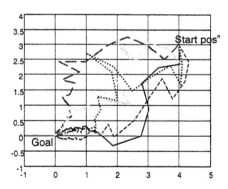

Figure 5: Biased RW navigation in a stable field

These simulations demonstrate that the biased random walking approach can form the basis for a robust and flexible navigational strategy. Therefore, it is natural to conclude that, in general, chemotaxis is inferior to biased random walking even though under specific conditions, such as in stable

diffusion fields and/or high concentration levels it can also be used with success.

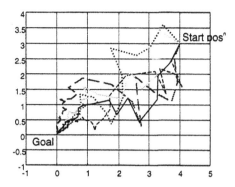

Figure 6: Biased RW navigation in a noisy field

5. CONCLUSIONS

The paper has investigated the potential use of a new field theoretic approach for autonomous navigation proposed by Kadar (1996). Potential fields are used in robust navigation, but they have various shortcomings. In particular, the optimality and plasticity of navigation in a chemical diffusion field has been discussed. The proposed conjugate field approach suggested that a biased random walking type of strategy is superior to chemotaxis (in terms of flexibility and robustness). The four conditions of computer simulations presented clearly demonstrated why biased random walking is a more flexible strategy in seeking a source of a chemical diffusion field. These findings are encouraging when designing robot navigators for possible industrial applications, where chemical and radio frequency signals can potentially be used in defining the field.

Other important points which need to be thoroughly investigated before a viable practical strategy can emerge include how the method performs with moving single and multiple targets and when the motion capabilities of the agents and targets are varied. These issues are currently being studied by the authors. In addition, the practical implementational aspects of such a solution for mobile robots have to be evaluated. For example mobile machines operate in unstructured and dynamic environments. Hence chemical diffusion fields are not sufficient for successful navigation on their own because they do not contain sufficient information about the environmental layout and solutions to address concerns such as obstacle avoidance need to be found. It is the authors intention to explore these aspects for practical robots in the near future.

6. REFERENCES

Beer, R. D. (1990). *Intelligence as adaptive behaviour: An experiment in computational neuroethology.* New York: Academic.

Berg, H. C., & Brown, D. A. (1972). Chemotaxis in Escherichia coli analysed by three-dimensional tracking. *Nature, 239*, 500-504.

Fosnaught, K. & Greenberg, E. P. (1988). Motility and chemotaxis of Spirochaeta aurantia: computer assisted motion analysis. *Journal of Bacteriology, 170*, 1768-1774.

Gibson, J. & Crooks, L. (1938/1982). A theoretical field-analysis of automobile-driving. In Reed, E., & R. Jones (Eds.), *Reasons for realism.* (pp. 120-136). Hillsdale, NJ: Lawrence Erlbaum & Associates.

Gibson, J. J. (1986). *The ecological approach to visual perception.* Hillsdale, NJ: Erlbaum. (Original work published 1979).

Kadar. E. E. (1996). *A field theoretic approach to the perceptual control of action.* Unpublished Dissertation. University of Connecticut.

Kleerekoper, H. (1969). *Olfaction in fishes.* Bloomington, IN: Indiana University Press.

Koshland, D. E. Jr. (1980). *Bacterial chemotaxis as a model behavioural system.* New York, NY: Raven Press.

Krogh, B. H. (1984). A generalised potential field approach to obstacle avoidance control. *International Robotics Research Conference.* Bethlehem. PA. August.

Lackie, J. M. (1986). *Cell movement and cell behaviour.* London: Allen & Unwin.

Lewin, K. (1938). *The conceptual representation and the measurement of psychological forces.* Durham, NC: Duke University Press.

Papi, F. (1992). *Animal homing.* New York: Chapman & Hall.

Payton, D. W. (1990). Internal plans: A representation for action resources. In Maes P. (Ed.). *Designing autonomous agents: Theory and practice from biology to engineering and back.* pp. 89-103. Brandford Books. MIT Press.

CURVATURE FUZZY MODELLING FOR A MOBILE ROBOT

J. Ruiz-Gomez, A. Garcia-Cerezo, R. Fernandez-Ramos

*Systems Engineering and Automation Department. University of Malaga. Spain.
ruizg@ctima.uma.es. Ph. 34-952132749. Fax 34-952133361*

Abstract: The automatic identification of the mobil robot RAM path curvature arising from an input-output data set is the aim of this work. The applied method comes from Takagi- Sugeno modelling and Mamdani inference methods, and generates the fuzzy rules that report a process behaviour by using an Artificial Intelligence classical automatic learnig algorithm as Quinlan ID3. The proposed method is implemented as a tool in C language, compatible with the Fuzzy Toolbox of MATLAB. *Copyright © 1998 IFAC*

Keywords: Fuzzy Modelling, Inductive Learning, Mobile Robot, Fuzzy Control, Non-linear systems, Identification.

1. INTRODUCTION

The aim of this work is to apply a software tool (Ruiz-Gomez and Garcia-Cerezo, 1997a) that implements a modelling method that by using Mamdani inference improves on intuition and also achieves an automatic knowledge acquisition from experimental data. This tool is considered to be useful to practitioners and people not expert in modelling who needs a simple way of describing the behaviour of some system.

In this paper the dynamical system to be modelled is the RAM mobile robot (Muñoz, V. *et al.*, 1994;

Fig. 1 RAM mobile robot.

Ollero *et al.*, 1992) shown in figure 1, developed at Malaga University for application on indoor and outdoor industrial environments.

Modelling, by means of fuzzy rules, a non linear process arising from a set of data that reflect past behaviour of a is an alternative approach to conventional I/O modelling of dynamic systems.
Fuzzy identification is adopted due to the difficulty of conventional mathematical modelling on certain systems, specially on non-linear ones, and due to the ability, settled by Wang (1992) and Buckley *et al.* (1993) to approximate any real function by means of a set of fuzzy rules.

Fuzzy identification has been treated from different view points: First approaches, based on Sugeno inference method, arise on 1985 (Takagi and Sugeno, 1985; Sugeno and Kang, 1986; Sugeno and Yasukawa, 1993). There are several inductive learning based methods: Sison and Chong (1994) approach applies ID3 to generate a fuzzy rulebase, and also allows to select the relevant inputs, though it does not build a sequence of models as the Sugeno approach; Batur *et al.* (1991) also use ID3 algorithm by means of a training data set, though the model is not a fuzzy one but a crisp rules model applied to discrete time systems; Tani *et al.* (1992) identify the premises structure and select the effective input variables by use ID3; Delgado and Gonzalez, (1993) work on an inductive learning process based on the frequency of appearance of certain patterns from raw data. Tejero and Sillero (1994) and Sillero (1997) worked with ID3 by using first not all the experimental data but a trial set and afterwards all the data. Other identification methods are based on I/O data approximation: García-Cerezo *et al.* (1996) use a least squares approximation technique; Abe y Lan (1995) developed a method for extracting fuzzy rules from numerical I/O data for pattern classification, and by activation hyperboxes defining input regions and a fuzzy inference net of four layers; Araki *et al.* (1991) proposed a modelling approach based on repeating parameter adjustment and monitoring an inference error, and Wang and Mendel (1991) combine fuzzy and numeric learning from numerical data; Harris (Mills and Harris, 1995; Bossley, Brown and Harris, 1995) has applied neurofuzzy modelling to underwater vehicles.

The paper layout is:after this introduction about fuzzy identification and related works, there is an overview on the basis needed for the proposed method (Takagi-Sugeno identification method and ID3 algorithm), followed by a description of the modelling method and two applications to an autonomous mobile robot modelling process.

2. MODELLING BASES: TAKAGI-SUGENO AND ID3 METHODS

One of the first fuzzy identification methods for dynamic systems was proposed by Takagi and Sugeno [16] based in Sugeno fuzzy inference method. Nevertheless the Mamdani inference method is more intuitive and easy to understand in natural language. And also is more suitable with the way of reasoning in fuzzy logic, more fitted for symbolic than numerical processing.

Rules, in Takagi-Sugeno approach are of the format:

$$\text{IF } x_1 \text{ is } \underline{A_1^l}, ..., \text{ and } x_k \text{ es } \underline{A_k^l}$$
$$\text{THEN } y = p_0^l + p_1^l * x_1 + ... + p_k^l * x_k \qquad (1)$$

The identification method tries to determine all elements: variables, x_ji ; fuzzy sets, $A_k^{l,}$, and consequent coefficients, p_m^n. The Takagi-Sugeno approach has some disadvantages that arise from the Sugeno inference method itself, as the consequent parameter computation accomplished by a Kalman filter. This estimation is restricted to consequents that are algebraic equations, as in Sugeno consequents, but can not be used for fuzzy consequents as Mamdani ones.

The first stage on this approach, the variables choosing process for rules premises and the partition of input ranges in fuzzy sets are not very structured, though Sugeno and Kang (1986) point out several possible criteria, as to use the system physical laws, heuristic knowledge or to choose highly correlated variables.

The second stage, the premises parameters identification, input spaces are divided into fuzzy sets and a tree to look for stables states is built. In the third stage, optimal consequent parameters, p_m^n are computed by a Kalman filter, a recursive algorithm for computing the parameters of an lineal algebraical equation. This identification process arises to several models generation, as several inputs ranges partitions are considered.

The approximation between the model and the set of experimental data is computed by means of a least squares index.

ID3 method was developed by Quinlan *et al.* (1979) as a classifying inductive learning method based on an entropy or information point of view. ID3 obtains rules from an initial set of examples described by objet-value-attribute. A classification process in the format of a decision tree can be constructed for any collection of objets. Each objet is described in terms of its attributes. An object is classified by means of a process that begins on the tree root node and ends in a all-of-the-same-class leaf. En each node, to choose the attribute that provides more information, Quinlan selects the attribute with the maximum information contents from all the experiences. To measure information content, ID3 uses entropy, approximating a class probability with relative frequency.

If the probability of message is p, the expected message information content is:

$$MI = -p \log_2 p \qquad (2)$$

And information gain on each level is:

$$\cdot I.G. = M(S) - B(S, A) \qquad (3)$$

where,

$$M(S) = -P \log_2 P - (1 - P) \log_2 (1 - P) \qquad (4)$$

M(S) is the measure of the information, P is the rate of objects belonging to the desired class, and B(S, A) the residual information of attribute A, expressed by:

$$B(S, A) = \sum_j \left(W_j * M(S) \right) \qquad (5)$$

where Wj is the rate of objects of every subset related to the original set before a node classification.

3. DATA FUZZIFICATION

As the measured data from a physical systems usually are discrete ones this data must be translated to fuzzy sets. Two important points on numeric-fuzzy conversion of data arise: the number of sets to split every variable range, and the way the intersection between two adjacent fuzzy sets is treated.

Every variable range has been divided in two equal sets, by the middle of the range value, with an overlap of 25 per cent between adjacent sets. Also it possible a second way that splits the variable range on the basis of critical points of the I/O curve (Ruiz-

Gomez and Garcia-Cerezo, 1997b) owing a derivative of zero value or not possessing derivative at all.
The second important feature is about how to select the frontier points to assign in every set. Here the Sison and Chong (1994) technique by means of crisp corresponding sets is followed.

4. RULE GENERATION

As ID3 is a learning from classes method, a classification process of all variables is needed. All numerical data are fuzzified. The raw data obtained by experimentation, give raise to a set of fuzzy experiences as these shown in the following table 1.

TABLE 1: Fuzzy Experiences

Experience #	Attribute x_1	Attribute x_2	Consequent y
1	vl	vl	vl
2	l	a	vl
3	g	a	g
4	a	l	m
5	...	...	..
...	..	...	...

a: all; vl: very little; l: little; m: medium; g; great;

Afterwards a link between "experiences" and rules must be established, to apply ID3 to a fuzzy rules generation process. As facts or "experiences" a set of data resulting in one measurement is considered, i.e. a row in a numerical data table. For every input variable, the consequents that reflect the action of every fuzzy set of the selected input are found.

The probability Pi is obtained dividing the number of experiences in the input related with a same consequent, by the total number of times the antecedent is in the table. After calculate the amount of information, a tree based on the attributes or input variables is buit. The tree development continues by the node with a greater entropy. In a leaf node all its elements belong to the same class. A model rule is given by the path from the root node to a leaf node. The sequence of models stops when the performances for a certain model achieves a desired value.
Model validation can be accomplished by an numeric index based on least mean squares error between experimental and generated by the model data, also by using an error histogram or last by viewing a 3D graphic of experimental and modelled surfaces.

5. CURVATURE MODELLING

This paragraph shows how the proposed method has been used to model a real system knowing only a table of measured I/O data. As a real system the dynamics of a mobile robot have been chosen. Input variables are the desired velocity and curvature values, and as output variable the actual path curvature has been taken, measured by taking the

diference between the lenght travelled by the left and right wheel and also the heading wheel angle.

Two models for this system have been developed: a first one based on a good approximation index, and a second looking for a minimum number of rules.

5.1 Fuzzy Model Based on the Best Error Index.

The aim is to obtain a very low error index, leaving free the number or rules, and by using the middle point range division method. Error index in this system can vay from 0 to 150. Let us suppose that a model with a low error index, and no matter how many rules, is wanted.

A model with a good error value (9.2733) has been obtained. The model has 9 rules.

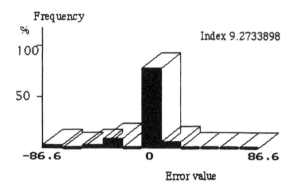

Fig. 2 First model error histogram.

On figure 2 about 90 % of deviations lay on sector of 0 value. The fuzzy sets resulting on partitioning the system inputs and output are shown in next fig. 3:

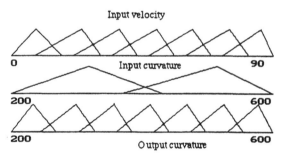

Fig.3 First model input an output fuzzy sets

The model report, Table 2 in Appendix, depicts all this first model features.

5.2 Fuzzy Model Minimizing Rule Number.

Looking for a low number of rules, a second model with only four rules have been obtained. But now the error index value have increased up to 19.249573. The index value is worse than in the first model as expected in a less rules model.

The software tool implemented also allows to confront experimental and modelled surfaces by using a 3-D plot. The report generated by the tool, on Table 3 in Appendix, shows all the important model features: error index, rules obtained and resulting fuzzy sets.

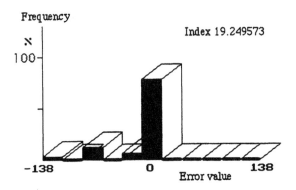

Fig. 4 Second model error histogram.

6. CONCLUDING REMARKS

This paper attempts to model the non-linear dynamics of an Autonomous Mobile Robot by a set of fuzzy rules by using a tool that implements an inductive learning technique.

The rules are induced from an experimental data set by the Quinlan ID3 algorithm and the performance of the models can be evaluated by a least squares index, an error histogram, and a graphic comparison between the experimental and the simulated model outputs.

As input data two variables have been taken, desired speed and curvature; and as output the actual measured curvature variable have been considered. But other input quantities could be chosen, e.g.: input voltages applied to left and right motors, sonar or laser information about the environment, and so on.

From the same data, two different fuzzy models have been achieved. The first is accomplished by minimising a least mean squares index, while on the second model a minimum amount of rules is looked for. The first model is a 9 rules one, and the second only 4. Both model get a good level of accuracy, as is shown in the results in Section 5, and on reports on Appendix.
This modelling technique does not require an explicit specification about the linearity or not linearity of the system.

An outlining feature is that no knowledge on modelling is required. The user only needs the software tool and a file of experimental raw data from the process to cope with modelling a dynamical process. These simple fuzzy models give an useful information for designing fuzzy controllers.

REFERENCES

Abe, S. and M-S Lan (1995). Fuzzy Rules Extraction Directly from Numerical Data for Function Approximation. *IEEE Trans. Syst., Man, Cybernetics*, **25**, No 1, 119-129.

Araki, S. et al.(1991). A Self-generation Method for Fuzzy Inference Rules. *Proc. IFES'91 Fuzzy Engineering towards Human Friendly Systems.* 1047-1058.

Batur, C., S Arvind and C-C. Chan (1991). Automated rule based model generation for uncertain complex dynamic systems. *Proc. IEEE Int. Symp. Intelligent Control*. Arlington. (VA). 275-279.

Bossley, K. M.; M. Brown and C. J. Harris (1995). Parsimonious Neurofuzzy Modelling. Techn. Report. University of Southampton. http://www.isis.ecs.soton.ac.uk/

Buckley, J. and Y. Hayashi (1993). Fuzzy input-output controllers are universal approximators. *Fuzzy Sets and Systems* **58**, 273-278.

Delgado, M. and A. González (1993). An Inductive learning procedure to identify fuzzy systems. *Fuzzy Sets and Systems* **55**, 121-132.

Garcia-Cerezo, A. J., M. J Lopez-Baldan and A. Mandow (1996). An Efficient Least Squares Fuzzy Modelling Method for Dynamic Systems. *Proceed. CESA'96. IMACS Multiconf. Symp. Modelling, Analysis and Simulation*. Lille. 885-890.

Mills, D. J. and C. J. Harris (1995). Neurofuzzy modelling and control of a six degree of freedom AUV. Techn. Report. University of Southampton. http://www.isis.ecs.soton.ac.uk/

Muñoz, V., A. Ollero, M. Prados and A. Simon (1994). Mobile Robot Trajectory Planning with Dynamic and Kinematic Constraints. *Proc. IEEE Int. Conf. Robotics and Automation*. San Diego (CA). May. 2802-2807.

Ollero, A., A. Simon, F. Garcia and V. Torres (1992). Integrated Mechanical Design and Modelling of a New Mobile Robot", *Preprints of the SICICA'92 IFAC Symposium on Intelligent Components and Instruments for Control Applications*, Malaga (Spain), May. 557-562.

Quinlan, J. R. et al. (1979). Discovering rules by induction from large collections of examples. In: Expert Systems in the microelectronic age (D. Michie, D., Ed.). Edinburgh University Press. Edinburgh.

Ruiz-Gomez, J. and A. Garcia-Cerezo (1997a). An Inductive Inference Tool for Fuzzy Modelling. *Proceed. IDEA'97. Intelligent Design in Engineering Application Symposium*, Aachen (Germany). Sept. 113-117.

Ruiz-Gomez, J. and A. Garcia-Cerezo (1997b). Inductive Learning Fuzzy Modelling. *Proceed. INCON'97. International Workshop on Intelligent Control*. Sofia (Bulgaria). Oct. 23-28.

Sison, L. S. and E. K-P Chong (1994). Fuzzy modeling by induction and pruning of decision trees. *Proc. 199 IEEE Int. Symp. Intelligent Control*. Columbus. (OH.). 166-171.

Sillero, J. L. (1997). *Modelado Borroso de Sistemas*. Master Thesis. ETS. Ing. Informatica. University of Malaga.

Sugeno, M. and G. T. Kang, (1986). Fuzzy modelling and control of multilayer incinerator. *Fuzzy Sets and Systems*, **18**. 329-346.

Sugeno, M. and T. Yasukawa (1993). A Fuzzy-Logic-Based Approach to Qualitative Modeling. IEEE Trans. Fuzzy Systems, 1, No 1. 7-31.

Takagi, T. and M. Sugeno (1985). Fuzzy Identification of Systems and Its Applications to Modeling and Control. *IEEE Trans. Syst., Man, Cybernetics*, **15**. 116-132.

Tani, T., M. Sakoda and K. Tanaka (1992). Fuzzy Modelling by ID3 algorithm and its application to prediction of a heather outlet temperature. *Proc. 1992 IEEE Int. Conf. Fuzzy Systems*, San Diego (CA), 923-931.

Tejero, V. and J. L. Sillero (1994). *Identificación de Sistemas con Control Borroso*. Master Thesis. E.U. Politécnica. Univ. de Málaga.

Wang, L-X. (1992). Fuzzy systems are universal approximators. *Proc. Int. Conf. Fuzzy Systems*, San Diego. 1163-1170.

Wang, Li-Xin and J. M. Mendel (1991). Generating fuzzy rules by learning from examples. *Proceed. 1991 IEEE Int. Symp. Intelligent Control*. Arlington.(VA). 263-269.

APPENDIX

TABLE 2. First model report.

Report: ifcurv74.inf

Model error index: 9.2733898

MFS Matrix (I/O fuzzy sets):
Input 1:

1	0	8.67	8.67	17.36
1	8.35	23.79	23.79	30.21
1	21.2	36.64	36.64	43.07
1	34.07	49.5	49.5	55.93
1	46.93	62.36	62.36	68.79
1	59.79	75.21	75.21	81.64
1	72.64	85.82	85.82	90

Input 2

1	200	316	316	432
1	368	516	516	600

Output:

1	200	234.6	234.6	269.1
1	245.1	297.7	297.7	326.3
1	302.3	354.9	354.9	383.4
1	359.4	412	412	440.6
1	416.	469.1	469.1	497.7
1	473.7	526.3	526.3	554.9
1	530.9	577.4	577.4	600

Model Rules (RLS Matrix):

9	10	10	10
10	1	1	1
10	1	2	4
10	2	1	1
10	2	1	2
10	3	1	1
10	3	1	2
10	3	2	4
10	3	2	7
10	7	0	4

TABLE 3 Second Model Report.

Report : ifcurv7.inf

Model error index: 19.249573

MATRIZ MFS (I/O fuzzy sets):
Input 1:

1	0	45	45	90

Input 2:

1	200	316	316	432
1	368	516	516	600

Output:

1	200	234.6	234.6	269.1
1	245.1	297.7	297.7	326.3
1	302.3	354.9	354.9	383.4
1	359.4	412	412	440.6
1	416.6	469.1	469.1	497.7
1	473.7	526.3	526.3	554.9
1	530.9	577.4	577.4	600

Model rules (RLS Matrix):

4	10	10	10
10	1	1	1
10	1	1	2
10	1	2	4
10	1	2	7

First column value 1 in inputs and output tables means the set in that row is active. The same meaning has values 10 in rules table.

EXTERNAL MODEL AND SYNCCHARTS DESCRIPTION
OF AN AUTOMOBILE CRUISE CONTROL

Bayart M.[#], Lemaire E.[#], Péraldi M-A.[*], André C.[*]

[#]*LAIL (URA CNRS 1440), Bât. P2, UFR IEEA,*
Université des Sciences et Technologies de Lille, 59655 Villeneuve d'Ascq Cedex - France,
Tel. : (33) 3 20 43 45 65 - FAX : (33) 3 20 33 71 89, e-mail : bayart@univ-lille1.fr
[*]*Laboratoire Informatique, Signaux, Systèmes (I3S)*
Université de Nice-Sophia Antipolis / CNRS, 41, bd Napoléon III - F - 06041 NICE Cedex
Tel. : (33) 4 93 21 79 56 - FAX: (33) 4 93 21 20 54, e-mail : {andre,map}@alto.unice.fr

Abstract : In this paper, the generic model developed for smart instruments and the synchronous model SYNCCHARTS are used to specify automated systems. The generic model provides us with an external description, which is the user's point of view, and the SYNCCHARTS model gives a behavioral model of the system. For those two models to provide complementary information upon a device, a method is proposed to obtain a coherent syncChart from a part of the external model. The classical example of an automobile speed cruise control system is used for illustration. *Copyright © 1998 IFAC*

Keywords : Cruise Control, Formal Specification, Intelligent instrumentation.

1. INTRODUCTION

Various models are used to describe automated systems : functional, behavioral, object-based, internal or external models (Robert, *et al*, 1993, Staroswiecki and Bayart, 1994). The external model, using the concept of service offered to users and an organization based on operating modes, has led to a generic model description in a formal language (Bouras, 1997) that allows to specify and to qualify smart instruments and hybrid systems (Bayart and Lemaire, 1997).

Adding a behavioral model like SYNCCHARTS (André, 1996) to the external model leads to a more complete description of the equipment and offers extended facilities for simulation and validation.
In this paper, we briefly present the external model concepts, followed by a short introduction to the SYNCCHARTS formalism. In the third part, we explain how to derive a SYNCCHARTS representation from a part of the external model in order to complete the equipment description. A cruise speed control system illustrates the proposed method.

2. THE EXTERNAL MODEL

The external model describes the device from the point of view of the services it is able to provide to external entities (operators, other field instruments, computers ...). It introduces the following notions (Staroswiecki and Bayart, 1994, 1996) :

A **service** is defined as a procedure whose execution results in the modification of at least one datum in the instrument data base, or/and at least one signal on its output interface.
Services are required by the **users** who intervene on the equipment during its whole life cycle, i.e. not only during its exploitation (supervision, maintenance, technical management) but through out its life cycle, from its conception to its dismantling (initialization, configuration,...).
The description of a service consists in the description of the result, which is produced by its execution, i.e. the outputs.

In order to define the obtained values, one will have to describe the computations which are done (algorithmic or sequential procedures, qualitative or fuzzy inferences,...), the variables on which they are applied (inputs) and the required resources (hardware, software). Moreover, before it can be executed, a service must verify some activation conditions. So, a service is described by :

<Service>::=<Inputs, Outputs, Procedure, Activation Condition, Resources>

According to the resources state whose estimation is given by the Fault Detection and Isolation (FDI) algorithms that are implemented in the intelligent instrument, several versions of services (nominal and degraded) may be designed.

The services executions can be either dependent (precedence, mutual exclusion,...) or independent and concurrent. Likewise, the service can have a limited duration or can end on the occurrence of simple or complex events (operator request, emergency alarm,...). Finally, the services are organized according to user operating modes.

A **User Operating Mode** (USOM) is a coherent sub-set of services. It contains at least one service, and each service belongs at least to one USOM. Moreover, in each USOM, there exists a notion of context. The context is the subset of services of the USOM that are implicitly executed (implicit request) as long as the system remains in the given USOM. The other services of the USOM are the (requestable) services.

From this description one can obtain a formal specification of the intelligent equipment. However, the external model is usually not sufficient to validate the instrument behavior. So, we propose to use the SYNCCHARTS formalism in order to complete and to be able to simulate and to validate the equipment running before realization.

3. THE SYNCCHARTS FORMALISM

"SYNCCHARTS" is an acronym for Synchronous Charts. SYNCCHARTS inherit from STATECHARTS (Harel, 1987) and ARGOS (Maraninchi, 1990). They are a new graphical representation of reactive behaviors based on the synchronous paradigm. They offer enhanced preemption capabilities and any syncChart can be translated into an equivalent Esterel program. Recall that Esterel (Boussinot and De Simone, 1991) is a powerful imperative synchronous language dedicated to reactive system programming.

Let us have a look at the hypotheses underlying the synchronous approach (André, 1996) :
- **Signals** : the system interacts with its environment through signals : input signals and output signals.
- **Global Perception** : we assume that all input and output signals are perceived simultaneously and that

this perception is objective. So the model deals with tuples of signals.
- **Logical Time** : there is no physical time, but logical instants.
- **Zero-Delay Hypothesis** : internal operations are supposed to be executed in zero-delay. So the output signals are synchronous with the inputs that cause them.
- **Broadcasting** : all signals are instantaneously broadcast.

In SYNCCHARTS, preemption is a first class concept. There are two types of preemption : suspension and abortion. Abortion can be either weak or strong : the strong one kills the process immediately and the weak one kills the process after it executes its current reaction.

SYNCCHARTS, like Esterel, deals with sequence, concurrency, preemption and communication in a fully deterministic way.

SYNCCHARTS is endowed with a mathematically defined semantics, fully compatible with the ESTEREL one. The main differences between SYNCCHARTS and STATECHARTS are a stricter semantics and a richer preemption management for the former.

Finally, SYNCCHARTS have their graphical representation. It is a state-based description of a reactive behaviors. It supports states, hierarchy of states, concurrency and transitions of several types. The basic block is the state or **star** (Fig.1) :

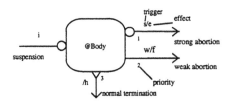

Figure 1 : The star of a syncChart.

Stars are interconnected to make a **constellation**. Only one star at a time can be active in a constellation, so that a constellation can be seen as a classical state-graph.

Each constellation has, at least, one initial star pointed to by an arrow. A parallel composition of constellations is a firmament or a **macro-state**. Concurrent constellations in a macro-state are delimited by dashed lines.

The structure of a syncChart is defined recursively: the body of a star can be a macro-state (or even an Esterel module, i.e., a textual description). This feature is convenient for supporting hierarchical descriptions.

A prototype SYNCCHARTS editor/compiler is now available[1], a commercial version is under

---

[1] http ://alto.unice.fr/~andre

development. The SYNCCHARTS compiler generates an Esterel program equivalent to the syncChart. Thus, SYNCCHARTS can use the rich software environment developed for Esterel (compilers, links to proof systems, interactive simulation, efficient code generation).

4. EXTERNAL MODEL TO SYNCCHARTS DESCRIPTION

4.1. General method.

We develop a method to obtain a SYNCCHARTS description from the external model.

First of all, each USOM of the external model is defined as a macro-state. In each of those macro-states, the different constellations are formed by the different services present in the USOM. In this case, all the services are concurrent. The transitions from USOMs to USOMs are still the same in terms of syntax. However, the designer must specify if the transition is a weak abortion, a strong one or a normal termination. Finally, the suspension within each USOM for each service must be established according to the given specifications.

The obtained representation could be the final one. However, due to the large number of services in complex systems, the representation might be illegible. In that sense, we propose a method to obtain a clearer internal representation. This method identifies subsets of services, which might be grouped in a same macro-state.

The first step of the method is to find the USOMs, which can be grouped according to the information provided by the external model.
Let M be the set of the user operating modes.
$$M = \{m_j ; j \in J\}$$

The set of USOMs is now augmented with the set S_{cm} of transition conditions. Let T be the set of transitions.
$$T = \{(m_i, t_{ij}, m_j) / m_i \in M, m_j \in M, t_{ij} \in S_{cm}\}$$
t_{ij} indicates the logical condition required by the change from m_i to m_j.

Table 1 shows the activation condition between USOMs.
Table 1 : activation condition between USOMs.

USOM \ USOM	m_1	...	m_j	...
m_1	ϕ			
...				
m_i			t_{ij}	
...				

From this table, we deduce the USOMs that can be grouped. The "grouping condition" of two modes m_i and m_j expresses the fact that the resulting aggregated model should remain deterministic. For each entry in Table 1, one has to check :

$\forall m_i \in M, m_j \in M, m_{ij}$ can be formed if and only if the following conditions hold :
- $\forall k \in J / t_{ik} \neq \phi, t_{jk} \neq \phi, t_{ik} = t_{jk}$,
- $\forall k \in J, \forall l \in J / k \neq l, \quad t_{ik} \wedge t_{jl} = 0$
$$\text{and} \quad t_{il} \wedge t_{jk} = 0.$$

Let K be the set of all the possible pairs m_{ij} of USOMs $\{m_i, m_j\}$.

The syncChart legibility is increased if a macro-state contains USOMs that possess common services since those services will be expressed only once in the syncChart. So, the set K of candidate pairs for aggregation can be ordered according to the number of common services they exhibit.

Let S be the set of services that the equipment can perform.
$$S = \{s_1, s_2, ..., s_n\}$$

Let Ls be the application, which associates to a USOM, the set of services, which are offered to users in this mode. P(S) is the set of subset of S.
$$Ls : \quad M \to P(S),$$
$$m_j \to Ls(m_j).$$

So, the best macro-state in the set K obviously corresponds to the pair m_{ij}, which shows the largest number of common services. One can validate the pair m_{ij} if :

$$|Ls(m_i) \cap Ls(m_j)| \geq |Ls(m_k) \cap Ls(m_l)|, \forall m_{kl} \in K$$

This application is a transitive one, so, more than two USOMs can be grouped if they verify, two by two, the above conditions. The new-formed macro-state can be used as a "new" USOM to obtain a hierarchical representation well-suited to the recursive definition of a syncChart.

For each obtained group of USOMs, the common services are then put as orthogonal constellations in the SYNCCHARTS formalism. The resulting transition is built with the expression, which includes all the transitions. Finally, some conjunction may be added to specify the fact that some USOM is only reachable from a particular USOM of the created macro-state.

As an example, the proposed design method of intelligent instrument is applied to the well-known example of the Automobile Cruise Control. This example provides several points of interest : on the one hand there is a continuous part relative to the control of the speed of the vehicle and, on the other hand a discrete part provided by the driver actions. This example of an Automobile Cruise Control was described in detail in Hatley (1987) and used for the illustration of the SART method. For our specification, we use the application case described in Calvez (1990) and we focus our study on the cruise control part.

4.2. Specification.

The Cruise Control is an additional device that allows a driver to assign a constant speed set point for long journeys. The vehicle speed is controlled by an action on the electric valve that commands the fuel injection of the motor. When the appropriate speed is reached, the driver can engage the regulation. When the regulation is on, the driver can, at any time, take back the control of the vehicle by acting on the brake or on the accelerator :
- on an action on the accelerator, the speed of the vehicle increases and, at the end of the acceleration the speed is again regulated by the cruise control at the previous set point,
- on an action on the brake, the cruise control is deactivated; the driver can reactivate it by pressing the "Resume" button.

At the setting up, a calibration procedure determines the conversion factor between the wheel impulsional coder and the calculator. This procedure allows the cruise control to adapt to any type of wheel.

The Cruise Control must react to several events : the operation of the driver and the coder signals. It must also make several calculations bound to the speed regulation and manage the mode changes. The Cruise Control is an hybrid system in comparison to purely reactive ones.

4.3. Description using the external model.

We give here a short presentation of the external model of the automobile cruise control detailed in Bayart and Lemaire, (1997). This takes into account the above-restricted specification.

The specification of the Cruise Control begins with the enumeration of all the services available to the driver. Let S be this set of services.
S is composed of seven services : Regulation Speed Capture, Speed Calibration, Regulation Valve Control, Manual Valve Control, Distance Elaboration, Speed Elaboration and *Clock*.
Those external services are deduced from the specification given in natural language. Some other services exist in the specification for the designer. This is the case of the "Clock" service which is

transparent to the users in the specification but which is necessary, for example, in the Speed Elaboration service.

The next step of the specification is the composition of the USOMs. In this case, five USOMs appear in the specification : Stop, Driver, Regulation, Brake in Regulation and Calibration. The composition of each USOM must be explicitly given. These can be done in a Backus Naur Norm form already used in Bouras, (1997) :

USOMs_list ::=
  (Stop::= ( )
    Driver::= (Distance_Elaboration,
      Man_Valve_Control, Speed_Elaboration,
      Clock)
    Regulation::= (Distance_Elaboration,
      Reg_Valve_Control, Reg_Speed_Capture,
      Speed_Elaboration, *Clock*)
    Brake_in_Regulation::= (Distance_Elaboration,
      Man_Valve_Control,   Reg_Speed_Capture,
      Speed_Elaboration, *Clock*)
    Calibration::= (Man_Valve_Control,
      Speed_Calibration)
)

When the set of USOMs, called M, is defined, the behavior of the Automotive Cruise Control has to be given. This can be done by defining the set of transitions T between the different USOM seen above. Here, a transition is considered as a disjunction of conjunctions of boolean terms. So, it is easy to verify the determinism of the transition graph (Atlee and Gannon, 1993). Furthermore, the connexity analysis of the graph and the lack of deadlock allow us to ensure its vivacity (Gondran and Minoux, 1990).

The obtained deterministic automata is given below (Fig. 2) with the set T of transitions between modes. In the transitions, the symbols used are logical and the part given after a "/" is a command that must be executed at the time the transition is taken.

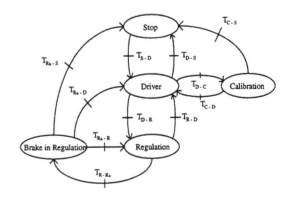

Figure 2 :   External   model   automata   of   the Automobile Cruise Control.

The transitions T_{i-j} labelled S for Stop, D for Driver, C for Calibration, R for Regulation and Rb for Brake in Regulation, give first the source USOM and second the destination USOM. This transitions are :
- T_{S-D} : (Motor Ignited)
- T_{D-S} : (¬Motor Ignited)
- T_{D-C} : Start Measurement KM / Caliber = 0
- T_{C-D} : Stop Measurement KM /

$$Caliber = pulses$$
- T_{C-S} : (¬Motor Ignited)
- T_{D-R} : (Regulation Engaged)∧(S>50km/h)∧ (¬brake)∧(Gear Lever Engaged)∧(Caliber ≠ 0) /

$$RS= S$$
- T_{R-D} : Regulation Disengaged / RS = 0
- T_{R-Rb} : (Brake)∨(¬Gear Lever Engaged)
- T_{Rb-R} : ((¬Brake)∧(Gear Lever Engaged)∧

$$∧(Resume)) ∨$$
((Regulation Engaged)∧(S>50km/h)∧ (¬brake)∧(Gear Lever Engaged) /

$$RS = S)$$
- T_{Rb-D} : Regulation Disengaged / RS = 0
- T_{Rb-S} : (¬Motor Ignited) / RS = 0

The deterministic automata gives us the behavior of the user operating modes of the Automotive Cruise Control.

The model proposed above is well fit to Hybrid Systems. It gives both a good and a clear representation of the general system behavior without describing it in details (i.e. giving a full internal description). In particular, it does not take into account the service organization. Let us now present the SYNCCHARTS representation to fit more closely to internal behavior and functionalities.

4.4. Application to the Cruise Control.

In order to obtain SYNCCHARTS description we use the method proposed above. First, we are looking for the pair of USOM that can be grouped : {Driver - Calibration}, {Driver, Regulation}, {Driver, Brake in Regulation} and {Regulation - Brake in Regulation}.

We are then looking for the subsets of services belonging to several USOMs. Here we can find three great subsets :
- { *Clock*, Distance Elaboration, Regulation Speed Capture, Speed Elaboration } in Regulation and Brake in Regulation.
- {*Clock*, Distance Elaboration, Speed Elaboration} which belong to the three USOMs : Driver, Regulation and Brake in Regulation,
- and {Manual Valve Control} which belong to three other USOMs : Driver, Brake in Regulation and Calibration,

The formation of the new syncChart proceeds as follows :

- From the first subset and from the pair {Regulation - Brake in Regulation}, we form a macro–state which can be called "Cruise Control Engaged".
- the first subset is also contained in the second one. Moreover all conditions hold to group the set of three USOMs { Driver - Regulation - Brake in Regulation}. In this case, the macro-state will contain the USOM Driver and the macro-state "Cruise Control Engaged" and it will be called "Running".

The grouping {Driver - Calibration}was not chosen because it takes into account less common services than the grouping { Driver - Regulation - Brake in Regulation} and because the pair {Calibration - Regulation} does not exist.

The obtained SYNCCHARTS representation from the external model is given below (Fig. 3). All the five USOMs described in the external model can be found in this representation.

For an efficient use of SYNCCHARTS some adaptations are desirable. Even if SYNCCHARTS may deal with conditions (predicates), event-driven descriptions are preferable. Thus, we introduce the following events :
- Ignition_ON (Ignition_OFF, resp.) is the event that causes the motor to enter (to leave, resp.) the state « Motor Ignited ».
- Regulation_ON (Regulation_OFF, resp.) is the event that causes the regulation to enter (to leave, resp.) the state « Regulation Engaged ».

For sake of clarity we also define the boolean function B = (S>50km/h)∧(¬brake)∧(Gear Lever Engaged).

The hierarchical description supported by SYNCCHARTS makes them different from automata, which are « flat » models. The use of SYNCCHARTS leads to a reduction of the number of arcs (e.g., transitions T_{Rb-D} and T_{R-D} in Fig. 2 are factorized in Fig. 3).

Last but not least, clever use of preemption may result in more concise description (e.g., note the self-loop arc on the « Cruise Control Engaged » macrostate. It is used to instantaneously restart the regulation with a new reference speed).

5- CONCLUSION :

The external model allows to describe an intelligent equipment from the point of view of the users. It provides a functional description, which is not sufficient to validate entirely the instrument behavior. In that sense, we propose to complete the external model with a SYNCCHARTS description. The interest of this method rests on a hierarchical representation of user operating mode and a clear presentation of concurrent services in each mode.

In this paper, we propose a method to derive a SYNCCHARTS description from the external model. At the present time, several SYNCCHARTS descriptions can be found from the same specification. Future work should aim at obtaining unique canonical description.

REFERENCES

André, C. (1996) "Representation and analysis of reactive behaviors : a Synchronous Approach", *Invited paper CESA'96*, pp. 19-29, Lille, France.

Atlee, J.M., and J.Gannon J. (1993). "State-Based Model Checking of Event-Driven System Requirements.", IEEE Transactions on Software Engineering, **Vol. 19 n°1**, pp. 24-40.

Bayart M., M. Staroswiecki M.(1993). "A Generic Functional Model of Smart Instrument for Distributed Architectures", *IMEKO, Intelligent instrumentation for remote and on-site measurement*, Brussels, Belgium.

Bayart, M., E. Lemaire (1997). "Intelligent Formal Description Applied to Hybrid Systems.", *IFAC-SICICA'97 :* , pp. 715-720, Annecy, France.

Bouras, A. (1997). "Contribution à la conception d'architectures réparties : modèles génériques et interopérabilité d'instruments intelligents.", PhD thesis, Université de Lille I, France.

Boussinot, F. and De Simone, R. (1991). "The Esterel language.", *Proceeding of the IEEE, 79*, pp. 1293-1304.

Calvez J.P. (1990). "Spécification et Conception des Systèmes", pp. 61-98, Masson, Paris.

Gondran, M. and M. Minoux (1990). "Graphes et algorithmes", Eyrolles, Paris.

Harel, D.(1987). "StateCharts, a Visual Formalism for Complex Systems", Sc. of Computer Prog., **8**, pp. 231-274.

Hatley, D.J.and I. Pirbhai (1987). "Strategies for Real-Time System Specification", Dorset House, NY.

Maraninchi, F.(1990). "ARGOS : un langage pour la conception, la description et la validation des systèmes réactifs.", PhD thesis, Université Joseph Fourier, Grenoble I, France.

Robert, M., M. Marchandiaux, M. Porte (1993) "Capteurs Intelligents et Méthodologie d'Evaluation", Hermès, Paris.

Staroswiecki, M. and M. Bayart (1994). "Actionneurs intelligents", Hermès, Paris.

Staroswiecki M. and M. Bayart (1996). "Models and Languages for the Interoperability of Smart Instruments", Automatica, **Vol. 32**, pp. 853-873.

Ward, P., Jensen and Al (1988). "ESML : An Extended Systems Modeling Language Based on the Data Flow Diagram", ACM Software Engineering Notes, **Vol. 13**, pp. 1-19.

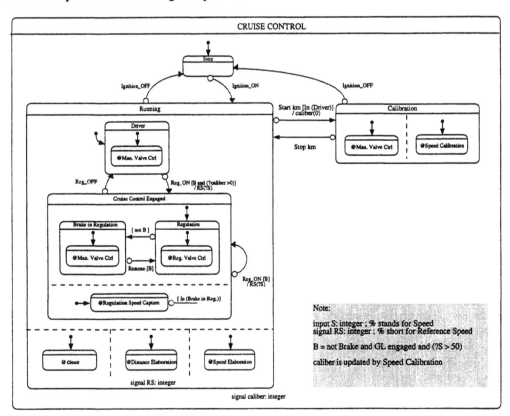

Figure 3: The syncChart of the Automobile Cruise Control.

PREDICTION OF JOURNEY CHARACTERISTICS FOR THE INTELLIGENT CONTROL OF A HYBRID ELECTRIC VEHICLE

C P Quigley and R J Ball.

Warwick Manufacturing Group,
Advanced Technology Centre,
University of Warwick,
Coventry CV4 7AL, UK,
Tel: +44(0)1203 523794
Fax: +44(0)1203 523387
E-Mail: c.p.quigley@warwick.ac.uk

Abstract: A hybrid electric vehicle is one which utilises both an internal combustion engine and electric motor for propulsion. This combination of power sources make its optimal control difficult. To enable optimised control of a hybrid electric powertrain it is desirable to have a priori knowledge of the characteristics for a given journey. Unfortunately this information is only known upon the completion of the journey and therefore it has to be intelligently estimated. This paper presents work that explores the possibility of prediction of journey characteristics upon journey departure. Usage data was collected from a number of vehicles with different usage characteristics over a period of one month each. The journey *distance* and *duration* were derived from this data. After ascertaining that there were predictable patterns inherent within the usage data, a fuzzy modelling approach was used to automatically generate rules for the prediction of journey *distance* and *duration* from departure time only. One conclusion drawn is that it is possible to predict the journey characteristics for some users whereas others prove more problematic. *Copyright © 1998 IFAC*

Keywords: Hybrid Vehicles, Prediction Methods, Automotive Control, Intelligent Control, Navigation Systems, Fuzzy Modelling.

1. INTRODUCTION

Recently there has been much research into new forms of vehicle propulsion, motivated by legislation intended to limit the polluting effects of vehicle exhaust emissions. Future legislation world-wide will introduce progressively more stringent limits on vehicle exhaust emissions. Ultimately, there will be a requirement for vehicle manufacturers to supply Zero Emission Vehicles (ZEVs) and Low Emission Vehicles (LEVs) for use as a form of private transport. At present electric vehicles are the only practical candidates as ZEVs, whilst hybrid electric vehicles (HEVs) currently form the most serious contenders as LEVs.

HEVs have a propulsion system which includes an internal combustion engine (ICE), and one or more electric motors (EM) and generators with an associated traction battery to supply power. Their powertrains can be configured in a variety of ways, but essentially narrow down to two main forms; parallel and series. The parallel drive HEV has both the ICE and EM connected directly to the vehicle

transmission and therefore gives rise to three possible modes of operation:-
- ZEV mode - vehicle is propelled solely by the EM.
- ICE mode - vehicle is propelled solely by the ICE.
- HEV mode - vehicle is propelled by a combination of EM and ICE.

The optimal control of the powertrain in ZEV and ICE mode is a relatively manageable problem to solve. However, optimal powertrain control in HEV mode is more problematic but has received attention in previous studies (Farrall and Jones, 1993; Farrall, 1992).

On the other hand the series drive HEV is a simpler design with only the EM connected directly to the vehicle transmission and therefore gives rise to only two possible modes of operation:-
- ZEV mode - vehicle is propelled solely by the EM.
- HEV mode - vehicle is propelled solely by the EM, but with the ICE used to recharge the battery via a generator.

Generally the control of the series drive HEV is easier than the parallel drive HEV. Both HEV configurations can recover lost energy under braking by regenerative methods (Wyczalek and Wang, 1992).

2. OPTIMISED HYBRID ELECTRIC VEHICLE CONTROL

One goal of HEV control is to use electrical energy and therefore the EM as much as possible to propel the vehicle, since this is the cheapest and cleanest source of power. One reason for this is the poor energy conversion efficiency of an ICE and fossil fuel combination when compared to an EM and battery combination. Hence recharging the vehicle battery via the domestic supply is often recommended or preferred. There are generally six strategies which can be used to control HEVs. Their application is dependent upon which of the two configurations of HEV is implemented.

For the parallel HEV, the three main control strategies are ICE assist (Kalberlah, 1991) which uses the EM as the primary propulsion system, EM assist (Farrall and Jones, 1993) which uses the ICE as the primary propulsion system, and Optimised. An optimised parallel HEV is a possible improvement on the EM assist parallel HEV and also uses the ICE as the primary propulsion system. It requires that the expected characteristics of the journey ahead to be predicted. If this is supplied, the battery would not necessarily need to be charged to its upper switching threshold once depleted. Instead it would only be charged to a level of SOC (state of charge) to allow the vehicle to complete the journey without further recharging. This results in the battery SOC depleted to its lower minimum threshold at the end of the journey. The advantage of this is that the net ICE use and net emissions are lowered and the performance losses of the EM Assist method are minimised. This method does not implement a ZEV mode.

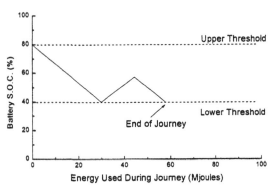

Figure 1 : A simplified example of the optimised series control strategy

For the series HEV, the three main control strategies are Thermostatic (Farrell and Barth, 1996), Load Following (Farrell and Barth, 1996) and Optimised (Farrell and Barth, 1996). The Optimised Series HEV is a compromise between the Thermostatic and Load Following strategies. The EM is used as the primary propulsion system until battery SOC is depleted down to a range between an upper and lower switching threshold. Within this range, the ICE is switched on and undertakes a limited load following strategy. Its strategy is limited by the ICE emissions (which limits the rate of change of the ICE output) and the ICE efficiency (which maintains the output above a minimum value). If the expected characteristics of the journey ahead were to be predicted then the vehicle would revert to ZEV mode once there is enough energy stored in the battery to complete the

remainder of the journey in this mode. Therefore the battery SOC is fully depleted at the end of the journey. As a result the HEV would automatically complete short journeys in ZEV mode. A simplified example of this control strategy is shown in figure 1. In this example, it has been predicted that the journey will be 60 km in length. The powertrain controller ensures that the battery SOC is depleted to its lower switching threshold at the end of the journey.

3. DATA COLLECTION

Both optimised parallel and series HEV control strategies require a priori knowledge of a given journey at journey departure. This is the motivation for the work presented in this paper. The aim of the work presented in this paper is to explore the possibility of the prediction of the journey parameters *distance* and *duration*. For this, experimental data is required that describes the usage characteristics of different vehicles in regular daily use. The *distance* and *duration* of a journey can be obtained simply from a speed / time profile of that journey.

3.1 DATA ACQUISITION EQUIPMENT

Typically a data acquisition system to record a speed / time profile would require a direct connection with the speedometer drive of the vehicle in order to fit some kind of transducer. Additionally this requires that the gear ratio between the speedometer and gearbox be ascertained. This usually varies between vehicle and vehicle manufacturer. The solution to this problem was to use a GPS navigation receiver connected via a serial communication link to a lap top personal computer. This then formed the basis of a data acquisition system that can record speed / time profiles and is independent of vehicle or vehicle manufacturer. Its configuration is shown in figure 2. The data acquisition system recorded the speed / time profile and the departure time for each journey.

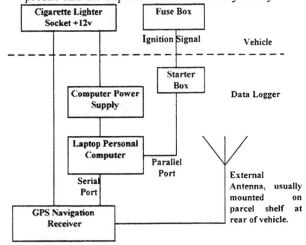

Figure 2 : Configuration of the GPS based data acquisition equipment for the collection of vehicle usage data.

3.2 EXPERIMENT DESIGN

A vehicle data logging programme was set up for the collection of such data. Careful consideration went into the selection of the ten subjects taking part in the programme. They were chosen so that the main vehicle user represents a subset of the UK driving population, and their selection is based on age and sex statistics from UK driving licence registrations (National Travel Survey 1989/91, 1993). They were also selected to give a spread of different occupations, different geographical location and different hours of work (shift, fixed hours, flexible time, part-time etc.).

Each subject had the GPS data logger installed in their car for approximately the same period of time (one month). This length of data collection period was primarily chosen due to the time and equipment availability constraints of the project. Due to problems with the data logging equipment, only data from eight vehicles was collected successfully. These are referred to as subject A to H inclusive throughout the rest of this paper.

3.3 DERIVATION OF JOURNEY PARAMETERS

The use of a GPS based system does have some error inherent in the speed / time profile data. This is due to loss in satellite reception caused by bridges, tree canopies, tall buildings etc. The most severe data loss is caused by the time to first fix (TTFF). This is the time from power on that the GPS system takes to acquire at least three satellites. For the particular system that was used this could take up to two minutes. This results in an error in the *distance* measurement for each journey. This error has been approximately compensated for by extrapolating the vehicle speed back, therefore recovering some of the lost distance measurement. Some journeys of less than two minutes *duration* recorded a *distance* of zero. This is also due to the TTFF. The journey *distance* is the integral of the speed / time profile. Figure 3 shows an example speed / time profile for an example journey. Journey *duration* is simply the length of the speed / time profile from *ignition on* to *ignition off*.

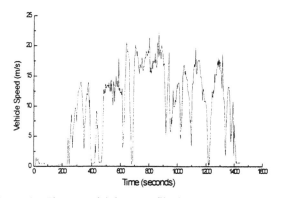

Figure 3 : The speed / time profile for a single journey

4. PRELIMINARY DATA ANALYSIS

Previous work by the authors had been carried out on data from subject A (Quigley et al, 1995; Quigley et al, 1996). It was found that predictable patterns were only inherent in week day journeys, due to the existence of commuting journeys. The aim was to ascertain whether *duration* and *distance* could be predicted from *departure time* only (the only information available at journey departure). The subject worked a five day week, Monday to Friday, and so it was expected that week day journeys might appear predictable and weekend journeys unpredictable. Figure 4 shows an example of such a predictable pattern for subject A's week day journeys. It can be seen that there is a significant peak on the 3-D histogram which represents the occurrence of many journeys at 8.00 a.m. with a *duration* of around 1200 seconds. Indeed weekend journeys did appear unpredictable. It was shown that 'eye balling' the data allowed simple rules for the prediction of journey parameters could be developed, e.g.

If (week day) and (*departure time* is around 8.00 a.m.) **then** (journey *distance* will be around 13 km)

Rules such as this have an inherent vagueness or fuzziness about them. This intuitively leads to the use of a fuzzy modelling approach for the automatic development of rules for the prediction of journey characteristics. This is presented in the next section. Similar 'eye balling' techniques with the other subjects showed that some had predictable patterns during the week day, but generally weekend journeys appeared unpredictable. Therefore the work presented in the remaining part of this paper considers week day journeys between the 8 subjects.

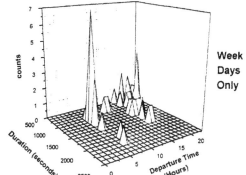

Figure 4 : 3-D histogram of *duration* versus *departure time* for subject A's week day journeys.

5. A PRELIMINARY FUZZY MODELLING APPROACH TO THE PREDICTION OF JOURNEY PARAMETERS

If a journey characteristic prediction capability were to be implemented in the powertrain controller of a HEV, the usage characteristics of that particular vehicle would need to be described by some method. The most obvious method would be to define its characteristics in the form of rules, as shown in the previous section. Ultimately an automatic method of rule generation will be required in the final system. This presents a number of problems:-
- There is no way of establishing the optimum number of rules required to describe the vehicle use characteristics.
- It is difficult to determine how to divide the *departure time* input space (i.e. do we separate the *departure time* into hour blocks, ten minute blocks etc.).

Fuzzy modelling (based on data clusters representing each rule) is a possible technique that can overcome such problems and could potentially be implemented in software in the powertrain controller of a HEV.

The fuzzy modelling method presented here is known as subtractive fuzzy clustering and is based on previous methods of cluster estimation (Chiu, 1994; Yager and Filev, 1994). The algorithm used was the MATLAB fuzzy logic toolbox implementation (Jang and Gulley, 1995). The algorithm itself is a fast one pass method for estimating the number of clusters in a set of input / output data. The number of clusters correspond to the number of rules that can be constructed from the data. Once the clusters have been found, they can be used to construct a Sugeno Fuzzy Inference System (FIS) (Sugeno and Kang, 1986). The algorithm has one parameter that must be sent to it, known as the cluster radius. The cluster radius specifies the range of each of the cluster centre's influence. There is no specific value of cluster radius that must be chosen for optimal fuzzy model generation. Therefore an incremental search must be carried out. This search must initially involve setting the cluster radius to a small value, and then on successive iterations making small increments to the cluster radius. There are three main operations

in this search for finding the optimum number of rules in the FIS. These are:-
- Find cluster centres.
- Use cluster centres to build a fuzzy inference system.
- Estimate success of predictions.

The fuzzy model or FIS that has the most success in making journey parameter predictions is then chosen as the optimal model. Further information on this technique can be found in previous work by the authors (Quigley et al, 1996).

The technique is demonstrated by the *duration* versus *departure time* results for subject A, since these are the least corrupt. However these are also very similar to the results obtained for *distance*. Figure 5 shows a scatter plot of the *departure time - duration* data collected from subject A. Observation of this input - output data set shows that there are perhaps two main clusters. One cluster is tight and well defined occurring around 450 minutes *departure time* and 1200 seconds *duration*. The other cluster is much sparser and occurs between 800 and 1400 minutes *departure time* and 100 to 1500 seconds *duration*. This data set was split into training and testing data sets using a 50:50 data partitioning method. The training set is used to build the FIS and the testing data is used to ensure that the FIS built can generalise well to previously unseen data.

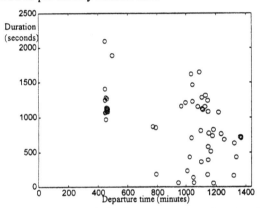

Figure 5 : The whole *departure time - duration* data set for subject A week day only journeys.

The performance of the various FISs developed were ascertained by examination of the RMS errors for both the training and testing data. To initially identify the FIS the cluster radius was varied (0.1 increments from 0.1 to 1.0) and the system performance evaluated as previously described. The FIS with the lowest testing RMS error was chosen as the best model. Figure 6 shows how during model identification the number of fuzzy rules generated, training RMS error and testing RMS error all vary with change in cluster radius. It can be seen that the lowest testing error is when the cluster radius is between 0.7 and 1.0. At these cluster radii values the clustering algorithm has indeed chosen 2 clusters (and thus 2 rules). It can also be seen that a small cluster radii value leads to many rules, which leads to the FIS over learning the training data. This is demonstrated by the fact that for lower cluster radii values the training RMS error tends to zero, whilst the testing RMS error tends to become increasingly large.

Figure 7 top shows how during FIS identification two input membership functions have been selected whose mean values correspond with the centres of the clusters of the input - output data (i.e. 450 and 1100 minutes *departure time*). The bottom of this figure shows the output response of the FIS with respect to *departure time*. This shows how the FIS has made the input and output non-linear mapping. Figure 8 top

shows the actual *duration* versus predicted *duration* of the training data set. The dashed diagonal line represents the ideal FIS performance when the predicted *duration* equals the actual *duration*. It can be seen that there are two main clusters of data. One of these clusters corresponds to a number of journeys of about 1200 seconds *duration* which have been predicted around 800 seconds *duration*. Thus these regularly occurring journeys have not been learnt correctly from the training data set. The second cluster corresponds to a number of journeys about 1100 seconds in *duration* predicted as 1200 seconds which is more reasonable. This journey in fact corresponds to the morning journeys that demonstrated the significant peak in the previous chapter and the tighter of the two clusters in figure 5. The bottom of figure 8 shows the actual *duration* versus the predicted *duration* of the testing data set. This again shows the appearance of similar clusters to those that occurred with the training data set. This confirms that the model that has been trained does indeed give a good prediction at around 1200 seconds.

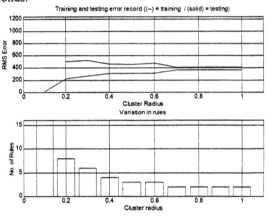

Figure 6 : The training / testing RMS errors / number of clusters found versus cluster radius for subject A.

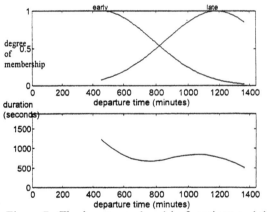

Figure 7 : The input membership functions and the output surface generated by the FIS.

Further information is given by observation of figure 9 which compares the training and testing data sets with respect to the FIS's output surface. Unfortunately, journeys in the second cluster occurring in the evening in both training and testing data sets range from about 100 seconds to 1500 seconds *duration* and the FIS has included journeys all in the same cluster during the model building process. As there are a number of different types of journey *duration* occurring around the same *departure time*, some additional information is needed in order to separate these journeys and predict their *duration* more accurately.

144

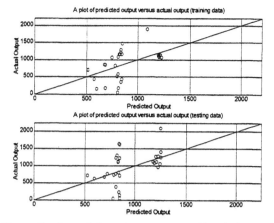

Figure 8 : Actual *duration* versus predicted *duration* of the training and testing data sets when presented to the fuzzy *duration* model for subject A.

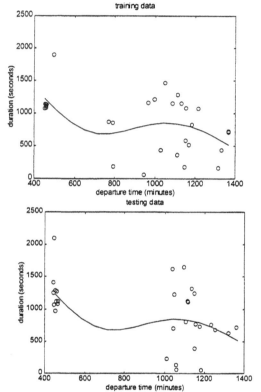

Figure 9 : Training and testing data sets with the FIS's output surface superimposed (represented by the broken line).

6. FUZZY MODELS OF REMAINING DATA SETS

The fuzzy modelling method applied to subject A week day journeys showed a reasonable amount of success since approximately 30 % of week day journeys were modelled quite well (see figure 9). Unfortunately, perhaps as expected, some were not modelled so well. In this section the same fuzzy modelling method is applied to the data taken from the remaining subjects. A *duration* and *distance* model with respect to *departure time* is developed for each of the subjects. This resulted in three types of performance. An example of each type of performance is shown in figure 10. Figure 10 shows two sets of error histograms for each performance type's fuzzy *duration* model. The top histogram for each type is its percentage errors for its week days only testing data. The bottom histogram is the actual

error for the same data set in seconds. The *distance* versus *departure time* histograms gave similar results.

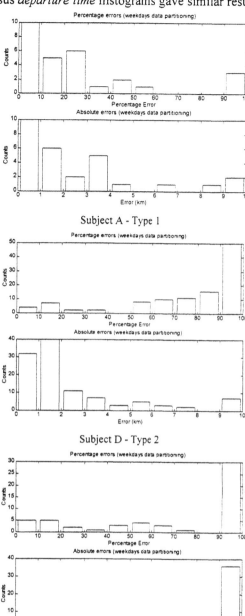

Subject A - Type 1

Subject D - Type 2

Subject C - Type 3

Figure 10 : Three types of error histograms for *departure time* versus *duration* FIS for the eight subjects.

The three main types are:-
1. The FIS performed reasonably for both percentage error and actual error, depicted by both distributions being left skewed (Subject A).
2. The FIS performed badly in terms of percentage error (right skewed distribution) but reasonably in terms of its actual errors (left skewed distribution) (Subjects B, D and F).
3. The FIS performed badly in terms of both percentage error and actual errors, depicted by both distributions being right skewed (Subjects C, E, G and H).

Type 1, only represented by subject A, represents a subject whose morning journeys and some afternoon journeys can be modelled reasonably well (as seen in figure 9). The morning journeys were the most successfully modelled. This was mainly due to them being defined by a tight cluster of data, as shown in figure 5. *Type 2*, represented by subjects B, D and F performed poorly. Further examination of type 2 performance found that these data sets contained loose clustering of data and therefore appeared unpredictable. Their low error in terms of seconds and kilometres was due to most journeys being of low *duration* and *distance*. *Type 3*, represented by subjects C, E, G and H, performed poorly also. *Type 3* also exhibited some loose clusters but sometimes had more than one tight cluster with respect to departure time (representing journeys of different types occurring at the same *departure time*).

7. CONCLUSIONS

The work presented in this paper has continued to explore the possibility of the prediction of journey *duration* and *distance* of a passenger car from *departure time* information only. Data collected from eight different vehicles in daily use over a period of one month each was used in the investigation. A preliminary investigation showed weekend day journeys to be apparently unpredictable. However week day journeys did show an element of predictability which allowed simple journey characteristic prediction rules to be developed heuristically. Fuzzy models for the prediction of the characteristics *duration* and *distance* from *departure time* only, exhibited three types of performance. *Type 1* performance, represented by 1 subject, performed well on week day morning journeys, but not so well on afternoon journeys. This was due to the morning journeys being defined by a tight cluster of data and afternoon journeys being defined by a loose cluster of data. *Types 2* and *3* performed poorly either due to loose clustering or more than one tight cluster with respect to departure time (representing journeys of different types occurring at the same *departure time*). Clearly, additional input information is needed to improve the prediction performance for subject A afternoon week day journeys and vehicle users of *type 2* and *3*.

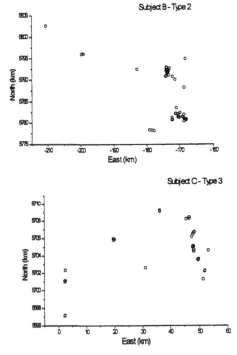

Figure 11 : Locations visited by Subjects B and C.

In practice, a single vehicle's usage pattern would probably consist of some unpredictable journeys. A way of providing additional input information with which to improve predictions is to include *place of departure* information as a pre-processor to the previously presented fuzzy modelling approach. Such information can be provided by a GPS receiver which could feasibly be incorporated into future HEVs economically (the cost of such technology has decreased significantly since project commencement). It has to be determined whether place of departure information would be a suitable pre-processor for other types of vehicle user also. Figure 11 shows a plot of the *place of departure* (expressed in displacement in km from latitude, longitude = 0 degrees) for subjects B and C, who were of *type 2* and *3* performance respectively. It can seen that there are at least two significant clusters of data representing locations regularly visited by subject B. For subject C, there are clearly at least seven locations that are regularly visited by this particular vehicle. These locations could potentially be used as a pre-processor for the prediction of subjects B and C's journey characteristics. A similar situation is observed with data from the other subjects, each with a different number of regularly visited locations. Future work will investigate the use of this information further.

ACKNOWLEDGEMENTS

The work was funded in part by the UK Engineering Physical Sciences Research Council (Grant No. GR/K35976). The authors would also like to acknowledge the support provided by Rover Group for this work.

REFERENCES

Chiu, S., (1994); "Fuzzy Model Identification Based on Cluster Estimation", *Journal of Intelligent & Fuzzy Systems*, **vol. 2**, no. 3, pp. 267 - 278.

Farrall, S. D. and Jones, R. P. (1993), "Energy management in an automotive electric/heat engine hybrid powertrain using fuzzy decision making", *Proceedings of 1993 IEEE International Symposium on Intelligent Control*, pp463 - 468.

Farrell, J.A. and Barth M.J. (1996); 'Hybrid electric vehicle energy management strategies', *Final report prepared for ISE Research*, 4909 Murphy Canyon Road, Suite 330, San Diego, CA 92123, USA, March 1996.

Kalberlah, A (1991); 'Electric hybrid drive systems for passenger cars and taxis', *SAE paper 910247*.

Jang, J.S.R., and Gulley, N. (1995), *Fuzzy Logic Toolbox - User's Guide*, The MathWorks, Inc.

National Travel Survey 1989/91 (1993), Transport Statistics Report, HMSO, London, September.

Quigley C.P., Ball R.J., Vinsome A.M., Jones R.P. (1995); 'Predictng the use of a hybrid electric vehicle', *Prodeedings of IFAC Workshop on Intelligent Components for Autonomous and Semi-Autonomous Vehicles, ICASAV'95*, Toulouse, France, pp. 139 - 144.

Quigley C.P., Ball R.J., Vinsome A.M., Jones R.P. (1996); 'Prediction of journey parameters for the intelligent control of a hybrid electric vehicle', *Proceeding of the International Symposium on Intelligent Control*, Dearborn, USA, pp. 402 - 407.

Sugeno, M. and Kang G.T. (1986); 'Fuzzy modelling and control of multilayer incinerator', *Fuzzy Sets and Systems*, **Vol. 18**, pp.329 - 346.

Wyczalek, F.A. and Wang T.C. (1992); 'Regenerative braking concepts for electric vehicles - a primer', *SAE paper 920648*

Yager R. and Filev D., (1994) 'Generation of fuzzy rules by mountain clustering', *Journal of Intelligent and Fuzzy Systems*, **vol. 2**, no. 3, pp. 209 - 219.

IMU : INTEGRATED MONITORING UNIT OF

THE SAVE DIAGNOSTIC SYSTEM

N. Hernandez-Gress, D. Estève

LAAS/CNRS 7, Avenue du Colonel Roche 31077 Toulouse Cedex 4 France.
hdez@laas.fr

and A. Bekiaris

TRD International 28, Alexandras Ave. 106 83, Athens Greece.
trnspcon@athena.compulink.gr

SAVE[1] European Programme

Abstract: In this paper, the diagnostic part of the SAVE system is described. It is composed of three main diagnostic subsystems *1)* behavioural diagnosis, *2)* physical diagnosis and *3)* critical diagnosis. Each subsystem uses different sensors on-board the vehicle and the information is placed in a hierarchical order and mixed to obtain a general diagnosis. A model (black-box), learned *off line*, is compared *on line* to the actual driver's performance. Owing to the inherent difficulty of this problem, the model is created by using Statistical and Artificial Intelligence algorithms (Neural Networks and Fuzzy Logic). Laboratory experiments were made and yielded a success rate in excess of 95%. On-line experiments have started using the SAVE demonstrators in which the system is able to differentiate between the driving behaviour of two persons with a success rate of 98% under the same real traffic conditions. *Copyright © 1998 IFAC*

Keywords: *Data Acquisition, Data Fusion, Neural Network models, Supervision, Decision Making, Real Time.*

1. INTRODUCTION

The aim of the Integrated Monitoring Unit (IMU) system within [SAVE] is to detect changes in the driving behaviour of the pilot caused by human-factors such as: intoxication (alcohol or drugs), fatigue, ill-health and other causes of impairment. If a driving deterioration is detected, the system must perform a number of actions like sending messages to a) the pilot, b) the environment and c) deciding in real time whether to activate the Automatic Control Device (ACD).

To carry out monitoring of the driver's behaviour in real time in a most efficient manner, diagnosis has been separated into three subsystems. These are :

1) *behavioural diagnosis* : obtained by sampling external behavioural variables such as vehicle speed and acceleration, pedal (clutch, brake and accelerator) position, wheel angle.

2) *physical diagnosis* : the driver 's eyelid motions are monitored by a movie camera mounted on the dashboard of the vehicle.

3) *critical diagnosis* : two different sensors are used, namely the grip force sensor and the head rest position sensor.

To produce the system's baseline, other subjective measures are needed such as the EEG, EOG, blood pressure or specific measurements. These « physical signals » are more direct and require physical contact with the driver, thereby precluding their use in real time. However they can be used to develop and test the system as they are well-correlated with the driver's hypovigilance of the driver. [Brookhuis 93].

---

[1] SAVE : System for effective assessment of the driver state and vehicle control in emergency situations.

The diagnostic system uses Multisensory Fusion by means of Statistical and Artificial Intelligence algorithms allowing the main characteristics to be determined and learning of a base-line (*driving model*) by Artificial Neural Networks and Fuzzy Logic. As a result some parameters are recorded on a personalised smart card. The base-line is compared with the actual driving characteristics; if any abnormal behaviour is detected some ergonomic actions are initiated such as messages to the pilot and/or to the environment or even the activation of the ACD system. Moreover, the system must be able of changing the driving model (if the pilot so requests) in order to customise it to the driver's characteristics and to reduce a number of false alarms.

To develop the IMU, several experiments have been conducted within the SAVE framework. The hybrid system is being tested using SAVE demonstrators. System performance is in excess of 95% when it comes to distinguishing between normal and abnormal (*fatigue, intoxication, inattention*) driving. The different algorithms are run on a PC-Platform, and the architecture has been developed to be fitted on-board the vehicle.

Section 2 describes the IMU subsystems, Section 3 develops the methodology retained and tested within SAVE. The results are discussed in Section 4 and finally some Conclusions and perspectives are commented.

2. THE IMU SUBSYSTEMS

Several subsystems operate within the IMU, *1) behavioural diagnosis 2) physical diagnosis* and *3) critical diagnosis*. Each one of these systems already produces a local diagnosis. The first diagnosis is based on mechanical sensors processed by Artificial Neural Networks and Fuzzy Logic. The second is a function developed by Renault [Artaud and Planque 97] which takes into account the blinking frequency and eyes closure duration. For the final diagnosis, some experiments have been made to investigate the appropriated CRITERIA (response time and characteristics of an ill-driver). Functional association is shown in Fig. 1. The end goal is to produce a final diagnosis as a fusion of the three diagnoses. The behavioural and physical diagnoses compute a driver's state every 30s and the critical diagnosis is obtained in 20ms.

In parallel with IMU, other systems are in operation. One of them is called the Hierarchical Manager. It acts as the brain of the overall SAVE system.

Second, the ergonomic communication with the driver and/or the environment which represent a very important part of the overall SAVE system managed by the Human Machine Interface and the Warning System (HMI/SWS) subsystems (developed by SIEMENS, IAT and TRD for SAVE). The Smart Card stores the driver's personal information and the customised parameters to provide a better decision system. An finally the so called Automatic Control Device (ACD) developed by CRF (FIAT) for SAVE and which function is to safely stop the car if a problem is detected.

As there is no ideal sensor capable of predicting the driver's difficulties [Hernandez 95], the fusion of several sensors is the approach retained to improve the diagnosis. A number of these sensors are already installed on commercial vehicles.

In the following paragraphs, the three subsystems are detailed from a Hardware point of view. The Software part will be detailed in Section 3.

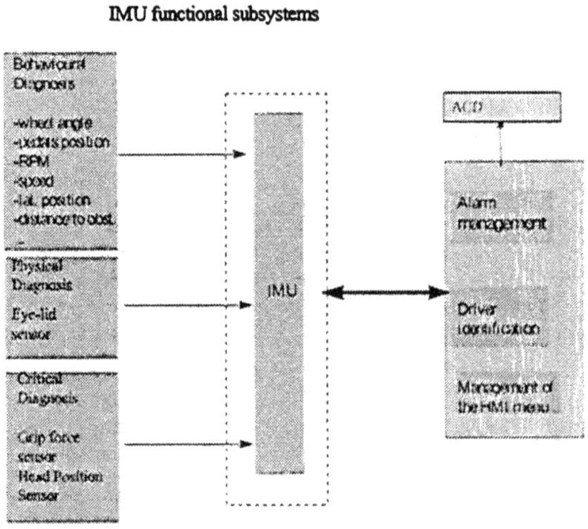

Figure 1 : Integrated Monitoring Unit functional Analysis.

2.1 Behavioural subsystem.

This subsystem performs a diagnosis using mechanical information of the vehicle. The sensors sampled on-board the vehicle, can be listed under two major headings:

- Driver's behaviour: pedal position (clutch, accelerator, brake) steering wheel angle, head position,...
- Vehicle behaviour: lane position, speed and acceleration, horizontal acceleration and relative speed and distance to obstacles, Revolutions per Minute,...

The dimensional space of these variables is reduced as some variables are correlated and/or non-independent, generating a small space of well-suited information. This reduction is completely dependent on the driving characteristics.

2.2 Eyelid sensor subsystem

In literature, many approaches have tried to correlate the blinking frequency and the eyelid closure duration to the to states of hypovigilance. These approaches have led to new functions to predict hypovigilance [Yamamoto et al 95]. For the author's point of view there is no an ideal sensor. However another approach is on-going to be tested, it consists to fusion the eyelid closure duration with the behavioural signals.

2.3 Critical Incident Diagnosis

This subsystem uses two sensors, the grip force sensor and the head position sensor. These sensor have developed under the SAVE framework, they consists to study the driver from a physical point if view.

The Behavioural and Physical Diagnosis take some time (frequency explained in the sequel) before achieving a relevant diagnosis. If the pilot encounters difficulties (i.e. ill-health) the system intervention, is needed over a few mili-seconds; this being designed to yield a fast diagnosis

3. Methodology

The proposed methodology has been presented in [Hernandez and Esteve 97]. Three major tasks must be performed 1) base-line generation 2) real time diagnosis and 3) evolution or customisation. The methodology is outlined in Fig. 2.

From a technical point of view, several classes must be considered (*normal, fatigue, intoxicated...*) and not only the detection of changes in the driver's behaviour, which is another possibility offered by the proposed methodology. As a result of the model generation, some parameters are extracted which are used by the real time task of the system. The parameters are stored on an appropriate smart-card.

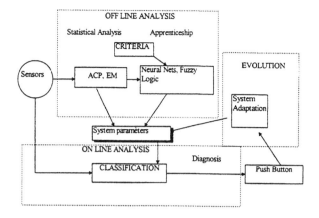

Figure 2 : General detection methodology.

3.1 Base-line generation

First, a base-line has to be created representing the model of the system. The statistical and Artificial Intelligence algorithms and methods are used off-line to produce some parameters which are stored in the smart card and used by the on-line part of the system.

This corresponds to a flexible and customised system. To perform the learning step, algorithms need a database which is created by sampling the variables of an experienced pilot and following the advice of an expert classifying the behaviour by means of EEG, EOG and visual analysis. Search for the hypovigilance levels to be detected and for the most important variables accounting for the behaviour known as CRITERIA and form a topic of research under SAVE.

3.2 Diagnosis

The parameters produced by the base-line generation are recorded on the smart card, for use during real time diagnosis. The actual state vectors represented by the variables are classified using the parameters. Filtering and generalisation using ANNs and FL are computed per class. The membership degree of the different classes is thus studied on a moving window to compute the evolution from one class to the other in relation to time and space.

The diagnosis is enhanced by maintaining the history of the system and keeping some diagnosis (in practice 50) and studying the time space relationship between them. As the driving skills vary greatly from one driver to another [Hernandez and Esteve 97], the evolution must be possible to allow the system parameters to be customised to the driver's characteristics.

3.3 Evolution

This approach also features a custom system proposed, since a learning phase has been provided prior to diagnosis. In practice, an individual map must already be available. Then, with the pilot's participation, the driver's personal driving characteristics are learned. The main objective is to adapt or customise the system to the driver's characteristics. In the case of a false alarm, the pilot dialogues with the system through a push button. The vector state (of variables) is recorded for off-line processing. These new vectors are processed to tailor the parameters to the user needs.

The system is designed so that vectors remain close to the actual classes. Thus only small changes can be taken into account, and noisy vectors can be filtered.

4. Detection Algorithms

4.1 Feature extraction

Two methods have been tested, namely, the so-called Principal Components Analysis (PCA) and the Independent Component Analysis (ICA). These methods yield a small number of variables from a set of initial variables. On the one hand, the PCA is a good visualization technique revealing the correlation between the initial variables and creating less new variables with lower or zero correlation. On the other, ICA is a non-linear method which minimizes the statistical dependence between components. As a matter of fact, the ICA concept may be viewed as an extension of the PCA which only imposes independence to second order statistics.

4.2 Learning

To learn the driver's characteristics, different methods have been tested like the binary unit based ANNs constructive algorithm « MOBCPSL » [Poulard and Hernandez 97a]. However in our case classes overlap. Another tested methodology detailed in [Poulard and Hernandez 97b] called « Multidimensional Fuzzy Sets » is shown in Fig. 4. In this method the activation function of neurons has been changed to obtain new fuzzy zones. However, a more efficient methodology is available with the Generalised Radial Basis Functions (GRBF) Fig. 5. The choice of methodology has taken into account the fuzzification required to handle overlapping classes. Both methodologies have been tested select the most suitable one.

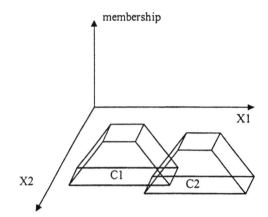

Figure 3 : Multidimensional Fuzzy Sets learning approach.

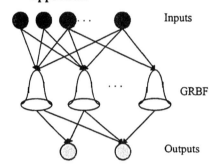

Figure 4 : Learning procedures a) MFS b) GRBF

4.3 Final decision

The use of a better diagnostic methodology is needed to reduce false alarms and the fuzzy logic concept is utilised to obtain the final decision. Given the complexity of the problem (overlapping classes), this symbolic methodology is very important to arrive at a better diagnosis, capable of decreasing the number of false alarms while increasing the system capability. Fuzzy Logic features advantages relative to binary logic.

5. Results

The methodology has been tested for several hypovigilance procedures. Here, only the inattention and intoxication (alcohol) issues are discussed.

5.1 Inattention experiment

The experiment was carried out using the TRC driving simulator. For the purposes of the experiment, twenty persons were enrolled in the study. These individuals were requested to drive under certain conditions: a) using a telephone or staying alert b) with or without following a car and c) in the morning or evening. These combined factors made up the complete experiment for one driver.

Data were sampled at 10Hz and recorded in ascii files.

At different intervals, a simple hand-held portable phone had to be answered. The telephone-task required looking up a telephone number. Subjects were instructed to give priority to safe driving over the telephone-task.

The parameters used were the following: gas pedal position, brake pedal position, clutch and pedal position, rotations/minute, speed, steering wheel position, lateral position on the road, time-to-collision.

Based on the PCA analysis, the most important variables were: (format: initial variable, correlation to the corresponding Principal Component (PC)):
PC 1 : (gas,-0.8); (rpm,-0.9); (velocity,-0.9);
PC 2 : (brake,+0.5); (thw,-0.7); (ttc,-0.6); (leaddis,-0.7);
PC 3 : (steer,-0.8); (latpos,+0.8);
PC 4 : (tlc,-1.0);
PC 5 : (clutch,-0.6);

The learning step has been computed over 5% of the overall data. Generalization (test of the NN with the overall data) makes use of the whole database (12754 and 12901 samples, respectively). Only three new variables were utilized in the learning step (3 Principal Components or 3 Independent new variables).

Performance of the ANN is shown in Table 1. Based on two different pre-processings: PCA and ICA.

	PCA	ICA
list02	74.4%	85.6%
list18	80.9%	90.0%

Table 1 : Performance using PCA and ICA pre-processing steps.

5.2 Intoxication (alcohol) experiment

The original data files contained 15 variables, of which only 6 have been used in our analysis. The variables are those containing the most pertinent information, and corresponding to the information generated by the vehicle. It is assumed that the information about the environment will not be available. These variables are:

(1) speed of the driver's vehicle, (2) distance to lead vehicle, (3) lane position, (4) steering wheel position, (5) brake position and (6) accelerator position.

Sampling rate is 10 Hz. and each data file has approximately 1 hour of data.

The scenario is that of an oval-shaped circuit with two long straight lines, four bends and two short straight lines.

For learning purposes, the first 8 minutes were used (5000 patterns) representing more than one lap track. Thus learning and test data were obtained.

The first step, consisted in building up the PCA (Principal Component Analysis). As a result, three components were found to exhibit the highest amount of information, and the variables taken into account were as follows:

Principal Component 1 : speed of the driver's vehicle 60%, distance to lead vehicle 70%, steering wheel 70%. Principal Component 2 : accelerator 80%. Principal Component 3 : lane position 80%.

Filtering has not been used due to the simplicity of data. We need more data to know whether or not isolated patterns are outliers.

Using the methodology, the performance of this diagnostic system reaches 100% but drops to 75% for the overall data.

The final diagnostic step is carried out, The ANNs are utilized to compute the firing strength of each rule in a rule base. The weight of each rule depends on the activation function (AF) of the Neural Network. For example we utilize the binary AF, then the utilization weights are either 0 or 1. Then, we have to find the consequent part of each rule which is a linear combination of input variables plus a constant term. The final output is the weighted average of each rule's output. We tune the parameters of the linear combination using the least squares method. The performance of such a system rises to 92%.

In order to compare the methodology with a normal Neural Networks method, the Principal Components are used to obtain a Neural Network using the Barycentric Correction Procedure [Poulard and Labreche 95] (BCP) and the results obtained over the same data reach approx. 86%.

6. Conclusions

In this paper, an on-board Monitoring Unit is described. A set of algorithms has been adapted which has yielded to theoretical developments. The most important problem, that is, how to model the driving behaviour under different driving behaviours has been addressed. The methodology has been developed and tested in laboratory using normal-abnormal data. Real time experiments making use of demonstrators placed in real traffic situations are performed to test the system.

In the author's opinion the originality of our method lies on the fusion of several sensors operated in conjunction with some point algorithms that are demonstrating the efficiency. Also this methodology can be adapted to match other process in which the mathematical model is difficult to establish.

The first prototype will be completed by the beginning of 1988. Plenty of work remains to be done to allow our system to be mounted on a commercial vehicle but, surely car crashes still have to be reduced.

Acknowledgements

The authors would like to thanks to the European programme SAVE for support. N. Hernandez would like to thanks CONACyT and ITESM for financial aid.

References

[Artaud and Planque 97] P. Artaud and S. Planque : « Validation tests of the complete drowsiness eyelid sensor based detector », SAVE European programme. July 97. Restricted.

[Brookhuis 93] K.A. Brookhuis : « The use of physiological measures to validate driver monitoring », Driving Future vehicles, Taylor and Francis, 1993.

[Hernadez 95] N. Hernandez-Gress : « Methodologie basée sur la fusion Multisensorielle et les réseaux de Neurones: Application à la Sécurité Active dans la conduite automovile » LAAS report 95307.

[Hernandez and Esteve 97] Neil Hernández-Gress, Daniel Estève, Driver Drowsiness Detection : Past Present and porspective work. Traffic Technology International June/July 1997.

[Poulard and Hernandez 97a] H. Poulard and N. Hernandez « Two efficient multiple output construction algorithms » LAAS report 97003 July 1997, submitted to Connection Science.

[Poulard and Hernandez 97b] H. Poulard and N. Hernandez « Multidimensional Fuzzy Sets : A novel Approach, » in Proccedings of Fuzzy Logic and Applications ISFL97 (Zurich, Switzerland).

[Poulard and Labreche 95] H. Poulard and S. Labreche : « A new threshold unit learning algorithm » LAAS report 95504 Dec. 1995.

[SAVE] EXTIT Telematics « System for effective assessment of the driver state and vehicle control in emergency situations TR 1047, » 1996-1998
http ://www.iao.fhg.de/Projects/SAVE/

[Yamamoto et al 95] Yamamoto K, Yoshikawai M, Higuchi S. Development of Driver's Alertness Monitoring System ACCV95.

AUTOMATIC VEHICLE CONTROL
IN EMERGENCY SITUATIONS:
TECHNICAL AND HUMAN FACTOR ASPECTS

Coda Alessandro, Antonello Pier Claudio, Damiani Sergio
Centro Ricerche Fiat
Strada Torino 50 Orbassano (T0) ITALY
Tel. ++39-11-9023 131 - Fax ++39-11-9023 083
Peters Björn
Swedish National Road and Transport Research Institute (VTI)
S-581 95 Linköping SWEDEN
Tel. ++46-13-204070 - Fax ++46-13-141436

Abstract: This paper presents the results from the first driving experience of the emergency control function as implemented in the European project (DGXIII) SAVE. Aspects related to both the technical and the user's point of view are discussed showing the results from real tests and from a simulation of the manoeuvre.   The emergency manoeuvre, performed by a vehicle in an automatic way, has the objective to improve safety: when a driver impairment is detected the system let the car stop in a safe way at the roadside. There is a need of a number of emergency control procedures and the paper describes the best control strategies to safely stop the vehicle without causing any problems to the driver and to the surrounding traffic. Actually a simple scenario is considered for the first prototype vehicle, i.e. straight road with low traffic, but the results can be applied also to different traffic and road conditions. The paper includes also a full description of the test vehicle currently under development at CRF, related driving simulator test performed within the project and finally a short description of further plans for test at the VTI driving simulator. *Copyright © 1998 IFAC*

Keywords: Autonomous vehicles - Image Processing - Intelligent cruise control - Vehicles - Vision

Introduction

Automatic Vehicle Control is considered to become a promising way to reduce, or even completely avoid road accidents. These ideas are not new and autonomous driving vehicles have been presented earlier (Ulmer, 1994).   However the actual technology does not guarantee a sufficiently precise scene reconstruction and obstacle detection to safely activate automatic control systems, substituting the driver, in all possible road / traffic conditions.

On the other hand, several car manufacturers are currently concerned with the development of various driver impairment detection systems. In most cases the work is focused to driver drowsiness detection systems (Kaneda et al., 1994). This was revealed during the latest ITS congress (e.g. Yamamoto et al., 1996, Shimotano et al., 1996).   The Advanced Safety Vehicle program in Japan promote this work (ASV, 1996).   There are a number of techniques

applied to detect driver impairment such as physiological data (heart-rate, EEG etc.) and driving performance data.   Most of these systems are intended to warn or alert the driver by different means (sound vibrations, menthol scent etc.) if an impaired status is detected.

SAVE (System for effective Assessment of the driver state and Vehicle control in Emergency situations - TR 1047) is a project within the Telematics Application Programme of the EC DGXIII. The SAVE project is concerned with the development of a system that detects driver impairment caused by drowsiness, illness, or drug abuse but it differs from the efforts described above in that it incorporates an *emergency control function* that will stop the vehicle if the driver is no longer capable to drive safely. This project is a direct continuation of the work that was done within the DETER project (Brookhuis, 1995).

153

Functional description of the Automatic Control Device

The emergency control function in the SAVE system is performed by the Automatic Control Device. The ACD will take control of the vehicle and automatically manoeuvre it to the roadside ad stop it in a safe and comfortable manner for driver and for others road users. This happens only when a driver impairment is detected and the driver does not respond adequately to alerts or warnings or after a direct request by the driver.

The main purpose of the ACD is therefore to safely and fast execute a stopping manoeuvre and place the vehicle away from the traffic by stopping it in the emergency lane. Others goals of the ACD system, are in no emergency situation, to provide information on the vehicle dynamic (speed, steering wheel, brake, throttle ...) and on the road and traffic state to others SAVE sub-systems (i.e. Integrated Monitoring Unit). In order to realise an ACD as outlined above the system needs to control a number of vehicle functions like acceleration/deceleration steering angle etc. but it also depends on information about the vehicle like current speed, steering wheel position, heading direction etc. and finally information about the surrounding environment like road geometry, vehicle position, headway and time-to-collision to other vehicles and objects etc..

The development of an ACD prototype - System Architecture

A prototype ACD will be developed and tested as a first step to a full working ACD. The prototype ACD is based on a simple road geometry recognition and co-operation between a dedicated lateral control and a standard Adaptive Cruise Controller.

Figure 1shows a block diagram, organised in three levels, of the ACD prototype.

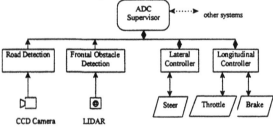

Figure 1 - ACD prototype architecture

The lower one is the **sensor and actuator level**. At this level there are the main blocks to collect information from vehicle status and from the road environment around the car. steer, throttle, brake. These blocks are at the same time sensors and actuators.

The higher level is named **supervisor level**. At this level, all the information from the vehicle and the road environment is available in a unique time/spatial reference system. Here the whole control strategy to perform the type of manoeuvre is planned and monitored. The task of this level is also to interface the ACD toward the others on-board systems and SAVE sub-systems.

At middle level (**processing level**) the processing work take place:

* the large, noisy, unstable information available, collected by the sensors at the lower level, is summarised in few, synthetic, stable parameters for the supervisor level;
* the complete control strategy coming from the supervisor level is delegate al this level by carrying out fine control and splitting it in longitudinal and lateral control to drive the actuators.

The **Road Detection** sub-system is used on structured road environment and is able to recognise the borders of the vehicle lane and to locate them with respect to the lane. The system is based on an opto-electronic sensor (b/w CCD camera), located behind the windshield at the centre of the vehicle, looking at the road in front of the car. The processing unit acquires images from the camera and analysing them, by means of particular image processing techniques, extracts a synthetic parametric description of the lane a head the vehicle. The lane marking parameters are: position, direction, curvature and marking type (continuous, broken); from these data it is possible to locate the car respect to the lane (car position, yaw angle).

The **Frontal Obstacle Detection** is able to recognise, locate and track any obstacle in front of the vehicle. The system is based on an opto-electronic sensor (scanned laser radar), located inside the front bumper of the car. The map produced by the sensor, is acquired by the processing unit that locate obstacles and describe them by means of position and relative speed.

At the **Lateral Controller** is delegate the task to keep the car in a opportune position inside the lane. The actuator, on the steering column, is an electrical power assistance system. It, besides assists the driver during normal driving phase, is used as slave device of the Lateral Controller during the automatic emergency manoeuvre.

At the **Longitudinal Controller** is delegate the task to control the speed of the car. Taking into consideration the will of the supervisor (recommended speed, acceleration, obstacles on lane), this sub-system controls active booster (brake) and drive by wire device.

Finally it is important to remark that the architecture presented is an open architecture; if in the future other sub-systems will be available, they could be added at the structure without problem. In particular other sensors (i.e. lateral obstacle detection, backward vehicle detection) could be installed on the car and contribute to a complete description of the environment around the vehicle.

Actual ACD Prototype Strategies

Starting from any vehicle condition (speed, trajectory etc.), the ACD is able to slow down the car and drive it on the right hand side of the road (emergency lane) before stopping it.

The ACD's *Supervisor* is responsible for the whole manoeuvre. It has to set parameters (accelerations, speeds, positions) for both controllers and verify the reached targets.

Moreover, it has some fundamental rules to manage both controllers. An example is that the anti-collision avoidance manoeuvre has a higher priority with respect to the lane change manoeuvre; if an obstacle is detected in front of the vehicle inside the *relative safe distance*, the longitudinal control has to decrease speed and stabilise it before the lateral shift could start.

At the beginning, the use of the brake is expected only to perform the obstacle avoidance function and to stop the car at the end of the manoeuvre when the speed is low and the car is already in the emergency lane. The combined usage of both controllers will be limited in order to avoid vehicle's stability problems.

A first simple strategy hypothesis is shown in the state diagram of Figure 2

In this schema the speed changes during steering intervention are minimised. The Supervisor uses the two controllers in separate way; when the speed is stable in the longitudinal domain, it changes lane position in the lateral domain.

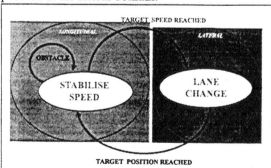

Figure 2 - Control strategy: simplified hypothesis

The actual implementation is based upon the following **hypothesis:**

- highway without junction, multi - lane with emergency lane (see example)

- Lateral controller with target lateral speed of 0.3 m/sec
- No obstacle on the trajectory
- Longitudinal control at constant speed or low deceleration (1 m/sec2)
  and with some **constraints**
- The vision system is able to recognise only one lane at time (left and right borders).
- The identification of a lane border is possible only if the car is inside the lane.
- A good identification of the lane border type (continuous/dashed) is possible only for the nearest border.

Starting from these a control based on the following state machine is implemented Figure 3.

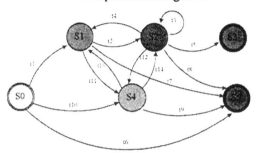

Figure 3 - ACD state diagram

The states are described in the following tables.

Table 1 - ACD control - State table

State description	
S0: STAND BY	waiting activation
S1: FIRST PHASE	go close to the right lane border
S2: SECOND PHASE	cross the lane border
S3: END	end of the manoeuvre - keep the left continuous border
S4: RECOVERY	recoverable problem - keep actual position
S5: ERROR	no recoverable problem: no lane visible

For any state is possible identify a car position in the lane as illustrated in Table 1.

The state diagram is completed by the following table describing events and actions related.

Table 2 - ACD control - Events and actions

Event & Action description		
	Events	*Actions*
t1	ACD manoeuvre activated (by driver or IMU) & right border is	- take the right border like reference - enable lateral controller - move the car close to the

	visible	right lane border
t2	the car is close to the right lane border	- cross the lane border
t3	the line is at the centre of the car	- leave the left border - the right border becomes the new left - search for a new right border (change lane)
t4	the car is in the new lane close to the left border & the left border is dashed & exist a right border	- change reference (take the new right border) - move the car close to the right lane border
t5	the car is in the new lane close to the left border & the left border is continuous.	- keep the car close to the left border MANOEUVRE FINISHED
t6	ACD manoeuvre activated (by driver or IMU) & both border are not visible	TBD - the lateral controller actually is disabled
t7 t8	both border are not visible	TBD - the lateral controller actually is disabled
t9	(both border are not visible) or (Time-out finished)	TBD - the lateral controller actually is disabled
t10	ACD manoeuvre activated (by driver or IMU) & the reference right border is not visible	- keep the car lined up to the other border - start the Time-out counter
t11 t12	the reference border is not visible	- keep the car lined up to the other border - start the Time-out counter
t13 t14	the reference border appear again	- check congruous situation - restart manoeuvre

The consequence of this possible solution is: to use both controller simultaneously; for the longitudinal domain there is no target speed setting but a target constant deceleration while the lateral controller resumes a constant lateral speed.

The lane change manoeuvre will appear with different trajectories as shown in Figure 5a and Figure 5b

The figures are relative to a highway with 3 lanes; in particular in the second one there is an assumption for the speed to be respected in the lane itself. The Supervisor requires to the longitudinal control to reach a speed accordingly to the lower in the lane. Only if this condition is reached the lane change will allowed to start. In this way a step by step strategy is applied. In the first picture, relative to the other solution, it is possible to implement a more continuous and smooth car trajectory.

EMERGENCY MANOEUVRE EXAMPLE

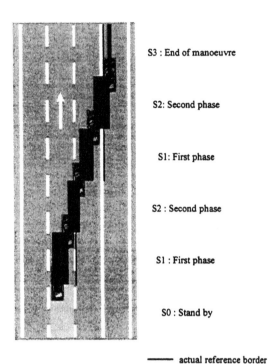

S3 : End of manoeuvre

S2: Second phase

S1: First phase

S2 : Second phase

S1 : First phase

S0 : Stand by

——— actual reference border

Figure 4- ACD control - States and relative car position in the lane.

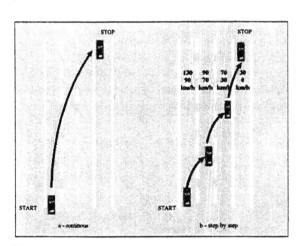

Figure 5- Different trajectories in emergency manoeuvre

Related to these strategies there are some technical and behavioural problems that need further research, such as:

- speed and acceleration technical limits;
- single step and whole manoeuvre requested time;

- behaviour of other drivers on the road with both hypothesis.

These arguments will be studied in depth on a simulator and by driving the prototype in real traffic.

Initial test with the ACD demo car

CRF has performed an initial test with the demo car on a real road. At an initial speed of 125 km/h the ACD performed a lane-change manoeuvre. Speed was allowed to decrease by releasing the accelerator. It was a straight road with two lanes and an emergency lane (hard shoulder). Lane width was 3.5 m and the car was placed in the left most lane when the manoeuvre was initiated. The target lateral speed was 0.3 km/h. The manoeuvre was completed in approx. 27 seconds. Figure 6 shows the raw lateral position data of the vehicle. The two lines represent data emerging from the two rows of lane marks defining the lane. The deviation between left and right data are probably caused by bad lane marks. As can be seen from figure the needed longitudinal length for moving the vehicle to the emergency lane (7.94 m) was approx. 700 m. The graphs represent the middle of the car. The lane-change angle was estimated to 0.65 °. The average lateral speed for the complete manoeuvre was 0.29 m/s^2.

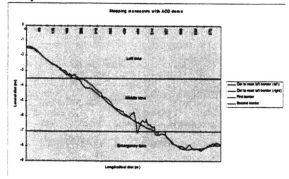

Figure 6- Lane-change manoeuvre on a real road with the SAVE demo controlled by the ACD

Lane changes and simulator experiments

There is a need for tests and experiments involving potential users/drivers in order to support the ACD development. A user analysis performed within the project revealed a somewhat negative attitude to the ACD function as it was described in the questionnaire (Bekiaris and Petica, 1996). This negative attitude was probably based on the imagination and expectations of *automatic driving* which is different at least in some respects from the ACD function of the SAVE system. Automatic driving in general could be viewed as an "auto-pilot" which takes over the primary driving task from the driver. In this case the driver will become more like a supervisor than an active controller of the vehicle. The ACD will of course perform an automatic driving manoeuvre *but* with a very specific purpose i.e. to safely *stop the vehicle*. The

ACD is activated in order to avoid an imminent accident when there is no driver who can safely control the vehicle any more.

Both the driver of a SAVE equipped car and drivers of surrounding vehicles will be influenced by the way the emergency manoeuvre is executed. If the driver is unconscious or for other reasons completely out of control then the manoeuvre should be executed without causing any road accidents and without causing any further problems to the driver. If the driver has manually activated the ACD then it could be expected that the driver would prefer a "smooth" and "normal" stop of the vehicle. Probably this is also a valid strategy in order to not disturb the surrounding traffic too much. So it seems logical to let the ACD stop the vehicle in a fashion that as much as possible resembles the way a non-impaired driver would do. So there is a need for data on normal lane changes that could be used for the emergency control strategy.

Van Winsum, Brookhuis and De Waard (1997) performed a driver simulator experiment where data on normal lane change manoeuvres were collected. As a result normative data on lane change behaviour such as lateral speed were defined and tabulated. From these data it can be derived that a mean lateral speed during lane changes is approx. 0.5 m/s. In this experiment there was no traffic that the subjects had to interact with also the lane change manoeuvres were not self-paced but subjects were told when to make lane-changes both to left and right. So it might be suspected that the lateral speed found in the experiment should be too high but this does not seem to be the case. Wallman (1975) found that average lateral speed during lane changes was 0.85 m/s, based on observations of real traffic on a three lane motor way. In this case the driving task was self-paced and should reflect normal driving behaviour today even if the observations were made 23 year ago.

The Swedish National Road Administration (SNRA) have issued some guidelines for road design (SNRA, 1994), where we can find some guidance on acceptable lateral speeds. In this document there is a section that deals with jerks due to road design. Jerks are defined as momentary changes in forces on the driver and the vehicle. This guideline says that acceptable lateral jerks should be below 0.45 m/s^3 and if above 0.8 m/s^3 it will be experienced as uncomfortable by the driver. They have also estimated longitudinal lengths needed to make lateral movements up to 7 m. From this figure we can see that at a speed of 50 km/h the driver will need 110 m of longitudinal travel to make lateral motion of 3.5 m in order to assure that is will not be experienced as uncomfortable. A lateral motion of

3.5 m would be enough for a normal lane-change. Table 2 gives the needed distances for some other speeds.

Table 2 Longitudinal distances needed to make lateral motions (3.5 m and 7m) under various speeds in order to avoid driver discomfort

	3.5 m lateral motion	7 m lateral motion
30	70	88
50	110	135
70	148	185
90	175	220
110	192	240

From these data we can calculate a lane-change angle and an average lateral speed and it seems that an acceptable target value for lateral speed during lane changes is approx. 0.5 m/s or any value below. These are of course rough estimations but should give some guidance for the ECS.

So it seems like the target lateral speed used in the SAVE demo car is rather low (0.3 m/s). This low value is though required initially as a precaution in order not to cause vehicle instability and to ensure safety and comfort for the test driver. But in the future, after field tests with the demo car, the target lateral speed can most likely be increased. There is no need to keep the lateral speed low as this will prolong the manoeuvre and will also require longer merging gaps when there is traffic.

A series of AHS (Automatic Highway Systems) experiments have been conducted in a driving simulator by Bloomfield et al. and the result of these studies can also be valuable input for the ADC and the development of the emergency control strategy (Bloomfield et al., 1995a), (Bloomfield, Buck, Christensen, & Yenamandra, 1995b), (Bloomfield, M., Carrol, & Watson, 1996b), (Bloomfield, Christensen, D, M, & A, 1996a). These AHS experiments have addressed a number of questions concerning the transfer of control between the human driver and the AHS system, how to get in and out of a platoon and finally comfortable distances to lead vehicles once in the platoon. Here we can only mention that the level of average lateral speed during lane changes seems to quite well correspond to the data found by Van Winsum, Brookhuis and De Waard (1997).

Lane changing manoeuvres with traffic
Next step in the development of the ACD and the emergency control strategy is to include traffic when the SAVE car performs an emergency stop. In a currently running experiment there is traffic in the adjacent right lane which the driver has to interact with when making lane changes. There is a need to define criteria for acceptable merging gaps in the traffic flow in the adjacent lane that allow for safe lane changes. These criteria should be included in the emergency control strategy.

The experiment aim at defining what are the best strategies and criteria for lane changes during the emergency manoeuvre. The results should give an indication on which are the best combined parameters to guide the vehicle e.g. deceleration, lateral speed, steering angle, etc..

The experiment consists of three phases. In the first phase the subjects will have to make lane changes where they have to merge in to a stream of cars. They will overtake a slower running stream of cars where the gaps between the cars will increase as the driver passes. The subject will be instructed to change lane when he/she judges the gap to be sufficient. This task will be repeated for different speeds. In a second phase the subject will drive as one of the drivers in the stream of car in the right line. They will then experience their own lane change manoeuvres and they will have to judge if the manoeuvre was critical or not. In a third and final phase they will experience the lane change manoeuvres as "impaired" drivers when the car makes the lane change manoeuvre and also this time judge how critical the manoeuvre was. The results of this experiment is expected to give some initial data for ACD controlled lane changes when there is surrounding traffic that has to be considered during the emergency manoeuvre.

Conclusions
The emergency manoeuvre has been defined on a simple scenario and tested on the road and in a simulator according to the following characteristics: extra urban well structured road, no junctions, low traffic density, good weather conditions. The human factor aspects are also considered to define the control strategies.

The main purpose of the ACD is therefore to safely and fast execute a stopping manoeuvre and place the vehicle away from the traffic by stopping it in the emergency lane.

The actual control strategy is realised with a supervisor and two different dedicated controllers: longitudinal and lateral.

The human interaction is considered as being part of the system. A driver simulator is used in order to define this type of interaction. Under this point of view in a future step, the simple scenario will be changed to introduce: more vehicles on the road, curves and junctions. It will be tested on a group of drivers (inside the SAVE car) to find out how they react to this automatic stopping manoeuvre. A second experiment will be set up to study the reaction of surrounding drivers will as a SAVE vehicle performs a stopping manoeuvre. The question of interest is if the SAVE vehicle should warn the surrounding traffic and if so by what means and to what extent. These driving simulator experiments are planned to take place in a near future at VTI. The implementation on the car are under the responsibility of the Fiat Research Centre.

References

Antonello P. C., Bozzo S., Damiani S "Driver steering task support system to improve vehicle lateral control" Centro Ricerche Fiat - XXVII FISITA Congress, Praha 17-21/6/1996.

ASV (Advanced Safety Vehicle) Program of the study group for promotion of ASV in Japan, Ministry of Transport, 1996.

Bekiaris A. and Petica S. (eds) (1996) Driver Needs and Public Acceptance of Emergency Control Aids, SAVE Project (TR 1047) Deliverable 3.1, Athens, Greece.

Bloomfield, J. R., Buck, J. R., Carrol, S. A., S, B. M., Romano, R. A., McGehee, D. V., & North, R. A. (1995a). Human Factors Aspects of the Transfer of Control from the Automated Highway System to the Driver (FHWA-RD-94-114). Minneapolis: Honeywell.

Bloomfield, J. R., Buck, J. R., Christensen, J. M., & Yenamandra, A. (1995b). Human Factors Aspects of the Transfer of Control from the Driver to the Automated Highway System (FHWA-RD-94-173). Minneapolis: Honeywell.

Bloomfield, J. R., Christensen, J. M., D, P. A., M, K. J., & A, G. (1996a). Human Factors Aspects of the Transferring Control from the Driver to the Automated Highway System with varying Degrees of Automation (FHWA-RD-95-108). Minneapolis: Honeywell.

Bloomfield, J. R., M., C. J., Carrol, S. A., & Watson, G. S. (1996b). The Driver's Response to Decreasing Vehicle Separations During Transitions into the Automated Lane (Working Paper FHWA-RD-95-107). Minneapolis: Honeywell Inc.

Brookhuis K. (1995) DETER (Detection, Enforcement & Tutoring for Error Reduction) Final report, Deliverable 20, TRC, Haren, the Netherlands.

Kaneda M., Iizuka H., Ueno H., Hiramatsu M., Taguchi M., and Tsukino M. (1994) Development of a drowsiness warning system, Nissan Research Centre, The 14th Int. Technical Conference on Enhanced Safety of Vehicles, Munich, 23 - 26 May, 1994.

Shimotani M., Satake T., Seki M., Nishida M., Terashita H., Ogawa K., Suzuki Y. (1996) On Board Drowsiness Detection & Alert Device Based on Image Processing, 3rd Annual World Congress on ITS, Orlando USA, October 14 - 18, 1996.

SNRA. (1994). Kapitel 3.3.6 Ryck i Vägutformning 94 (Chapter 3.3.6 Jerks in Roaddesign 94) (Publication 1994:049): SNRA (Swedish National Road Adminstration), Borlänge; Sweden.

Ulmer B. (1994) Autonomous Automated Driving in Real Traffic, First World Congress on Applications of Transport Telematics and Intelligent Vehicle-Highway Systems, Paris , France, 30 November - 3 December 1994.

Van Winsum W., Brookhuis. K. De Waard, Dick (1997) Characteristics of lane change behaviour, Internal SAVE project report, Soesterberg, The Netherlands

Wallman, C.-G. (1975) Simulering av trafikflöden i planskilda korsningar (Simulation of Traffic Flow in Graded Separated Intersections), Doctoral Thesis, Chalmers Technical Highschool, Gothenburg, Sweden.

Yamamoto K., Yoshikwa M., Higushi S. (1996) Development of Driver's Alertness Monitoring System, 3rd Annual World Congress on ITS, Orlando USA, October 14 - 18, 1996.

SMART VIDEO SENSOR FOR DRIVER VIGILANCE MONITORING.

A. Giralt, S. Boverie

SIEMENS Automotive SA - B.P.1149 av. du Mirail - 31036 TOULOUSE Cedex France.

Abstract:
Monitoring the car driver's vigilance is a complex problem. It requires a non intrusive system able to measure and interpret symptoms independently of driver's physical characteristics, way of driving and environment. Within the European Program SAVE a multi-sensor approach for improving the driver impairement diagnostic is considered. Among the information which can be used to predict driver impairement the observation of eyelid patterns seems to be one of the most relevant. Siemens is developing a non intrusive, real time, on board video-image processing system for measuring: the driver's eyelid pattern i.e. the eyelid sensor. This paper gives an overview about the SAVE program. It describes in section 3 the main HW and SW characteristics of the eyelid sensor. Finally, the results of preliminary evaluations performed for car day time real-driving conditions and for night conditions on driving simulator are presented.
Copyright © 1998 IFAC

Keywords: Video sensor, Image processing, automotive systems, safety

1. Introduction.

Improvement of safety in road traffic corresponds to an increasing demand from a majority of users as well as to a social and economical necessity.

Among the functions that impact on traffic safety the drivers physical and psychological state and more espicifically driver impairement play an important role. The assessment of the driver vigilance state has attracted wide interest both in basic research and in solution to the development of monitoring systems.

The term "driver impairment" encompasses all the situations in which the driver's alertness is diminished, and therefore when the driving task cannot be maintained at an adequate level of performance. It is a consequence of stress, fatigue, alcohol abuse, medication, inattention, effects of various diseases. Driver impairment in connection with these states is the first cause of accidents on European motorways. According to Vallet (1991), « it is generally a loss of alertness which is the principal cause of fatal accidents (34%) followed by punctures (14%), lack of attention to meteorological conditions (12%), non respect of distance between vehicles (11%), loss of control through speeding, if a monofactorial analysis of accidents is possible ». Similarly, Smiley and Brookhuis (1987) stated that about 90% of all traffic accidents can be attributed to human failure, for instance, fatigue, inattention, drowsiness at the wheel.

To monitor driver's vigilance is a complex problem which requires a user accepted system able to measure and interpret symptoms independently of driver's physical characteristics, way of driving and environment.

The SAVE@ project (System for effective Assessment of driver state and Vehicle control in

---

@ The SAVE Consortium gathers representatives of all the actors of automotive field, including end users (FIA: International Federation of Automobile-clubs), car manufactures (Fiat, Renault), on-board electronics suppliers (Siemens), road safety institutes and research laboratories (AVV, CEBHA-CNRS, HUSAT, IAT, INRETS, LAAS-CNRS, TNO/TM, TRC, TRD, VTI). The prime contractor is the TRD company. SAVE is a 3 years project, which started in Jan 1996

Emergency situations), is a direct result both of the critical situation arising from the number of accidents caused by driver impairment and the parallel development of new ITT driving technologies (see below). This project is partly funded by the European Union's Transport Telematics programme under contract number TR 1047.

The aim of the SAVE project is to diminish the severe consequences of road drivers impairments.

We consider SAVE as a system to be an essential part of future technology, which will improve road safety, particularly by decreasing the number of accidents due to driver impairment. It is planned to be a non-obtrusive system which will monitor driver alertness in order to:

- **warn** the driver in case of impairment
- **assist** him by safely parking the car in case of emergency, through the use of an Automatic Control Device.

The SAVE system will operate thanks to various integrated systems. (For further details, see Bekiaris 1996).

A. System monitoring driver alertness (IMU)

- Several sub-systems will be integrated into this main unit so as to control the driver's performance. Thus, depending upon the driver's degree of alertness (and implicitly, the degree of danger of the situation

This system will detect and assess in real-time various driver states and driver impairment due to alcohol, medication, stress, fatigue, inattention, illness or any other relevant cause.

Moreover, it is planned that an integrated system of all these data will provide the analysis of the real driver behaviour ; it will be able to compare it to the driver's usual driving style. Detection is based on non intrusive techniques, including:

- driver's profile identification and recognition of serious deviations from it,
- detection of relevant road-safety aspects violation,
- measurement of steering wheel force,
- observation of the driver's face, considering in particular the eyelid pattern.

B. System which "takes the steering wheel" in case of emergency (Automatic Control Device ACD)

In case of emergency, and only if the driver is not responding or responding erroneously to driver warning system indications (see C), the previous system will inform this automatic control device. Once informed, it will take charge of the situation : it will perform an automatic manoeuvre. In parallel, the driver himself could initiate this automatic operation, by means of a "panic button". ACD will stop the vehicle safely and it will guide it to the roadside in order to park it in a safe location. To decide to conduce the manoeuvre, there will be a supervisor which processes the relevant information in order to adopt the most suitable strategy.

C. SAVE Warning System (SWS)

A Telematics network will make it possible to warn surrounding drivers and emergency services. This system could warn both the driver as well as other road users so that they keep a safe distance from the vehicle in danger. On one hand, in case an emergency is identified by **IMU**, a **pre-emergency warning** system will inform the driver and the surrounding traffic. On the other hand, an **emergency warning system**, based on a multimedia concept, will warn the driver <u>prior to ACD activation</u>, warn surrounding traffic before and during its operation; and once the vehicle has reached a safe, <u>stationary</u> position to call for help both through the surrounding traffic and an emergency control centre (for more details, see Boverie *et al*, 1997)

In conclusion, SAVE could be considered as a preventive as well as an accident avoidance technology.

The contribution of Siemens to SAVE project concerns the Integrated Monitoring Unit and the SAVE Warning System.

- IMU: Siemens define, develop and test a non-intrusive eyelid sensor prototype.
- SWS: Siemens prototypes the Driver Warning Device according to preliminary specifications which take into account the inputs coming from SAVE consortium partners.

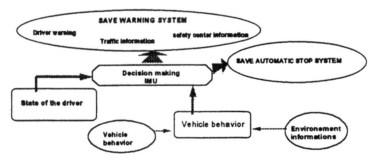

Figure 1.1 SAVE general concept description

162

2. The driver monitoring system

The objective of the IMU is to detect and assess in real time that a driver impairement is imminent, giving the possibility to the driver, through adapted warning to take a decision and to react to the situation (if possible).

One of the main purpose of this study was to find and to design adapted systems which would be able to provide information to predict driver impairement situation. Figure 2 presents a cause to effect diagram which shows the different level of observation we can get. Four level can be considered, for each one different measurements, observations can provide information on the driver state. Unfortunately, in no way one measurement by itself is able to present 100% of reliability in the sense of the impairement prediction. Depending on the measurement type this lake of reliability can be due to several factors:

- The sensor reliability.
- The environemental conditions; for example in the case of the eyelid movement sensing; this measurement suppose that the driver's face is correctly located in the field of the camera, when the driver is looking on the side of the road or in general when he moves the head out of the field of the camera (which situations are current in driving conditions) no measurement can be performed.
- the relevance of the measured phenomena; for example in the case of eyelid movement analysis statistical analysis show that the relevance of the phenomenum is around 90 to 95%.

The basic idea of SAVE is to fuse different informations, provided by non-intrusive sensors in order to improve as much as possible the reliability of the driver state diagnosis. The selected informations are based on:

- hand pressure on the steering wheel
- steering wheel movement
- pedal movement
- position on the vehicle in the lane
- eyelid movement (Which is one of the most reliable information). This sensor proposed by Siemens Automotive combines a smart image sensor and a powerful processor for real time scene analysis and local features interpretation.

The following sections present the eyelid sensor system in more details and its preliminary evaluation results.

3. Eyelid sensor prototype description

3.1 Background and overview:

Eyelid pattern (eye blink, slow eye closure) appears to be one of the most relevant symptom for detecting driver's drowsiness. Wierville uses the measure of the proportion of time that the driver eyes are 80 to 100 % closed to asses drowsiness (Wierville, 1994). To measure eyelid pattern two systems are currently being developed:

- the first one is an intrusive one based on a infrared emitter/sensor mounted on a eyeglass frame. The device emits an infrared beam at the eye and measure the reflected light
- the second one which is actually developed by Siemens is a non intrusive system based on a distant on-board video sensor and image processing (similar approach is explored by Renault for trucks on night driving conditions and by Nissan) (Giralt, 1995) (Lavergne, 1996), (Kaneda, 1994).

Eyelid Sensor is composed of three main elements:

- an IR video image sensor which is placed on the dashboard in front of the driver in order to obtain images of the driver's face
- a distant image processing unit which localizes and tracks the eyes and measures the degree of eye opening on the digitized image

an infrared light which is activated during low illumination driving conditions (night, ..) (Fig3).

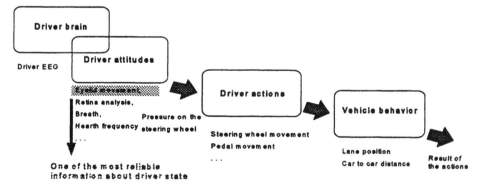

Figure 2: Cause to effect diagram for driver state prediction

3.2 System main specifications:

The Eyelid Sensor main specifications derive from and comply with the two following crucial constraints:First the necessity of users acceptance implying non-intrusiveness and very limited cooperation, if any; Second, to guarantee a sufficient level of validity within realistic operational limits.

3.2.1 Acceptability:

Indeed car user's acceptability entails necessarily that the system must not be intrusive or require any action from the driver. The driver must not have to identify himself, to initialize the system, by image centering or by staying still for a while, to wear any tag like special glasses even for a very short period. We have made the reasonable assumption that only a few menial tasks such as (possibly the only one) to clean once in a while the optical lens would be acceptable. May be for professional drivers going for a long trip a learning phase of a few seconds could be accepted. The driver would have to stay still or wear special glasses. This phase would increase the system reliability and would reduce Its complexity. In both case the system must be felt by the user like an help for safe driving an not a mandatory system.

3.2.2 Eye opening measurement validity and accuracy:

Eye opening measurement must be reliable

- independently of the size of driver, his facial characteristics, (skin color, hair cut, ...)
- for every stable position of the head (head on the side, head backward, ..)
- for falling asleep positions (head nodding)
- when the driver performs long duration actions (smokes, chews gum, ..). In the case of brief actions like specific head movements (the driver looks at the rear mirror, at the radio, at the dashboard,) measurement is not relevant and it can be discarded.
- when the driver's wears glasses. Glasses is one of the most difficult problem: during day light reflections can mask the eyes. Some sun glasses are totally opaque to infrared making measurements impossible.

A typical eye blink of an alert person is characterized by a 10mm amplitude, a 80ms closing duration and a 200ms opening duration [Guitton, 91] [Stern, 84]. The duration depends mainly on the level of alertness, a factor of at least 1.5 differentiates a drowsy driver from an alert one [Regouby, 94]. The amplitude depends mainly on eyes characteristics, drowsiness level, sight orientation, head inclination, ..for a majority of persons vision can still be possible with a 2 to 3mm opening). Of course indication of the direction of sight would be very interesting for inattention detection but it requires a much more complex system.

4. Algorithm outline:

The algorithms were developed in collaboration with two research laboratories of Toulouse: the LAAS (Laboratoire d'Analyse et d'Architecture des Systemes) and the LEN7 (Laboratoire d'Electronique de l'Ecole Nationale Supérieure d'Electrotechnique d'Electronique d'Informatique et d'Hydraulique de Toulouse), they take advantage of previous algorithm developed by the Siemens Research center for general purpose applications (Fouchet, 1995), (Scotto, 1996).

A first level in our approach processes the global scene for robust eye localization adapted to driver situations and morphologies and to beat whenever is relevant in a bounded sub-image (represented by dark rectangles centered on the eyes on the figure 3). An edge filtering function is used to enhance regions of interest (eyes, hair, mouth, ..). Possible eyes positions are selected by a morphological analysis of previously selected regions. The final eyes position are choosen by template matching. Real time processing is not required but processing speed should be high enough to insure accurate positioning of the bounded sub-image

A second level independently processes the pixels within the bounded sub-image. The algorithms here yield to locate accurately the eye (tracking) for measurement of the eye opening degree in real time (40 ms), the result are validated by the first level situation assessment (Fig 4). This interplay between the two systems aim to achieve the essential features of a high validity rate for true detection but even more a very false low rate for false detection.

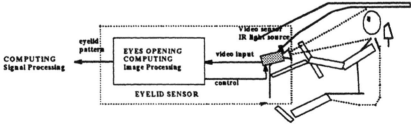

Figure 3 : Eyelid sensor system overview

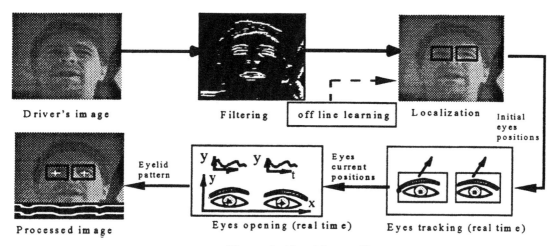

Driver's image   Filtering   off line learning   Localization   Initial eyes positions

Processed image   Eyes opening (real time)   Eyes tracking (real time)

Eyelid pattern   Eyes current positions

Figure 4: Algorithm outline

5. System characteristics

The camera is a CCIR monochrome standard, interlaced scanning, 1/2" sensor, 752 (h) x 582 (v) pixel definition which is located on the dashboard in front of the driver, under the helmet so It looks at the driver's face through the steering wheel. Special care must be taken to protect the lens from direct sun light.

It's rotated 90° , so driver's hair is seen on the left of the screen (this configuration increases accuracy of eyelid measurement).

It's inclination is around 25 °, the focal length is 8.5 mm one and localization distance is 65 cm. This values have been determined so no mechanical adaptation to driver's position and morphology is necessary.

- two infrared lighting unit stuck to the camera is activated for low illumination conditions.
- image processing is performed by a system based on a multi DSP TMSC40 board plugged in a PC.

6. Evaluation results:

The purpose of the evaluations carried out with the Eyelid Sensor prototype is to evaluate its performances regarding eye localization, and next the global performance of the system.

6.1 Methodology for eye localization performance analysis

Tests have been performed on 10 different subjects on car driving situations at daytime (at 3 p.m.) on a sunny afternoon in a Renault 21 car and more then 20 subjects on laboratory on low illumination environment. These 30 persons have been chosen in order to be a representative sample of face characteristics. Glasses, beard, mustache as well as different color of the skin have been taken into account. All the tests were performed without any adaptations of the system characteristics to driver's morphology and seating position, in other words

camera position, camera incline, focal length, localization distance and lens aperture were not modified. Likewise no software parameters have been modified.

6.2 Eyes localization results

The analysis of the video tape provided interesting and quite encouraging results summarized below. Subjects without glasses:

Detection rate	96 %
False detection rate	1%
No detection	3 %

Subjects with glasses:

Detection rate	65 %
False detection rate	1%
No Detection	34%

6.3 Methodology for the global evaluation of the eyelid sensor

Tests have been performed on vehicle simulator, by night driving conditions. All the tests were performed without any adaptations of the system characteristics to driver's morphology and seating position. Because of the huge amount of work needed for quantitative analysis of the eyelid pattern the analysis has been performed on a restricted number of subjects. It consisted in a manual comparison between a video display of the driver's image and eyelid pattern datas recorded by the eyelid system. For each driver four consecutive driving sequences have been considered. For each sequence the correlation between eyelid time closure automatically measured - time closure manually calculted - and simultaneous closure of the 2 eyes have been calculated. False closure time detection considers the case where the difference between closure time manually measured and automatically measured is more than 20%

6.4 Results of the global evaluation of eyelid sensor

Figure 5 a-b show a sample of the results obtained for various drivers. They globally show a very low false detection rate and a correlation between real and measured closure time between 50 and 70%

Most important reasons for non detection of closure time are: head movement, abnormal positions, hands on the face

False alarms are mainly generated by large blinking. Correlation based on simultaneous blinking of both eyes reduce drastically the false alarm rate but also the detection rate. In addition, more detailed results are given in figure 6 for subject number one. Figure 6a shows, for each eye the correlation between eyelid sensor measurement and manual measurement (for right and left eyes) and next the cross correlation for both eyes.

Figure 6b shows the drastic reduction of the false alarm rate brought by the 2 eyes cross correlation, for this subject, false alarm rate is reduced to zero.

7. Product cost analysis:

The cost of the electronic components integrated in car is a major and permanent trend. From 1980 to present the amount has more than been multiplied by a factor of 5, increasing from $300 to $1600 with an expected figure of $2000 for the year 2000 [BPA,]. This constant raise is generated by more and more novel and highly sophisticated electronic systems which provide the capacities to solve the ever growing constraints of reliability, confort, security, cleanness, silence and fuel economy.

First applications appeared in the years 70 and were dedicated to engine control, later on security related system (Airbag, ABS), power management, on-board driver information, ...were integrated in the vehicle. No doubt that the systems devoted to active security such as vigilance monitoring, anticollision.. will presently follow.

Video system and image processing appear to be a quite different story since almost nothing exist but a few rear-vision systems that equip some busses and trucks. Yet it is easy to contend that a number of very interesting applications do exist ranging from outside perception systems (e.g. obstacles detection, enhance visibility, lateral vision, overtake checker, ..) and inside ones (vigilance monitoring, anti theft, ..).

There is unfortunately, a major drawback: the cost! Nowadays a $100 low-complexity vision system is a conceivable price. Such system would have a CCD/CMOS sensor of 256x256 resolution, a 25 MIPS Digital Signal Processor and a 512 Kbytes image memory, and an approximate size of 250 cm3.

The effort to lower vision system cost are focused on two major components:

- CCD sensor by integration on the same chip of a CCD/CMOS sensor , an Analogic to Digital Converter and pre processing image functions allows high perspectives of cost reduction by means of miniaturization and hybrideless components.
- processor by development of dedicated image processors. A 25 MIPS Digital Signal Processor should reach a $2 to $3 (high quantity) by the year 2000.

These development directions are part of the Electronic Eye project which aim to:

- design a modular picture-processing architecture based on a Vision Instruction Processor which combines biologic inspired and conventional processing.
- design a low cost system architecture based on new smart sensor and VLSI systems according to the real-time requirements and neuronal signal processing.
- to demonstrate the efficiency (reliability, real-time, cost, ..) of these vision systems in industrial applications.

The Electronic Eye consortium gathers various teams of the Siemens Research Center (ZFE), many German university and industries, among them Siemens Automotive, which provide specific applications. We strongly believe that today concurrent efforts from industry and research center, in particular the large cooperative project Electronic Eye will achieve decisive results that will bring down the cost of automobile oriented vision systems to market level breakpoint for use in cars in the year 2005.

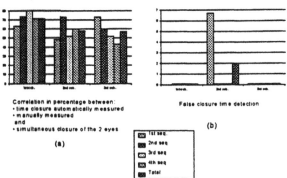

Figure 5 Evaluation of the performances of eyelid sensor for different subjects

166

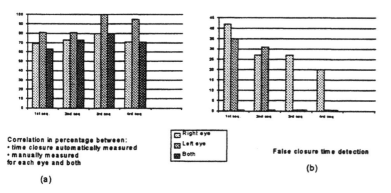

Figure 6 Evaluation of the performances of eyelid sensor for subject1 Comparison of the performances for each eye

8. Conclusion.

In this paper we have presented the SAVE European program activity and more espicifically the work performed by Siemens on the eyelid sensor development. The evaluation results performed up to now have shown interesting and promising results.

Nevertheless, it is now clear that 100% of detection rate would never be reached by such sensor. This can be easily explained by many unmanageable factors; outside the reliability of the observed symptom by itself, the detection of the eyes should be critical in many situations:

abnormal position of the driver face with respect of the video sensor observation field, transient movements of the driver, head movements, . . .

But the equivalent problems have also been encountered with other sensors or other observed symptoms.

This why a good diagnosis of driver state should be based on a fusion of several informations coming from different sensors; This is the objective of the SAVE program to fuse these informations.

Acknowledgments.

Part of this work has been supported by SAVE European project and founded by the European community.

References:

BPA Technology and Management. Report N° 573.

Brookhuis, K.A. (1995). Integrated systems. Results of experimental tests, recommendations for introduction. Report DETER Deliverable 18 to the European Commission. Haren: Traffic Research Centre, University of Groningen.

Boverie S. Bekiaris E, (1997). Strategy for pre- and post emergency actions, SAVE deliverable n°8.1.

Fouchet, (1995) "Developpement d'algorithmes de traitement d'images pour la localisation des yeux de conducteurs routiers", Rapport LAAS.

Giralt, (1995)"Technologie vidéo pour la détection de vigilance des conducteurs routiers", SIA.

Guitton, (1995) "Upper Eyelid Movements Measured With a search Coil During Blinks and Vertical Saccades", Invest Ophtalmol Vis Sci, Vol. 32, N° 13, 91.

Kaneda, (1994)"Development of a drowsiness warning system" ESV, Munich.

Lavergne, (1995)." Results of the feasibility study of a system for warning of drowsiness at the steering wheel based on analysis of driver eyelid movements", E.S.V.

Regouby C., (1994) "Evaluation de la baisse de vigilance par analyse du mouvement palpébral", Rapport de stage, Siemens Automotive.

Scotto Di Rinaldi (1996)"Détermination d'une chaine d'acquisition d'images vidéos et développement d'algorithmes de traitement d'images pour l'extraction du mouvement palpébral de conducteurs routiers", Rapport ENSEEIHT.

Stern, (1984)"The Endogenous Eyeblink", Psychophysiology, Vol. 21.

Vallet M. (1991) "Les dispositifs de maintien de la vigilance des conducteurs de voiture" In: "Le maintien de la vigilance dans les transports", Caen: Paradigme.

Wierville, (1994). "Overview of research on driver drowsiness definition and driver drowsiness detection" ESV, Munich.

POSE ACQUISITION THROUGH LASER MEASURES IN STRUCTURED ENVIRONMENTS

J.C. Raimúndez Álvarez, E. Delgado Romero

Dpto. de Ingeniería de Sistemas y Automática, E.T.S.I.I. Universidad de Vigo.
Lagoas Marcosende, s/n. 36200 Vigo (Spain).
Fax: +34 (86) 812201. Phone: +34 (86) 812244. E-mail: cesareo@centolla.aisa.uvigo.es

Abstract: The desirable autonomy in a mobile robot, force to identify its position and orientation ,"pose", in relation to a global reference system previously defined. . Without "a priori" assumptions, i.e. a region for start of search, the problem of pose acquisition has a combinatorial feasibility potential that can represent a serious challenge to naive attempts of solution. Even in the simple case of knowledge of the most probable starting region, myopic procedures inspired in steepest descent do not succeed in many trials. This paper presents a procedure of pose acquisition based on Evolutive Strategies and afterwards maintaining the pose using a myopic procedure here presented as a result of potential minimisation between measurement points and a CAD map segments. The switch from pose acquisition to pose correction is made depending on the closeness measure obtained at each step. If the closeness value exceeds a given bias, the procedure of pose acquisition follows and thereafter the pose correction takes place. *Copyright © 1998 IFAC*

Keywords: Mobile robots, Laser scanner, Evolutive Strategies, Pose acquisition, Pose correction.

1. INTRODUCTION

To know the starting point in any path following task, with confidence, is necessary. Structured environments means "a priori" knowledge about the surroundings main features. Those features must be detectable in some degree to the robot sensors on board.

Laser scanning offers a polar span centred in the laser focus and reaching each obstacle apart, according to a straight issuing from the focus. Depending on the cadence of laser shots in the polar span, rich set of information can be extracted containing data about the relative pose of the conveying platform.

Assuming a surroundings description given by a map whose features are described by straight segments, here cited from now on as CAD (stems

for Computer Aided Design) map in absolute coordinates, and scanning it horizontally from a constant height, the initial pose acquisition problem can be formulated now as: find a rigid coordinates transformation (a rotation and a translation) over the CAD map such that the laser range data its the closest possible, to some partially occluding map segments. The rotation and translation found are the local to global coordinate transformations needed (Ingemar, 1989).

2. EVOLUTIONARY STRATEGIES

Simulated evolution is based on the collective learning processes within a population of individuals, in the quest for survival (Bäck T. et al, 1997). Each individual represents a search point in the space of potential solutions to a given problem.

There are currently three main lines of research strongly related but independently developed in simulated evolution: Genetic Algorithms (GA), Evolution Strategies (ES), and Evolutionary Programming (EP). In each of these methods, the population of individuals is arbitrarily initialised and evolves towards better regions of the search space by means of a stochastic process of selection, mutation, and recombination.

These methods differ in the specific representation, mutation operators and selection procedures. While genetic algorithms emphasise chromosomal operators based on observed genetic mechanisms (e.g., crossover and bit mutation), evolution strategies and evolutionary programming emphasise the adaptation and diversity of behaviour from parent to offspring over successive generations.

2.1 Evolutionary Computation and Mathematical Programming

In the classical approach to optimisation methods there are many procedures iterative ones and systematic others, to search for stationary points in a response surface. Those points thereafter will be qualified as local maxima or minima or even saddle. Depending on such severe hypothesis as convexity, qualification can be global.

In general, optimisation procedures require finding a set $x \in \Omega \subset \Re^n$ such that certain quality criterion $f: \Re^n \to \Re$ is maximised (minimised). The solution to the global optimisation (maximisation) problem can be stated as:

$$\text{find a vector } x^o \in \Omega \text{ such that } \forall x \in \Omega \Rightarrow$$
$$f(x) \leq f(x^o) = f^o$$

Here Ω is the set of feasible solutions, f the objective function, and x^o one of the solution points. In real-world situations the objective function and the restrictions g_i which qualifies $\Omega \subset \Re^n$ are often not analytically treatable or even not given in closed form, when the classical procedures based mainly in the model regularity, fail to apply (Raimúndez J.C., 1997). The main contributions in the evolutionary computation approach are:

Model regularity independence.
Population search $\times$ individual search (classical).
General meta-heuristics.

2.2 Evolutionary Meta Heuristics.

Evolution is the result of interplay between the creation of new genetic information and its evaluation and selection. A single individual of a population is affected by other individuals of the population as well as by the environment. The better an individual performs under these conditions, the greater is the chance for the individual to survive for a longer while and generate offspring, which inherit the parental genetic information.

Evolutionary algorithms mimic the process of neo-Darwinian organic evolution and involve concepts such as:

$t \equiv$ Time or epoch.
$\# \equiv$ Individual.
$P(t) \equiv$ Population.
$\Phi \equiv$ Fitness.
$o \equiv$ Variation, Selection, etc.

A simple evolutionary algorithm follows:

```
t ← 0
initialise P(t)
evaluate Φ(P(t))
while not terminate
      P'(t) ← variation P(t)
      evaluate Φ(P'(t))
      P(t +1) ← select P'(t) ∪ Q
      t ← t +1
end
```

Q is a special pot of individuals that might be considered for selection purposes, e.g. $Q = \{\varnothing, P(t), ...\}$. An offspring population $P'(t)$ of size λ is generated by means of variation operators such as recombination and/or mutation from the population $P(t)$. The offspring individuals $\#_i(\pi_k) \in P'(t)$ are evaluated by calculating their fitness represented by $\Phi(f)$. Selection of the fittest is performed to drive the process toward better individuals.

2.3 Evolution Strategies

In evolution strategies the individuals consist of two types of parameters: *exogenous* π which are points in the search space ($\pi \equiv x$), and *endogenous* σ which are known too as *strategic parameters*. Then $\# = \#(\pi, \sigma)$. Variation is composed of *mutation* and *self-adaptation* performed independently on each individual. Thus:

$$\#(\pi', \sigma') \leftarrow \text{mutate } (\#(\pi, \cdot)) \cup \text{adapt } (\#(\cdot, \sigma))$$

where mutation is accomplished by

$$\pi_i' = \pi_i + \sigma \cdot N_i(0,1)$$

and adaptation

$$\sigma_i' = \sigma_i \cdot \exp\{\tau' \cdot N(0,1) + \tau \cdot N_i(0,1) \}$$

for $\tau' \propto (\sqrt{2n})^{-1}$ and $\tau \propto (\sqrt{2\sqrt{n}})^{-1}$

$N(0,1)$ indicates a normal density function with expectation zero and standard deviation 1.

Selection is based only on the response surface value of each individual. Among many others are specially suited:

Proportional. Selection is done according to the individual relative fitness

$$p(\#_i) = \frac{\Phi(f(\pi_i))}{\sum_k \Phi(f(\pi_k))} \qquad (1)$$

Rank-based. Selection is done according to indices that correspond to probability classes, associated with fitness classes.

Tournament. Works by taking a random uniform sample of size $q > 1$ from the population, and then selecting the best as a survival, and repeating the process until the new population is filled.

(λ, μ). Uses a deterministic selection scheme. μ parents create $\lambda > \mu$ offspring and the best ν are deterministically selected as the next population $[Q = \varnothing]$.

$(\lambda + \mu)$. Selects the ν survivors from the union of parents and offspring, such that a monotony course of evolution is guaranteed $[Q = P(t)]$.

3. FITNESS EVALUATION

The fitting quality between the data points obtained through laser scanning and the CAD map is evaluated by means of an index that in some way represents the disclosure between them. It is recommended to use a norm of a kind such as the mean distance from points to their closest segments in order to attenuate the importance of data out of standard deviation.

The pseudo code representing the closeness or so called fitness can be represented as:

```
distmean = 0;
for all points Γ(k), { k − 1, numpoints}
{
    distmin = ∞;
    for all segments S(i), { i = 1, numsegments}
    {
        Distance calculation dist = dist (P(k),S(i));
        if dist ≤ distmin then distmin = dist;
    }
    distmean+ = distmin;
}
distmean /= numpoints;
```

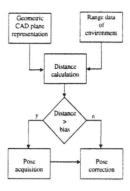

Fig. 1. Positioning estimation module

4. POSE ACQUISITION AND CORRECTION.

The positioning estimation module can be summarised as shown by figure 1.

4.1 Pose acquisition.

The distance function may have several relative minima and at least an absolute minimum. It is also a continuous function but not differentiable, which implies the impossibility of using methods based on gradient calculus, Newton type, etc. It could be used a Simplex type method (Nelder& Mead), but it does not solve the problem of the local minima. Therefore, is propounded the utilisation of Evolutive Strategies.

Likewise, distance is function of α, δ_x, δ_y where:

$$\text{dist} = f(\alpha, \delta_x, \delta_y) \qquad (2)$$

$\alpha \equiv$ planar rotation angle.
$\delta_x \equiv$ linear translation in x axis.
$\delta_y \equiv$ linear translation in y axis.

So, are established the following equivalencies:

chromosome $\Rightarrow \{\alpha, \delta_x, \delta_y\}$
fitness $\Rightarrow \{distmean\}$

Fitness represents the objective function to maximise. Generally, within the evolutive process, this function gets values in the interval [0,1]. So, it is considered:

$$fitness = \frac{1}{1 + distmean} \qquad (3)$$

As example, a CAD plane formed by fifteen straight-line segments is chosen. It is fired the laser scanner that provides thirty-four measuring points. Later on are showed results of some simulations in order to obtain the improve parameter of tuning under these conditions.

Table1 Simulation results of pose acquisition

Individuals	Epochs	Cross	Fitness	Time
20	109	1	0.95	2m
10	167	1	0.96	1m10s
6	400	1	0.86	2m
10	74	0	0.97	30s
15	91	0	0.97	40s

The final result were obtained using the following characteristics:

Search space:
angle: $-\pi/2$ a $\pi/2$
x co-ordinate translation: -30 a 30
y co-ordinate translation: -30 a 30
Number of individuals: 10
Number of epochs: 74
Cross probability: 0
Mutation probability: 0.01

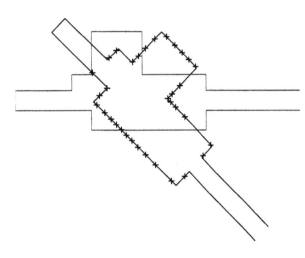

Fig.2 Pose acquisition through Evolutive Strategies.

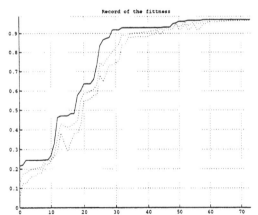

Fig. 3. Record of the fitness.

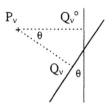

Fig. 4. The rigid CAD plane movement.

4.2 Pose correction.

In order to support the matching when the distance between the range data and the CAD plane is smaller than a certain bias, it is used a fast iterative procedure robust enough, inspired in the elastic potential energy associated to the CAD plane. So, it is considered as a rigid body connected to the measuring points (P_v) through lineal springs. According to the figure 4, the CAD plane potential energy will be:

$$V(q) = \frac{1}{2} \sum_{v=1}^{v=n} \langle f_v, P_v - Q_v \rangle \qquad (4)$$

$$f_v = k(P_v - Q_v) \qquad (5)$$

with k constant in every points.

The rigid CAD plane movement implies:
$Q_v{}^\circ \equiv$ point position before movement.
$Q_v \equiv$ point position after movement.
all the points Q_v will accomplish:

$$Q_v = R(\theta) \, Q_v^o + T \qquad (6)$$

where:

$$R(\theta) = \begin{pmatrix} \cos(\theta) & -\sin(\theta) \\ \sin(\theta) & \cos(\theta) \end{pmatrix} \qquad (7)$$

$$T = \begin{pmatrix} t_x \\ t_y \end{pmatrix} \qquad (8)$$

and then:

$$V(\theta,T) = \frac{1}{2} k \sum_{v=1}^{v=n} \langle P_v - R(\theta) Q_v^o - T, P_v - R(\theta) Q_v^o - T \rangle$$

$$V(\theta,T) = \frac{1}{2} k \sum_{v=1}^{v=n} \{ \langle P_v, P_v \rangle + \langle R(\theta)Q_v^o, R(\theta)Q_v^o \rangle + \langle T,T \rangle - 2(\langle P_v, R(\theta)Q_v^o \rangle + \langle P_v, T \rangle - \langle R(\theta)Q_v^o, T \rangle) \}$$

$$V(\theta,T) = \frac{1}{2} k \sum_{v=1}^{v=n} \{ \langle P_v, P \rangle + \langle Q_v^o, Q_v^o \rangle - 2(\langle P_v, R(\theta)Q_v^o \rangle + \langle P_v, T \rangle - \langle R(\theta)Q_v^o, T \rangle) \} \qquad (9)$$

Using the property:

$$\frac{d}{d\theta} R(\theta) = S\, R(\theta) \quad \text{with} \quad S = \begin{pmatrix} 0 & -1 \\ 1 & 0 \end{pmatrix}$$

Are obtained the partial derivatives to calculate the minimum potential energy point:

$$\frac{\partial V(\theta,T)}{\partial \theta} = 2\sum_{v=1}^{v=n} \left\{ \langle SR(\theta)Q_v^o, P_v \rangle + \langle SR(\theta)Q_v^o, T \rangle \right\} \quad (10)$$

$$\frac{\partial V(\theta,T)}{\partial T} = 2\sum_{v=1}^{v=n} \left\{ nT + R(\theta)Q_v^o - P_v \right\} \quad (11)$$

If it is considered:

$$\mu_p = \frac{\left(\sum_{v=1}^{v=n} P_v\right)}{n} \quad \text{and} \quad \mu_Q = \frac{\left(\sum_{v=1}^{v=n} Q_v^o\right)}{n} \quad (12)$$

$$tan(\theta) = -\frac{\sum_{v=1}^{v=n} \langle Q_v^o, SP_v \rangle + n\langle \mu_Q, S\mu_P \rangle}{\sum_{v=1}^{v=n} \langle Q_v^o, P_v \rangle + n\langle \mu_Q, \mu_P \rangle} \quad (13)$$

$$T = \mu_P - R(\theta)\,\mu_R \quad (14)$$

With these expressions it is constructed an iterative procedure that adjusts the results obtained former by the evolutive algorithm. Each iteration consist:

1. Determining the Q_v^o points on the CAD plane at a smaller distance of the P_v. (Figure 5)
2. Calculating θ_k, T_k according to the expressions (13) (14). So, determining the rigid CAD plane movement at instant k:

$$Q_v^k = R(\theta)\,Q_v^0 + T \quad (15)$$

3. Continue until $\sum_{v=1}^{v=n} \left\| P_v - Q_v^k \right\| \leq Error$.

A simulation of this iterative procedure that can be applied as already said, when the range data are next to CAD plane can be seen in the figure 6.

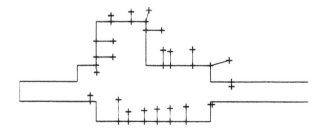

Fig 5. Points on the CAD plane at a smaller distance of the measuring points.

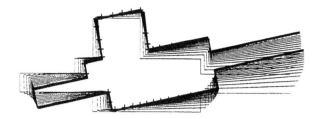

Fig. 6. Pose correction.

4. CONCLUSIONS

Pose acquisition is a first necessary step in any planned movement strategy. Their satisfactory accomplishment depends on the sensoring features on board. The laser range finder when used in structured environments can build an accurate and quick polar trace of the surrounding obstacles. Matching a CAD plane against this polar profile in local coordinates gives an answer to the pose acquisition. The matching process with its algorithm implementation must guarantee convergence in reasonable time.

This pose acquisition has been programmed in C⁺⁺ for PC Pentium at 133 MHz and has attained good results with simulated cases in which were included ambiguous features, like symmetries. The results obtained in the range of seconds to the minutes were accepted as good enough for the task. Once the pose known, a fast and deterministic algorithm is used for close following step by step. The results depend on accurate geometric setting of the laser and frame of attachment to the robot, discarding the need for using a Kalman filter instead.

ACKNOWLEDGEMENT

The authors would like to acknowledge CICYT for funding the work under grant TAP-96-1184-C04-03

REFERENCES

Bäck T., Hammel U., Schwefwel H.P (1997).
    Evolutionary Computation: Comments on the History and Current State.
    IEEE Trans. on Evolutionary Computation, **1** n°1, April.
Ingemar J. Cox (1989)
    Blanche: Position Estimation for an Autonomous Robot Vehicle.
    Autonomous Robot Vehicle- Springer Verlag.
Raimúndez Álvarez, J.C. (1997)
    Estrategias evolutivas y su aplicación a la síntesis de controladores.
    Doctoral Thesis Universidade de Vigo.

CORRECTION OF ODOMETRIC ERRORS IN MOBILE ROBOT LOCALIZATION USING INERTIAL SENSORS

L. Jetto, S. Longhi, S. Zanoli

Dipartimento di Elettronica e Automatica, Università di Ancona
v. Brecce Bianche, I-60131 Ancona, Italy

Abstract. A basic requirement for an autonomous mobile robot is its capability to elaborate the sensor measures to localize itself with respect to a coordinate system. The low cost positioning system here proposed corrects the partial odometric information, and the consequent long-term drift, exploiting the measures coming from an optical fiber gyroscope. The fusion algorithm is based on a suitably defined extended Kalman filter. *Copyright © 1998 IFAC*

Keywords. Wheeled mobile robots; Localization systems; Sensor fusion, Inertial sensors.

1. INTRODUCTION

An accurate determination of location is a fundamental requirement when dealing with control problems of mobile robots. Two different kinds of localization exist: relative and absolute. Relative localization is important because of its outdoor applicability and of the relatively low cost of sensors (Cox and Wilfong, 1990; Barshan and Durrant-Whyte, 1995; Bury and Hope, 1995; Crane et al., 1995; Rintanen et al., 1995; Schönberg et al., 1995; Vaganay et al., 1993). The most widely used internal sensors are the optical incremental encoders. These sensors are generally fixed to the axis of the driving wheels or to the steering axis of the vehicle, at each sampling instant the position is estimated on the basis of the encoder increments along the sampling interval. A drawback of this method is that the errors of each measure are summed up. This heavily degrades the position and orientation estimates of the vehicle, especially for long and winding trajectories (Wang, 1988). Other typical internal sensors are gyroscopes and accelerometers which provide angular rate information and velocity rate information, respectively. The information provided by these inertial sensors must be integrated to obtain absolute estimates of orientation, position and velocity. Therefore, like for the odometers, even small errors in the singular measure may give rise to unbounded errors in the absolute measure.

The problem of a continuous growth in the inte-grated measurement error can be overcome by periodically correcting the internal measures with the data provided by absolute sensors like sonar, laser, GPS, vision systems (Curran and Kyriakopoulos, 1993; D'Orazio et al., 1993; Freund and Dierks, 1995; Jarvis, 1995; Mar and Leu, 1996; Talluri and Aggarwal, 1992; Zhuang and Tranquilla, 1995). To reduce the frequency of these costly corrections it is necessary to improve the performance of the internal sensors as discussed e. g., in (Borestein and Feng, 1996) and (Barshan and Durrant-Whyte, 1995). A systematic calibration method for the reduction of odometry errors has been proposed in (Borestein and Feng, 1996), the accuracy of low cost internal sensor system like gyroscopes, accelerometers and tilt sensors, is analyzed in (Barshan and Durrant-Whyte, 1995). The experimental results reported in this last reference showed that gyroscopes are significantly more useful than accelerometers for improving the navigation system performance.

Based on the foregoing discussions, the purpose of this paper is to perform a fusion of odometric and gyroscopic data to obtain an accurate low cost internal sensor equipment. The data fusion is performed through a suitably defined Extended Kalman Filter (EKF).

The use of Kalman filtering techniques requires to formulate the problem in a space-state framework. This has been here done exploiting the knowledge of the kinematic model of the mobile robot and of

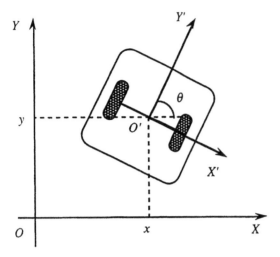

Y Y'

θ

y

O'

X'

O x X

Fig. 1. The scheme of the unicycle robot.

the measure equipment.

The paper is organized in the following way. The robot model and the sensor device equipment are described in Section 2. The adaptive algorithm is reported in Section 3. Section 4 contains a detailed description of the experimental tests performed on the mobile base.

2. THE SENSOR EQUIPMENT

2.1 Odometric measures

Consider an unicycle-like mobile robot with two driving wheels, mounted on the left and right sides of the robot, with their common axis passing through the center of the robot (see Figure 1). Localization of this mobile robot in a two-dimensional space requires the knowledge of coordinates x and y of the midpoint between the two driving wheels and of the angle θ between the main axis of the robot and the x-direction. The kinematic model of the unicycle robot is described by the following equations:

$$\dot{x}(t) = \nu(t)cos\theta(t), \qquad (1)$$
$$\dot{y}(t) = \nu(t)sin\theta(t), \qquad (2)$$
$$\dot{\theta}(t) = \omega(t), \qquad (3)$$

where $\nu(t)$ and $\omega(t)$ are, respectively, the displacement and angular velocities of the robot, and are expressed by:

$$\nu(t) = \frac{\omega_r(t) + \omega_l(t)}{2}R, \qquad (4)$$
$$\omega(t) = \frac{\omega_r(t) - \omega_l(t)}{d}R, \qquad (5)$$

where $\omega_r(t)$, $\omega_l(t)$ are the angular velocities of the right and left wheels, respectively, R is the wheel radius and d is the distance between the wheels.

If the sampling period $\Delta t_k := t_{k+1} - t_k$ is sufficiently small, the increment on the position and orientation of the mobile robot on a time interval Δt_k in which the velocities $\omega_r(t)$ and $\omega_l(t)$ are constant can be expressed as (Wang, 1988):

$$x(t_{k+1}) = x(t_k) + \bar{\nu}(t_k)\Delta t_k \frac{sin\frac{\bar{\omega}(t_k)\Delta t_k}{2}}{\frac{\bar{\omega}(t_k)\Delta t_k}{2}}cos(\theta(t_k)$$
$$+\frac{\bar{\omega}(t_k)\Delta t_k}{2}), (6)$$

$$y(t_{k+1}) = y(t_k) + \bar{\nu}(t_k)\Delta t_k \frac{sin\frac{\bar{\omega}(t_k)\Delta t_k}{2}}{\frac{\bar{\omega}(t_k)\Delta t_k}{2}}sin(\theta(t_k)$$
$$+\frac{\bar{\omega}(t_k)\Delta t_k}{2}), (7)$$

$$\theta(t_{k+1}) = \theta(t_k) + \bar{\omega}(t_k)\Delta t_k. \qquad (8)$$

where $\bar{\nu}(t_k)\Delta t_k$ and $\bar{\omega}(t_k)\Delta t_k$ are:

$$\bar{\nu}(t_k)\Delta t_k = \frac{\Delta q_r(t_k) + \Delta q_l(t_k)}{2}, \qquad (9)$$
$$\bar{\omega}(t_k)\Delta t_k = \frac{\Delta q_r(t_k) - \Delta q_l(t_k)}{d}, \qquad (10)$$

and $\Delta q_r(t_k)$ and $\Delta q_l(t_k)$ are the incremental measures on the interval Δt_k of the encoders attached to the right and left wheels of the robot.

Under the hypothesis of sufficiently small sampling period Δt_k, equations (6) – (10) introduce an estimation of the position and orientation of the mobile robot at time t_{k+1}. An analysis of the accuracy of this estimation procedure has been developed in (Wang, 1988). The encoder errors introduced incremental errors in the above estimation procedure, which especially affects the estimate of the orientation θ, and reducing its applicability to short trajectories.

2.2 Fiber optic gyroscope measures

The operative principle of a Fiber Optic Gyroscope (FOG) is based on the Sagnac effect. The FOG is made of an fiber optic loop, fiber optic components, a photo-detector and a semiconductor laser. The phase difference of the two light beams traveling in opposite directions around the fiber optic loop is proportional to the rate of rotation of the fiber optic loop. The rate information is integrated to provide the absolute measurements of orientation. A FOG does not require frequent maintenance and have a longer lifetime of the conventional mechanical gyroscopes. In a FOG the drift is also low. The fiber optic gyroscope HITACHI mod. HOFG-1 was used for measuring the angle θ of the mobile robot. In this gyroscope the angular measure is available in a digital form by a RS232 serial line. The main characteristics of this FOG are reported in the Table 1.

Rotation Rate	- 60 to + 60 deg/s
Angle Measurement Range	-360 to +360 deg
Random Walk	≤ 0.1 deg/√h
Zero Drift (Rate Integration)	≤ 1 deg/h
Non-linearity of Scale Factor	within ± 1.0 %
Time Constant	Typ. 20 ms
Response Time	Typ. 20 ms
Data Output Interval	Min. 10 ms
Warm-up Time	Typ. 6.0 s

Table 1. Characteristics of the HITACHI gyroscope mod. HOFG - 1.

3. ESTIMATION OF ROBOT LOCATION

The proposed EKF for the fusion of odometric and gyroscopic measures, providing on line estimates of robot position and orientation, is derived in this section.

Denote with $X(t) := [x(t) \quad y(t) \quad \theta(t)]'$ the robot state and with $U(t) := [\nu(t) \quad \omega(t)]'$ the robot control input. The kinematic model of the robot can be written in the compact form of the following stochastic differential equation

$$dX(t) = F(X(t), U(t))dt + d\eta(t), \quad (11)$$

where $F(X(t), U(t))$ is obtained by (1) – (3) and $\eta(t)$ is a Wiener process such that $E(d\eta(t)d\eta(t)^T) = Q\,dt$ representing the model inaccuracies (parameter uncertainties, slippage, dragging). Its weak mean square derivative $d\eta(t)/dt$ is a white noise process $\sim N(0, Q)$. Assuming a constant sampling period $\Delta t_k = T$ and denoting t_{k+1} by $(k+1)T$, the following sampled measure equation can be associated to equation (11):

$$Z((k+1)T) = C\,X((k+1)T) + v(kT), \quad (12)$$

where $Z(kT)$ is the vector containing the odometer measures and the inertial measure of the orientation and $v(kT)$ is a white sequence $\sim N(0, R)$. The measure vector $Z(kT)$ is composed of four entries $Z(kT) = [z_1(kT) \; z_2(kT) \; z_3(kT) \; z_4(kT)]^T$ where $z_1((k+1)T) = x_{odom}((k+1)T) = x((k+1)T) + v_1((k+1)T)$, $z_2((k+1)T) = y_{odom}((k+1)T) = y((k+1)T) + v_2((k+1)T)$, $z_3((k+1)T) = \theta_{odom}((k+1)T) = \theta((k+1)T) + v_3((k+1)T)$ are the measures provided by the odometric devices with $x_{odom}((k+1)T)$, $y_{odom}((k+1)T)$ and $\theta_{odom}((k+1)T)$ given by (6) – (8), and $z_4((k+1)T) = \theta_{gyro}((k+1)T) = \theta((k+1)T) + v_4((k+1)T)$ is the angular measure provided by the FOG.

By definition of the measurement vector it follows that the matrix C has the following form:

$$C = \begin{bmatrix} 1 & 0 & 0 \\ 0 & 1 & 0 \\ 0 & 0 & 1 \\ 0 & 0 & 1 \end{bmatrix}. \quad (13)$$

Linearization of model (11), (12) about the current estimate $\hat{X}(t)$ and discretization with the period T results in the following Extended Kalman Filter (EKF) (where explicit dependence on T has been dropped for simplicity of notation),

$$\hat{X}(k+1, k) = \hat{X}(k, k)$$
$$+ B_d(k)B(k)F(\hat{X}(k, k), U(k-1))$$
$$+ B_d(k)(U(k) - U(k-1)), \quad (14)$$

$$P(k+1, k) = A_d(k)P(k, k)A_d^T(k) + Q_d, \quad (15)$$

$$K(k+1) = P(k+1, k)C^T S(k+1)^{-1}, \quad (16)$$

$$S(k+1) = [CP(k+1, k)C^T + R], \quad (17)$$

$$\hat{X}(k+1, k+1) = \hat{X}(k+1, k)$$
$$+ K(k+1)\Gamma(k+1), \quad (18)$$

$$\Gamma(k+1) = [Z(k+1) - C\,\hat{X}(k+1, k)], \quad (19)$$

$$P(k+1, k+1) = [I - K(k+1)C]P(k+1, k), \quad (20)$$

where:

$$A_d(k) = e^{A(k)T},$$

$$A(k) = \left[\frac{\partial F(X(t), U(t))}{\partial X(t)} \right]_{\substack{X(t) = \hat{X}(k,k) \\ U(t) = U(k-1)}}$$

$$= \begin{bmatrix} 0 & 0 & -\nu(t)\sin\theta(t) \\ 0 & 0 & \nu(t)\cos\theta(t) \\ 0 & 0 & 0 \end{bmatrix}_{\substack{\theta(t) = \hat{\theta}(k,k) \\ \nu(t) = \nu(k-1)}}$$

$$B(k) = \left[\frac{\partial F(X(t), U(t))}{\partial U(t)} \right]_{\substack{X(t) = \hat{X}(k,k) \\ U(t) = U(k-1)}}$$

$$= \begin{bmatrix} \cos\theta(t) & 0 \\ \sin\theta(t) & 0 \\ 0 & 1 \end{bmatrix}_{\theta(t) = \hat{\theta}(k,k)} \quad (21)$$

$$B_d(k) = \int_{kT}^{(k+1)T} e^{A(k)((k+1)T - \tau)}d\tau,$$

$$Q_d = \int_{kT}^{(k+1)T} e^{A(k)((k+1)T - \tau)} Q e^{A^T(k)((k+1)T)}d\tau. \quad (22)$$

The covariance matrices Q and R are assumed to have the following form:

$$Q = \sigma_\eta^2 I_3, \quad (23)$$

$$R = diag[\sigma_{v,1}^2, \cdots, \sigma_{v,4}^2]. \quad (24)$$

The diagonal form of Q understands the hypothesis that model (11) describes the true dynamics of the three state variables with nearly the same degree of approximation and with independent errors. Its value is considered to be constant over each sampling period. The diagonal form of R means that no correlation is assumed between the measurement errors introduced by the sensors. Equations (22) and (23) imply :

$$Q_d = \sigma_\eta^2 \bar{Q}, \tag{25}$$

$$\bar{Q} = \int_{kT}^{(k+1)T} e^{(A(k)((k+1)T-\tau)} e^{(A^T(k)((k+1)T-\tau)} d\tau. \tag{26}$$

The above EKF can be implemented once estimates of Q_d and R are available. The values of σ_η^2 and $\sigma_{v,i}^2$, $i = 1, \cdots, 4$, in the EKF equations were chosen on the basis of first simulation results. If no "a priori" reliable estimates of noise statics can be obtained, adaptive filtering procedures should be used (Jetto et al., 1997).

4. EXPERIMENTAL RESULTS

The experimental tests have been performed on the LabMate mobile base in an indoor environment. This mobile robot is realized with two driving wheels, as reported in Figure 1, and the odometric data are the incremental measures that at each sampling interval are provided by the encoders attached to the right and left wheels of the robot. These measures are directly acquired by the low level controller of the mobile base. The gyroscopic measures on the absolute orientation have been acquired in a digital form by a serial port on the computer.

The control algorithm is based on a proper discrete-time implementation of the control algorithm proposed in (De Luca and Oriolo, 1994) and based on the kinematic inversion approach. The continuous-time control algorithm is explicitly represented by the following equations:

$$\nu(t) = K_\nu(\dot{y}_d(t)cos\theta(t) - \dot{x}_d(t)sin\theta(t)) + C\nu e_t(t), \tag{27}$$

$$\omega(t) = K_\omega(\theta_d(t) - \theta(t)) + C\omega e_n(t), \tag{28}$$

where

$$e_t(t) = (y_d(t) - y(t))cos\theta(t) - (x_d(t) - x(t))sin\theta(t)), \tag{29}$$

$$e_n(t) = (y_d(t) - y(t))sin\theta(t) + (x_d(t) - x(t))cos\theta(t)). \tag{30}$$

and $x_d(t)$, $y_d(t)$ and $\theta_d(t)$ represent the desired planned trajectory. The EKF has been implemented on a personal computer with Windows 3.1 system by the developing environment described in (Bonifazi et al., 1996). In this development system, the planned trajectory has been computed considering the non-holonomic and environment constraints according to the algorithm proposed in (Conte et al., 1996). The system is connected directly with the

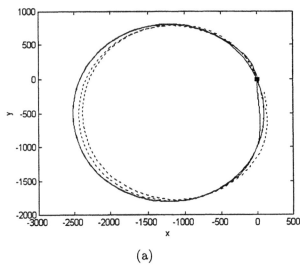

(a)

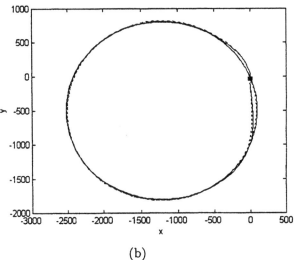

(b)

Fig. 2. The plain path is the planned trajectory, the dashed path is the realized trajectory; the unit of x-, y-axis is $m10^{-3}$.

robot low level controller and the Fiber Optic Gyroscope by two standard serial ports RS232. Relatively long trajectories have been considered in all the experimental tests to check the performance of the proposed localization algorithm in real situations.

Figure 2 illustrates an example of the developed experiments. Part (a) of this figure represents the realized trajectory with localization only deduced by odometric measures. The planned trajectory is a twice repeated closed curve computed by the developing environment described in (Bonifazi et al., 1996). The black path is the planned trajectory and the dashed path is the trajectory realized only with odometric measures. In this case, at the end of the test, the robot is out of the planned path with errors of 10 cm, 15 cm and 12 $degrees$ on the x, y and θ, respectively. Part (b) shows the same test with the localization based on the fusion of gyroscopic and odometric measures realized through the proposed EKF. In this case, at the end of the test, the

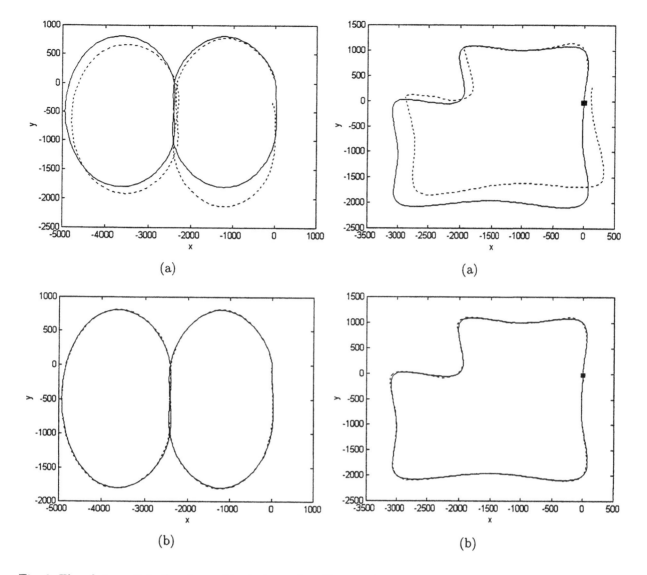

(a)

(b)

(a)

(b)

Fig. 3. The plain path is the planned trajectory, the dashed path is the realized trajectory; the unit of x-, y-axis is $m10^{-3}$.

Fig. 4. The plain path is the planned trajectory, the dashed path is the realized trajectory; the unit of x-, y-axis is $m10^{-3}$.

robot is out of the planned trajectory with errors of 0.9 cm, 1.3 cm and 2 $degrees$ on the x, y and θ, respectively. The plot clearly evidences the reduction of odometry errors obtained with the introduction of the gyroscopic measures.

In the second experiment the planned trajectory is composed by a set of orientation changes. The planned trajectory is a closed curve with symmetric orientation changes. Experimental results are reported in Figure 3. Part (a) of this figure represents the realized trajectory with localization only deduced by odometric measures. The black path is the planned trajectory and the dashed path is the trajectory realized with the odometric measures only. In this case, at the end of the test, the robot is out of the planned trajectory with errors of 6 cm, 28 cm and 15 $degrees$ on the x, y and θ, respectively.

Part (b) shows the same test with localization based on the proposed EKF. In this case, at the end of the test, the robot is out of the planned trajectory with errors of 1.1 cm, 6 cm and 1 $degrees$ on the x, y

and θ, respectively.

Figure 4 contains the results of an experiment with changes on the orientation of 90 $degrees$. Parts (a) and (b) of this figure have the same meaning than in the previous figures. Plot (a) shows that at the end of the test the robot is out of the planned trajectory with errors of 18 cm, 35 cm and 20 $degrees$ on the x, y and θ, respectively. Part (b) shows that at the end of the test the robot is out of the planned trajectory with errors of 4 cm, 7 cm and 3 $degrees$ on the x, y and θ, respectively.

5. CONCLUSIONS

This paper proposed a new method for the accurate localization of a mobile robot by a low cost internal sensor equipment. The underlying idea is to improve the odometric sensor performance exploiting the measure provided by inertial sensors like gyroscopes. The fusion algorithm is based on a EKF

179

incorporating all the available information, namely the kinematic model of the robot and the measure equipment. The performed experiments confirmed that a substantial reduction of odometric errors can be really attained, so that high performances of the localization algorithm are obtainable over relatively long trajectories.

The use of absolute measures can further improve the accuracy of the localization algorithm by reducing the absolute error and/ or allowing the mobile robot to track longer trajectories.

Future research activities in this area will be addressed to this last issue: the proposed localization algorithm will be integrated with the measures provided by a set of sonar or by a Differential Global Position System (DGPS). Hence, the results here presented can be seen as the first step towards the definition of an algorithm for the localization of a mobile robot operating in a out-door environment.

REFERENCES

Barshan, B. and H.F. Durrant-Whyte (1995). Inertial navigation systems for mobile robots, *IEEE Trans. on Robotics and Automation*, vol. 11, no. 3, pp. 328-342.

Bonifazi, M., F. Favi, T. Leo, S. Longhi and R. Zulli (1996). A developing environment for the solution of the navigation problem of mobile robots with non-holonomic constraints, *Proceedings of the 4th IEEE Mediterranean Symposium on New Direction in Control Automation*, Krete, Greece, pp. 107112.

Borenstein, J. and L. Feng (1996). Measurement and correction of systematic odometry errors in mobile robots, *IEEE on Trans. Robotics and Automation*, vol. 12, pp. 869-880.

Bury, B. and J. C. Hope (1995). Autonomous mobile robot navigation using a low-cost fiber optic gyroscope, *Proc. of the 2nd IFAC Conference on Intelligent Autonomous Vehicle (IAV 95)*, Espoo, Finland, pp. 37-42.

Conte, G., S. Longhi and R. Zulli (1996). Motion planning for unicycle and car-like robots, *International Journal of Systems Science*, vol. 27, no. 8, pp. 791-798.

Cox, I.J. and G.T. Wilfong (1990). *Autonomous Robot Vehicles*, Springer-Verlag, Berlin.

Crane III, C. D. , A. L. Rankin, D. G. Armstrong II, J. S. Wit and D. K. Novick (1995). An evaluation of INS and GPS for autonomous navigation, *Proc. of the 2nd IFAC Conference on Intelligent Autonomous Vehicle (IAV 95)*, Espoo, Finland, pp. 208-213.

Curran, A. and K.J. Kyriakopoulos (1993). Sensor-based self-localization for wheeled mobile robots, *Proceeding of the IEEE Mediterranean Symposium on New Direction in Control Automation*, pp. 8-13.

De Luca, A. and G. Oriolo (1994). Local incremental planning for nonholonomic mobile robots, *Proceeding of the IEEE International Conference on Robotics and Automation*.

D'Orazio, T. , M. Ianigro, E. Stella, F.P. Lovergine and A. Distante (1993). Mobile robot navigation by multi-sensory integration, *Proceeding of the IEEE Mediterranean Symposium on New Direction in Control Autom.*, pp. 373-379.

Freund, E. and F. Dierks (1995). Map-based free navigation for autonomous vehicles, *Proc. of the 2nd IFAC Conference on Intelligent Autonomous Vehicle (IAV 95)*, Espoo, Finland, pp. 185-190.

Jarvis, R. (1995). An all-terrain intelligent autonomous vehicle with sensor fusion based navigation capabilities, *Proc. of the 2nd IFAC Conference on Intelligent Autonomous Vehicle (IAV 95)*, Espoo, Finland, pp. 25-31.

Jetto, L., S. Longhi and G. Venturini (1997). Development and experimental validation of an adaptive estimation algorithm for the on-line localization of mobile robots by multisensor fusion, *Proc. of the 5th Symposium on Robot Control, SYROCO'97*, Nates, France, pp. 165-172.

Mar, J. and J.-H. Leu (1996). Simulation of the positioning accuracy of integrated vehicular navigation systems, *IEE Proc.-Radar, Sonar Navig.*, vol. 143, no. 2, pp. 121-128.

Rintanen, K., H. Mäkelä, K. Koskinen, J. Puputti, M. Sampo and M. Ojala (1995). Development of an autonomous navigation system for an outdoor vehicle, *Proc. of the 2nd IFAC Conference on Intelligent Autonomous Vehicle (IAV 95)*, Espoo, Finland, pp. 220-225.

Schönberg, T., M Ojala, J. Suomela, A. Torpo and A. Halme (1995). Positioning an autonomous off-road vehicle by using fused DGPS and inertial navigation, *Proc. of the 2nd IFAC Conference on Intelligent Autonomous Vehicle (IAV 95)*, Espoo, Finland, pp. 226-231.

Talluri, R. and J.K. Aggarwal (1992). Position estimation for an autonomous mobile robot in an outdoor environment, *IEEE on Trans. Robotics and Autom.*, vol. 8, no. 5, pp. 573-584.

Vaganay, J., M.J. Aldon and A. Fournier (1993). Mobile robot attitude estimation by fusion of inertial data, *Proceeding of the IEEE International Conference on Robotics and Automation*, pp. 277-282.

Wang, C.M. (1988). Localization estimation and uncertainty analysis for mobile robots, *Proceedings of the Int. Conf. on Robotics and Automation*, pp. 1230-1235.

Zhuang, W. and J.M. Tranquilla (1995). Effects of multipath and antenna on GPS observables, *IEE Proc.-Radar, Sonar Navig.*, vol. 142, no. 5, pp. 267-275.

CONSTRUCTIVE RADIAL BASIS FUNCTION NETWORKS FOR MOBILE ROBOT POSITIONING

Manuel R. Arahal* Manuel Berenguel*

* Departamento de Ingeniería de Sistemas y Automática de la
Universidad de Sevilla, Escuela Superior de Ingenieros. Camino
de Los Descubrimientos, E-41092. E-mail: arahal@cartuja.us.es

Abstract: This paper presents a comparison of several methods for constructing neural classifiers in the framework of mobile robot positioning. The task of positioning a mobile robot using a ring of sonars or laser sensors is used as a testbed for different radial basis function neural network constructive algorithms that produce topological maps of the environment. The experiments performed using two different experimental systems show the advantages and drawbacks of the compared algorithms. Copyright © 1998 IFAC

Keywords: Neural networks, Mobile robots, Navigation, Self organizing systems, Learning algorithms.

1. INTRODUCTION

Mobile robots are often equipped with environment sensors such as sonars or laser rangers that provide a measure of the distance to obstacles to be used for map-building. Most of the work presented here is valid for other types of sensors; however, in the tests conducted just sonars and lasers have been used. Positioning consists in producing an estimate of the robot localization using the measurements of the sonars and/or the laser. Map-based positioning can be used to generate representations of the environment that can change with time in order to improve positioning accuracy through exploration. However, techniques for map-based positioning are often considered to consume extensive processing time, especially geometric maps. Topological maps are based on the observed features and their geometric relationships (Kortenkamp and Weymouth, 1994). Unlike geometric maps, the absolute position with respect to the frame of reference is not used. The problem of place recognition with topological maps is viewed as the task of matching recorded sensory maps and the actual sensed data.

In this paper, previously visited places are used as landmarks for the positioning problem. Measures from a ring of sonars or from a laser ranger can be arranged in a vector $\mathbf{x}$ of n_x signals

$\mathbf{x} = [x_1, \cdots, x_{nx}]^T$. This vector will be used as input to a neural network that will provide an estimation of the position of the robot in the map. In the following, it will be considered that there is a set of n_y landmarks, or privileged positions in the environment, $\mathbf{L} = \{l_1, \cdots, l_{ny}\}$. These particular places or landmarks can be selected manually or automatically during exploration. Also, the number n_y of landmarks can grow with time to improve accuracy or decrease to merge local features into more global areas.

Given a vector of distance measurements $\mathbf{x}$, and the set $\mathbf{L}$, the task of the neural positioner consists in producing a binary valued output vector $\mathbf{y} = [y_1, \cdots, y_{ny}]$ that indicates the closeness of the robot to any of the landmarks. More precisely, if the robot is close to landmark i, the output of the positioner should be a vector with all elements equal to -1 except the i-th element $y_i = 1$. In the cases when the robot is not close to any of the landmarks, the output vectors ought to be a null vector.

The above exposed problem of robot positioning based in a number of environment signals, is just a pattern classification problem and has been considered in this way in some previous works (Courtney and Jain, 1994). There are many approaches in the literature to tackle this problem. In this paper, constructive RBF solutions are considered as proposed in (Zimmer *et al.*, 1994). The objective is to analyze different constructive algorithms for radial basis function (RBF) networks in the framework of the mobile robot positioning problem. Algorithms that construct neural networks during training are superior to non-constructive ones because the size of the network does not have to be given. Among a number of different algorithms found in the literature, the following have been selected: the Non Constructive Hybrid Model (NCHM) due to Moody and Darken (Moody and Darken, 1989), Platt's Resource Allocation Network (RAN) (Platt, 1991), and the Incremental RBF Networks (IRBFN) of Fritzke (Fritzke, 1994).

In the next section, the neural framework is presented and the different algorithms exposed. The sensors used in the experiments are described in section 3 together with a description of the sonar ring and laser sensor. Section 4 is devoted to present and compare the results of the tests. Some conclusions are presented at the end.

2. NEURAL FRAMEWORK

The neural structure used in this paper correspond to the RBF type. The RBF nets are one-hidden layer feedforward networks with linear output nodes (see figure 1). The neurons in the hidden layer receive as inputs the whole input vector and perform a nonlinear function of this vector and a inner vector called center. The most used basis function is a gaussian of the distance from the input vector to the center. If the input vector is denoted as $\mathbf{x}$, and the center of nodes as $\mathbf{c}$, then, the contribution of node i to the output is,

$$n_i(\mathbf{x}) = e^{-\frac{\|\mathbf{x}-\mathbf{c}_i\|^2}{2\sigma_i^2}} \qquad (1)$$

where σ_i is the width of the $i-th$ basis function. The output of the network is the weighted sum of the outputs of all nn nodes:

$$NN(\mathbf{x}) = \sum_{i=1}^{nn} w_i n_i(\mathbf{x}) \qquad (2)$$

The properties and capabilities of RBF nets have been studied in detail by many authors (Poggio and Girosi, 1989) and compared with other approximation schemes (Haykin, 1994)

For the purpose of pattern classification, a network with nx inputs, nn hidden nodes and ny outputs will be used. The number of nodes is the only structural parameter, and has to be fixed *a priori* when using a non-constructive algorithm. Constructive procedures begin with network with

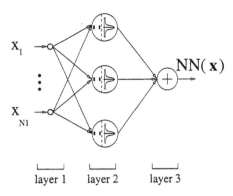

Fig. 1. A radial basis function network with three nodes in the hidden layer.

a small number of nodes, as training progresses, the procedure adds more nodes if needed.

The learning procedure for RBF networks can be data-driven for all parameters or just for some of them being the rest selected by heuristics. There is a large number of different algorithms for the selection of the parameters, most of them fall in any of the types listed below.

- Fixed centers and variances. The centers of the basis functions can be selected randomly or regularly spaced in the input space (Sanner and Slotine, 1992) or being a subset of the sampled input data. The widths are fixed according to the distance between centers. In this situation, training consists in modifying the output weights using the pseudoinverse method or any other algorithm. It has to be noted that the problem is linear in the parameters.

  Another approach is to select the centers according to the probabilistic density in which input vectors appear. The k-neighbor rule can be used to accommodate centers in such a way that regions with more dense data receive more basis (Moody and Darken, 1989).

- Adaptation of all parameters. The centers, widths and output weights are adapted using a gradient-descent rule (Wettschereck and Dietterich, 1992). This has the drawback that local minima can appear.

- Growing networks. A problem with the former approaches is that the number of basis functions has to be chosen *a priori*. A more effective way is to let the learning procedure decide to add and/or delete nodes. The procedure is driven by the error in the identification, achieving a meaningful allocation of resources. Among many reported algorithms, the following references can be given (Kadirkamanathan *et al.*, 1991),(Platt, 1991), (Fritzke, 1994).

In the following, the algorithms that will be compared in the tests are presented.

2.1 Non Constructive Hybrid Algorithm

The model proposed in (Moody and Darken, 1989) consists on placing the centers of the RBF using a self-organizing unsupervised algorithm off-line. The number of nodes has to be chosen *a priori*; then, the algorithm will allocate more nodes in regions of the input space with denser data. Next, a nearest neighbor heuristic is used to determine the widths of the nodes. Having fixed the centers and variances, only the output weights have to be selected. This is in most cases a problem linear in the parameters, allowing fast supervised adaptation algorithms to be used. The algorithm can be summarized as follows.

(1) Gather a set of input-output pairs $(\mathbf{x}, \mathbf{y})$.
(2) Use an unsupervised clustering algorithm to find a set of nn centers that represent the input data. To this end, a number of competitive learning procedures are of use. In essence, the clustering algorithm begins with centers placed at random positions, and, after the presentation of each input vector $\mathbf{x}$, moves the closest node n in the direction of $\mathbf{x}$: $\mathbf{c_n} \leftarrow \mathbf{c_n} + \alpha(\mathbf{x} - \mathbf{c_n})$, being α the learning rate.
(3) Determine the variances of the RBF. For every node i, the variance σ_i is calculated using the distance to the nearest neighbor j, $\sigma_i = \kappa\|\mathbf{c_i} - \mathbf{c_j}\|$. Some overlap is allowed, being determined by the constant κ.
(4) Initialize the vector of output weights $\mathbf{w}$ with small random values.
(5) For each input-output pair in the data set,
   □ Compute the classification error $\mathbf{e} = \mathbf{y} - \hat{\mathbf{y}}$
   □ Adjust the output weights as
   $\mathbf{w} \leftarrow \mathbf{w} - \eta\frac{\partial \mathbf{e}^2}{\partial \mathbf{w}}$.
(6) Repeat last step until end of training cycles.

This approach achieves a solution many times faster than backpropagation does with feedforward networks of various hidden layers, and using less nodes. However, there are some problems. Firstly, the unsupervised placement of centers can allocate far more resources than needed in regions having dense data and small variations in the output vector. This first problem can be alleviated using an extended input-output metric for the clustering phase (Saha and Keeler, 1990), and allowing the centers to be moved in the training phase (Wettschereck and Dietterich, 1992). Secondly, the number of nodes can be too small or too large for the given task, producing either poor classification of bad generalization, respectively.

2.2 Resource Allocating Network

To avoid some of the problems of the previous approach, Platt (Platt, 1991) introduced a strategy for allocating new units in RBF networks. The algorithm decides to place a new unit or to adapt the existing units based on the distance of the current input to the set of centers and on the output error. If the error is not too large, gradient descent is used to adjust the parameters; otherwise, and if the input vector is far from the centers of the network, a new node is placed in the location of the input vector. The algorithm works entirely on line and can be described as follows:

(1) Begin with a network with no nodes.
(2) For each input-output pair $(\mathbf{x_i}, \mathbf{y_i})$.
   □ Compute the classification error $\mathbf{e} = \mathbf{y} - \hat{\mathbf{y}}$, and the nearest node to the input $\mathbf{c_n}$.
   □ If $\mathbf{e} > \epsilon$ and $\|\mathbf{x_i} - \mathbf{c_n}\| > \delta$, then allocate a new node nn with, $\mathbf{c}_{nn} = \mathbf{x}$, $\mathbf{w}_{nn} = \mathbf{e}$, and $\sigma_{nn} = \kappa\|\mathbf{x_i} - \mathbf{c_n}\|$
   □ Else adjust the output weights as
   $\mathbf{w} \leftarrow \mathbf{w} - \eta\frac{\partial \mathbf{e}^2}{\partial \mathbf{w}}$, and the center of the best matching unit n as $\mathbf{c_n} \leftarrow \mathbf{c_n} + \alpha(\mathbf{x} - \mathbf{c_n})$, being η and α the learning rates.

The parameter ϵ is selected to be the desired accuracy of the network. The minimum distance δ is decreased with time to increase the resolution of the classification given by the network. An exponential decay is normally used $\delta = max(\delta_{max}\exp(-t/\tau), \delta_{min})$.

By selecting the RBF centers according to both the error and the distance criterion, the number of nodes grows sublinearly with the number of patterns. However, the insertion of new nodes can be affected by the presence of noise. To alleviate this problem, Fritzke (94) proposes the use of accumulated error to decide where to insert new units.

2.3 Incremental Networks

The approach presented in (Fritzke, 1994) is also a constructive algorithm for RBF networks. The units in the network are tied by neighboring relationships represented by edges. The initial network consists of two units with randomly selected parameters connected by an edge. Each unit has a gaussian function associated and there is a certain overlap between them. Training produces centers to move to regions with dense data and output weights to adapt according to the delta rule. In addition, edges have a counter associated whose value (referred to as age) indicates its usefulness to represent neighborhood information, and units have a local error variable that are used to show regions where the classifications are often erroneous. Both informations are used to set the parameters of new units according to the following algorithm that is applied for each input-output pair $(\mathbf{x_c}, \mathbf{y_c})$, $c = 1, 2, \cdots$.

(1) Increase observation counter c by one.

(2) Compute the classification error $\mathbf{e} = \mathbf{y} - \hat{\mathbf{y}}$, and the node closest to the input n_1.

(3) Increase the error variable of unit n_1 with the value of $\mathbf{e}^T\mathbf{e}$.

(4) Reduce the value of the error variable of all units multiplying them by ν with $0 < \nu < 1$.

(5) Move the center of the nearest node a fraction α_1 in the direction $(\mathbf{x} - \mathbf{c_{n_1}})$, and the centers of its neighbors a fraction α_2, with α_1 and $\alpha_2 < \alpha_1$ the learning rates.

(6) Update output weights as $\mathbf{w} \leftarrow \mathbf{w} - \eta\frac{\partial \mathbf{e}^2}{\partial \mathbf{w}}$, being η the learning rate.

(7) Update edges as follows
   □ Connect the nearest node n_1 and the second nearest node n_2 by and edge and set its age variable to zero.
   □ Increase age of all edges emanating from n_1.
   □ Remove edges whose age counter exceeds a constant a_{max}.
   □ Remove units with no edges.

(8) If counter c reaches c_{max}, insert a new unit in the following manner
   □ Place new center between units u_1 and u_2, being $u_1 = argmin_i(ev(i))$ the unit with maximum error variable, and u_2 its neighbor with maximum error variable.
   □ Replace original edge between u_1 and u_2 by new edges from the new unit and set its ages to zero.
   □ Halve the error variable of units u_1 and u_2.
   □ Set the error variable of the new unit to the mean of the values of u_1 and u_2.

3. DESCRIPTION OF THE EXPERIMENTS

The sensors used for the experiments are the SENSUS 200^{TM}, a ring of sonars incorporated to the NOMAD 200 Mobile Base^{TM}, and the SICK^{TM} laser scanner that was mounted on top of a LABMATE^{TM} robot. The ring of sonars consists of 16 independent sonars of Polaroid transducer. These devices determine the distance to objects based on the time of flight between transmission of ultrasonic pulses. The range of the SENSUS 200 is from 6 to 255 inches. The the firing rate can be set from 4ms to 1024 ms.

The laser scanner provides the distance to the environment in a 180° opening angle with an angular resolution of 0.5°, and a maximum range of 50m. The distance to objects is determined based on the fime of flight.

In order to obtain a data set, a number of sonar and laser measurements have been recorded placing the robot in different positions in an indoor environment. A number of those positions have

Fig. 2. The sensor systems used in the experiments. On the left, the sonar ring SENSUS 200^{TM} of the NOMAD 200 Mobile Base^{TM}. On the right the SICK^{TM} laser scanner.

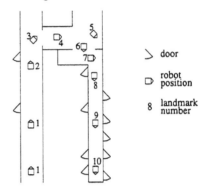

Fig. 3. Landmarks used in the experiments.

been selected as landmarks as shown in figure 3. It has to be noted that some landmarks (1, 10) cover a wide area, such as an aile or dead end, while others (5,7) correspond to narrow regions around a significative place such as a corner.

In figure 4(a), several measurements of the ring of sonars corresponding to landmark number 10 are shown. It can be seen that noise and slight movements of the robot make the positioning problem a difficult task. Using the laser sensor, noise is reduced, but the problem of a non-perfect positioning of the robot remains, as can be seen in figure 4(b). The data set consists thus in a series of sonar measurements $\mathbf{x}_s$ and laser measurements $\mathbf{x}_l$, along with an indication of the landmark corresponding to that position (if any). Some of the data pairs $(\mathbf{x}, \mathbf{y})$ are used to train a neural classifier, while the rest is used to test its performance.

4. EXPERIMENTAL RESULTS

For all the three algorithms, a process of trial and error was performed to find the optimal set of parameters. The reported results correspond to these parameters and have been divided in two sections, one using data from the ring of sonars and the other using the laser.

4.1 Results with the sonar

The objective of the algorithms is to achieve a hundred percent in the classification task with the smaller number of RBF nodes and with the

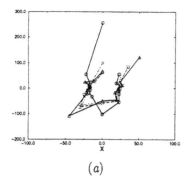

(a)

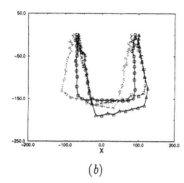

(b)

Fig. 4. *(a)* Several measurements from the ring of 16 sonars corresponding to the same landmark. *(b)* The same using the laser sensor.

faster training time. Table 1 shows the number of nodes, the number of iterations, percentage of correct classifications (PCC), the floating point operations (flops) during training and CPU time [1] for the three algorithms using the data from the sonar ring. The results are an average over ten runs, since some algorithms use random initial values. In the case of the NCHM algorithm, the reported 12000 training cycles correspond to 10000 iterations of the SOFM and 2000 for the gradiential adaptation of output weights. The parameters used in the experiments are:

Algorithm	Parameters	
NCHM	$\sigma = 1.12$	$\alpha = 0.2$
	$\eta = 0.1$	
RAN	$\kappa = 0.5$	$\epsilon = 0.4$
	$\delta_{max} = 1$	$\delta_{min} = 0.05$
	$\tau = 15$	$\eta = 0.1$
	$\alpha = 0.01$	
IRBFN	$\alpha_1 = 0.05$	$\alpha_2 = 0.005$
	$\eta = 0.5$	$\kappa = 0.5$
	$e_{max} = 10$	$c_{max} = 20$
	$\nu = 0.99$	

From table 1 it is clear that RAN performed better than the other two algorithms in this particular problem. However, it is interesting to check the sensitivity of the algorithms to variations in the parameters, since the time spent in the election of such parameters is of great importance for

practical purposes. Table 2 shows the effect of the variation of some parameters (with respect to the previously reported nominal values) in the performance of the classification and in the number of nodes of the resulting network.

Algorithm	Nodes	Cycles	PCC	MFlops	Time
NCHM	40	12000	100	12	55 s
RAN	20	6	100	0.8	1 s
IRBFN	32	10	100	6	10 s

Table 1. Classification results using data from the sonar ring.

Algorithm	Parameter	Variation	PCC	Nodes
NCHM	σ	+20%	96	-
		−20%	97	-
	α	+50%	100	-
		−50%	88	-
	nn	−10%	87	-
		−20%	68	-
RAN	κ	−50%	89	21
		+50%	97	22
	δ	+200%	88	18
		−200%	100	23
	η	+200%	92	20
		−200%	100	22
IRBFN	κ	−50%	85	30
		+50%	80	34
	α_1	+50%	98	32
		−50%	97	34
	α_2	+50%	100	33
		−50%	95	34

Table 2. Analysis of the sensitivity of the algorithm to variations in the parameters.

4.2 Results with the laser

As in the previous section, the comparison of the algorithms have been carried out considering the number of iterations, percentage of correct classifications, number of neural nodes and CPU time. Table 3 shows the results averaged over ten runs, since some algorithms use random initial values that affect their performance.

Algorithm	Nodes	Cycles	PCC	MFlops	Time
NCHM	35	12000	100	17	70 s
RAN	17	7	100	1	2 s
IRBFN	25	60	100	9	30 s

Table 3. Classification results using data from the laser.

The parameters used in the experiments are:

---

[1] Corresponding to a Pentium-S 200 MHz.

Algorithm	Parameters	
NCHM	$\sigma = 1$	$\alpha = 0.25$
	$\eta = 0.05$	
RAN	$\kappa = 0.5$	$\epsilon = 0.4$
	$\delta_{max} = 0.3$	$\delta_{min} = 0.01$
	$\tau = 20$	$\eta = 0.15$
	$\alpha = 0.01$	
IRBFN	$\alpha_1 = 0.2$	$\alpha_2 = 0.015$
	$\eta = 0.2$	$\kappa = 0.5$
	$e_{max} = 35$	$c_{max} = 80$
	$\nu = 0.99$	

From table 3 it is clear that RAN performed better than the other two algorithms in this particular problem. Table 4 shows the effect of the variation of some parameters with respect to the previously reported nominal values.

Algorithm	Parameter	Variation	PCC	Nodes
NCHM	σ	+20%	97	-
		−20%	98	-
	α	−50%	98	-
		+50%	99	-
	nn	−10%	87	-
		−20%	73	-
RAN	κ	−30%	99	16
		+30%	100	18
	δ	+300%	100	17
		−300%	100	17
	η	+200%	90	16
		−200%	95	18
IRBFN	κ	−50%	87	25
		+50%	83	27
	α_1	+50%	99	25
		−50%	97	28
	α_2	+50%	100	27
		−50%	98	29

Table 4. Analysis of the sensitivity of the algorithm to variations in the parameters.

5. CONCLUSIONS

Some conclusions can be drawn from the obtained results. First, RBF with constructive algorithms have been shown as a valuable tool for building qualitative topologic maps for mobile robot navigation. Among the different constructive algorithms tested, the RAN algorithm showed superior characteristics for this particular problem. This is due to the fact that RAN uses a distance criterion to insert new units and to set its output weights conveniently. The IRBFN algorithm uses cumulated error variables to insert new units, but its output weights need to be adjusted for a number of cycles before producing the correct output.

Finally, it can be noted that the best results are achieved with fewer neural nodes using the laser sensor. This is due to the fact that the sonar sensors produce noisier measurements, making the classification problem more difficult.

6. REFERENCES

Courtney, J. and A. Jain (1994). Mobile robot localization via classification of multisensor maps. In: *Proceedings of the IEEE International Conference on Robotics and Automation*. San Diego, CA. pp. 1672–1678.

Fritzke, B. (1994). Fast learning with incremental rbf networks. *Neural Processing Letters* **1**(1), 2–5.

Haykin, S. (1994). *Neural networks a comprehensive foundation*. Macmillan College Publishing Company.

Kadirkamanathan, V., M. Niranjan and F. Fallside (1991). Sequential adaptation of radial basis function neural networks and its application to time-series prediction. In: *Advances in Neural Information Processing Systems* (D. S. Touretzky, Ed.). Vol. 3. Morgan Kaufmann, San Mateo. pp. 721–727.

Kortenkamp, D. and T. Weymouth (1994). Topological mapping for mobile robots using a combination of sonar and vision sensing. In: *Proceedings of the Twelfth National Conference on Artificial Intelligence (AAAI-94)*. pp. 642–647.

Moody, J. and C. Darken (1989). Fast learning in networks of locally-tuned processing units. *Neural Computation* **1**, 281–294.

Platt, J.C. (1991). Learning by combining memorization and gradient descent. In: *Advances in Neural Information Processing Systems* (D. S. Touretzky, Ed.). Vol. 3. Morgan Kaufmann, San Mateo. pp. 715–720.

Poggio, Tomaso and Federico Girosi (1989). A theory of networks for approximation and learning. Technical report. Artificial Intelligence Laboratory, Massachusetts Institute of Technology (MIT). Cambridge, Massachusetts.

Saha, A. and J.D. Keeler (1990). Algorithms for better representation and faster learning in radial basis function networks. In: *Advances in Neural Information Processing Systems* (D. S. Touretzky, Ed.). Vol. 2. Morgan Kaufmann, San Mateo. Denver 1989. pp. 482–489.

Sanner, R.M. and J.-J.E. Slotine (1992). Gaussian networks for direct adaptive control. *Neural Networks* **3**(6), 837–863.

Wettschereck, D. and T. Dietterich (1992). Improving the performance of radial basis function networks by learning center locations. In: *Advances in Neural Information Processing Systems* (D. S. Touretzky, Ed.). Vol. 4. Morgan Kaufmann, San Mateo. pp. 1133–1140.

Zimmer, U. R., C. Fisher and E. von Puttkamer (1994). Navigation on topologic feature-maps. In: *Proc. of the IIZUKA '94*.

LOCALIZATION AND PLANNING IN SENSOR-BASED NAVIGATION FOR MOBILE ROBOTS

Juan C. ALVAREZ   **Jose A. SIRGO**   **Alberto DIEZ**   **Hilario LOPEZ**
Department of Electrical Engineering
Area of Systems Engineering and Automation
University of Oviedo

Abstract. Most approaches to automatic motion planning can be classified depending on the amount of information available to the robot each planning period. In the paradigm of motion planning with incomplete information, only a subset of the work environment is known by the robot each instant as given by its sensors, and no other previous knowledge of it is supposed. This is called Sensor-Based Motion Planning (*SBMP*). As the basic model of *SBMP* assumes no errors in localization, this paper address the question of what is the effect of position estimation errors in these algorithms performance. We found two types of problems in the experiments. Then, an study of the relation between position error and movement direction is presented. It leads to a different approach to mobile robot localization based in the coupling between planning and positioning: the robot motion planner takes into account the uncertainty in localization. Simulations show the possibilities of the proposed solution, and give hints to explain its limits and to suggest future research.
Copyright © 1998 IFAC

Key Words. Sensor-based motion planning, Localization, Robotic Navigation, Mobile Robots

1. INTRODUCTION

In motion planning with complete information, the work environment and the robot motion characteristics are completely known a priori. It puts the problem inside the field of computational geometry, where exact solutions are known (see (Latombe, 1991) for a review). In the paradigm of motion planning with incomplete information, only a subset of the work environment is known by the robot each instant as given by its sensors, and no other previous knowledge of it is supposed. Again exact algorithmic solutions are known, which are fast enough to work in real-time and guarantee convergence. This is called Sensor-Based Motion Planning (*SBMP*) (Lumelsky and Stepanov, 1986). A lot of work had been done in heuristic strategies to solve *local* obstacle avoidance problems with sensors (Koren and Borenstein, 1991), but in SBMP we are concern with approaches where a *global* convergence criteria is assured.

Position estimation must be part of any robot navigation system, either map-based or sensor-based (Lumelsky and Skewis, 1990). Localization problems come from the imprecision of relative position measurements, like dead reckoning or inertial navigation. These errors grow unboundly, making necessary to use, sooner or later, some global positioning method.

Robot navigation implies, at least, motion planning and localization. In short, the common method to deal with these two problems is to isolate them: first, to increase the localization precision with global localizing ("map matching" or "landmarks"); and then, to apply the planning solution supposed perfect odometry. But this 'two-steps' approach can not free planning from localization errors, because of the noisy nature of the sensory information used for positioning. Even with complete information, map matching is difficult, and landmark navigation is sensitive to factors like visibility and relative movement of observer and beacons (Motwani *et al.*, 1995).

The experimental implementation of Sensor-Based Motion Planning algorithms in real robots has to deal with two kind of errors with source in the sensors: *localization errors* and *sensorial-skill* (an operation which depends on the sensors accuracy, like "Detecting a Hit-Point" or "Moving Around an Obstacle") accuracy. As the basic model of SBMP assumes no errors in localization, we will first address the question of what's the effect of position estimation errors in these algorithms performance.

2. EFFECTS OF LOCALIZATION ERRORS IN PLANNING

Sensor-Based motion planning is very sensible to localization errors, as detected in the experiments even with simple environments (Alvarez *et al.*,

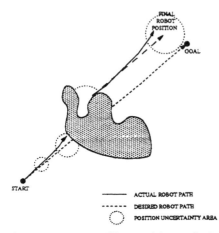

Fig. 1. Offset type error. The goal is reached, but far away from the selected one.

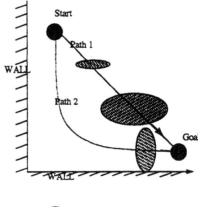

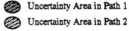

Fig. 2. The path from S to G will depend on the model of global localization. In the case of a corner, a curved line path *seems to be better in the sense of minimum uncertainty area.*

1997). There are two kind of elemental problems related to this in SBMP: "offset" and "loops". We call an *offset* error to odometry errors that do not avoid finishing the mission, but produces a result different to the one expected. As shown in Figure 1, the uncertainty in position, modeled in this case as a circle, leads to an offset. The robot odometry gives the Goal position, but we can be far from it. In the figure the error come from messing the Leave-Point, but even without obstacles we suffer this odometry errors. More dangerous is the *loop* error: the robot stays in an loop around an obstacle, because it can not find the original Hit-Point (or Leave-Point) because of odometry errors. This problem is specific of SBMP. Even if the sensory skills are error–free, odometry problems can produce that. Both effects had been found in practice in our laboratory. When the working environment complexity grows, the possibility of success decreases greatly.

They only can be solved reducing the dead-reckoning error or using any global positioning methods. A measure of our experimental odometry precision was done with the UMBmark test (Borenstein and Feng, 1996). It showed that even in experiments with trajectory lengths under 6 meters, the odometry errors are big enough to demand a global positioning aid. As the nature of sensor–based planning demands a positioning method as fast as possible, positioning based in map matching or costly computer vision based methods are avoided. For indoor navigation, our intention is to use the "natural features" of a room, as walls and corners, to reduce the uncertainty in position to levels small enough to accomplish every mission inside it.

Figure 2 resumes the main idea the coupling between planning and localization. The "optimal" path (in the sense of minimal position uncertainty) is not the straight line, because of the landmark used to positioning (a corner). The main

idea is "to move while controlling the uncertainty". If the Planner wants to move along a path where the uncertainty will grow too much, the Localizator should have something to say, giving the Planner a chance to select a better path. If there were only one solution to the geometrical problem (one path from S to G), our only option will be to follow it until either the goal (success) or the uncertainty limit (target unreachable) were reached. But if there are more possibilities to move in a provable correct way, then we should move along the rute of minimal increase of the position error.

3. RELATED WORK

The explicit conexion between planning and localization had not been very explored. In known environments, the main result is reported in (Takeda and Latombe, 1996), where a pre-processing of the map leads to a mapping between configuration and uncertainty. This criteria is used when selecting the path, and both computations are made off-line.

However, related issues are being studied under the name of "self-localizing" and "landmark recognition". The "self-localizing" problem deals with how to move with the purpose of decreasing the localization uncertainty, given a landmark (Lee and Zhao, 1995). A "good movement" will go in the direction of the descendant gradient of the map $E_r = f(x, y, \theta)$, which indicates how the uncertainty in localization E_r grows in a given workspace. This idea leads to move always in the direction of the landmark position whenever the uncertainty is proportional to the distance to the corner. Adding more landmarks to the workspace makes E_r more complicated. In this case, the better movement should lead the robot to a zone with

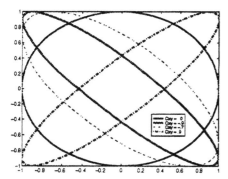

Fig. 3. Relation between numerical values and signs of P_k and their reflection in the uncertainty area

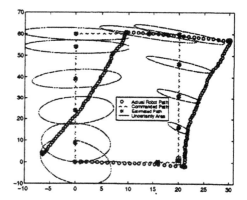

Fig. 4. The robot is commanded to follow a rectangular trip within 200 control periods. The desired path is (0,0)-(20, 0)-(20, 60)-(60,0). Actual Positions, and Uncertainty areas (each 15 control) periods are plotted.

landmarks, where it has more options to stay localized. The other connected problem is related to "exploratory" movements in order to extract relevant features from the workspace (Bauer and Rencken, 1995). This add a new third goal to the robot motion planning: 1) moving to reach the goal (planning), 2) moving to keep localized (self-localizing), and 3) moving to extract new landmarks (exploration).

In conexion with that there are different problems related to management of growth of uncertainty:

- Given a desired path, to decide when to do global positioning
- Given a desired path, to decide what beacons to use (Komoriya *et al.*, 1992)
- Given an goal, to decide when we need exploration to identify a new beacon (Bauer and Rencken, 1995)
- Given a goal, and and a set of landmarks, to decide what move takes us toward the goal with minimum uncertainty

4. UNCERTAINTY AND DIRECTION OF MOVEMENT

To address the localization problem it is necessary to define two operations related to uncertainty representation: 1) how to represent and propagate it, and 2) how localization reduces it (Smith and Cheeseman, 1986). The position uncertainty of a robot can be represented as an "uncertainty area". This area can be related to the variances of the error in the two Cartesian coordinates, σ_x and σ_y and their mutual relation σ_{xy}, or in form of the Covariance Matrix

$$P = \begin{pmatrix} \sigma_x & \sigma_{xy} \\ \sigma_{xy} & \sigma_y \end{pmatrix} \qquad (1)$$

Figure 3 shows the relation between numerical values and signs of P_k and their reflection in the uncertainty area. The physical meaning of the cross-covariance term σ_{xy} is not evident. As a measure

of this area, we will use the product of minor and mayor axis $l_M l_m$, and the width of the distribution of the estimated position error S_a along the path (Komoriya et al., 1992).

How this area varies with movement can be modeled with difference stochastic equations. Let us assume the following model of a differential gear robot

$$x_{k+1} = x_k + \frac{u_{rk} + u_{lk}}{2}\cos\theta_k \qquad (2)$$

$$y_{k+1} = y_k + \frac{u_{rk} + u_{lk}}{2}\sin\theta_k \qquad (3)$$

$$\theta_{k+1} = \theta_k + \frac{u_{rk} - u_{lk}}{l} \qquad (4)$$

where u_r and u_l are the rotational speeds of the two driving wheels, and l the distance between them. Rewriting the state space variable expression, and modeling the noise sources as an stochastic processes

$$x_{k+1} = f_k(x_k, u_k + v_k) + w_k \qquad (5)$$

with

$$x_k = \begin{pmatrix} x_k \\ y_k \\ \theta_k \end{pmatrix} \qquad u_k = \begin{pmatrix} u_{rk} \\ u_{lk} \end{pmatrix} \qquad (6)$$

with v_k the error related to the inputs, and w_k noise corresponding to the states measure. We can estimate the position each control period with $\hat{x}_{k+1} = f_k(\hat{x}_k, u_k)$, and the states covariance P_k can be updated using a linearized model

$$\Delta x_{k+1} = \frac{\partial f_k}{\partial x_k}\Delta x_k + \frac{\partial f_k}{\partial u_k}\Delta v_k + \Delta w_k \qquad (7)$$

$$= A_k\Delta x_k + F_k\Delta v_k + \Delta w_k \qquad (8)$$

$$P_{k+1} = A_k P_k A_k^T + F_k V_k F_k^T + W_k \qquad (9)$$

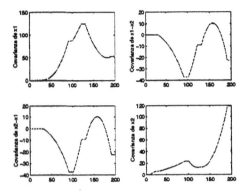

Fig. 5. The covariance matrix in the previous journey

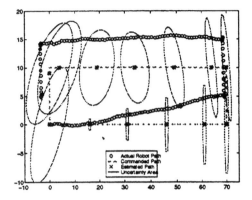

Fig. 6. Different path for the same task than in Fig. 4

with

$$A_k = \begin{pmatrix} 1 & 0 & -\frac{u_{rk}+u_{lk}}{2}\sin\theta_k \\ 0 & 1 & \frac{u_{rk}+u_{lk}}{2}\cos\theta_k \\ 0 & 0 & 1 \end{pmatrix} \quad (10)$$

$$F_k = \begin{pmatrix} \frac{1}{2}\cos\theta_k & \frac{1}{2}\cos\theta_k \\ \frac{1}{2}\sin\theta_k & \frac{1}{2}\sin\theta_k \\ \frac{1}{l} & -\frac{1}{l} \end{pmatrix} \quad (11)$$

In a workspace without landmarks, position uncertainty grows unbounded. This growth is because the previous uncertainty P_k expands, and a new uncertainty adds because of the inputs V_k and W_k. Both effects are added in a compounding operation , mathematically expressed as a matrix sum. In Figure 4 we see a simulation of a robot journey and the growth in the uncertainty in position using the previous model. The commanded path starts in $(0,0)$ and moving to $(20,0)$, $(20,60)$ and $(0,60)$ in straight lines. Actual positions are plotted in every 200 control period. The noise values were

$$W = \begin{pmatrix} 0.01 & 0 & 0 \\ 0 & 0.01 & 0 \\ 0 & 0 & 0.0001 \end{pmatrix}$$

$$V = \begin{pmatrix} 0.0001 & 0 \\ 0 & 0.0001 \end{pmatrix}$$

In Figure 5 the area uncertainty covariances are plotted for the whole journey. They vary depending on the direction of movement. Final values in $(0,0)$ were

$$P_{200} = \begin{pmatrix} 47 & -22 \\ -22 & 113 \end{pmatrix}$$

The uncertainty area grows following equation (9), therefore there are two components of uncertainty:

1) Even if no external noise is considered (V_k and W_k are both zero) the uncertainty area changes with movement:

$$A_k P_k A_k^T =$$

$$\begin{pmatrix} 1 & 0 & -p \\ 0 & 1 & q \\ 0 & 0 & 1 \end{pmatrix} \begin{pmatrix} \sigma_x & 0 & 0 \\ 0 & \sigma_y & 0 \\ 0 & 0 & \sigma_\theta \end{pmatrix} \begin{pmatrix} 1 & 0 & 0 \\ 0 & 1 & 0 \\ -p & q & 1 \end{pmatrix}$$

$$= P_k + \sigma_\theta \begin{pmatrix} p^2 & -pq & -p \\ -pq & q^2 & q \\ -p & q & 0 \end{pmatrix}$$

with

$$p = \frac{u_{rk}+u_{lk}}{2}\sin\theta_k \qquad q = \frac{u_{rk}+u_{lk}}{2}\cos\theta_k$$

In this case, and although P_k volume does not change, the area only is constant if there are no uncertainty in orientation $\sigma_\theta = 0$ (or if the robot model is such as $A_k = I \ \forall k$). Therefore, as shown in Figure 4, moving in the x axis direction (which makes $p = 0$) only increase uncertainty σ_y.

2) The effect of $F_k V_k F_k^T + W_k$ makes P_k volume grow each control period. Again, this growth in independent of the direction of movement too, but the area changes because of F_k. The movement in the x axis direction increases σ_x and σ_θ because the V part.

It is difficult to obtain an analytical model to compute the relation between directions of movement and uncertainty growth. A numerical simulation similar to the one in Figure 4, presented in Figure 6, shows a different rute (200 control periods "long"), with four 90 degrees turns too. The error conditions are the same, but the final result was:

$$P_{200} = \begin{pmatrix} 60 & 53 \\ 53 & 105 \end{pmatrix}$$

Then, to fulfill the same task, and although the path length is the same in both of them, the second rute leads to bigger uncertainty area level. As we can see in Figure 7, in both journeys the uncertainty grows, but at different rates. This measure could give us a criteria to decide when to follow a path or no, depending on if the uncertainty cross a certain level.

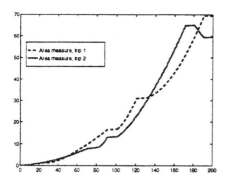

Fig. 7. Comparison of the uncertainty growth in both previous tasks

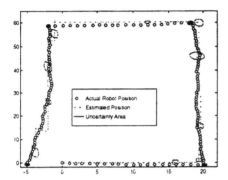

Fig. 8. The covariance matrix with landmarks

5. UNCERTAINTY REDUCTION WITH LANDMARKS

In order to control the growth of uncertainty landmarks provide us with an additional source of position information. Fusing both data (vg. with the Extended Kalman filter as fusion method), we can obtain a better estimation

$$z_k = H_k x_k + q \qquad (12)$$
$$\hat{x}_k^+ = \hat{x}_k^- + K_k[z_k - H_k \hat{x}_k^-] \qquad (13)$$
$$P_k^+ = P_k^- - K_k H_k P_k^- \qquad (14)$$
$$K_k = P_k^- H_k^T [H_k P_k^- H_k^T + Q]^{-1} \qquad (15)$$

where x_k is the real state, $\hat{x}_k^+$ the estimated one after the measurement z_k, and $\hat{x}_k^-$ the estimated before localization. When the relation between measures and states is non linear, a linealization around the mean value takes to the Extended Kalman Filter

$$z_k = h_k(x_k) + q \qquad (16)$$
$$h_k(x_k) = h_k(\hat{x}_k) + H_k(x_k - \hat{x}_k) \qquad (17)$$
$$z_k - h_k(\hat{x}_k) + H_k \hat{x}_k = H_k x_k + q \qquad (18)$$

with H_k the Jacobian matrix

$$H_k = \left(\frac{\partial h_k}{\partial x_k} \right)_{x_k = \hat{x}_k}$$

In Figure 8 the distance to a vertical wall in $X = 0$

task than in previous examples. The result is an smaller uncertainty area.

REFERENCES

Alvarez, Juan C., A. Shkel and V. Lumelsky (1997). Accounting for mobile robot dynamics in sensor–based motion planning: the safe–scan algorithm. Technical Report RL-97007. Robotics Laboratory, University of Wisconsin-Madison.

Bauer, R. and W. D. Rencken (1995). Sonar feature based exploration. In: IEEE/RSJ/GI Int. Conf. on Intelligent Robots and Systems. range.

Borenstein, J. and L. Feng (1996). Umbmark-a method for measuring, comparing and correcting dead-reckoning errors in mobile robots. IEEE Trans. Robotics and Automation.

Komoriya, K., E. Oyama and K. Tani (1992). Planning of landmark measurement for the navigation oa a mobile robot. In: IROS. pp. 1476–1481.

Koren, Y. and J. Borenstein (1991). Potential field methods and their inherent limitations for mobile robot navigation. In: IEEE Int. Conf. on Robotics and Automation. pp. 1398–1404.

Latombe, Jean-Claude (1991). Robot Motion Planning. Kluwer Academic Publishers.

Lee, Sukhan and Xiaoming Zhao (1995). A new sensor planning paradigm and its application to robot self-localization. In: IEEE/RSJ/GI Int. Conf. on Intelligent Robots and Systems.

Lumelsky, V. and T. Skewis (1990). Incorporating range sensing in the robot navigation function. IEEE Trans. Robotics and Automation 20(5), 1059–1069.

Lumelsky, V. J. and A. A. Stepanov (1986). Dynamic path planning for a mobile automaton with limited information on the environment. IEEE Trans. Autom. Control.

Motwani, Rajeev, L. Guibas and P. Raghavan (1995). The robot localization problem. In: Algorithmic Foundations of Robotics (J-C. Latombe K. Goldberg, D. Halperin and R. Wilson, Eds.). A. K. Peters (Boston).

Smith, Randall C. and Peter Cheeseman (1986). On the representation and estimation of spatial uncertainty. Int. J. Robotics Research.

Takeda, Haruo and Jean-Claude Latombe (1996). Sensory uncertainty field for mobile robot navigation. In: IEEE Int. Conf. on Robotics and Automation. pp. 335–340.

FUSION OF SENSORS APPLIED TO A COLLISION AVOIDANCE SYSTEM

J. Pontois (1) (2), P Deloof (1), A.M. Jolly-Desodt (2), D. Jolly (2)

(1) Institut National de Recherche sur les Tranports et leur Sécurité
20 rue Elisée Reclus, 59650 Villeneuve D'Ascq cedex, France.
Tel : 33 (0)3 20 43 83 94. Fax : 33 (0)3 20 43 83 59.
E-mail jerome.pontois@inrets.fr

(2) Laboratoire I³D, Bat P2,
Université des Sciences et Technologies de Lille,
59655 Villeneuve d'Ascq cedex, France.

Abstract : This article describes a multisensor system developed in order to avoid road crashes. The system deals with data issued from sensors placed on an experimental vehicle in a real environment. Indeed, actually no technology allows to conceive an ideal sensor responding to the safety conditions in the different situations met in road driving. Our objective consists in establishing a clever fusion of data issued from different existing sensors. The experimental vehicle is equipped with different sensors (a microwave anti-collision radar, a laser telemeter and a linear stereo vision system). The goal of this project is to design a hardware and software system which can prevent road crashes. The integration of the multisensor system aims at fusing different information sources in order to take safer decisions. A lot of heterogeneous and asynchronous data must be taken into consideration for the study and achievement of an on-board driving aid. Information given by different sensors pass through a Controller Area Network. The first step is the research of a common spatio-temporal referential for the different data. The sensors limitations are taken into account and the system takes advantage from each sensor. The sensor set allows the reconstruction of a local road map as accurately as possible.
Copyright © 1998 IFAC

Keywords : multisensor integration, range finders, vision, networks, Kalman filters, multitarget tracking,

1. INTRODUCTION

Road accidents are often due to multiple crashes and to the bad visibility with foggy or wet weather. To find a solution for this type of accident, we have to resort to anticollision systems which are used to warn the driver about an undergoing of a limited distance from an eventual obstacle. This safety distance corresponds to the necessary braking distance in case of a sudden stop of the front vehicle. Since twenty years, a lot of researches have been conducted in different countries on the means to reduce the number of collisions. It appears that a satisfactory solution to multiple crashes risks would be to fit out each vehicle with an interval regulation system comparable to some anti-collision systems used in public transport. This regulation would be based on a measure of the distance which separates the vehicle from the following one, and on the respect of a security distance which permits the emergency braking without collision The principle of the device consists in detecting, as accurately as possible, all the obstacles which would appear suddenly in front of the vehicle (Deloof, et al., 1994).

The specificity of the demonstrator is to use jointly different information sources which come from different sensors. Each available sensor has its own limit in the sense that it works in a satisfactory way only under certain conditions. Its performances are particularly limited by atmospheric conditions (rain, fog...) and the visibility (night or day drive).

The sensor set is directed so as to examine the front of the vehicle. Some measurements can be redundant, having the same sight zones. This is the case with the road on which the vehicle is driving, it is the way where all the beams are concentrated. In other cases, some information may be complementary.

The first part of this article describes the perception system which is made up of three sensors : a microwave radar, a linear stereo vision system and a laser telemeter. The data coming from this system pass through a CAN network to the fusion module. The network shall be introduced in the second part. The dynamic characteristics of the road environment imply the necessity of checking the data timing coherence and carrying out the tracking of a potential obstacle. This issue and the superior fusion module will be described in the last paragraph.

2. DESCRIPTION OF THE SENSORS

2.1 The microwave radar

A pulsed amplitude modulation (PAM) radar mock-up, at frequencies of 60 GHz, is installed in front of the car. The computation of the distance between the two cars is based on an approximation of the propagation time. An electromagnetic impulsion is emitted. After reflection on an obstacle, the radar picks up the reflective wave.

The beam aperture is 2°. The basic principle for scanning is a virtual defocusing of the microwave lens allowing a beam rotation of +/- 3°. We obtain 8° in front of the car with 0.14 of step. This radar can detect obstacles up to 150 meters but over 25 meters (fig 1).

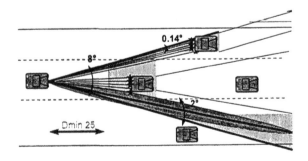

Fig. (1) Schematic radar scene.

2.2 The linear stereo vision system

The main idea is to use two linear cameras instead of classical matrix camera. However, the peculiarity of linear cameras, which do not restore the vertical dimension, needs a particular arrangement : otherwise the two cameras would not shoot the same scene. This would make 3D pictures impossible (fig 2).

A specific stereo system, using two linear cameras installed behind the car radiator grill, has been developed by the "Centre d'Automatique de Lille" to capture pairs of linear images sequences (Bruyelle, et al., 1994).

A specific calibration procedure has been developed to adjust the parallelism of the two optical axes in a common plane of view attached to the two cameras. The two axes define an « optical plane » which intersects the planar surface of the road about 100 metres in front of the car (fig 2).

Points of interest from each linear image are extracted by means of a recursive differential operator proposed by Deriche. After derivation, a gradient magnitude image is obtained. This image indicates the amplitude of the reponse of the Deriche operator and the sign of the derivative of the signal. Before proceeding, it is necessary to select pertinent local extrema among these remaining edge points. This is achieved by splitting the gradient magnitude signal into adjacent intervals where its sign remains constant. At this stage of the procedure, the edge points extracted from left and right linear images are matched to find the distance from the obstacle (Burie and Postaire, 1994).

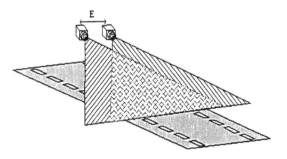

Fig. (2) Principle of linear stereo vision .

2.3 The laser telemeter

The telemeter, like the radar, calculates the distance to the obstacle using an approximation of the propagation time. The telemeter used, has two beams, one in the horizontal position and the other one for short distances (fig 3). The beam aperture is 1.1° for the directive beam and 4° for the wide one. The telemeter detects obstacles up to 120 meters (150 meters with an optional external equipment). The telemeter gives only the information of the distance to the nearest objet, after a mean on 200 measures.

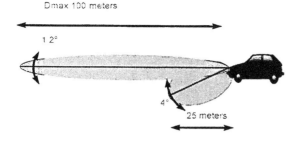

Fig. (3)   Laser Range finder .

3. THE NETWORK

The information given by each sensor passes through a Controller Area Network or CAN bus towards the final decision module. The CAN network is a serial data bus specifically conceived for geographically compact and moderate data bandwidth applications. On top of the three sensors named before, a lock sensor and a speed sensor are also connected on the bus (fig 4). A software allows to record in real-time the data present on the bus.
A video recording allows to compare subsequently the data with the real road scene.

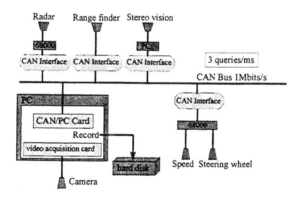

Fig. (4)   Network architecture

3.1 The functions

The network allows the transmission of data on the bus. After a local data processing on each sensor, the information is dated by a local timer on the interface card. The processing time is then calculated in order to readjust the data in time. Each local clock is synchronised by a general clock every two minutes. The error messages and the scan of the present sensor on the network allow a fast control of the technical failure. If the production of data is faster than the reading, the oldest data are deleted. In the same way, if the reading is faster than the production, the most recent data are read again.

4. FUSION

Data fusion is hierarchical. The first level of fusion starts at the signal production. Each sensor carries out local treatments to transmit only high level semantic data. Objects representing obstacles are built from radar and telemeter echoes or the video signal. These objects are described with a specific formalism : the longitudinal and transversal distance, the obstacle size and the acquisition instance. To merge information from the different sensors , they must be in a same spatio-temporal referential. The fusion system must include a module to readjust the data in time. A prediction and an adjustment are applied to each obstacle by a Kalman filtering. Therefore each obstacle can be tracked in time.
Sensors have a wide common sight zone. Thus, it is imperative to match obstacles when they are detected by several sensors.

A map of the road is obtained by mixing the complementarity and redundancy characteristics of each sensor.
At last, a decision module informs the driver of the road danger state.

4.1 The different observation zones

Figure 5 points out the observation zones of the different sensors. Zone [1] is the common zone of the three sensors. In zones [2] and [3], two sensors can give information: on the first zone the telemeter and the radar, on the other the telemeter and the stereoscope. In zones [4] and [5] objects are seen by only one sensor, on the first zone by the stereoscope and on the other by the radar. Zone [6] is the blind zone, where no observation is possible.

The observation zones allow to widen the sighting of one sensor but also to predict the potential obstacle movements. Thus, appearance and disappearance of obstacle can be predicted. If an obstacle comes in front of the car, it is first seen by the radar. Then its moving is used to anticipate the appearance in the others zones. In the same way, disappearance of obstacles, while coming near to the blind zone, can be anticipated. The knowledge of the different zones improves the occurrence control by the creation, the elimination and the tracking of potential obstacle.

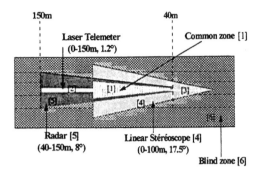

Fig. (5) Observation zones

4.2 Data timing coherence and tracking of a potential obstacle

Each sensor describes the scene by a list of obstacles. This list constitutes the sensor map. There is one map per sensor. Prediction and correction are carried out with the Kalman filter (Chui and Chen, 1990). After each measurement, this procedure prediction/correction is applied for each obstacle and for each sensor to update the sensor map (fig 6).

4.3 Fuzzification

The fusion of different maps is realised by an aggregation of fuzzy numbers. The sensor gives different estimations of the distance from the obstacle. Each obstacle is defined by its co-ordinates X and Y and some other parameters (time, speed, width, ...) but also some a-priori data like the measurement precision. This distance can be expressed by a triangular fuzzy number.

4.4 Tracking

A very large variety of fusion operators has been proposed in the literature (Bloch, 1996, Dubois, 1994).
The final distance obtained by the set of three sensors can be calculated as the fuzzy arithmetic mean of the three fuzzy distances. However, it is possible that one of the sensor gives information related to another vehicle. So, if we take into account this information it would make the result of fusion erroneous. In this case a compatibility test has to be implemented to see which sensor must not be taken in account. The fusion will then use only information coming from sensors than does not disagree. For example, if two objects are separate from more than 5 meters, we consider that they must

be different. In some cases, it will not be appropriate to take into account all sensors.

The method for aggregating fuzzy number was developed by Yager (1996) from the concept of OWA (Yager, 1988). This method is able to include the concept of compatibility and partial aggregation for a reinforcement of the information. We can also use the notion of external trust given to some information using information coming from the environment and the climatic conditions. The ideas presented in introduction and 4.1 about the conditions of good working of the sensors can give us a subjective evolution of this trust. So, taking in account at the same time the compatibility of information and the external trust we can define a global coefficient of trust which characterise each source. This allows us to give greater or lesser reliability to the data (Delmotte, et al., 1995). The aggregation of the different sensor maps gives a global map. At each new measures, the obstacles position is predicted from the previous global map by a Kalman filter and aggregated with the positions given in the three sensor maps (fig 6). The global map is then updated at each new measurement.

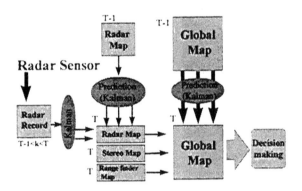

Fig. (6) The fusion architecture

5. CONCLUSION

Too many accidents are still uncontrollable. Accidents are rarely due to a single cause but generally, they result of a succession of circumstances which , taken individually, would not be able to cause an accident. The speed is the principal factor of the road crashes. The drivers estimate badly the environment danger and come in front of the difficulty with an excessive speed. The other causes are the tiredness and the atmospheric and visibility conditions. The anticollision system has to give a pertinent assistance to the driver in the critical situations, in order to avoid that these situations lead to accidents. We automatically have to acquire all the information which are necessary to the detection of an accident risk, and to present these information, after doing a processing under a form adapted to the driver needs, that is to say that they have to be as complete and synthetic as possible.

Collision avoidance on the road entails numerous problems linked to the real-time aspect of the system and the dynamic environment of the road. The moving vehicle is surrounded by other moving vehicles. At any time information about the road is needed, even if no new information is available. Therefore it is necessary to predict the behaviour of the obstacles and give the most accurate and reliable information as possible.

REFERENCES

Bar-Shalom Y, (1988).Tracking and Data Association Academic press, inc.

Bloch Isabelle, (1996). Information combination Operators for data fusion : A comparative review with classification. IEEE Transactions on systems, Man, and cybernetics. Part A : Systems and Humans, vol. 26, N°.1.

Bruyelle J L, Burie J C and Postaire J G, (1994). A new vision system for obstacle detection. 2nd congress of Mecatronique, Takamatsu, Japon, Vol 2, P 747-750, 1-3 November 1994.

Burie J C, Postaire J G, (1994). A coarse to fine matching procedure for linear stereo vision. International Federation of Automatic Control, 7th Symposium on Transportation Systems: Theory and Application od Advanced Technology, Tianjin, Chine, p 164-169, 24-26 august 1994.

Chui C K, Chen G, (1990). Kalman Filtering with real-time Applications. Second Edition, 1990, Spring-Verlag.

Delmotte F, Dubois L, Desodt A-M, Borne P, (1995). Using trust in uncertainty theories. Information and systems engineering. Vol 1, Number 3,4 Decembre 1995. IOS press.

Dellof P G, Rolland P A, Haese N, (1994). La prévention des collisions routières par radar anticollision micro-onde, Synthèse des études effectuées au GRRT de 1989 à 1993.
Synthèse INRETS n°25 juillet 1994.

Dubois D, Prade H, (1994). Possibility theory and data fusion in poorly informed environnements. Control Eng Practice, Vol 2, n° 5, p 577-583.

Yager R, (1988). On ordered weighted averaging aggregation operators in multicriterion decision making. IEEE Trans Systems Man Cybernet vol 18, p 183-190.

Yager R, A Kelman (1996). Ffusion of fuzzy information with considerations for compatibility, partial aggregation, and reinforcement. International Journal of approximate reasoning vol 15, p 93-122, 1996.

UNMODELLED OBSTACLE AVOIDANCE IN STRUCTURED ENVIRONMENTS USING FEEDFORWARD NEURAL NETS

E. Delgado Romero, J.C. Raimúndez Álvarez

Dpto. de Ingeniería de Sistemas y Automática, E.T.S.I.I. Universidad de Vigo
Lagoas Marcosende, s/n. 36200 Vigo (Spain).
Fax: +34 (86) 812201. Phone: +34 (86) 812610. E-mail: emma@centolla.aisa.uvigo.es

Abstract: An unmodelled obstacle elusion procedure for wheeled mobile vehicles is presented. It is shown the adoption of two neural nets: the first one, for path following and the second, for obstacle avoidance. This module is presented as a subsystem attached to a global supervisory module. The modelled robot is of LABMATE style. The sensors used, are inexpensive POLAROID sonars. The simulation furnishes good advice for number of sensors and arrangement, as well as for time step discretization. *Copyright © 1998 IFAC*

Keywords: Mobile robot, sonar sensor, neural networks, supervised training, unmodelled obstacle avoidance, path following.

1. INTRODUCTION

In the quest for autonomy in mobile robots, several tasks must be defined and specified. Depending on the surroundings, like indoor structured environments, routines as: following a wall or a path, passing a door, avoiding an unmodelled obstacle, etc. include the relevant functions needed to attain a goal.

Those tasks can be modelled using several approaches such as Fuzzy logic or Neural nets. Neural nets as black box approximators, offer the advantage of quick implementation when compared with traditional Fuzzy logic (Patrick, 1994). Neural nets use backpropagation over desired data whereas Fuzzy Logic needs rules, sets and scaling. The results obtained using Neural Nets are good enough for prototyping and simulation having as by products, the final number and arrangement of sensors, etc.

2. ROBOT AND SENSOR MODELS

The modelled robot is of LABMATE style whose kinematic equations are:

$$\left. \begin{array}{l} rb(k+1) = rb(k) + \Delta rb \\ x(k+1) = x(k) + step * \cos(rb) \\ y(k+1) = y(k) + step * \sin(rb) \end{array} \right\} \underline{X}(k+1) = f(\underline{X}(k), \underline{U}(k)) \quad (1)$$

where:

$\underline{X}(k) \equiv$ The state vector at instant k given for the vehicle's position $(x(k), y(k))$ and its orientation $(rb(k))$, in relation to the global reference system.

$\underline{U}(k) \equiv$ Actuation, that implies a translation (step) and a rotation (Δrb).

The mobile robot is equipped with POLAROID sonar sensors that provide information about the environment. It is important to emphasise the following aspects related with them (Leonard, 1992):

2.1 Sonar sensor

In the arguments advanced by Brown (Brown, 1985), all solid surfaces appear as acoustic reflectors. Also, because the acoustic wavelengths are generally quite long, most of them perform as acoustic mirrors.

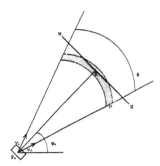

Fig. 1. Geometry of the sensor

Consequently, surfaces that are not orthogonal to the direction of propagation reflect signal energy away from the source.

The responses from the detected surfaces, take the form of circular arcs in Cartesian coordinates (sequences of headings over which the range value measured is the distance to the target). The figure 1 illustrates the geometry of the sensor where:

$Ps \equiv$ Position of the sensor in the global frame.
$\varphi_s \equiv$ Orientation of the sensor in the global frame.
$\theta \equiv$ Visibility angle.
$C \equiv$ Visibility cone $\equiv Ps + l_i.v_i + l_d.v_d \quad | \; l_i,l_d > 0.$
$MN \equiv$ Segment $\equiv M + (N-M)\gamma \; | \; 0 \le \gamma \le 1.$
$\sigma \equiv$ Measuring uncertainty.
$R \equiv$ Reference. Point of the segment that returns signal.

2.2 The environment seen by a sonar sensor

The environment is modelled through straight oriented segments, here cited from now on as CAD plane (CAD stems for Computer Aided Design). This plane represents the surroundings where the robot moves.

Therefore, the problem will be reduced to the study of the detectability of every segment composing any solid surface. A segment is detected by a sensor if it accomplishes the following conditions:

The segment is not occluded.

The segment is visible: If at least one of the ends of the segment is within the visibility cone and in addition, when the sensor receives echo of any internal point of the segment (surface that are orthogonal to the direction of propagation).

The segment is at a lower distance than the maximum range of the sensor.

Every sonar reading generates a range value (one return per time step). If the segment is detected by the sensor, the reading will inform about the minimum distance between them. Otherwise, the sensor will return the maximum range.

$$h_i\big(X(k), R\big) = \sqrt{(R_x - Ps_x)^2 + (R_y - Ps_y)^2} \quad (2)$$

2.3 Sensorial modelling at the vehicle.

The required number of sensors, and arrangement were established through simulations. Lastly, the use of nine sensors with a symmetric distribution, five at the front side of the vehicle and two in both lateral sides is adopted. It was assumed that the car always moves forward.

3. NEURAL NETWORKS

The backpropagation learning rule can be used to adjust the weights of any feedforward network in order to minimise the sum-squared error of it. This type of learning relies on the generalisation property that it makes possible to train a network on a representative set (the training data file) of input/output pairs that associate the desirable output values to the specific input values. In both designed nets, is adopted the use of the backpropagation techniques with momentum and adaptive learning rate, that increase the speed and reliability of it.

3.1 Neural network for the pursuit

This neural net must provide the necessary actuation for the following of a prefixed trajectory with variable slope.

Generation of the training data file. A representative input/output data table is constructed by the algorithm of pursuit on a straight line. With the definitions (see fig. 2):
XY $\equiv$ Global reference system.
X'Y' $\equiv$ Reference system attached to the robot.

$$d(k) = P_r(k) - P_a(k) \quad (3)$$

$$d(k+1) = d(k) + V_x \quad (4)$$

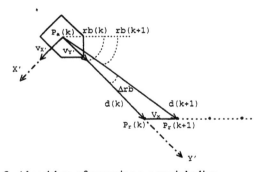

Fig. 2. Algorithm of pursuit on a straight line.

200

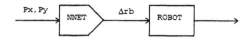

Fig. 3. Inputs/output of the NNET for path following

$$px = d(k+1)'* v_{X'} \qquad (5)$$

$$py = d(k+1)'* v_{Y'} \qquad (6)$$

$$\Delta rb = rb(k+1)-rb(k) | \quad -\pi \leq \Delta rb \leq \pi \qquad (7)$$

where:
$v_{X'} \equiv$ Versor in the X' direction.
$v_{Y'} \equiv$ Versor in the Y' direction.
$P_a \equiv$ Absolute robot's position.
$rb \equiv$ Absolute robot's orientation.
$P_r \equiv$ Absolute trajectory's position.

$px \equiv$ Projection of the distance to the X' axis.
$py \equiv$ Projection of the distance to the Y' axis.
$\Delta rb \equiv$ Actuation on the robot's steering angle.

Every vector entering the NNET, will include the projections px and py, and will associate its corresponding normalised actuation on the robot's steering angle, that will be the explicit NNET output (Fig 3). The other component of the actuation, the translation, will appear as a training parameter in an implicit form and will admit variations depending on the distance between the robot and the reference trajectory.

This way, a training data file with 80 input/output vectors is formed, as the fig. 4 shows.

Pursuit network architecture. The network has two layers, with sigmoidal transfer functions. The number of nodes at the distribution and outside layers corresponds to the inputs (2) and output (1) of the NNET. It was satisfactory the choice of two nodes for the hidden layer.

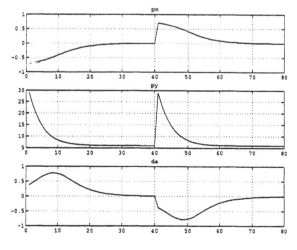

Fig. 4. Training data file.

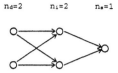

Fig. 5. Pursuit network architecture

Simulations. The following graphics show the NNET learning capacity, comparing the output data of the training file with the output of NNET when is excited for the same input vectors (Fig. 6) and testing how the neural network, trained off-line, acts in the following of a prefixed trajectory with variable slope (Fig. 7).

3.2 Neural network for obstacle avoidance

This neural net must provide the necessary actuation in the robot's steering angle in order to avoiding every obstacle that appears in its planned trajectory.

Generation of the training data file. The training data set for this network is obtained from several paths generated bordering a regular polygonal obstacle, and eluding it. The paths have clearly two phases. The first one, leaving the planned path and smoothly avoiding the obstacle; the second surrounding it.

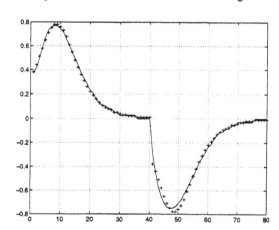

Fig. 6. The NNET learning capacity. Output data of the training file (points) and the NNET.

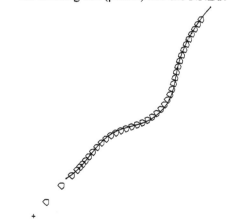

Fig. 7. Simulation

Fig. 8 Inputs/output of the NNET.

The data generation was performed using nine sonar sensors. After submitting the mobile robot to each training path, the range data of the nine sensors and the differences between these values and the obtained at the instant before, are normalised and filed for further network training. These are the input data. The output at each step is given by the normalised increment in the robot's steer angle. Therefore, every vector of the data set has 18 inputs and 1 output, making a training data file with 198 input/output vectors (see fig. 8).

$$z_k = \left\{ h_i(k) \mid 1 \le i \le 9 \right\} \quad (8)$$

$$z_k - z_{k-1} = \left\{ h_i(k) - h_i(k-1) \mid 1 \le i \le 9 \right\} \quad (9)$$

for h_i given by the equation 2

Obstacle avoidance network architecture. The best results were obtained with a net of three layers with sigmoidal transfer functions. The number of neurones at the distribution and outside layers corresponds to the inputs (18) and output (1) of the NNET. It was satisfactory the choice of three neurones for the first hidden layer and two neurones for the second one.

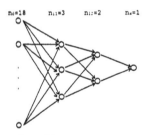

Fig. 9 Network architecture

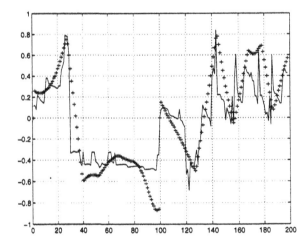

Fig. 10 The NNET learning capacity. Output data of the training file (points) and the NNET.

The results of the learning process are showed in the figure 10 where are compared the output training data and the output of the NNET.

Simulations. For checking the net obtained, a planned path (a straight beginning from initial pose) is built and over it, is generated an irregular random polygonal object. Using the sensorial collected data and its variations, the network generates the incremental angle actuation. Figure 11 illustrates the robustness of the net, which admits variations upon the training velocity (with the consequent change in the sample period); noise in the sensorial measures and the presence the random obstacles in volume, position and orientation.

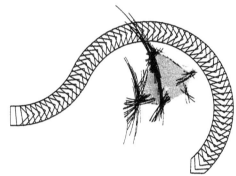

Fig. 11 Simulation

4. UNMODELLED OBSTACLE ELUSION PROCEDURE

This module is presented as a subsystem attached to a global supervisory module and it is established under the following assumptions:

A CAD plane of reference. This plane represents the surroundings in which the robot moves, and it is described by straight segments as already said. The non-modelled objects such as chairs, boxes, etc. will be considered as unmodelled obstacles.

A sufficient knowledge of the robot's pose, acquired through accurate sensing.

Previous determination of planned trajectories.

The flux diagram of the figure 14, shows the interactions between the basic elements over which is built the system of unmodelled obstacle avoidance. The conditions are explained:

Free path: The planned path. If the theoretical measures calculated over the CAD from actual pose and the real world measures moving average have an error greater than a minimum previously defined through simulations, then this event is understood as the presence of an unmodelled obstacle, thus firing the elusive procedure.

Collision. The planned path avoids CAD collisions maintaining a secure margin. The chance of collisions thus, appears during elusive manoeuvres. In the case of dangerous approximation to the CAD plane, the elusion procedure can be deactivated by the supervisory module, replanning the path.

Bordering. When during the elusion procedure the front central sensors lose object contact but receive contact measures from lateral sensors.

Lost. When during the elusion procedure the front and lateral sensors lose object contact, the robot is considered lost.

Watching. When bordering, a virtual sensor placed on the robot, heading the frontal regions with aperture and range previously defined, watches ahead for planned path detection. When this occurs (the planned path is detected), the path following net is activated.

5. CONCLUSIONS

The adoption of feedforward neural nets for complex control tasks is growing "pari pasu" with the development of powerful training methods. The great advantage from the neural net side is due to the general methodology used for design. Backpropagation can be blended with other techniques (Raimúndez, 1994) to improve the training process.

Neural nets, as universal approximators, can model with accuracy and good level of robustness level, complex behaviours like those presented.

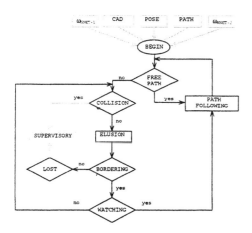

Fig. 12 Flux diagram.

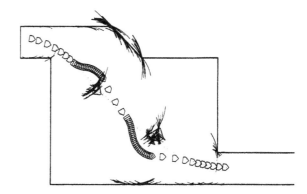

Fig. 13 Simulation

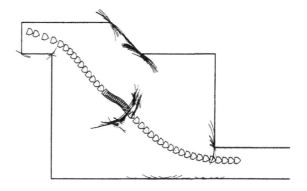

Fig. 14 Simulation

ACKNOWLEDGMENT

The authors would like to acknowledge CICYT for funding the work under grant TAP-96-1184-C04-03.

REFERENCES

Brown M. (1985).
  Feature extraction techniques for recognizing solid objects with an ultrasonic range sensor.
  IEEE Journal of Robotics and Automation, RA-1 (4): 191-205.

Leonard J. and Durrant-White H. (1992)
  Directed Sonar Sensing for Mobile Robot Navigation.

Raimúndez J.C. (1994)
  Utilisation of neural networks in environment perception and guidance of mobile robots.
  AIRTC'94, Valencia (Spain).

Reignier P. (1994)
  Fuzzy logic techniques for mobile robot obstacle avoidance.
  Robotics and Autonomous Systems **12**, 143-153.

MOVING OBSTACLE AVOIDANCE ALGORITHM FOR MOBILE ROBOTS UNDER SPEED RESTRICTIONS.

A. Cruz†, V. Muñoz†, A. García-Cerezo†, A. Ollero‡.

†Dpto. de Ingeniería de Sistemas y Automática. Universidad de Málaga. Plaza El Ejido
s/n, 29013 Málaga (Spain).
Phone: (+34) 5 213-14-07; Fax: (+34) 5 213-14-13; E-mail: victor@ctima.uma.es
‡Dpto. de Ingeniería de Sistemas y Automática. Universidad de Sevilla. Avenida Reina
Mercedes s/n, 41012 Sevilla (Spain).
Phone: (+34) 5 455-68-71; Fax: (+34) 5 455-68-48; E-mail: aollero@cartuja.us.es

Abstract:This paper presents a moving obstacle avoidance algorithm for mobile robots that work under some kind of speed limitations. The proposed method takes the information about these speed constraints and the moving obstacles in its environment, and provides a safe speed profile which allows to build a trajectory that bypasses such obstacles. The algorithm applies three steps: i) speed planning, ii) mobile obstacles avoidance, and iii) speed profile generation. The method has been successfully tested in the RAM-2 mobile robot. *Copyright © 1998 IFAC*

Keywords: Autonomous mobile robots, robot navigation, obstacle avoidance, trajectory planning, robot dynamics, robot kinematics.

1. INTRODUCTION

The control motion system of a mobile robot drives its course by using the position and speed references (trajectory) computed by the planning system. Hence, this trajectory should be defined in such a way that provides a safe navigation with no collisions. Therefore, the obstacle avoidance is defined as the main function of the planning process.

The approach used for performing the above process depends on the movement state of the obstacle. Some methods contemplate the obstacle's size and trajectory as entries to the local planner algorithm. Classic planning strategies can be adapted for this problem, such as visibility graphs (Gil de Lamadrid, 1994), configuration space schemes (Fujimura and Samet, 1990) and potential fields (Kyriakopoulos and Saridis, 1993).

The path-velocity decomposition is an efficient method for avoiding mobile obstacles by computing a safe speed function (Kant and Zucker, 1986). However, this methodology must be modified in order to provide the following features:

- First order continuity: The speed function and its first derivative should be continuous in order to reduce the speed tracking error.

- Admissibility from the point of view of the vehicle's kinematic and dynamic behaviour: The speed reference must be defined in such a way that allows the tracking system to follow the trajectory. Therefore, the kinematic and dynamic models of the robot must be taken into account in the speed function definition.

The method, for avoiding moving obstacles, proposed in this paper covers the above points by using a piecewise cubic function, which is defined by considering other kind of speed restriction imposed by the vehicle's physical characteristics.

These limitations (environmental and physical speed restrictions) are commented in section 2. Section 3 is devoted to the previous work where the method presented in this paper leans on (Muñoz, 1995). This work involves two stages: a speed planning process, which provides a speed planning that takes into account physical and operational speed limits, and a speed profile generation process, which builds the speed function needed to obtain the vehicle's trajectory. Section 4 details how the proposed method copes with the avoidance of moving obstacles. The speed planning made considerating only kinematic and dynamic constraints is combined with the information about the mobile obstacles. In this way, a set of safety zones representing that speed planning, is built. Whenever a hazardous situation is detected, the original planning is modified in order to get a safe navigation. Implementation and experiments on the RAM-2 mobile robot are showed in section 5, and finally, section 6 presents the conclusions of this work.

2. SPEED PLANNING PROBLEM.

A robot path is defined as a set of evenly spaced postures $Q=\{q_1,...,q_m\}$, which are to be executed by the path tracking algorithm. A posture q_i is composed of five basic elements: x_i, y_i, θ_i, κ_i and s_i. The first two elements are position components, the third is the heading with respect to a global work frame, the fourth is the curvature component, and the last one is the distance along the path from the starting posture to the current one.

In order to convert a path Q into a trajectory $\tilde{Q}$ it is necessary to append a speed component to each posture of the path. In other words, the trajectory conversion process must turn each $q_i=(x_i,y_i,\theta_i,\kappa_i,s_i)$ into $\tilde{q}_i =(x_i,y_i,\theta_i,\kappa_i,s_i,v_i)$, where v_i is the posture speed component. This transformation is made by the definition of a parametric arc length speed function V(s). Such a curve is defined in the space-speed plane (Shiller and Gwo, 1991), where the upper speed limits for each posture q_i of the path Q are represented. These limits are obtained by taking into account the speed constraints introduced by the vehicle features and operational speed limitations. Thus, V(s) is specified in such a way that it preserves all the posture speed limits, in order to obtain a speed profile with good tracking conditions. That means that V(s) must lie inside a safety area of the space-speed plane defined by the speed limits functions (see Fig. 1.).

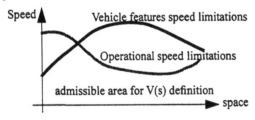

Fig. 1. Speed limitations in the space-speed plane.

The speed constraints considered at speed planning

time are shown in table 1.

Table 1: Speed limits classification

Type	Constraint
Physical	Mechanical (ME)
	Kinematics (KI)
	Dynamic (DY)
Operational	Goal Point (DG)
	Known mobile obstacles (KM)

The physical speed restrictions group is due to the kinematic and dynamic behaviour of the mobile robot (Prado et al., 1994). These restrictions impose a top speed and acceleration according to the pecularities of the vehicle and the path to be followed.

The second group presents the external speed limitations, which arise because of performing the task in a real environment. In this way, some situations force to set a safe speed value to the vehicle in order to synchronize with other elements in the working environment, or even to stop before getting into a zone with collision hazard.

3. PREVIOUS WORK: SPEED FUNCTION V(s).

The speed function V(s) definition is made in two steps: speed planner and speed generation processes.

3.1. Speed planner process.

This stage chooses a set of control path postures $C=\{q^1, ..., q^p\}$ which divides the path into a set $S=\{S_1,..., S_{p-1}\}$ of path segments, where S_i is composed of the path postures sequence between q^i and q^{i+1}. The choice of the elements which will be belong to C is made depending on the nature of the speed limitations.

A top speed v_i is assigned to each member q^i of C by using the minimum speed value provided by the speed limitations introduced by vehicle's features and operational speed constraints. This operation sets up a speed control set $V=\{0, v_1,...,v_{p-2}, 0\}$ for the path Q, whose first component (always null) is the starting speed for S_1 and the remaining components v_i are speed boundary conditions between segments S_{i-1} and S_i. Sets C and V resulting from the speed planning process, can be represented in the space-speed plane, and will be used by the speed profile generator process for building V(s).

3.2. Speed generation process.

The $V(s)$ is modelled through a spline curve defined as a set of space-time functions $\sigma_i(t)$. The ith component of this curve is assigned to the path segment S_i, and it is determined by the parameters set $\Pi_i = \{v_i, v_{i+1}, s_i, t_i\}$, where v_i and v_{i+1}, components of the set V, are the starting and ending speed assigned to the current path segment; s_i is its length, and t_i the navigation time.

The method evaluates $\sigma_i(t)$ by using two cubic functions $^1\sigma_i(t)$ and $^2\sigma_i(t)$, which are obtained from $\sigma_i(t)$:

$$
\begin{aligned}
^1\sigma_i(t) &= {}^1\alpha_{i0}t^3 + {}^1\alpha_{i1}t^2 + {}^1\alpha_{i2}t + {}^1\alpha_{i3} \\
^2\sigma_i(t) &= {}^2\alpha_{i0}t^3 + {}^2\alpha_{i1}t^2 + {}^2\alpha_{i2}t + {}^2\alpha_{i3}
\end{aligned}
\tag{1}
$$

Function $^1\sigma_i(t)$ covers the first half of the segment and $^2\sigma_i(t)$ the second one. Let ^it_m be $t_i/2$ and $t \in [0, {}^it_m]$. If the acceleration is null at the beginning and end of the path segment S_i and the second derivatives of both functions $^1\sigma_i(t)$ and $^2\sigma_i(t)$ are continuous at their joints, then, $^1\sigma_i''(0)=0$, $^2\sigma_i''(^it_m)=0$ and $^1\sigma_i''(^it_m)={}^2\sigma_i''(0)$. The first function is defined by the set $\{v_i, {}^1v_m, {}^1s_i, {}^it_m\}$ and the second one by $\{{}^1v_m, v_{i+1}, {}^2s_i, {}^it_m\}$ where ^1v_m is the average between v_i and v_{i+1}. These functions are detailed in matrix form as follow:

$$
M \times \begin{bmatrix} ^1\alpha_{i0} \\ ^1\alpha_{i1} \\ ^1\alpha_{i2} \\ ^1\alpha_{i3} \end{bmatrix} = \begin{bmatrix} 0 \\ ^1s_i \\ v_i \\ ^iv_m \end{bmatrix} \quad M \times \begin{bmatrix} ^2\alpha_{i0} \\ ^2\alpha_{i1} \\ ^2\alpha_{i2} \\ ^2\alpha_{i3} \end{bmatrix} = \begin{bmatrix} ^1s_i \\ ^2s_i \\ ^iv_m \\ v_{i+1} \end{bmatrix} \tag{2}
$$

where M is defined by the expression:

$$
M = \begin{bmatrix} 0 & 0 & 0 & 1 \\ (^it_m)^3 & (^it_m)^2 & ^it_m & 1 \\ 0 & 0 & 1 & 0 \\ 3(^it_m)^2 & 2(^it_m) & 1 & 0 \end{bmatrix} \tag{3}
$$

The values for the elements of these sets are:

$$
^it_m = \frac{s_i}{v_i + v_{i+1}} \qquad ^iv_m = \frac{v_i + v_{i+1}}{2}
$$

$$
^1s_i = \frac{s_i(5v_i + v_{i+1})}{6(v_i + v_{i+1})} \qquad ^2s_i = s_i - {}^1s_i \tag{4}
$$

Thus, the relationship between the starting and ending speeds for keeping the ith path segment top acceleration constraint is defined by:

$$
\text{if}(v_i < v_{i+1}) \text{then}
$$
$$
(v_{i+1} \le \sqrt{v_i^2 + {}^iA_t s_i}) \qquad \text{else}(v_{i+1} \ge \sqrt{v_i^2 - {}^iA_t s_i}) \tag{5}
$$

Finally, the speed profile $V(s)$ is defined in the space-speed plane by using expression (6).

$$
V(s) = \bigcup_{i=1}^{p} \frac{d}{dt}\sigma_i(\sigma_i^{-1}(s)) \tag{6}
$$

4. MOVING OBSTACLE AVOIDANCE.

The approach for avoiding mobile obstacles presented in this paper lays on the path-velocity decomposition, proposed by Kant and Zucker (1986). This decomposition works in the so called path-time space $s \times t$, where moving obstacles, as well as the searched speed planning, can be represented.

In the $s \times t$ space, obstacles are modelled as rectangles. These rectangles arise from the set of crossing points between the mobile obstacle's trajectory and the vehicle´s path. A crossing means that there is a time interval when a moving obstacle is extending over the vehicle's path. This situation defines a forbidden rectangular region which vertex determined by the maximum subsegment of the vehicle´s path occupied by the obstacle and the time needed to cross it.

Obviously, a rectangle is not the best solution to represent a crossing between the vehicle and a circular moving obstacle. Furthermore, the solution assumes that the crossing between the obstacle´s trajectory and the vehicle´s path is orthogonal, and this is not always possible. In fact, there are other solutions that fit better in this situation (for example, using an ellipse for modelling the obstacle, instead of a rectangle). However, figuring such others shapes out turns more difficult and expensive: the better a solution is found, the harder its computation becomes. So, a choice must be made between accuracy and efficiency.

Once all the crossings between the vehicle and the moving obstacles have been determined and represented in the $s \times t$ space, the speed planning begins. As it was stated in section 2, the speed planning is made in the speed-space phase plane. So, this planning has to be mapped into the path-time space, where the mobile obstacles are represented. The speed planning is rendered in this space by $V(t)$ function, obtained as follows:

$$
V(t) = \bigcup_{i=1}^{p} \sigma_i(t) \tag{7}
$$

where $\sigma_i(t)$ is the composition of $^1\sigma_i(t)$ and $^2\sigma_i(t)$ as shown in expression (1). $V(t)$ is defined as a safe planning when this function does not touch any rectangular region, in other case the speed planning must be modified. Let OB_i be a moving obstacle in the $s \times t$ space. The $V(t)$ function provides the maximum speed allowed by the vehicle's kinematics and dynamics for any time t. Therefore, for avoiding the collision with OB_i, the robot must reduce its speed and let the moving obstacle pass the collision point before the vehicle reaches it. This action means that the region OB_i is below the replanned $V(t)$ curve.

However, testing the collisions beetwen $V(t)$ and all OB_i is a time-expensive process. Instead of this, the proposed method uses a set of hull-convex areas, called the safety zones Z_i, which encloses every $\sigma_i(t)$ belonging to $V(t)$. Therefore, the proposed algorithm checks the collisions between every Z_i and OB_j, and whenever this situation is detected the safety zone Z_i will be moved properly in the $s \times t$ space. The $V(t)$ function must be rebuilt after this operation.

4.1. Safety Zones

The building of the safety zone Z_i assigned to $\sigma_i(t)$ is based on the following lemma:

Lemma 1: *The shape of the speed profile $\sigma_i(t)$ assigned to S_i is always either concave or convex.*

Proof: *Let $^1\sigma_i(t)$ and $^2\sigma_i(t)$ be the components of $\sigma_i(t)$. Their second derivatives are defined in the following expresion:*

$$^1\sigma''_i(t) = 6\,^1\alpha_{i0}t$$
$$^2\sigma''_i(t) = 6\,^2\alpha_{i0}t + 2\,^2\alpha_{i1} \tag{8}$$

where $^1\alpha_{i0}$ is obtained by solving expression (2) :

$$^1\alpha_{i0} = \frac{-(v_i - v_{i+1})(v_i + v_{i+1})^2}{s_i^2} \tag{9}$$

In the same way,:

$$^2\alpha_{i0} = \frac{(v_i - v_{i+1})(v_i + v_{i+1})^2}{6s_i^2}$$
$$^2\alpha_{i1} = \frac{-(v_i - v_{i+1})(v_i + v_{i+1})}{2s_i} \tag{10}$$

Concavity or convexity of these functions is studied by taking into account the sign of their second derivatives, which depends on the relationship between the starting and ending speed of S_i.

An increasing speed segment $(v_i < v_{i+1})$ proves that both derivatives are positive, which implies that functions $^1\sigma_i''(t)$ and $^2\sigma_i''(t)$ are concave. On the other hand a decreasing speed segment $(v_i > v_{i+1})$ forces that derivatives are negative, and both subsegments are convex.

By taking into account the above lemma, the safety zone Z_i can be modelled as a triangle in the $s \times t$

space as it is shown in Fig. 2.

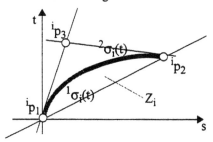

Fig. 2. The safety zone Z_i assigned to $\sigma_i(t)$.

This figure presents the safety zone as a shady triangle defined by the following lines:

- Line segment from the starting point $^ip_1$ to the ending point $^ip_2$.
- Tangent line to $^1\sigma_i(t)$ at $^ip_1$ (slope=v_i).
- Tangent line to $^2\sigma_i(t)$ at $^ip_2$ (slope=v_{i+1}).

So, Z_i is defined by $^ip_1$, $^ip_2$ and the intersection point $^ip_3$ between the tangent lines.

4.2. Time-shifting the safety zones.

As it was stated before, the robot must decrease its speed when a collision between a safety zone Z_i and a forbidden region OB_j is detected. In $s \times t$ space, this means that Z_i will be shifted in such a way that OB_j will be left beneath Z_i. A change on a safety zone Z_i implies a change on its attached $^1\sigma_i(t)$ and $^2\sigma_i(t)$ functions. Therefore, the values that define both functions, i.e., $\{v_i, {}^iv_m, {}^1s_i, {}^it_m\}$ and $\{{}^iv_m, v_{i+1}, {}^2s_i, {}^it_m\}$, are also modified. This situation increases the navigation time t for travelling S_i , so a new either starting or ending speed will be computed. The relationship between speeds and time (expression (4)) shows the calculation of these values, depending on the increasing or decreasing speed of the segment S_i:

$$v_i < v_{i+1} \Rightarrow v_{i+1} = \frac{s_i}{\tilde{t}_i} - v_i$$
$$v_i > v_{i+1} \Rightarrow v_i = \frac{s_i}{\tilde{t}_i} - v_{i+1} \tag{11}$$

where $\tilde{t}_i = t_i + \Delta t_i$ is the new navigation time for S_i. Secondly, it is necessary to take into account the side effects on the following segment S_{i+1} (if $v_i < v_{i+1}$) or the previous segment S_{i-1} (if $v_i > v_{i+1}$).

In this way, the algorithm for avoiding the collision between a safety zone and an obstacle is implemented below:

```
Function Ṽ =AvoidMobileObstacle(OB_j,S,V,Π_i)
    let Z_i=ComputeSafetyZone(v_i,v_{i+1},s_i)
    while Collision(OB_j,Z_i) do
        let t_i=t_i+STEP
        if v_i<v_{i+1},
            v_{i+1}=(s_i/t_i)-v_i;
            modify next segment S_{i+1};
        else
            v_i=(s_i/t_i)-v_{i+1};
            modify previous segment S_{i-1};
        end if
    end while
End Function.
```

As it is shown, the parameters needed for that algorithm are the obstacle OB_j, the set of path segments S, the speed control set V, and the parameters set Π_i of the current segment S_i. The first operation is the calculation of the safety zone Z_i for the ith segment. Once Z_i is built, the next step performs two actions inside a loop: i) modifies the navigation time t_i of S_i; and ii) computes the new values of the starting or ending speed, taking into account the side effects on the contiguous segments. The loop finishes when no collision with obstacle OB_j is detected. The result of the algorithm is the modified speed control set Ṽ, that assures the avoidance of the obstacle OB_j in segment S_i.

The value STEP stands for the increase of the time needed to traverse S_i avoiding the mobile obstacles in its neighbourhood. This value is fixed by a rule obtained from experimentation, and it is based on the distribution of the obstacles on the s × t space.

4.3. Final algorithm.

The integration of the proposed method for avoiding mobile obstacles inside the speed planning and generation processes is described in the following algorithm.

```
Procedure Path2Trajectory(Q,OT)
    Q̃ ={}
    {S}=DividePathIntoPathSegments(Q)
    {V}=ComputeSpeedControlSet(S)
    {OB}=ComputeObstaclesSet(S,OT)
    For each S_i belonging to S do
        For each OB_j belonging to OB do
            {V}=AvoidMobileObstacle(OB_j,S,V,Π_i)
        end loop
    end loop
    For each S_i belonging to S do
        Compute ^1σ_i(t) and ^2σ_i(t)
        σ_i(t)=Composition(^1σ_i(t),^2σ_i(t))
        V_j(s)=σ_i'(σ_i^{-1}(s))
        Q̃ = Q̃ ∪ {S_i, V_i(s ∈ S_i)}
    end loop
End Procedure
```

The avoidance algorithm is shown in bold type. It is necessary to consider that moving a safety zone Z_i to avoid a forbidden region OB_j might cause collisions between previous safety zones Z_k with k=0,...,i-1 and rectangles OB_l with l=0,...,j-1; so, the search for collisions must be restarted if the function *AvoidMobileObstacle* modifies the previous speed planning.

5. IMPLEMENTATION AND EXPERIMENTS

The speed planning and profile generator algorithms have been integrated in the intelligent control architecture of RAM-2 (see Fig. 3.) (Ollero et al., 1993). .

Fig. 3. RAM-2 autonomous mobile robot.

Figures 4 to 6 show the results of the proposed method with the mobile robot RAM-2. The studied situation involves an environment with a moving obstacle. Figure 4 shows the path to be tracked by the robot in solid line and the trajectory followed by the obstacle in dotted line.

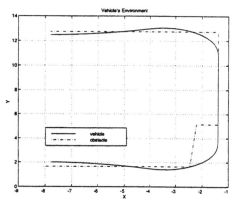

Fig. 4. Robot path and the moving obstacle.

The original speed planning provided by the speed planning process is represented, on space-speed plane, with a dotted line in figure 5. However, the moving obstacle crosses the robot path several times along its trajectory. This information is used for replanning the original speed profile, and obtaining a safe speed planification for the mobile robot (solid line) which avoids the collision.

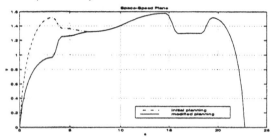

Fig. 5. Original speed planning and its modification for avoiding the mobile obstacle.

The above result is achieved via path-time representation (see Fig. 6.).

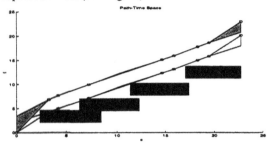

Fig. 6. Path-time representation.

This figure details the moving obstacle as a set of dark rectangles, and also contains the safety zones for every segment of the path as light shady triangles . The method detects the collision between the first safety zone and the first obstacle, and eliminates this situation. The result of this action is shown as the set of dark shady triangles in Fig. 6. The new speed planning avoids the mobile obstacle and also considers the robot motion constraints due to its physical features.

6. CONCLUSIONS.

This paper presents a moving obstacle avoidance algorithm for mobile robots that work under some kind of speed limitations. The proposed method takes the information about these speed constraints and the moving obstacles in its environment, and provides a safe speed profile which allows to build a trajectory that bypasses such obstacles. The algorithm applies three steps: i) speed planning, ii) mobile obstacles avoidance, and iii) speed profile generation.

The method modifies the speed planning when is required due to a collision situation. Moreover, it does not handle a visibility graph on the $s \times t$ space, and therefore does not use any kind of graph-search algorithm. The result is a speed profile which avoids the moving obstacles and takes into account other kind of speed constraints (i.e. kinematic and dynamic limitations). This feature provides the real time execution for speed

replanning which can be also made while the robot is tracking the path.

Finally, working in the presence of mobile obstacles demands a complete knowledge about them, i.e., their sizes and trajectories. Though this situation is usually real in multi-robot systems, the navigator must use a local speed planner which modifies the current speed reference using feedback control (Mandow et al., 1996). This action solves the uncerties about the obstacle's trajectory, that the planner does not consider in planning time.

7. ACKNOWLEDGEMENTS

This work has been done within the framework of the projects TAP96-1184-C04-02 and TAP96-0763 of the C.I.C.Y.T. (Spain)

8. REFERENCES.

Fujimura, K. and Samet, H. (1990). Motion planning in a Dynamic Domain. *Proc. IEEE Int. Conf. on Robotics & Automation*, pp. 324-330

Gil de Lamadrid, J. (1994). Avoidance of Obstacles with Unknown Trajectories: Locally Optimal paths and Periodic Sensor Readings. *The International Journal of Robotics Research*, Vol 13,No 6, pp. 496-507.

Kant K., Zucker S. (1.986). Toward Efficient Trajectory Planning: The Path-Velocity Decomposition. *The International Journal of Robotics Research*, Vol 5, No. 3.

Kyriakopoulos, K. J. and Saridis, G. N. (1993). An Integrated Collision Prediction and Avoidance Scheme for Mobile Robots in Non-Stationary Environments. *Automatica*, Vol. 29, No. 2, pp. 309-322.

Mandow A., Muñoz V., Fernandez R., García-Cerezo A. (1997). Dynamic Speed Planning for Safe Navigation. 1997 *Proc. IEEE International Conference on Intelligent Robots and Systems*.

Muñoz V. (1995). Trajectory Planning for Mobile Robots. *Ph. D. Thesis*. University of Malaga (Spain).

Ollero A., A. Simón, F. García, and V. E. Torres (1993). Integrated Mechanical Design of a New Mobile Robot. *Proc. SICICA '92*. Pergamon Press.

Prado M., Simon A., Muñoz V., Ollero A. (1994). Autonomous Mobile Robot Dynamic Constraints due to Wheel-Ground Interaction. *Proc. of EURISCON'94*. Vol 1, pp 347-360. Malaga (Spain)

Shiller Z., Gwo Y. (1991). Dynamic Planning of Autonomous Vehicles. *IEEE Transactions on Robotics and Automation*, Vol 7 No. 2.

NOMAD: A DEMONSTRATION OF THE TRANSFORMING CHASSIS

E. Rollins, J. Luntz, B. Shamah, and W. Whittaker

Robotics Institute, Carnegie Mellon University

Abstract: During the Summer of 1997 - Nomad - a planetary-relevant mobile robot, was driven via satellite link for more than 125 miles in the Atacama Desert of Chile by novice operators in North America, demonstrating many technologies relevant to robotic exploration of the planets. An innovative "transforming" chassis was demonstrated which uses a simple linkage to change the footprint of the vehicle from a stowed to a deployed position. This linkage also enables both double Ackerman and point-turn steering. This paper presents details on the design of the transforming chassis including kinematic analysis used in low and high level control.
Copyright © 1998 IFAC

Keywords: Steering, Vehicle Suspension, Mechanisms, Kinematics

1. INTRODUCTION

Interest in the exploration of the planets and moons of our solar system is growing. Robotic explorers are ideal for long-duration exploration of such harsh and unknown environments. These environments present challenges for communication, sensing, autonomy, power, and locomotion systems. In order to develop robotic explorers qualified for these challenges, it is necessary to first test them in analogous missions on the earth. Harsh environments such as the deserts and polar regions of Earth offer accessible proving grounds for such experiments.

During June and July of 1997 *Nomad*, a lunar rover prototype developed at Carnegie Mellon University, made the longest teleoperated cross-country traverse ever accomplished by a robot - 138 miles through the Atacama Desert in northern Chile. Nomad was driven via satellite for most of its journey by team members and the general public at Pittsburgh's Carnegie Science Center, the University of Chile, NASA Ames Research Center, and even from their own homes through Pittsburgh Public Television using the telephone. This field experiment successfully demonstrated several technologies slated by NASA to be in-

cluded in future planetary exploration missions. Among these technologies were a new imaging device - the *panospheric camera*, a high-performance antenna pointing device for high bandwidth communication, remote geological investigation, and safeguarded and autonomous modes of driving.

In order to accomplish its journey, Nomad needed a locomotion system capable of tackling the planetary-analog terrain of the Atacama desert while carrying all of the necessary equipment for the rest of the task. The special demands of the terrain combined with geometric constraints on the robot required a new chassis, the motivation, design, and kinematic analysis of which is the focus of this paper.

2. PREVIOUS RELEVANT ROVER DESIGNS

In 1970, the unmanned Soviet rover Lunakhod traversed 10km of the Lunar surface, collecting data. Lunakhod, with a fixed wheel base and skid steering, was constructed with eight self-contained electrically powered wheel modules. The idea of self-contained wheel modules has become the standard for planetary-relevant vehicles. Later, the LRV rover also made a lunar excursion, driven by American astronauts. This rover had four-wheel Ackerman steering and was expandable from a stowed position to a deployed position. Astronauts manually unfolded "outriggers" holding the wheel and steering modules, providing a larger, more stable wheelbase than would have fit in the landing module. More recently, JPL's Rocky series of micro-rovers (Gat *et al.* 1994, Miller *et al.* 1992), which lead to the production of the Martian rover Sojourner, used four-wheel explicit steering on a six wheel rocker-bogie suspension. The suspension allows the Rocky rovers to tackle large obstacles with relatively smooth body motions. Sandia's rover, Ratler (Purvis and Klarer 1994), used a four wheel bogie suspension with skid steering.

Nomad combines the concepts of Lunakhod's wheel modules and Ratler and Rocky's bogie suspensions with a new linkage to provide both explicit steering, as used by LRV and Rocky, and an automatic version of LRV expanding wheelbase. Nomad's *transforming chassis* automatically expands the wheelbase from a stowed to a deployed position with the same actuators used for steering.

3. MOTIVATION

Two factors motivated the unique design of the transforming chassis of Nomad, both involving volumetric constraints. The first is a constraint on the overall size of the robot, and the second is a constraint on parts within Nomad.

The design constraint for the size of Nomad was generated from a proposed plan to explore the moon with two large rovers transported by a Saturn V rocket. The payload faring constrains each rover's footprint to 72" × 72".

The Moon's surface is bombarded by a constant shower of micrometeorites. creating a deep, loose layer of fine dust that covers much of the surface. A lunar vehicle's wheels must be sufficiently large to "float" on rather than sink into the soil. In fact, such sinking ended the mission of the Soviet rover Lunakhod. A set of equations was developed (Bekker 1956) which empirically represent the behavior of wheels in soils of varying composition. These equations were applied (Apostolopoulos 1997) to the problem of Nomad's wheels floating in lunar soil, and determined optimal wheel size of 30" diameter and 18" width, along with parameters such as balancing sinkage, traction, and locomotive power draw.

Four wheels of this size occupy a significant portion of Nomad's lower volume. This forces other components to be placed higher raising the vehicles center of gravity and reducing its stability. Further volume would be occupied by the wheels

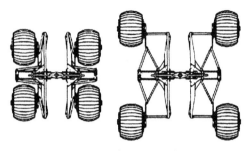

Fig. 1. Nomad's transforming chassis in stowed (left) and deployed(right) positions.

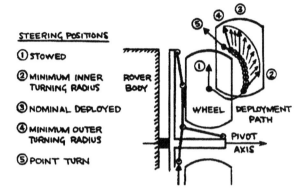

Fig. 2. Positions and steering range for Nomad's transforming chassis

as they are rotated by a standard explicit steering mechanism. It is important that as much of this lower volume be used for heavy components to keep a low center of gravity.

This motivated an investigation of alternative steering mechanisms that could steer explicitly without sacrificing internal space by moving the wheels away from the body as they steered. A four bar linkage with this property was developed which had the additional property of producing two positions in which the wheels point straight ahead. For Nomad, the first such position is the original, *stowed* position. The second such position is the *deployed* position about which the wheels are steered. Figure 1 shows this linkage in the stowed and deployed positions. Figure 2 shows the range of steering angles around the deployed position including the point turn position.

Nomad's *transforming chassis* enables explicit steering while keeping a low center of gravity and expanding the footprint of the robot into a deployed position, increasing stability. Nomad can exit a lunar lander by driving straight ahead and using skid-steering in the stowed position until it is clear of lander and can transform into the more stable and mobile deployed position for the remainder of its mission.

4. SYSTEM DESIGN

The *transforming chassis* along with the *averaging suspension system* and the self contained *wheel*

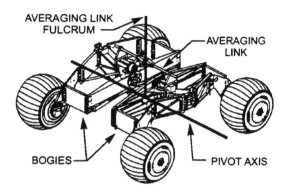

Fig. 3. Nomad's averaging system, consisting of bogies and an averaging link

modules make up the locomotion system of Nomad. These systems are described below.

4.1 *Wheel module*

Each of Nomad's wheels is a self contained system consisting of a tire, hubs, axle, motor and harmonic drive. All of these components are sealed inside a aluminum tire and are physically independent from the rest of the robot. The tires have grousers to improve traction.

4.2 *Averaging System*

The averaging system acts as a suspension smoothing the motions of the robot's body relative to the motions of its wheels. The two wheels on each side of the rover are attached through the steering system to a bogie which pitches relative to the body about a central axis (see Figure 3). This allows all four wheels to rest on the ground regardless of the terrain. With the pivot placed in the middle of each bogie, the vertical excursion of the pivot is the average vertical excursion of that bogie's two wheels. A similar averaging is experienced across the rover's body. Therefore the center of the body lifts by an amount equal to the average vertical excursion of all four wheels. The pitch of the body is fixed at the average of the pitches of the two bogies by the averaging link. This pitch, roll, and vertical averaging reducing the effects of rough terrain on the rover body and its components. The following photograph shows the right front wheel perched on a large rock (Figure 4) while the other wheels remain on flat ground.

4.3 *Steering system*

Nomad can steer by three methods: skid steering, double Ackerman, and point turning. Skid steering is used only to position the robot for deployment, and is considered a backup mode for

Fig. 4. Nomad perches on a rock, demonstrating the averaging suspension and the stability-enhancing large footprint

use in the case that steering motors fail. Point turn mode is best used for reversing direction when progress is blocked. The preferred mode of steering for Nomad is double Ackerman. This is accomplished by two pairs of four-bar linkages - one pair on each of the two bogies. A push-rod is attached to one axis of each output link. These rods are driven by two racks which are pulled in opposite directions by a single pinion placed between the two linkages. The axle of each wheel module is rigidly attached perpendicular to each of the four output links so that the steering angle is equal to the angle of the output links.

The sizing of the transforming chassis linkage was accomplished by graphical methods with the following goals:

- Accommodate wheels 30" dia. × 18" width
- Enable point turns
- Limit steering actuator loading
- Minimize volume occupied by mechanism
- Limit non-deployed size to 72" × 72"
- Maximize deployed footprint size
- Limit minimum turning radius to no larger than one non-deployed vehicle width

With an overall length limit of 72", accommodating two wheel diameters on the side leaves 12" of space between. To accommodate cabling, support and actuation between the wheels, another 6" is consumed. This leaves 3" each for linkage element behind and in front of wheels. Similarly, accommodating an electric generator in the center of the rover leaves 9" on each side for linkage elements. Turning radius depends both on steering angle and wheel position (see Figure 5). Therefore, minimizing turning radius was an iterative process, requiring the simultaneous consideration of both wheel position and steering angle at the extreme steering position. Similar considerations were necessary to enable point turns. To enable a turning radius of 72", the resulting steering angle

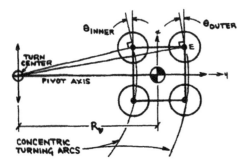

Fig. 5. Nomad's turning radius is a function of wheel position and steering angle

of the inner pair of wheels at the extreme position is 33°. To enable a point turn at the other extreme of turning, the resulting angle is 49°.

In the first stages of deployment much of the motion is sideways sliding rather than forward rolling, putting large loads on the steering actuators. In addition, one part of the linkage starts at an angle near a singularity, further increasing loads on the actuators. This situation improves with the increase of this starting angle. However, a large starting angle for this link widens the mechanism in the stowed position, taking up more volume from the body. A balance between these two constraints gives a satisfactory mechanism volume without overloading the steering actuators. Finally, larger links provide a greater deployed footprint, so the links were scaled to be as large as possible while still accommodating the other requirements. The stowed footprint is 72" × 72", while the deployed footprint becomes 96" × 96", greatly improving Nomad's stability.

5. KINEMATIC ANALYSIS

High-level control algorithms as well as human operators request the center of the robot to steer at a particular turning radius. The turning radius is a function of both the position and steering angle of the wheels. This, combined with the fact that the steering actuators act through a set of linkages, complicates the functional relationship between actuator commands and high-level commands. In order to implement control of Nomad's steering, a kinematic analysis of the linkage is necessary.

First, the case where all wheels lie in a plane is examined, and the relationship between actuator position and turning radius is established. Then, the case where the wheels rest on uneven terrain is considered by projecting the wheels positions into a plane and then calculating the turning radius.

5.1 *2-Dimensional*

Following is an analysis of one linkage set. This is mechanically linked to a mirror image linkage

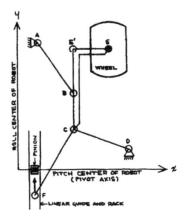

Fig. 6. Schematic of one quarter of Nomad's steering system

on the same side of the rover, and this system is duplicated on the other side. Figure 6 shows a schematic of the links that make up the four bar linkage that steers one wheel. The motion of point F is constrained in the x direction by a linear rail and is actuated by a rack and pinion in the y direction. The linkage is mirrored about the x axis such that this pinion drives both racks. Thus the motion of the two linkages is mechanically linked and the steering angles that are produced are equal. The steering mechanisms on each side of the vehicle are independent, and each must be positioned so that the two turning arcs are concentric, producing double-Ackerman steering. The turning center can also be placed at the center of the vehicle, enabling a point turn.

Turning radius commands from autonomous or user control must be transformed into actuator inputs (motions of point F) and wheel velocities. This is accomplished by calculating the turning radius as a function of the actuator input, computing the appropriate wheel velocity, and using lookup tables to reverse the calculations. The transformation must be done separately for each side of the vehicle since the inner and outer steering radii differ and a single steering radius for the center of the vehicle must be determined.

Using trigonometric relationships, wheel position E and steering angle θ are computed as follows:

- Given input F_y, fixed position F_x, fixed point D, and link lengths CD and FC, determine position of C from triangle CDF.
- From the position of C, fixed point A, and link lengths AB and BC, find position B from triangle ABC.
- Calculate steering angle, θ, from the relative positions of B and C.
- Using θ, the position of B, and link lengths BE' and $E'E$, determine position E.

Figure 5 shows the relationship between turning radius, wheel position, and steering angle. Equa-

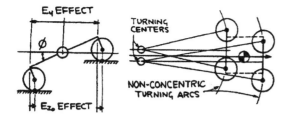

Fig. 7. Out-of-plane motions of Nomad's wheels move the turning centers of both pairs of wheels

tions for calculating the vehicle turning radius R from both the inner (+) and the outer (−) wheel pairs are:

$$R = \frac{E_y}{\tan \theta} \pm E_x$$

When the radius calculated from the outer and inner sides matches the desired vehicle turning radius, the appropriate inner (−) and outer (+) wheel velocities, v are calculated based on the desired vehicle velocity, $v_{vehicle}$.

$$v = \frac{v_{vehicle}}{R} \sqrt{E_y{}^2 + (R \mp E_x)^2}$$

Physical limits of the steering mechanism restrict the Ackerman turning radius to a 72" minimum - one (non-deployed) vehicle width.

5.2 3-Dimensional

The above analysis describes the rover as it functions on flat ground. As it travels over uneven terrain, the system does not exist in a plane and must be considered as three bodies, rotating about the y-axis of the rover. The rover body is considered ground and bogie angle is defined as the angle of rotation of the bogie with respect to the rover body (see Figure 7). Likewise, turning radius is considered in the plane of the body.

The averaging mechanism forces the bogies to have equal and opposite bogie angle ϕ. With steering commands for the rover being given with respect to the global coordinate system, the positions and angle of each wheel must be related to the body coordinate system. Thus the effect of bogie angle ϕ on wheel position E as it is projected into the rover body plane is that of reducing its distance from the pivot axis - E_x remains the same, but E_y is shortened. Furthermore, the wheel centers and the pivot axis are not actually in the same plane but are separated by the vertical distance E_{z_o}. Unlike the symmetric ($\cos \phi$) effect of bogie angle ϕ with E_y on the wheel center location, the effect of bogie angle ϕ with E_{z_o} is to move both projected wheel locations in the same direction ($\sin \phi$) (see figure 7). The adjusted projected position, E' of the wheel center is:

$$E'_x = E_x \qquad E'_y = E_y \cos \phi + E_{z_o} \sin \phi$$

The calculations above consider E as the position of the center point of the wheels, but the point at which the wheel contacts the ground is relevant to steering. The projection of these points into the body plane is coincident in the planar case, but not necessarily in the 3-dimensional case. Nomad, however, has no way of determining the location of the contact point, so the nominal case is assumed where the wheels contact the ground on a plane parallel to the body plane and tangent to the wheel, regardless of bogie angle. The projection of such a contact point is coincident with the projection of the wheel center, requiring no additional transformation to the position of E.

Since the wheels no longer steer about an axis normal to the body plane, an additional correction factor is necessary to adjust the steering angle, θ:

$$\tan \theta' = \cos \phi \tan \theta$$

The new effective steering center for a pair of wheels lies on a line shifted by the same amount the E'_y was shifted. The radius can be calculated from the 2-D case, using the adjusted wheel positions and steering angles. For the inner (−) and for the outer (+) wheel pairs, the new steering radius is:

$$R' = \frac{E_y + E_{z_o} \tan \phi}{\tan \theta} \pm E_x$$

Also, because the wheel "rolls" sideways on the ground as the bogie angle changes when the wheel is steered, the contact point is no longer at the radius of the wheel. The new contact point is closer to the wheel axis, so the wheel velocity must be adjusted. Since the contact surface of the tire is spherical, the adjusted wheel velocity is

$$v' = \frac{1}{\cos \phi} v$$

Due to the fact that the projections of the wheel positions are no longer symmetric about the pivot axis but the effective steering angles of the two wheels on each side are still equal, the turning center of each side no longer lies on the pivot axis (see figure 7). Furthermore, the direction in which the turning center is transformed is opposite between the two sides, making it impossible to generate concentric steering arcs. Because of this mismatch, the moving rover wheels slip, making it impossible to accurately predict the true turning center. As a compromise the average turning center could be considered for steering commands (see figure 7), adding a dimension to the lookup table to accommodate adjustments for bogie angle.

At the bogie angle's physical limit of $\phi = 25°$ the difference in the pivot axis and the averaged turning center is fairly small. This leads to an error of less than 10% in the vehicle turning radius.

Because of this small error, and the fact that this adjustment would be based on many assumptions about soil mechanics and weight distribution between the wheels, three dimensional effects are ignored.

6. PERFORMANCE

The primary goal of traversing 125 km under the direction of distant operators, operating through satellite connections was met. Two design flaws prevented a perfect performance from the locomotion system. Rocks occasionally became wedged between the inner hub of the wheel and the wheel-link, causing severe abrasion and in some cases punctures to the hub and damage to bearing seals. The actuation assembly in the wheel modules was not sufficiently fastened and vibrations allowed the assembly to become misaligned. This led to a chain of failures resulting in gear wear and several broken bolts. Even with these events, the rover traversed 138 miles.

The transforming chassis performed extremely well, experienced no failures, and showed no signs of impending trouble. Its wide wheel base provided the extremely stable platform necessary for traversal of the desert environment. Nomad traversed down slopes as steep as 38°, up slopes as steep as 22°, across slopes at 33°, and over discrete obstacles up to 22" high (see figure 4). Although there were some obstacles it could not surmount, nothing short of vertical walls higher than 2 feet were shown to be a problem to the stability of the rover.

7. CONCLUSIONS

The in-wheel propulsion was mechanically simple and it placed heavy elements like motors and gearheads low, dropping the center of gravity and increasing stability. The averaging linkage distributed loads among all the wheels and smoothed body angles and excursions, even under extreme conditions. Explicit steering eliminated the side loads, extreme power draw, and most of the slippage common in skid-steered systems. The extremely wide wheelbase Nomad achieves through this transforming chassis gives it stability that belies its stowed size. The extreme slopes and obstacles it encountered in the Atacama were dispatched with confidence by the sprawling locomotion.

The biggest problem with the transforming chassis is its weight. While some weight could have been removed through structural optimizations, the structure needed to be strong enough to withstand the high forces experienced in deployment.

When the chassis is deployed and simply performing steering maneuvers, the forces in the mechanism drop by half an order of magnitude. Thus the mechanism must be sized to accommodate the loads that it experiences only during deployment, which only occurs once during a planetary exploration mission.

Since the transformation from compact to deployed position would occur only once during a planetary mission, spring assistance combined with explosive bolt deployment could aid in the initial stages of the motion, reducing the worst case design loads on linkage members and actuators. Further improvement can be gained through reduction of the number of parts and reduction of parts' weight in the form of improvements in materials and optimization of shape for loading conditions. These methods, however, promise only incremental improvements. Revolutionary changes are necessary if this is to be a useful configuration for planetary exploration. Current efforts focus on the development of new linkages that give the same or better performance but do not experience the same side loads during deployment. Parts count reduction, materials improvements, the addition of crab steering mode, more compact mechanisms and thus improved utilization of body volume, deployment expansion in the vertical direction to improve ground clearance and reduce compacted volume, variable averaging amplifying motions in the middle range and truncating motions in the extremes, and the investigation of springs in the suspensions are all targets for research leading to the development of a next generation of Nomad.

8. REFERENCES

Apostolopoulos, D. (1997). Analytical Configuraion of Wheeled Robotic Locomotion. PhD thesis. Robotics Institute, Carnegie Mellon University.

Bekker (1956). *Theory of Land Locomotion.* University of Michigan Press. Ann Arbor.

Gat, E., R.S. Desai, R. Ivlev, J. Loch and D.P. Miller (1994). Behavior control for robotic exploration of planetary surfaces. *IEEE Journal of Robotics and Automation* 10(4), 490–503.

Miller, D.P., R.S. Desai, E. Gat, R. Ivlev and J. Loch (1992). Reactive navigation through rough terrain: Experimental results. In: *Proceedings of the 1992 National Conference on Artificial Intelligence.* San Jose CA.

Purvis, J. and P. Klarer (1994). Ratler: Robotic all terrain lunar exploration rover. In: *Proceedings of the Sixth Annual Space Operations, Applications, and Research Symposium.* Johnson Space Center, Houston TX.

A FORCE CONTROLLED ROBOT FOR AGILE WALKING ON ROUGH TERRAIN

José A. Gálvez, Pablo González de Santos and Manuel Armada

Instituto de Automática Industrial - CSIC
Departamento de Control Automático
La Poveda, 28500 Arganda del Rey, Madrid, Spain
Tel.: +34 1 8711900. Fax: +34 1 8717050
E-mail: jagalvez@iai.csic.es

Abstract: Force controlled legged vehicles are worldwide researched for their intrinsic advantages in locomotion over different types of unknown irregular soil. The work presented in this paper is intended to be a contribution to improve walking machines' stability, agility and adaptability over uneven terrain. A design for feet and ankles with an integrated sensor system that enables the walking robot to determine the local surface orientations of the ground under each foot is described. Such information can then be used by the robot's control system for the real-time computation of more favourable foot force distributions. *Copyright © 1998 IFAC*

Keywords: Force control, many-degrees-of-freedom systems, mobile robots, redundancy control, sensor systems, walking.

1. INTRODUCTION

The most evolved terrestrial animal species have developed legged locomotion systems over millions of years of evolution. This quality gives us a distinctive ability to traverse unstructured rough terrains and adapt our movements to a greater extent of situations or contingencies than wheeled and tracked locomotion systems can do.

In the technological world, the continuous development of Microelectronics, Sensory Perception, Automatic Control and Computer Sciences is bringing about that walking machines are steadily growing in significance. It can be expected that, within some years, legged robots will prevail over wheeled or tracked vehicles in such tasks as autonomous planetary exploration or, more generally, those jobs where the ability to robustly and flexibly traverse different kinds of irregular, possibly unknown, terrain is required (Bares and Whittaker, 1993; Sukhatme and Bekey, 1995). The

same applies for environments that were primarily designed for humans, with stairs or varied obstacles that can be overcome by legs, but not by wheels. There is a need for research on improved concepts and techniques to reduce the gap between animal walking and robot walking.

One of the topics related to legged locomotion that has been intensively researched in the last two decades is force based control. Some important functions of force control are: 1) optimization of the walking machine's stability, traction and adaptability; 2) avoidance of the risk of foot slippage; 3) smoothing of the robot's motion, by actively damping the mechanical impacts of possible brusque footsteps; 4) improvement of energy efficiency, and 5) identification of mechanical properties of the underground.

For the purpose of investigating these concepts, a prototype robotic leg (see figure 1) was designed and constructed at the *Instituto de Automática Industrial*

Fig. 1. Prototype robotic leg.

(I.A.I.) embodying the essential features of the four-legged machine that is being under construction at the moment. It is an insect-like leg with the peculiarity that it incorporates an articulated foot. The leg possesses three rotary joints actuated by three DC-servomotors with planetary gearheads; the second and third rotary joints also include special spiroid gears. A sensor system has been conceived and integrated at the leg's ankle which, in conjunction with appropriate force adaptive algorithms (Gardner, 1992), will give the robot a distinctive ability to blindly, rapidly and nimbly walk over unknown uneven terrains.

Gardner (1992) presented a general modeling and formulation of the force distribution problem for statically stable walking machines in which the feet contact the ground at arbitrary inclinations. Then he developed a computationally efficient technique for the solution of the problem. This was an important extension of previous techniques since the primary purpose of legged vehicles is locomotion on rough terrain and Gardner's formulation was the most generalized one for a force-controlled walking machine on rough terrain: Gardner formulated the equilibrium equations in the local support coordinate systems oriented according to the local surface normals at the zones of contact between *each* foot and the ground. Then he outlined an approximate solution to solve them.

The computer simulations that Gardner realized to demonstrate his technique indicated that it performs favourably when compared to the widespread

pseudo-inverse method and a computationally excessive optimal scheme. However, his approach was merely theoretical and, to the authors' best knowledge, it was not implemented in a real walking robot. In order to put Gardner's method into practice, a sensor system would be required that would enable the robot to determine the arbitrary orientations of the local footholds (of each leg). Nothing about this matter is mentioned in Gardner's work.

Nevertheless, the referred determination of the footholds' orientations can be accomplished with great simplicity if the robot's feet are articulated and have flat or concave soles, so that the feet adapt to the local inclinations of the ground (see figure 3). In the design presented in this paper, two angle sensors are integrated at the robot's ankles, which consist of a universal joint (i.e. two passive rotary degrees of freedom). In this manner, the orientations of the surface normals (i.e. two parameters) at the local footholds can be measured. This simple mechanical design -already implemented at the prototype leg- is described in section 3. The information gained by the conceived sensor system can then be used by the robot's control system for the real-time computation of the ideal commanded forces according to Gardner's approach.

2. FORCE DISTRIBUTION IN STATICALLY STABLE WALKING MACHINES

The force control of a walking robot involves the solution of the problem of the force distribution at its footholds. A statically stable legged robot supports itself with at least three legs contacting the ground. Each leg must possess three degrees of freedom to permit arbitrary foot placement. Hence, the six degrees of freedom of moving the machine's body through space are controlled using at least nine actuators. This means that six equilibrium equations but nine or more free unknowns are available. The force distribution problem is therefore statically indeterminate: for any kinematic situation of the robot, there exists an infinite number of foothold force distributions (or, correspondingly, actuator torque combinations) that satisfy the equilibrium equations. Among them, the control system may choose the force distribution that is considered most appropriate according to some criteria. This section addresses the determination of this most appropriate force distribution.

Figure 2 (Gardner, 1992) depicts a generalized representation of a walking machine on uneven terrain, by which the foot reaction forces neither necessarily align with the z-axis nor with the surface normals at the footholds.

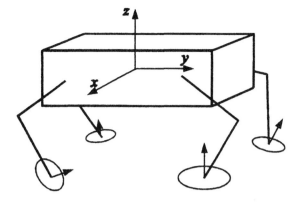

Fig. 2. A legged machine on uneven terrain. Footholds' orientations are represented as ellipses and foot reaction forces as arrows.

To support and drive the machine, a certain net force and moment must be achieved by the set of forces that the soil exerts on each foot. On a first intuitive approach, the "useful" forces applied by the ground on each foot are those forces which contribute to the robot's support and drive without causing the feet to "fight each other". These forces are the so-called *equilibrating forces*, whose accurate definition can be found in (Kumar and Waldron, 1988 and 1990).

Yet, by the interaction of each foot with the soil, another type of force components can be generated that, on this intuitive approach, could be considered as "useless": these are the forces that cause the feet to "fight each other", that is, those forces that exist *locally* at each foot, but cancel out when all of them are added to obtain the net or global force and moment exerted on the machine. These forces are termed *interaction forces*. Again, a precise formulation can be found in (Kumar and Waldron, 1988 and 1990). Figure 3 shows an example where the interaction forces are expected to have high values.

The interaction forces cause "useless" internal mechanical loads in the machine and worsen its energy efficiency, because the actuators are compelled to overcome higher forces than those that are actually needed to support and drive the robot. Thus, on a first approach, it seems that the force control system should pursue the minimization of these interaction forces, which have been described above as "useless" from some particular points of view. Such a strategy is achieved by methods based on the Moore-Penrose pseudo-inverse, which are described somewhere else (Klein, *et al.*, 1983; Salisbury and Roth, 1983).

Nevertheless, the above described scheme has an important drawback: it usually leads to foot forces for which the risk of leg slippage is high, i.e. the ratio of the force components tangential to, and normal to the ground at the feet is high (Kumar and

Waldron, 1989; Klein and Kittivatcharapong, 1990). In technical words, this method is unable to ensure that the friction cone constraints are met.

An improvement was presented by Kumar and Waldron (1988): they decompose the system of six equations of equilibrium into two subsystems: a subsystem of three equations involving the forces and torques perpendicular to the axis of the *load wrench* (a six-component vector including the components of the net force and moment with opposite signs) and another one involving the forces and torques parallel to this axis. Then the force distribution is solved imposing the hypothesis that there should be no interaction forces between the feet. The method is computationally more efficient and faster and, furthermore, permits the superimposing of interaction forces in order to optimize the contact conditions. It was supposedly implemented in the well-known *Adaptive Suspension Vehicle* (ASV).

The risk of slippage may be further reduced by the method presented by Lehtinen (1994). He makes use of the decomposition presented by Kumar and Waldron (1988) and applies the pseudo-inverse first to the first aforementioned subsystem of three equations and then to the second one. The method was implemented in the walking robot MECANT.

The two control strategies applied in the ASV and the MECANT pursue a *global* approximate minimization of the risk of slippage: they do not descend to analyze the *local* situation of each foot, what would be the most complete approach to the problem. As was already named in the introduction, such an approach was presented by Gardner (1992). His method may be more advantageous for the purpose of the light robot (aprox. 30 kg) developed at the I.A.I. The feet of this robot must conform to the soil to a greater extent than the feet of the ASV or of the MECANT: these are very heavy machines (they weigh respectively more than three tones and more than one tone); on soft terrain it is the soil that

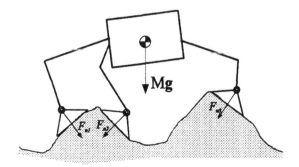

Fig. 3. A walking machine on extremely irregular terrain. The interaction forces -whose sum equals zero and cause the feet to "fight each other"- are here useful to avoid foot slippage.

conforms to these heavy machines rather than the opposite.

Gardner's mathematical solution can be found in the referred paper and is not reproduced here. The essence of his technique was outlined in section 1 of the present paper. His method goes a step further away from the objective of minimizing the interaction forces. Figure 3 shows an example where Gardner's technique would perform notably better than the previous techniques, since the action of "squeezing" the terrain, that the robot executes at the

expense of increasing the interaction forces, is here desired.

3. SENSOR SYSTEM

In order to implement Gardner's technique in a walking machine, a sensor system is needed that, in addition to permit force feedback, also enables the robot to sense the local orientations of the ground under each foot. Such a system has been designed at the I.A.I.; its implementation in the prototype leg is shown in figure 4.

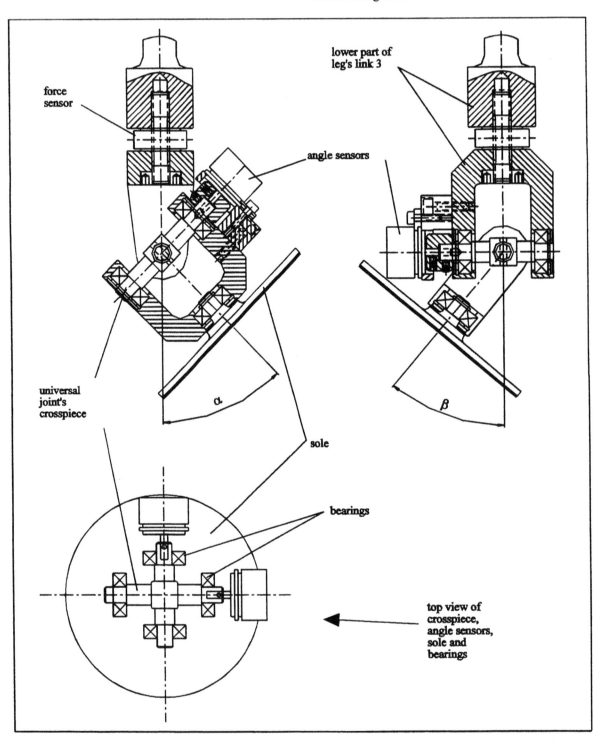

Fig. 4. Technical drawing of the prototype foot and ankle, showing the integrated sensor system.

Two rotary potentiometers are connected to each of the axes of a universal joint (the robot's ankle) to measure the two angles that define the orientation of the foot, which by this design corresponds with the orientation of the ground surface under the foot.

At present it is widely accepted that the realization of robot force control requires that force feedback should be introduced into the control circuit (Whitney, 1987; Gorinevsky and Schneider, 1990; Lehtinen, 1994). By hydraulically actuated robots, the easiest and most reliable way to do this is by measuring the pressures from the hydraulic actuators, as done in the ASV and the MECANT (Song and Waldron, 1989; Lehtinen, 1994). By most electrically actuated robots, the realization of force control requires the integration of external force sensors. Although the output torque of dc servomotors is linearly related to the armature current, force feedback can not be appropriately provided by motor current measurement. The measurement accuracy is affected by any friction in the motor bearings, associated gears and joint bearings. These friction can change a lot as the lubricants are heated during operation of the robot and also depends on the actual load supported by the joint and various other kinematic parameters. Besides, actuator inertia seen at the foot is increased by the square of the gear ratio.

In the walking robot presented in this paper, the force sensors are quartz transducers for the three force components. Due to the drift of the charge amplifier, these sensors allow only for dynamic or so-called *quasi-static* measurements. The task is successfully accomplished by resetting each time the foot is known to be off the ground, producing reliable output for a sufficient period of time when the foot is in support phase.

With this sensor system, reflex control has been successfully applied in the prototype leg to detect foot contact with the ground (sudden increase of normal force component), foot slippage (sudden drop of tangential force components) or disadvantageous footstep (too high inclination measured by the potentiometers or overcoming of acceptable kinematic limits). It can also be used to smooth the otherwise abrupt raise of the force when a foot impacts the soil at the end of the lowering run of the leg.

The performance of a smooth force redistribution during the so-called joint support phase, which separates the transfer phases of different legs (i.e. when all legs are on the supporting surface), is described in (Gorinevsky and Schneider, 1990). When the force sensor detects that the foot touches the ground, the leg lowering ceases and the leg goes to the support phase. With the change of the set of supporting legs, the calculated ideal force distribution vary abruptly and therefore should not be tracked; otherwise it would give rise to mechanical impact stresses and control instabilities. Hence, during the joint support phase the foot forces are smoothly redistributed until another leg (or set of legs in the case of robots with more than four legs) goes to the transfer phase. The commanded vector force for the leg that goes to the transfer phase is obviously zero. So, when a leg is lifted, the new set of supporting legs is already loaded with the new set of ideal commanded forces.

This concept of the smooth force redistribution during the joint support phase is doubly beneficial for the robot presented in this paper, since it allows the control system to acquire the information of the new foothold's orientation before the foot actually has to exert the corresponding ideal vector force, giving time for this force to be calculated.

4. SUMMARY

Walking machines have the potential to perform the superior mobility that biological systems exhibit in unknown rough terrain. Because a large number of degrees of freedom have to be coordinated and controlled, legged locomotion poses laborious and challenging problems for researchers to work on.

One of the aims of this work is to contribute to the efforts to optimize the performance of walking robots by designing one in which inspiring algorithms could be implemented. In order to put these algorithms, formulated by Gardner (1992), into practice, the development of an appropriate mechanical and sensor system is required. Such a system was designed and constructed at the I.A.I. and has been presented in this paper.

Various situations by which the introduction of reflex control in legged locomotion may be advantageous have been successfully applied and tested in a prototype robotic leg. A quadruped robot based on this prototype leg is being built at the moment.

5. ACKNOWLEDGEMENTS

This work was supported by the *Comisión Interministerial de Ciencia y Tecnología* (CICYT), Spain, under Grant TAP94-0783.

REFERENCES

Bares, J. and W. Whittaker (1993). Configuration of Autonomous Walkers for Extreme Terrain. *Int.*

J. of Robotics Research. **Vol. 12**, No. 6, pp. 535-559.

Gardner, J.F. (1992). Efficient Computation of Force Distributions for Walking Vehicles on Rough Terrain. *Robotica*, **Vol. 10**, pp. 427-433.

Gorinevsky, D.M. and A.Yu. Schneider (1990). Force Control in Locomotion of Legged Vehicles over Rigid and Soft Surfaces. *Int. J. of Robotics Research*, **Vol. 9**, No. 2, pp. 4-23.

Klein. C.A., K.W. Olson and D.R Pugh (1983). Use of Force and Attitude Sensors for Locomotion of a Legged Vehicle over Irregular Terrain. *Int. J. of Robotics Research*, **Vol. 2**, No. 2, pp. 3-17.

Kumar, V. and K.J. Waldron (1988) "Force Distribution in Closed Kinematic Chains," *IEEE Trans., J. on Robotics and Automation*, **Vol. 4**, N. 6, pp. 657-664.

Kumar, V. and K.J. Waldron (1989). Actively Coordinated Vehicle Systems. *Trans. of the ASME, J. of Mechanisms, Transmissions and Automation in Design*, **Vol. 111**, pp. 223-231.

Kumar, V. and K.J. Waldron (1990). Force Distribution in Walking Vehicles. *Trans. of the ASME, J. of Mechanical Design*, Vol. 112, pp. 90-99.

Lehtinen, H. (1994). Force Based Motion Control of a Walking Machine. *Ph.D. Dissertation*, Technical Research Centre of Finland.

Salisbury, J.K. and B. Roth (1983). Kinematic and Force Analysis of Articulated Mechanical Hands. *ASME J. of Mechanisms, Transmissions and Automation in Design*, **Vol. 105**, pp. 35-41.

Song. S.-M. and K.J. Waldron (1989*). Machines that Walk*, The MIT Press, Cambridge, Massachussetts.

Sukhatme, G.S. and G.A. Bekey (1995). Mission Reachability for Extraterrestrial Rovers. *Proc. IEEE Int. Conf. on Robotics and Automation*.

Whitney, D.E. (1987). Historical Perspective and State of the Art in Robot Force Control. *Int. J. of Robotics Research*, **Vol. 6**, No. 1, pp. 3-14.

OUTDOOR NAVIGATION OF MICRO-ROVERS

S. Pedraza*, A. Pozo-Ruz*, H. Roth**, K. Schilling**

(*) University of Malaga. Dep. Ingenieria de Sistemas y Automatica.
Plaza El Ejido s/n, 29013-Malaga, Spain.
Phone +34 5 2131418, Fax +34 5 2131413, e- mail:salpedro@ctima.uma.es

(*) Hochschule für Technik und Sozialwesen, Fachhochschule Ravensburg-Weingarten,*
Institut für Angewandte Forschung, Postfach 1261, D-88241 Weingarten, Germany,
Tel.: +49-751-501 627
Fax: + 49-751-49240
e-mail: roth@fbe.fh-weingarten.de

Abstract: A series of micro-robots (MERLIN) has been designed and implemented for a broad spectrum of indoor and outdoor tasks on basis of standardised functional modules at the Fachhochschule Ravensburg-Weingarten. In particular a flexible modular vehicle control system based on a CAN-bus has been developed. It supports easy changes of sensors (e.g. range sensors on basis of infrared or ultrasonic, gyroscope, cameras, odometry, inclinometer, GPS) and motion control (tracked and wheeled locomotion) system. Power can be provided via tether or via primary batteries. Another interchangeable subsystem concerns telecommunications. Here solutions ranging from tether to analogue/serial radio data transmission are available. Because of the modular structure these different vehicles are ideal test platforms for use in education and training in the area of robotics and telematics, addressing also tele-operations via INTERNET.

The special emphasis of this paper is to describe and demonstrate by experiments the MERLIN's capabilities for outdoor navigation based on a sensor system with Differential GPS and gyros, a distributed on board control and a radio link for remote control and monitoring of sensor data. *Copyright © 1998 IFAC*

Keywords: Mobile robots, autonomous vehicles, outdoor navigation.

1. INTRODUCTION

A series of micro-robots (MERLIN: Mobile Experimental Robots for Locomotion and Intelligent Navigation) has been designed and implemented for a broad spectrum of indoor and outdoor tasks on basis of standardised functional modules. This development was carried out in a joint development project by the Steinbeis Transferzentrum ARS and Fachhochschule Ravensburg-Weingarten. In particular a flexible modular vehicle control system based on a CAN-bus has been developed. It supports easy changes of sensors (e.g. range sensors on basis of infrared or ultrasonic, gyroscope, cameras, odometry, inclinometer, GPS) and motion control (tracked and wheeled locomotion) system (Figure 1). Power can be provided via tether or via primary batteries. Another interchangeable subsystem concerns telecommunications. Here solutions ranging from tether to analogue/serial radio data transmission are available. Because of the modular structure these different vehicles are ideal test platforms for use in education and training in the area of robotics and telematics, addressing also tele-operations via INTERNET (Schilling, et al.,1997a). But also in commercial applications, in particular the sensor system implementation and test for the European Mars-Rover MIDD (Schilling, et al., 1997b), this platform was successfully employed.

The special emphasis of this paper is to describe the MERLIN's capabilities for outdoor navigation based on

- a sensor system, which is composed of three small ultrasonic sensors, a gyroscope, an inclinometer and a GPS-sensor working in differential mode (DGPS),
- distributed on-board control based on microcontrollers, which communicate by a high speed CAN-bus,
- a radio link for remote control and monitoring of sensor data from a host PC in an interactive graphical environment for Windows-95,

and demonstrate by experimental results the use of DGPS and gyro data obtained in real time for navigation.

Fig. 1: Different locomotion principles of MERLIN

2. THE MERLIN VEHICLE

MERLIN is an all-terrain micro-robot based on a modular concept wrt. the locomotion principles, the sensor system, the power supply and the tele-operations capabilities. Distributed on-board control implemented on micro-controllers, which communicate by a high speed controller area network (CAN) bus is used. This gives the possibility for an easily exchange of components and for using similar components in different vehicle types. The architecture of the control System is given in Figure 2: Three boards based on 80C592 Philips 8-bit micro-controllers are employed to control the vehicle's sensors and the motor. The main controller is responsible for the overall control of the vehicle, for communication with a host PC and for the interface to the GPS.

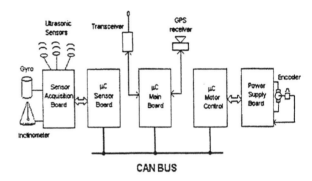

Fig. 2: MERLIN system architecture.

2.1 Drive System.

MERLIN is available with two locomotion principles:

- Wheeled locomotion, powered by one DC motor driving the rear axis through a gear-reduction. Depending on the reduction ratio the vehicle can drive with a speed of up to some 5 m/s. A servo-motor controls the steering angle of the two front wheels to a maximum angle of ±20°.
- Tracked locomotion, powered by two DC motors, one for each track. The maximum speed is below 1 m/s, but climbing and off-road capabilities are much better than for the wheeled vehicle. Steering is performed by different control of the two motors.

Every motor is equipped with an encoder for measuring the rotations and enabling an underlying control of the motor angular velocity by the motor micro-controller. The motors are controlled by PWM signals connected to the power board, which contents a PWM driver.

2.2 Sensor System.

The sensors in the MERLIN robot can be divided into two categories:
- internal sensors for the low-level control of the vehicle (e.g. optical incremental encoder for velocity control of the motor) and for measuring the state of the robot (e.g. battery voltage),
- external sensors for obstacle detection, modelling of the environment and position estimation and navigation of the robot in a global co-ordinate system.

For obstacle detection three small Polaroid ultrasonic transducers of 10.7 mm diameter are placed in front of the vehicle. The visible zone is formed by a cone of 20 degrees with a range between 20 cm and 200 cm and a resolution of some 2 to 5 cm.

For navigation and position estimation different sensors are used:
- a gyroscope determining the orientation by integrating the angular velocity of the gear angle,
- encoders in two wheels estimating the distance by integrating the wheel speeds,
- a GPS-sensor working in differential mode,
- a camera providing visual information to the operator during tele-manipulation,
- a digital incremental inclinometer.

The sensor board is connected to the sensor data acquisition system, which contains the electronic interface for the ultrasonic sensors, the gyro and the inclinometer. After an acquisition cycle is completed, the data are sent to the main micro-controller board. According to Figure 2 the GPS receiver is connected directly to the main controller. It is placed on the MERLIN robot working in differential mode. In use is a low-cost GPS, which uses 6-channels for tracking up to eight GPS satellites automatically, selecting the satellites to optimise position data. The GPS receiver can be connected to a backup battery to save the GPS configuration data when the main power supply is turned off. This receiver can be configured for three different serial communication protocols: Standard Interface Protocol (SIP), ASCII Interface Protocol (AIP) and NMEA 0183 (Marine Industry Standard ASCII Protocol for information transfer between marine navigation equipment). The NMEA protocol has been used in this application. Some technical data of the GPS receiver are shown in Table 1.

A GPS antenna of reduced dimensions is placed on the MERLIN. This antenna receives the GPS satellite signals and passes them to the receiver. Therefore, the vehicle must have a relatively unobstructed view to the sky. The received GPS signals are of very low power (approximately 140 dB), so the active antenna includes a pre-amplifier that filters and amplifies the GPS signal before delivery to the receiver.

Table 1: Technical characteristics of the GPS receiver.

Number of channels	6
Power supply voltage	9 to 32V
Size	127 x 102 x 28 mm
Weight	260 g
Communication port	2 DB9 serial port
Protocols available	SIP, AIP, NMEA 0183
Position accuracy in DGPS mode	2 to 5 meters
Backup optional battery	3.5 to 14 V

In order to improve the accuracy of a GPS, differential GPS (DGPS) have been introduced for civil applications, although this precision depends on the employed GPS receiver (Weber and Tiwari, 1995). By placing a reference base station receiver at a precisely surveyed location and comparing measured GPS satellite ranges with calculated ranges based on the satellite data transmissions and the known geographic position of the receiver, errors in the range measurements can be computed. The calculated error data is then transferred to another GPS receiver and applied to the second receiver's solution. Because the distance between the two GPS receivers is very small in comparison to the distance from satellites in space, the resulting solution will eliminate the errors that are common to the two receivers.

Some companies provide DGPS corrections over a FM link transmitting differential error data from the reference base station to several users. This technique, that has been employed in this application (Figure 3) is widely used nowadays. The Star Track LWRX (Figure 3) is a receiver that collects, through its antenna, the error data for correcting the position from a base station placed near to Frankfurt (Germany), and send them to the GPS receiver. Because the antennas of the Star Track LWRX and the GPS receiver on the vehicle are both active, they must be separated at least two meters for a correct working of the system.

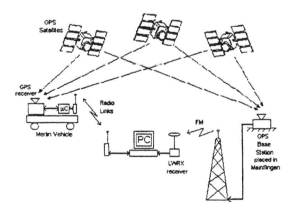

Fig. 3: Principle of the Differential GPS

2.3 Remote Control and Sensor Monitoring.

A radio transceiver module of small dimensions *(54x32x13.5* mm) connected to the control system on-board of MERLIN permits to implement a radio link with the remote host PC with a range of 30 meters in-building and 120 meters outside (Radiometric Limited, 1995). The data are packed and sent at a maximum rate of 40 Kbits/s in half duplex mode.

The main micro-controller board manages the communications to the remote host through the transceiver. Control commands, sensor data request commands and error correction data for DGPS are received, as well as sensor data are sent by the transceiver. On the other hand, the main board communicates with the GPS through the serial port, receiving information about position and sending the error correction data.

A graphical interactive software has been developed using Delphi 2.0 running under Windows 95 on a Pentium PC to remote control and monitor the sensors of MERLIN. Figure 4 represents the main window of the program. It is divided into two parts:
* the control section, placed in the middle of the window,
* and the sensor section, at the top and the bottom of the screen.

The MERLIN remote control can be done in two ways: through buttons touched with the mouse, or by a joystick connected to the game port of the PC. The Sensor section represents graphically the measurements obtained from ultrasonic sensors, gyroscope and inclinometer, as well as the information from GPS, consisting of the position relative to a determined reference point (set by the ‚RESET DATA' button), the number of satellites used for position measurement, and the operation mode of the receiver (single or differential mode). Sensor data can be saved into a file for post-analysis and graphical representation.

3. EXPERIMENTAL RESULTS

One of the main purposes of the MERLIN rovers are to perform tasks in outdoor environments. Especially for supporting a tele-manipulating operator by position estimation and navigation the measured data of the gyro, which provide after integration the orientation of the rover, and the GPS-data are of interest. To give an impression of the sensors accuracy different test scenarios have been performed. The sensor data are obtained each second and they are stored into a file. In Figure 5 the data of the DGPS and the gyro are displayed while the MERLIN has been driving on a closed path in an outdoor environment.

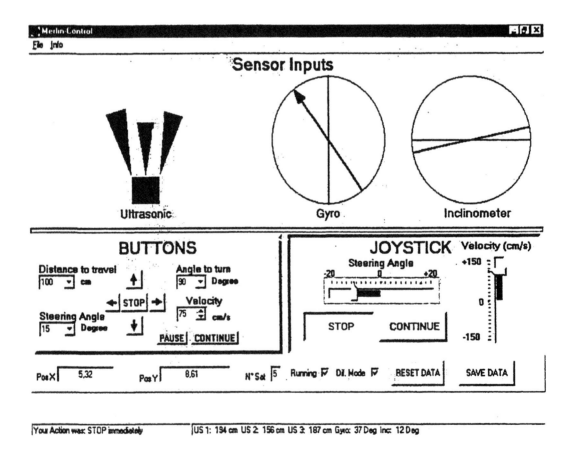

Fig. 4: Display of the remote control user interface.

In Figure 5a the measured data of the DGPS are displayed. The wheeled MERLIN starts at the initial position (0/0) and is tele-controlled by the operator via joystick on an off-road parcour. After some 80 m the rover returns to the same place, where it starts. However the measurement of the DGPS shows a different position. Thus a measurement error at the final point of (0.5 m/ 2 m) occurs.

At each location measured by the DGPS a gyro measurement of the angular velocity has been taken. These data have been numerically integrated in the sensor controller. At the starting point (0/0) the orientation of the vehicle was -90°, which is to be seen on the direction of the line in Figure 5b. Each gyro data has been represented as a orientated line and placed in the point provided by the GPS at the same time. At the final point the vehicle had in reality the same orientation as at the beginning. The integrated gyro data, which contain measurement errors because of drift and bias as well as errors from the integration algorithms, deliver according to Figure 5b an orientation error of some 10°.

In future works also the measurement data of the encoder inside the wheels shall be measured. Integrating these values and combining them with gyro measurements via Kalman filtering should deliver a position estimation alternative to the DGPS data.

4. CONCLUSIONS

In this paper the modular concept of the MERLIN rovers developed at the Fachhochschule Ravensburg-Weingarten in joint co-operation with the Steinbeis Transfercentre ARS has been described. The main topic was to demonstrate the position accuracy and outdoor navigation capabilities based on a differential GPS sensor and on gyro data.

ACKNOWLEDGEMENTS

The authors thank for support by the „Deutsche Akademische Austauschdienst (DAAD)" of the project „Applications of Fuzzy Logic in robotics and in space" within the bilateral Spanish-German „Acciones Integradas"-programme together with the Universidad de Sevilla and Universidad de Malága. The GPS-work has been performed during a research visit of the Spanish scientists, which have been supported by the „Fundación Ramón Areces" and the „Ministerio de Educacion y Cultura Espanol".

REFERENCES

Chansik Park, Ilsun Kim, Jang Gyu Lee, Gyu-In Jee and Choon Shik Kim (1995). A Satellite Selection Criterion Incorporating the Effect of Elevation Angle in GPS Positioning. *Proc. IFAC International Conference on Intelligent Autonomous Control in Aerospace' 95*. Beijing, (P.R. China) pp. 244-249.

National Marine Electronics Association (1992). „NMEA-0183 Standard for Interfacing Marine Electronic Device Version 2.00".

Radiometric Limited (1995). „RCP preliminary data sheet". Issue 2,21 th.

Schilling, K., H. Roth, R. Lieb (1997a). Teleoperations of Rovers - from Mars to Education. *Proceedings IEEE International Symposium on Industrial Electronics*. Guimaraes, Portugal, Vol. 1, p. 257 - 262.

Schilling, K., L. Richter, M. Bernasconi, C. Jungius, C. Garcia-Marirrodriga (1997b). Operations and Control of the Mobile Instrument Deployment Device on the Surface of Mars. *Control Engineering Practice* 5, p.837 - 844.

Weber, L. and A. Tiwari (1995). DGPS Architecture Based on Separating Error Components, Virtual Reference Station and FM Subcarrier Broadcast. *Proc. 1995 ION Annual Meeting*.

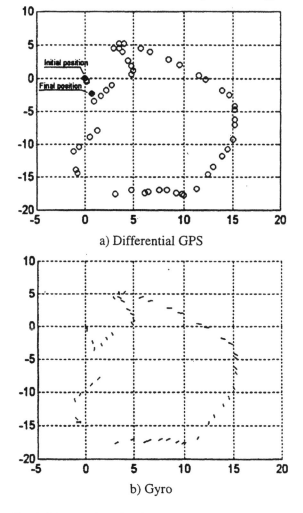

a) Differential GPS

b) Gyro

Fig. 5: Measured navigation sensor data.

CAD / CAM CONCEPT OF A CRASH-PROOFED, LIGHT-WEIGHTED WHEELCHAIR TO BE USED AS DRIVING AND TRANSPORTATION SEAT BY DISABLED USERS

Dr E. Bekiaris(*), Dr A. Vosinis(*), Mr. M. Zeroli(**)

*(*)Transeuropean Consulting Unit of Thessaloniki S.A.
28, Alexandras Ave., Athens, 106 83, GREECE
Tel: +30 (1) 82 53 777-9, Fax: +30 (1) 82 53 780
e-mail: trnspcon@compulink.gr*

*(**) FIAT Auto, Orbassano (TO), ITALY
Tel: +39 (11) 9029510, Fax: +39 (11) 9012904
e-mail: michele.zeroli@auto.fiat.it*

Abstract: An innovative wheelchair is developed within TRANSWHEEL (DE3013) project, with enhanced safety, comfort and maneuverability. It will be used as driver's or passenger's seat within private cars and other transportation means. Its safety against crash and its comfort will be first provided by using a standard car seat on the wheelchair mainframe. Then, a computer crash simulation will follow of CAD/CAM models of the wheelchair, the surrounding compartment of the car and their connecting mechanisms, together with the ergonomic design of the wheelchair. Comparison to actual crash-tests will lead to the final design of the initial wheelchair concept. *Copyright © 1998 IFAC*

Keywords: Computer-aided design, Computer-aided manufacturing, Structural optimization, System analysis, System architectures, Systems concepts, Systems engineering.

1. INTRODUCTION

The number of wheelchair users in Europe, with disabilities in their limbs, is up to 13 millions today. The variety of functions that human legs can perform is desirable to be done by a wheelchair also. This is a difficult goal to achieve, as a wheelchair has to function in different environments (buildings, transportation means and the way to them). In addition, it is necessary to support and to safely transport the disabled people, as well as to be comfortable and easy in maneuver execution (e.g. within a vehicle or at home and at the work place).

Nowadays, there exists no such multipurpose wheelchair but more than 700 kinds of them, each one suitable for some specific of the afore-mentioned tasks. Safe use of a wheelchair in transportation means is assured by transferring the user to a standard seat and separately storing the wheelchair. These operations are tiresome and even impossible to perform autonomously for many wheelchair users. In addition, such automatic storage systems costs are rather high. On the other hand, to drive from a wheelchair, even in these few countries that this is legal, means high danger in case of impact, high torque and limited visual fields, because the wheelchair is about 25cm higher than a standard car seat. Its comfort is also limited. Furthermore, its ingress / egress to / from a standard car is currently not feasible.

The solution of the above problems would require special design and construction of vehicles and buildings, if the existing types of wheelchairs remain. However, this would be costly and could not solve these problems under existing environments. Therefore, a new wheelchair design, taking into account the needs of motor disabled car users, is intented within the context of TRANSWHEEL (DE3013) project, funded by the TIDE Programme of the EEC/DG XIII.

2. STATE OF THE ART

The up to now work on transportable wheelchairs has focused on the safety and control specifications of them. Already an existing EU project (TELAID - V2032) has concluded a guideline about the safety of a wheelchair, that is in accordance to existing regulations.

Moreover, ECE Reg.17 as well as Directives 74/408 and 81/577 require that "such a wheelchair type driver seat should be able to withstand a longitudinal deceleration in both directions of 20g for at least 30 seconds and its back should withstand a rearward force reproducing a moment of 530 Nm around the 'H' point (the 'H' point being the pivot point between the legs and the torso on a mannequin), as any original driver seat does".

Many impact tests have been performed, merely by research institutes (e.g. Petzäll and Olsson, 1995). Various types of wheelchairs have been tested, including a dummy and different restraint systems to the vehicle. The results have been observed on the dummy, the restraint system and the wheelchair itself (Bertocci, et al., 1996; Kallieris, et al., 1981). As it can be concluded from these and similar tests (Naniopoulos, et al., 1997;), the head and thorax damages of the dummy have been acceptable, according to FMVSS 208 regulation. On the other hand, none of the tested restraint system configurations has satisfied the safety standards as do the corresponding ones of the modern cars.

Another configuration that has been tested by Permobil is with fixing plates, which has successfully satisfied the TSVFS 1985:22 regulation (Wenäll, 1989). However, these configurations do not allow the wheelchair to be lowered while tied in a vehicle. This is desirable in order to reduce the moments in case of impacts and enhance the visual field of the occupant.

The various tested types of existing wheelchairs cannot withstand successfully impact tests with decelerations of 25 - 27g (g being the gravity acceleration), whilst the vehicle seats can do. Solutions proposed by various manufacturers (e.g. improvement of the wheelchair webbing system, conventional reinforcing means of the wheelchair base combined with a standard car seat) have led to heavy and rather costly solutions. In addition, such a wheelchair is unsuitable for use outside of the vehicle.

On the other hand, relevant EU projects are directed towards the control and the self-navigation of the wheelchair as well as the reliable motoring of it. Also, an inventory with the safety needs of elderly and disabled people has been developed within these projects, but not with practical solutions.

3. TOWARDS AN ITERATIVE DESIGN CONCEPT TO PERFORM WHEELCHAIR CRASH-TEST SIMULATIONS

The design conception of a wheelchair through laboratory crash-tests is very costly. Computer simulations with existing powerful tools are often used for the design procedure. Thus, within the context of the TRANSWHEEL Project, CAD/CAM software will be used, capable of reproducing the 3D model of the wheelchair, the dummy and the vehicle compartment around them, with the fixing system. Examples of considered computer programs are I-DEAS, CADDS5 and HYPERMESH. Such computer programs have already been tested and can model complicated geometries. An application example from the automotive industry is shown at Figure 1.

On the other hand, the strength of the wheelchair will be checked by a corresponding elastic analysis software (e.g. NASTRAN, ANSYS) and its crash safety will be analyzed by a specialized computer tool (e.g. RADIOSS, LS-DYNA, DYTRAN). In the case that the strength and impact analyses prove insufficiency of the wheelchair, redesign must be performed. However, if no such problems exist, a structural optimization will be attempted, in order to minimize its weight and its construction cost. The afore-mentioned design process is shown at Figure 2. The begin of the design process will be the computer simulation of a crash-test carried out by FIAT on 15/9/1997, including the MEYRA "Optimus" wheelchair as driver's seat and two dummies (driver at the wheelchair and passenger at a standard car seat) in a FIAT Scudo (see Table 1, #1 crash-test). Thus, the correct use of the chosen software will be ascertained by comparison of the computed results to the crash-test measurements (Figures 3 and 4). The weak points of the construction will be identified by measuring the wheelchair deformations. A first indication of such points is shown at Figure 5. The drawing of the "Optimus" wheelchair is provided for comparison at Figure 6. Then, a new seat, based on the Lancia Kappa seat, will replace the existing one of the tested wheelchair. This is a standard car seat and it has already been verified to be crash-proofed. Therefore, this modification will assure the a priori crash-proof design of the wheelchair seat and will allow for the redesign and check only of the wheelchair frame and its tie-down system against impact loads. Even if the wheelchair shows no failure from the crash-test, its redesign is necessary because the frame must be lower for extra security and visibility reasons.

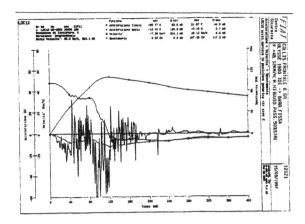

Fig. 4. Acceleration (o), mean acceleration (◁), velocity (□) and displacement (x) versus time at wheelchair front base.

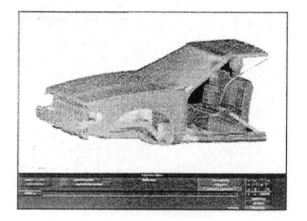

Fig. 1. Car compartment model by HYPERMESH.

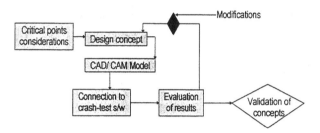

Fig. 2. The design process of the TRANSWHEEL prototype.

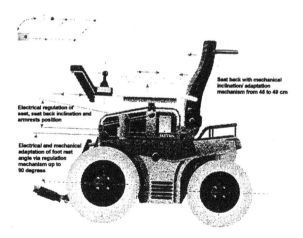

Fig. 5. Wheelchair and dummy after the crash-test.

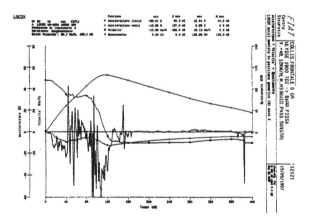

Fig. 3. Acceleration (o), mean acceleration (◁), velocity (□) and displacement (x) versus time at wheelchair rear base.

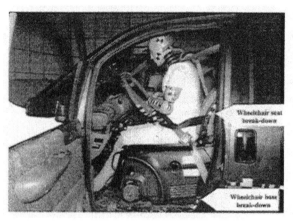

Fig. 6. MEYRA "Optimus" wheelchair

This new design encompasses a series of intelligent components, such as the impact safety of the wheelchair, described above. Additional safety characteristics are foreseen independently of the crash-proofed design of the wheelchair. The tie-down system has the inherent property to distribute the loads from the wheelchair to more than one anchoring points of the car floor. In addition, the wheelchair, once positioned as driver's or passenger's seat within a car, will be automatically locked by a hook to the tie-down system in a safe manner which will not permit its movement forwards or backwards along the longitudinal axis of the vehicle. Figure 7 presents a first concept by GUIDOSIMPLEX of the tie-down parts that will be assembled on the wheelchair. Figure 8 presents the corresponding part fixed on the car floor, where the locking hook is seen. A safety aspect of an other kind is the prevision to charge the wheelchair batteries through a connection to the car batteries. This, of course, assures its use as an autonomous transportation means for short distances as well as the electrical supply to its instruments. Finally, the comfort and adaptation of the wheelchair as driver's and passenger's seat are guaranteed by the use of a height adjustment mechanism and of a standard car seat on the wheelchair frame. The height adjustment mechanism is being developed by MEYRA and will assure the right height position for driving as well as the wheelchair frame locking lower than the wheels level. The latter operation is especially important for the occupant's safety too, because it allows for the correct function of the wheelchair tie-down system and it minimizes the turn over torque in case of impact. The seat, based on the Lancia Kappa one, will be suitably modified to fit several positions of the human body and to memorize them as well. Both adaptation and comfort features are easily operated through touch buttons with the help of integrated electrically powered components. Figure 9 presents these features at the lowest wheelchair position.

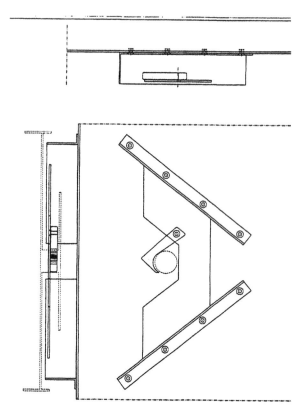

Fig. 8. Parts of tie-down system fixed on the car floor. First concept by GUIDOSIMPLEX.

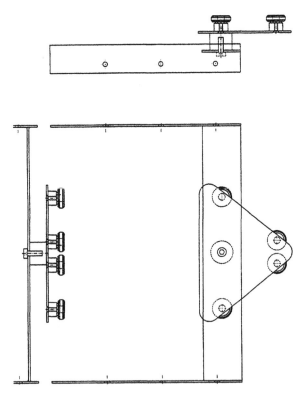

Fig. 7. Parts of tie-down system put at the wheelchair base. First concept by GUIDOSIMPLEX.

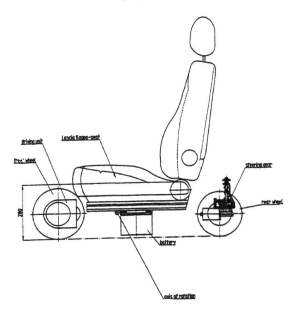

Fig. 9. Wheelchair at its lowest position.

4. VERIFICATION OF SIMULATION PROCESS BY LABORATORY CRASH-TESTS

Each designed configuration by the above procedure will be constructed as a prototype by MEYRA. It will then be tested against crash loads. A series of crash-tests, to be performed by FIAT, are already planned (see Table 1), for a duration of 24 months, starting at September 1997. As the design procedure is estimated to conclude within 4 - 5 design cycles, more than one crash tests will be performed for each designed wheelchair. The first computer modelling of the set "car compartment-wheelchair-dummy" is expected to be ready about December 1997 and the crash simulation of a first new designed configuration is expected for March 1998.

A simultaneous calculation by computer software will be performed. Comparison of the computer analysis results with the crash-test ones will define the points of modification at a next design cycle or will ascertain the final design of the wheelchair.

5. CONCLUSIONS

The configurations of existing wheelchairs show the need of new design concepts, so as to achieve safer ones, with enhanced possibilities of multi-use within vehicles and outside of them. The TRANSWHEEL design procedure intents to a crash-proofed wheelchair, with lower frame, enhanced comfort, adaptability and optimized values of weight and cost. The suggested process is based on a interactive numerical-experimental scheme. Laboratory crash-tests are the starting and the verification points of the design process. Computer geometrical simulation and strength analysis (both elastic one and dynamic one with plastic deformations) are its core. The reliability of the candidate computer tools, the experience of the automotive company that will carry out the crash-tests as well as the specialization of the companies that will construct the wheelchair and its anchoring system within a car, make sure that the project will provide the market with an ameliorated product.

REFERENCES

Bertocci, G., K. Digges and D. Hobson (1996). Shoulder belt anchor location influences on wheelchair occupant crash protection. *J. of Rehabilitation Research and Development,* **Vol. 33 No 3,** p. 279-289.

Kallieris, D. et al., (1981). Behavior and response of wheelchair, passenger and restraint systems used in buses during impact. *Proceedings, 25th Stapp car crash Conf., September 1981,* **SAE paper 811018.**

Naniopoulos, A., A. Gateley and E. Bekiaris (1997). *User needs in relation to various transportation modes and system requirements, 108-135.* Aristotle University of Thessaloniki, Thessaloniki, GREECE.

Petzäll, J. and A. Olsson (1995). Wheelchair Tie-Downs and Occupant Restraint Systems for Use in Motor Vehicles. *Proc. of the 7th Int. Conf. On Mobility and Transport for Elderly and Disabled People, Reading, UK,* **Vol. 2,** 130-137.

Wenäll, J., (1989). Test of seat fixture for Permobil electric wheelchairs in accordance with TSVFS. In: *Road and Traffic Institute Test Reports,* (VTI Ed.), Report No. 56124, VTI, Sweden.

Table 1: Plan of crash-tests to be performed by FIAT

N°	DESCRIPTION	STANDARD	TARGET	GOAL	REQUIRED MATERIALS
1	Frontal collision vs. rigid barrier (0°, with dummies)	USA Std. 208		Recording of vehicle deceleration	1 complete vehicle (if curves are not recoverable from previous tests; by FIAT)
2	Rearward collision	ECE32-34		Recording of vehicle deceleration	Same vehicle of test 1
3	Frontal collision simulation (HYGE sled; 95% dummy)	USA Std. 208	Fulfil standard	Check of seats anchorages and dummy's movement	
4	Rearward collision simulation (HYGE sled; 50% dummy)	ECE32-34	Fulfil standard	Check of seats anchorages and dummy's movement	–N° 6 seat kits (TRANSWHEEL seat by MEYRA)
5	Frontal collision simulation for luggage retention	DIN 75410	Fulfil standard	Check of seats anchorages	–N° 5 driver belt kits with pretensioners (by FIAT)
6	Frontal collision simulation (HYGE sled; 5% dummy)	USA Std. 208	Fulfil standard	Check of seats anchorages and dummy's movement for air-bag interference	–N° 4 steering wheel systems (by FIAT) –N° 4 dashboards (by FIAT) –N° 3 air-bag systems (by FIAT) –N° 1 body (by FIAT) with structural modifications (by GUIDOSIMPLEX)
7	Frontal collision simulation (HYGE sled; 50% dummy)	USA Std. 208	Fulfil standard	Check of seats anchorages and dummy's movement for air-bag interference	
8	Frontal collision simulation (HYGE sled; 95% dummy)	USA Std. 208	Fulfil standard	Check of seats anchorages and dummy's movement for air-bag interference	
9	Seats and seat anchorages	ECE 17	Fulfil standard	Check of seat-back and anchorages resistance for official approval	–N° 2 bodies (by FIAT) with structural modifications (by GUIDOSIMPLEX) –N° 1 seat (by MEYRA)
10	Belts anchorages	ECE 14	Fulfil standard	Check of belts' anchorages for official approval	
11	Internal lay-out	ECE 25	Fulfil standard	Geometrical checks and energy dissipation for official approval	
12	Internal lay-out	Internal standard	Fulfil standard	Check of seat's general installation	N° 1 vehicle (by FIAT) complete of a seat (by MEYRA) and handling system (by GUIDOSIMPLEX)
13	Frontal collision vs. rigid barrier (0°; with dummies)	USA Std. 208	Fulfil standard	Check of seat's performance during crash	Same vehicle of test 12

234

SENSOR SUPPORTED DRIVING AIDS FOR DISABLED WHEELCHAIR USERS

Klaus Schilling, Hubert Roth, Robert Lieb, Hubert Stützle
FH Ravensburg-Weingarten
Postfach 1261, D-88241 Weingarten
Tel. ++49-751-501 739, Fax ++49-751-48523
e-mail schi@ars.fh-weingarten.de

Abstract: Disabled people using electrical wheelchairs usually suffer from strong motoric handicaps. In order to assist them in their mobility, there has a sensor system been developed to warn them of obstacles in the way and to assist them in navigation to return to their home. The related low-cost technology uses sensor data fusion of a ranging system, active markers, a gyro and a GPS-system. With the same sensor configuration also additional useful capabilities for a hospital environment, like convoy driving of several wheelchairs, or autonomous driving and return are described in this paper. *Copyright © 1998 IFAC*

Keywords: Navigation, obstacle detection, autonomously guided vehicle, sensor data fusion, range sensors, GPS

1. INTRODUCTION

Usually an electric wheelchair is controlled via joystick directly by its user, without any further data processing. The radius for actions for the disabled is thus determined by his actual capabilities with respect to perception of the environment and to his motoric capabilities. As handicapped people with multiple disabilities (physical and psychical) are at present not able to control their environment without additional human support, it would be desirable to enable them by modern sensor and control technology to get a higher degree of independence.

Several approaches exist to detect standing obstacles ([1],[2]), but the problematic areas like smaller obstacles (e.g. kerbstones) or descending stairs and wholes can not yet be reliably detected.

Similar technical problem areas, related to the requirements of an „intelligent" wheelchair, have been addressed in the field of industrial transport robots. In the area of autonomous transport robots the Steinbeis Transferzentrum ARS has extensive experience concerning sensorics, navigation, obstacle avoidance and docking ([3],[4],[5]). These technologies are transferred in this project to the wheelchair application. The industrial use provides

also insight into maintenance, commercial and security aspects.

Fig. 1 : The sensor supported electric wheelchair INRO

2. OBJECTIVES AND PROJECT FRAMEWORK

The objective of this project INRO (= *In*telligenter *Ro*llstuhl) is to enable safe and intelligent control of an electric wheelchair by supporting the user with respect to navigation and obstacle detection by sensorics. The control has to adapt to the current capabilities of the user and should avoid accidents due to decreased motoric and orientation capabilities of a user. By telecommunication links, help requests are transmitted to the sick ward, giving information about the actual wheelchair location and some first information about the actual problem. Thus the safety in mobility of disabled people is increased, improving safety in unaided transport, indoors and outdoors.

For elderly or disabled people, using wheelchairs, often the problem occurs, that on a trip the orientation or the control over the wheelchair is lost, due to suddenly occurring handicaps (physical or psychical). The care of disabled or elderly people is guided by the principle to enable -as far as possible- autonomous mobility without involvement of nursing staff. Therefore the project aim is to develop and implement intelligent control strategies for wheelchairs, allowing disabled or elderly people to successfully perform trips to given aim points. Thus there are two specific requirements for the control addressed:

1. *The vehicle should be capable to repeat -in extreme cases autonomously- a defined route indoors or outdoors*
   Typical examples might be transports between rooms for therapy, dinning halls etc. The input of routes should be easily defined by the transported people as well as by the nursing staff. The vehicle should be able to reach given targets and to return autonomously to the starting point of the trip.

2. *Uncontrolled or dangerous actions of the user are to be identified and prevented, nursing staff is to be informed about the actual situation*
   Obstacles in the way are to be identified and the user has to be warned to avoid them. In case of immediate danger, the computer stops the driving motors. The classification of dangerous actions is based on criteria from experienced nursing staff and computer specialists to implement a related rule base. Information about the vehicle location and characteristic parameters to describe the user's and the vehicle's status are to be send to the nursing staff to enable quick actions for help. These telecommunication and user action classification topics will be described in a companion paper.

The navigation and control functions must work reliably at varying environments (different lightning conditions, wet/dry path etc.) and require robust, adaptive concepts for sensor and control system.

Two standard electrical wheelchairs (cf. Fig. 1) have been equipped with additional sensors and a notebook PC for sensor data processing. An interface electronic has been implemented to connect 2 serial ports of the PC to the regular wheelchair display and to the drive motor control.

The users have been intensively involved in this project. There has been close contacts with several local centres for handicapped people and a large wheelchair producer to ensure the relevance of results for practical use.

3. THE OBSTACLE AVOIDANCE SYSTEM

To detect larger obstacles, 5 ultrasonic sensors cover the near range area of 2 m in front of the wheelchair. When an obstacle is detected, a warning by sound is activated and the location of the obstacle is indicated on the driver's display. In order to detect also smaller obstacles, it would be useful to place the ultrasonics as low as possible, but then - in particular in the usual outdoor environment - there increase noisy reflections from an uneven floor. Frequent false warnings let the driver ignore these inputs. As consequence in our system the ultrasonics are placed in an height of 55 cm, but can thus only be employed to detect larger standing obstacles and to assist in navigation.

Fig. 2 : Ultrasonic system to detect obstacles in front of the wheelchair, viewn from the top. Ultrasonic beams are indicated in grey.

Complementary we use for the obstacle detection an active marking system, consisting of a laser, projecting stripes or grids in front of the vehicle, and a CCD-camera to detect the deviation of the stripes against the expected position (cf. Fig. 3). This active marking system is optimised to detect smaller, but still dangerous obstacles, like kerbstones. In addition this sensor system is also capable to detect descending stairs and dangerous wholes.

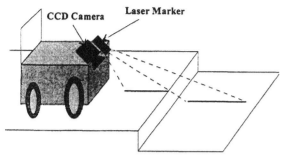

Fig. 3 :   The principle of the active marking system, consisting of a camera and a stripe laser

To allow reliable detection under different light conditions in the indoors and outdoors environment, an appropriate combination of the laser type and filters are to be employed, as well as substraction of the light background of the environment.

4. THE NAVIGATION SYSTEM

The navigation system is used to determine the actual position of the wheelchair. For this purpose a combination of sensors is employed, related to different accuracy requirements, indoors and outdoors.

Outside of buildings, the GPS (Global Positioning System) is suitable to provide a reference of the current position. In case of commercially available low-cost differential GPS (DGPS), the position can be determined to an accuracy of 5 m. If less accuracy is required, with a standard GPS the position can still be determined to 100 m accuracy. Thus this is quite sufficient to direct an user back home, in case he looses orientation in a city.

In the context of this project the coordinates of frequently visited objects (like the user's home, supermarkets, bank etc.) are stored and the user can ask the system to guide him to one of these targets. Then from the GPS data the actual direction and distance towards the target will be calculated and displayed on the monitor.

Fig. 4 :   The DGPS mounted on the wheelchair

Indoors usually a higher accuracy is required to access a specific room and the GPS signals can not be well received in-house. Thus a different navigation method, based on dead reckoning and a ranging system, is used there. From a defined start point (e.g. the house entrance or the room entrance) the values of the motor encoders are integrated to determine the actual position.

This inexpensive approach suffers from a drift proportional to the travelled distance according to accumulated noise. Floors with varying surfaces (and related different friction), or rough and uneven surfaces further decrease the accuracy. It is therefore necessary to recalibrate after certain distances the dead reckoning system in relation to reference landmarks.

For the project INRO we used characteristic profiles of natural landmarks, detected by the ultrasonic ranging system, also used for obstacle avoidance. Further technical details and performance results related to industrial transport robots have been published in [6]. An in-house environment is usually structured enough to provide sufficient natural landmarks, such as doors, edges, walls. Nevertheless this approach degrades, when many obstacles like persons walking around, are present. Then it would be advisable to use in addition artificial markers (optical or magnetic transponders) as references.

Although the main application of the dead reckoning system is in-house, it is also used outdoors together with fixed magnetic transponders to allow autonomous driving in a limited undangerous area, e.g. in a hospital park. The dead reckoning system allows the hospital personnel to teach a specific path in the hospital park, just by driving this path along, and to store it within the on-board computer. Here the odometry data, the reference transponder locations, gyro measurements for curve characteristics, ranging system data at natural landmark locations and GPS measurements are stored and combined by sensor data fusion to document reliably with a high degree of redundancy the path. Thus the wheelchair is able to repeat this path autonomously. With activated obstacle avoidance system, the wheelchair can thus drive safely the disabled people around, or can transport them to pre-programmed targets, as initiated by the hospital personnel.

5. CONVOY DRIVING

When going outdoors, normally one person from the hospital staff takes care of one disabled. Thus it might be much more efficient, if several wheelchairs would follow autonomously one after the other a guide. This function has been implemented by using the ranging system from obstacle avoidance to keep a fixed safe distance to the wheelchair in front and to adjust the direction towards this reference.

The distance between two wheelchairs is in the order of 1 to 1.5 m, such that detections of other obstacles don't disturb this function.

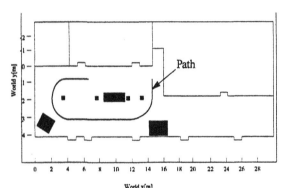

Fig. 5 : The path used for the convoy driving test, including fixed obstacles.

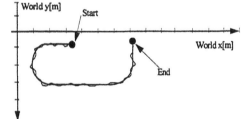

Fig. 6 : The reference path and the path taken by the vehicle following the guide wheelchair.

5. CONCLUSIONS

There has been developed a sensor supported wheelchair, allowing to assist disabled users in obstacle avoidance and navigation tasks. With these „intelligent" wheelchairs, elderly people or people with multiple disabilities will be able to explore their environment to a much greater extend, indoors and outdoors.

Due to the increased mobility of the handicapped people, the need for nursing staff might decrease and therefore allow to intensify the care in other fields. The increase in mobility due to an intelligent, sensor-supported wheelchair would allow the user to get easier in contact with other people or to take to a larger extend part in social life.

Acknowledgements
The authors want to thank the German Ministry for Science and Research (BMBF) for the financial support of the project „Intelligent Wheelchair (INRO)".

REFERENCES

[1] Miller, D., M. Slack, Design and Testing of a Low-Cost Robotic Wheelchair Prototype, Autonomous Robots 2 (1995), p.77 - 88.

[2] Pires, G., N. Honório, C. Lopes, U. Nunes, A.T. Almeida, Autonomous Wheelchair for Disabled People, Proceedings "IEEE International Symposium on Industrial Electronics", Guimaraes 1997, Vol.3, p. 797 - 801.

[3] Schilling, K., H. Roth, Homing Guidance Schemes for Autonomous Vehicles,
    in: A. De Carli, M. La Cava (eds.), Motion Control for Intelligent Automation, Pergamon Press 1992.
    p. 319 - 324.

[4] Schilling, K., H. Roth, B. Theobold, Fuzzy Control Strategies for Mobile Robots, Proceedings EUFIT`93,
    p. 887 - 893.

[5] Schilling, K., J. Garbajosa, M. Mellado, R. Mayerhofer, Design of Flexible Autonomous Transport Robots for Industrial Production, Proceedings "IEEE International Symposium on Industrial Electronics", Guimaraes 1997, Vol. 3, p. 791 - 796.

[6] Schilling, K., R. Lieb, H. Roth, Indoor Navigation of Mobile Robots Based on Natural Landmarks, Proceedings "3rd IFAC Symposium on Intelligent Components and Instruments for Control Applications", Annecy 1997, p. 527 - 530.

[7] Virtanen, Ari, A Drive Assistant, Horizons 1/96, VTT Finland, p. 17 - 18.

APPLICATION OF A NEURAL NETWORK TO THE IDENTIFICATION OF THE DYNAMIC BEHAVIOUR OF A WHEELCHAIR

L. Boquete, R. García, M. Mazo, R. Barea, I. Aranda

Electronics Department. Alcala University. Spain.
28801. Alcalá de Henares. Madrid.
E-mail: boquete@depeca.alcala.es

Abstract: The problem of identifying the dynamic behaviour of a wheelchair is solved by means of a new model of recurrent neural network. Once the architecture of the model has been defined, the formulae are obtained for the adjustment of its coefficients by the gradient descent technique; the conditions to be met by the gain factors are deduced to assure the convergence of the learning process and its performance is verified by identifying the real behaviour conditions of a wheelchair. *Copyright © 1998 IFAC*

Keywords: Neural networks, System identification, Radial base function networks, Robotics, Lyapunov stability.

1. INTRODUCTION

A great effort has been made in the past years in research into the application of neural networks for the identification and control of linear and non-linear systems (Kawato *et al.*, 1987; Narendra and Parthasarathy, 1991). One of the main advantages in using neural networks is that these networks are universal identifiers and they learn by means of examples. One of the models of neural networks most used is the radial base network (RBF), formed by base functions where each neuron responds to the inputs that are close to certain parameters of these functions. This type of network is a universal function approximater, with a high convergence speed, as it is formed by a single layer of neurons and the adjustable parameters depend linearly on the output and for this reason they can be used in systems which operate in real time.

Various recurrent radial basis neural networks are on record. For example Ciocoio's model (Ciocoiu, 1996) uses FIR/IIR filters as synaptic connections, the latter also being able to store time relations. Obradovic's model (1996) uses the outputs as new inputs so that the output at any time depends on the input at this moment and the previous output.

This paper describes the use of a model wherein each neuron has a FIR filter feedback (Fig. 1) (Boquete *et al.*, 1997). The main advantage of the proposed model is its learning speed - when using the gradient descent algorithm - and the fact that the network can identify systems with memory using the simplest identification scheme: parallel-parallel.

This article has been organised into the following sections: Section 2 describes the new model for the neural network and also gives the expressions for adjusting the coefficients of its filters. In Section 3 a stability study of the system is made applied to the model described. Some practical tests are made in Section 4 and the main conclusions on this work are summarized on in Section 5.

2. NEURAL NETWORK MODEL

The neural network used is a local recurrent model, i.e., it allows only backwards connections from a neuron's output to its own input. Figure 1 shows the feedback from the output of each neuron to its own input using a $F_{im}(z)$ filter, so that each neuron is capable of storing previous states. The neural network output are:

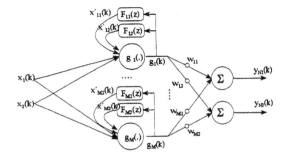

Fig. 1. Neural network model.

$$y_{Np}(k) = \sum_{i=1}^{M} w_{ip}.g_i(k) \qquad p = 1,2 \qquad (1)$$

The neuron output in the moment k is in turn given by:

$$g_i(k) = e^{-\frac{(x_1(k) + x'_{i1}(k) - C_{i1})^2 + (x_2(k) + x'_{i2}(k) - C_{i2})^2}{\sigma^2}} \qquad (2)$$

x'_{im} (m=1,2) is the filter output connecting the neuron output with its own input. Its formula is:

$$x_{im}'(k) = \sum_{j=1}^{S} a_{imj} \cdot g_i(k-j) \qquad i = 1,..M; \qquad (3)$$

For the sake of simplicity, it is supposed that the filter's order of the different neurons is the same (S). The adjustable parameters of the proposed model are the coefficients of all the filters (a_{imj}) and the synaptic connections w_{ip}. The adjustment of the various parameters must be made in the opposite direction to the error function variation with respect to each parameter. The formula to be applied is:

$$\theta(k+1) = \theta(k) + \Delta\theta(k) = \theta(k) - \alpha.\frac{\partial E(k)}{\partial \theta};$$

$$\alpha > 0; \qquad \theta = w_{ip}, a_{imj}; \qquad (4)$$

Instant error is defined as:

$$E(k) = \frac{1}{2}(y_{N1}-y_{d1})^2 + \frac{1}{2}.(y_{N2}-y_{d2})^2 \qquad (5)$$

where $y_{dp}(k)$ is the desired output. The adjustment of weights w_{ip} is carried out according to formula 6:

$$\Delta w_{ip}(k) = -\alpha.\frac{\partial E(k)}{\partial w_{ip}(k)} = -\alpha.(y_{Np}(k)-y_{dp}(k)).g_i(k) \qquad (6)$$

The increase of the coefficient a_{imj} is:

$$\Delta a_{imj} = -\alpha.\frac{\partial E}{\partial a_{imj}} =$$

$$= -\alpha.[(y_{N1}-y_{d1}).w_{i1} + (y_{N2}-y_{d2}).w_{i2}].\frac{\partial g_i}{\partial x'_{im}} \cdot \frac{\partial x'_{im}}{\partial a_{imj}} \qquad (7)$$

complying with:

$$\frac{\partial g_i(k)}{\partial x'_{im}(k)} = -2.[\frac{x_m(k)+x'_{im}(k)-C_{im}}{\sigma^2}].g_i(k) = g'_{im}(k) \qquad (8)$$

and:

$$\frac{\partial x'_{im}(k)}{\partial a_{imj}} = g_i(k-j) + \sum_{t=1}^{S} a_{imt}.\frac{\partial g_i(k-t)}{\partial a_{imj}} \qquad (9)$$

where:

$$\frac{\partial g_i(k-t)}{\partial a_{imj}} = \frac{\partial g_i(k-t)}{\partial x'_{im}(k-t)} \cdot \frac{\partial x'_{im}(k-t)}{\partial a_{imj}} =$$

$$= g'_{im}(k-t).\frac{\partial x'_{im}(k-t)}{\partial a_{imj}} \qquad (10)$$

3. STABILITY ANALISIS

To ensure convergence of the learning process, Lyapunov's second convergence criterion is used (Ogata, 1987) (Slotine and Li, 1991). The following Lyapunov function is defined:

$$V(k) = \frac{1}{2}.(y_{N1}(k)-y_{d1}(k))^2 + \frac{1}{2}.(y_{N2}(k)-y_{d2}(k))^2 =$$

$$= \frac{1}{2}.e_1(k)^2 + \frac{1}{2}.e_2(k)^2 \qquad (11)$$

If the vector **W** contains all the adjusting coefficients of the network:

$$W(k+1) = W(k) - \alpha.\frac{\partial E(k)}{\partial W(k)} =$$

$$W(k) - \alpha.\frac{\partial E(k)}{\partial y_{N1}(k)} \cdot \frac{\partial y_{N1}(k)}{\partial W(k)} - \alpha.\frac{\partial E(k)}{\partial y_{N2}(k)} \cdot \frac{\partial y_{N2}(k)}{\partial W(k)} = \qquad (12)$$

$$W(k) - \alpha.e_1(k).\frac{\partial y_{N1}(k)}{\partial W(k)} - \alpha.e_2(k).\frac{\partial y_{N2}(k)}{\partial W(k)}$$

and:

$$\Delta W(k) = -\alpha.e_1(k).\frac{\partial y_{N1}(k)}{\partial W(k)} - \alpha.e_2(k).\frac{\partial y_{N2}(k)}{\partial W(k)} \qquad (13)$$

In 13 it has been borne in mind that:

$$\frac{\partial E(k)}{\partial y_{Np}} = e_p(k) \qquad (14)$$

Defining:

$$\Delta e_p(k) = e_p(k+1) - e_p(k) \qquad (15)$$

The change in the function can be expressed as follows:

$$\Delta V(k) = V(k+1) - V(k) = \Delta e_1(k).[e_1(k) + \frac{1}{2}.\Delta e_1(k)] + \Delta e_2(k).[e_2(k) + \frac{1}{2}.\Delta e_2(k)] \qquad (16)$$

If the **W** vector contains the adjusting coefficients of the network:

$$\Delta e_1(k) \approx [\frac{\partial e_1(k)}{\partial W(k)}]^T.\Delta W(k)$$

$$\Delta e_2(k) \approx [\frac{\partial e_2(k)}{\partial W(k)}]^T.\Delta W(k) \qquad (17)$$

As:

$$\frac{\partial e_p(k)}{\partial W(k)} = \frac{\partial y_{Np}(k)}{\partial W(k)} \qquad (18)$$

It results:

$$\Delta e_1(k) = -e_1(k).\alpha.[\frac{\partial y_{N1}(k)}{\partial W(k)}]^T.\frac{\partial y_{N1}(k)}{\partial W(k)} - e_2(k).\alpha.[\frac{\partial y_{N1}(k)}{\partial W(k)}]^T.\frac{\partial y_{N2}(k)}{\partial W(k)} \qquad (19)$$

In a similar way:

$$\Delta e_2(k) = -e_2(k).\alpha.[\frac{\partial y_{N2}(k)}{\partial W(k)}]^T.\frac{\partial y_{N2}(k)}{\partial W(k)} - e_1(k).\alpha.[\frac{\partial y_{N1}(k)}{\partial W(k)}]^T.\frac{\partial y_{N2}(k)}{\partial W(k)} \qquad (20)$$

The variation of the energy function is:

$$\Delta V(k) = -e_1(k)^2\alpha.\|\frac{\partial y_{N1}(k)}{\partial W(k)}\|^2 - e_2(k)^2\alpha.\|\frac{\partial y_{N2}(k)}{\partial W(k)}\|^2 - 2.e_1(k).e_2(k).\alpha.(\frac{\partial y_{N1}(k)}{\partial W(k)})^T.\frac{\partial y_{N2}(k)}{\partial W(k)} + \qquad (21)$$

$$\frac{1}{2}.[e_1(k).\alpha.\|\frac{\partial y_{N1}(k)}{\partial W(k)}\|^2 + e_2(k).\alpha.[\frac{\partial y_{N1}(k)}{\partial W(k)}]^T.\frac{\partial y_{N2}(k)}{\partial W(k)}]^2 + \frac{1}{2}.[e_2(k).\alpha.\|\frac{\partial y_{N2}(k)}{\partial W(k)}\|^2 + e_1(k).\alpha.[\frac{\partial y_{N2}(k)}{\partial W(k)}]^T.\frac{\partial y_{N1}(k)}{\partial W(k)}]^2$$

Equation 21 can be expressed as follows:

$$\Delta V(k) = -\alpha.[e_1(k).\frac{\partial y_{N1}(k)}{\partial W(k)} + e_2(k).\frac{\partial y_{N2}(k)}{\partial W(k)}]^2 + \frac{1}{2}.\alpha^2.[e_1(k).\|\frac{\partial y_{N1}(k)}{\partial W(k)}\|^2 + e_2(k).[\frac{\partial y_{N1}(k)}{\partial W(k)}]^T.\frac{\partial y_{N2}(k)}{\partial W(k)}]^2 + \frac{1}{2}.\alpha^2.[e_2(k).\|\frac{\partial y_{N2}(k)}{\partial W(k)}\|^2 + e_1(k).[\frac{\partial y_{N2}(k)}{\partial W(k)}]^T.\frac{\partial y_{N1}(k)}{\partial W(k)}]^2 \qquad (22)$$

A negative increment in the Lyapunov function is obtained if, for instance, the first positive term is bigger than or equal to the second positive term:

$$[e_1(k).\frac{\partial y_{N1}(k)}{\partial W(k)} + e_2(k).\frac{\partial y_{N2}(k)}{\partial W(k)}]^2 > \alpha.\|\frac{\partial y_{N1}(k)}{\partial W(k)}\|^2.[e_1(k).\frac{\partial y_{N1}(k)}{\partial W(k)} + e_2(k).\frac{\partial y_{N2}(k)}{\partial W(k)}]^2 \qquad (23)$$

To obtain negative increment in V(k):

$$0 < \alpha < \frac{1}{\|\frac{\partial y_{N1}}{\partial W}\|^2} \qquad 0 < \alpha < \frac{1}{\|\frac{\partial y_{N2}}{\partial W}\|^2} \qquad (24)$$

These expressions can be applied to any neural model with two outputs and should be specified for the model proposed in this article.

3.1 Application to the proposed model.

All the adjustable parameters of the neural network should be considered limited between +1 and -1. In the proposed model:

$$W = [a_{111}, ..., a_{M2S}, w_{11}, ..., w_{M2}]^T \qquad (25)$$

The number of the sipnasis coefficients (w_{ip}) is 2M and the number of feedback filters coefficients is 2MS. For each one of the sipnasis coefficients:

$$\|\frac{\partial y_{Np}}{\partial w_{ip}}\|_{max} = \|g_i(k)\|_{max} = 1 \qquad (26)$$

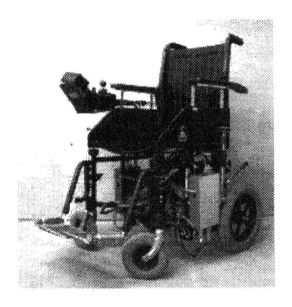

Fig. 2. Wheelchair.

The variation with respect to one of the coefficient a_{imj} belonging to the filter m of the ith neuron is:

$$\left\| \frac{\partial y_{Np}}{\partial a_{imj}} \right\|^2 = \left\| \frac{\partial y_{Np}}{\partial g_i} \cdot \frac{\partial g_i}{\partial x'_{im}} \cdot \frac{\partial x'_{im}}{\partial a_{imj}} \right\|^2$$

$$< \left\| w_{ip} M_d \cdot \frac{\partial x'_{im}}{\partial a_{imj}} \right\|^2 < \left\| M_d \cdot \frac{\partial x'_{im}}{\partial a_{imj}} \right\|^2 \qquad (27)$$

Being:

$$M_d = m\acute{a}x \left\| \frac{\partial g_i(k)}{\partial x'_{im}(k)} \right\| = \frac{\sqrt{2}}{\sigma} . e^{-\frac{1}{2}} \qquad (28)$$

From equations 9 and 10:

$$\frac{\partial x'_{im}(k)}{\partial a_{imj}} = g_i(k-j) + \sum_{t=1}^{S} a_{imt} . g'_{im}(k-t) . \frac{\partial x'_{im}(k-t)}{\partial a_{imj}} \qquad (29)$$

When a large number of sampling cycles (k→∞) have been carried out, the following condition is met:

$$\left\| \frac{\partial x'_{im}(k)}{\partial a_{imj}} \right\| < 1 + SM_d + (SM_d)^2$$

$$+ (SM_d)^3 + (SM_d)^4 + ... = \frac{1}{1 - SM_d} \qquad (30)$$

The following condition is to be met to ensure that the geometric series is limited:

$$SM_d < 1 \quad \Rightarrow \quad M_d < \frac{1}{S} \qquad (31)$$

It should be noted that this condition implies fixing a

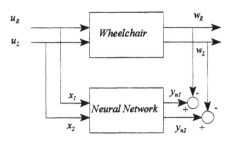

Fig. 3. Identification model used.

minimum value to the parameter "σ" used in the radial basis functions. If this value is very small the derivative of such functions is very high and therefore the previous series may not be limited, giving rise to an error in the learning process.

In practice, the limit values of the learning factors are:

$$0 < \alpha < \frac{1}{2M + \left(\frac{M_d}{1 - M_d . S} \right)^2 . S . M . 2} \qquad (32)$$

4. PRACTICAL TESTS

The described model was used for the identification of the dynamic behaviour of a wheelchair (Fig. 2) The aim is to obtain a good behaviour model that is also adaptive: network training must be effected while the chair is in operation: a parallel-parallel identification model is therefore most suitable (Fig. 3) applying two input signals u_R and u_L (each one corresponding to the voltage of the right or left wheel). Two readings are thereby obtained, each indicating the speed of the right or left wheel (w_R w_L).

In the first sampling cycle of the tests carried out, all the neural network coefficients have random values (different in each experiment). In this case it is possible to observe the time needed for the network to identify the wheelchair correctly.

The tests involved applying different signals to each of the chair's wheels (Fig. 4 and 5), distinguishing the speed of each wheel and the output given by the corresponding network. Figures 4.c and 5.c show the average quadratic error trend for each experiment, defined as:

$$MSE(k) = \frac{1}{k} \sum_{x=1}^{k} e^2(x) \qquad (33)$$

In the two experiments the following parameters were used to effect the identification: M=10; S=2; σ = 2,2, α = 0,007.

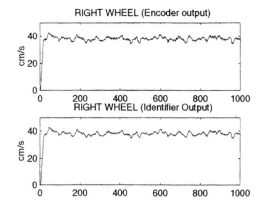

Fig. 4.a. Example 1

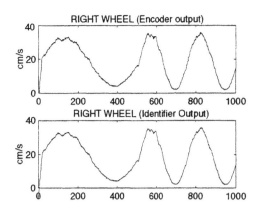

Fig. 5.a. Example 2

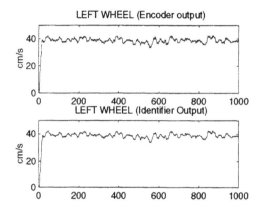

Fig. 4.b. Example 1

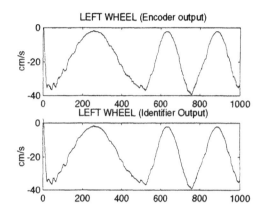

Fig. 5.b. Example 2

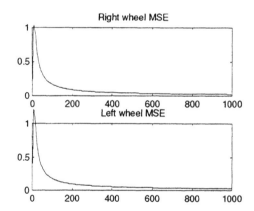

Fig. 4.c. MSE for example 1

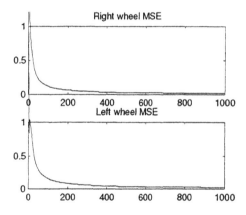

Fig. 5.c. MSE for example 2

5. CONCLUSIONS

A real physical system was identified using a new model of recurrent neural network. After obtaining the formulae for the adjustment of network coefficients, the necessary conditions were found for assuring convergence in the learning process. The tests carried out prove that the proposed model works well: minimum identification error and high convergence speed.

The final aim of this line of research is to implement a direct control system (Maeda and Figueiredo, 1997; Li & Häußler, 1996) to control the motion of the wheelchair even in those cases in which its working conditions vary, i.e., an increase in weight, a change in friction conditions, loss of battery power, etc.

ACKNOWLEDGMENTS

The authors would like to express their gratitude to the "Comision Interministerial de Ciencia y Tecnologia"

(CICYT), for their support through the project TER96-1957-C03-01.

REFERENCES

Boquete, L., R. Barea, R. Garcia, M. Mazo and J. A. Bernad (1997). Identification of Systems Using Radial Basis Networks Feedbacked with FIR Filters. In: *Lectures Notes in Computer Science. Computational Intelligence. Theory and Applications.* International Conference, 5th Fuzzy Days, pp. 46-51, Springer, Dormund, Germany

Ciocoiu, I. B. (1996) Radial Basis Function Networks with FIR/IIR Synapses. *Neural Processing Letters* 3, 17-22.

Kawato, M., Y. Uno, M. Isobe, and R. Suzuki (1987). A Hierarchical Model for Voluntary Movement and its Application to Robotics. In: Proc. 1987 IEEE Int. Conf. Neural Network, Vol. IV, pp. 573-582.

Maeda, Y. and R. J. P. de Figueiredo (1997). Learning Rules for Neuro-Controller via Simultaneous Perturbation. *IEEE Transactions on Neural Networks*, 8, n° 5, pp. 1119-1130.

Li, Y. and A. Häußler (1996). Artificial Evolution of Neural Networks and its Application to Feedback Control. Centre for Systems and Control and Dep. of Electronics and Electrical Engineering. Univ. of Glasgow.

Narendra, K. S., K. Parthasarathy, (1991). Identification and Control of Dynamic Systems Using Neural Networks. *IEEE Transactions on Neural Networks*, 1, n° 1, pp. 4-27.

Obradovic, D. (1996) On-Line Training of Recurrent Neural Networks with Continuous Topology Adaptation. *IEEE Transactions on Neural Networks*, 7. n° 1.

Ogata, K. (1987) *Discrete-Time Control Systems.* Prentice-Hall International Editions.

Slotine, J. J. and W. Li, (1991) *Applied Nonlinear Control.* Prentice Hall, Inc.

COOPERATIVE TASKS IN MOBILE MANIPULATION SYSTEMS[1]

O. Khatib, K. Yokoi, A. Casal

Robotics Laboratory
Department of Computer Science
Stanford University, Stanford CA 94305

Abstract.
Cooperative mobile manipulation capabilities are key to new applications of robotics
in space, underwater, construction, and the service environments. This article discusses the ongoing effort at Stanford University for the development of mobile manipulation capabilities to aid humans in a variety of material handling tasks. The work
presented in this paper focuses on the extension of the Augmented Object and the
Virtual Linkage models for fixed base manipulation, to mobile cooperative manipulation systems. We propose a new decentralized control structure for cooperative tasks
which is suitable to the more autonomous nature of mobile systems and prove its
efficacy in simulation. *Copyright © 1998 IFAC*

Keywords. Cooperative manipulation, decentralized control, mobile manipulation

1. INTRODUCTION

A central issue in mobile manipulation is cooperative operations between multiple vehicle/arm systems (Uchiyama
and Dauchez, 1988)(Adams *et al.*, 1995). Such systems
are usually characterized by a large degree of actuator force redundancy. Our approach to controlling such
systems is based on two models: The *augmented object* (Khatib, 1987*b*)(Khatib, 1987*a*) and *virtual linkage*
(Williams and Khatib, 1993). The *augmented object* describes the system's closed-chain dynamics and allows
for motion control specification of the entire system at
the object level. The *virtual linkage* model makes use of
the system's actuator force redundancy to allow for the
characterization and control of object internal forces.
Together, these models provide a way to control both
motion and forces at the object level in systems of multiple cooperating robots.

In this paper [1], we propose the extension of these mod-

els to mobile vehicle/arm systems and describe a new
strategy for the decentralized control of cooperative operations. The implementation of this decentralized control structure has been successfully proven in simulation
and results are presented here.

In order to provide an experimental testbed for the integration of the above mobile manipulation strategies,
we have designed and built two holonomic mobile manipulator platforms (Khatib *et al.*, 1997) (Pin and S.M.,
1994), SAMM (Stanford Assistant Mobile Manipulator),
shown in Figure 1. The aim of the SAMM project is to
develop a mobile "robotic assistant" to aid humans in a
variety of operations involving transportation and manipulation of materials.

2. COOPERATIVE MANIPULATION

We use the *augmented object* model to provide a description of the dynamics at the operational point for
a multi-arm robot system. The simplicity of the result-

---

[1] This work has been supported by NSF grant IRI-9320017.

Fig. 1. Stanford Mobile Platforms

ing equations is due to an additive property that allows us to obtain the system equations of motion from the dynamics of the individual mobile manipulators. The *augmented object* model is

$$\Lambda_\oplus(\mathbf{x})\ddot{\mathbf{x}} + \mu_\oplus(\mathbf{x}, \dot{\mathbf{x}}) + \mathbf{p}_\oplus(\mathbf{x}) = \mathbf{F}_\oplus \qquad (1)$$

with

$$\Lambda_\oplus(\mathbf{x}) = \Lambda_{\mathcal{L}}(\mathbf{x}) + \sum \Lambda_i(\mathbf{x}) \qquad (2)$$

$$\mu_\oplus(\mathbf{x}) = \mu_{\mathcal{L}}(\mathbf{x}) + \sum \mu_i(\mathbf{x}) \qquad (3)$$

$$\mathbf{p}_\oplus(\mathbf{x}) = \mathbf{p}_{\mathcal{L}}(\mathbf{x}) + \sum \mathbf{p}_i(\mathbf{x}); \qquad (4)$$

where $\Lambda_{\mathcal{L}}(\mathbf{x})$ and $\Lambda_i(\mathbf{x})$ are the kinetic energy matrices associated with the object and the ith effector, $\mu_{\mathcal{L}}(\mathbf{x})$ and $\mu_i(\mathbf{x})$ are the centrifugal and Coriolis vectors for the object and the ith effector, and $\mathbf{p}_{\mathcal{L}}(\mathbf{x})$ and $\mathbf{p}_i(\mathbf{x})$ are the gravity vectors for the object and the ith effector, respectively. $\mathbf{F}_\oplus$ also has the same additive property shown above for $\Lambda_\oplus(\mathbf{x})$, $\mu_\oplus(\mathbf{x}, \dot{\mathbf{x}})$ and $\mathbf{p}_\oplus(\mathbf{x})$.

Object manipulation requires accurate control of internal forces. We have proposed the *virtual linkage* (Williams and Khatib, 1993), as a model of object internal forces associated with multi-grasp manipulation. In this model, grasp points are connected by a closed, non-intersecting set of virtual links (Figure 2.)

For an N-grasp manipulation task, the *virtual linkage* model is a $6(N-1)$ degree of freedom mechanism that has $3(N-2)$ linearly actuated members and N spherically actuated joints. By applying forces and moments at the grasp points we can independently specify internal forces in the $3(N-2)$ linear members, along with $3N$

internal moments at the spherical joints. Internal forces in the object are then characterized by these forces and torques in a physically meaningful way. The relationship between applied forces, their resultant and internal forces is

$$\begin{bmatrix} \mathbf{F}_{res} \\ \mathbf{F}_{int} \end{bmatrix} = \mathbf{G} \begin{bmatrix} \mathbf{f}_1 \\ \vdots \\ \mathbf{f}_N \end{bmatrix}; \qquad (5)$$

where $\mathbf{F}_{res}$ represents the resultant forces at the operational point, $\mathbf{F}_{int}$ the internal forces and $\mathbf{f}_i$ the forces applied at the grasp point i. $\mathbf{G}$ is called the grasp description matrix, and relates forces applied at each grasp to the resultant and internal forces in the object. Furthermore, $\mathbf{G}$ can be written as

$$\mathbf{G} = [\ \mathbf{G}_1 \mathbf{G}_2 \ ... \ \mathbf{G}_N\ ] \qquad (6)$$

with

$$\mathbf{G}_i = \begin{bmatrix} \mathbf{G}_{res,i} \\ \mathbf{G}_{int,i} \end{bmatrix}; \qquad (7)$$

where each $\mathbf{G}_i$ represents the contribution of the ith grasp to the resultant and internal forces felt by the object and where $\mathbf{G}_{res,i}$ is the contribution of $\mathbf{G}_i$ to the resultant forces on the object and $\mathbf{G}_{int,i}$ to the internal ones. The inverse of the grasp description matrix, $\mathbf{G}^{-1}$, provides the forces required at the grasp points to produce the resultant and internal forces acting at the object:

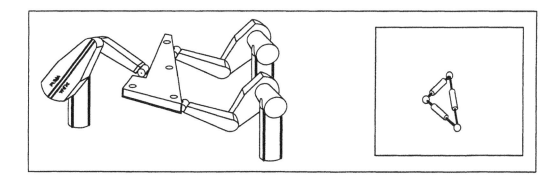

Fig. 2. The Virtual Linkage

$$\begin{bmatrix} \mathbf{f}_1 \\ \vdots \\ \mathbf{f}_N \end{bmatrix} = \mathbf{G}^{-1} \begin{bmatrix} \mathbf{F}_{res} \\ \mathbf{F}_{int} \end{bmatrix} \qquad (8)$$

with

$$\mathbf{G}^{-1} = \begin{bmatrix} \overline{\mathbf{G}}_1 \\ \vdots \\ \overline{\mathbf{G}}_N \end{bmatrix} \quad \text{and} \quad \overline{\mathbf{G}}_i = [\, \overline{\mathbf{G}}_{res,i} \ \ \overline{\mathbf{G}}_{int,i} \,]; \quad (9)$$

where $\overline{\mathbf{G}}_{res,i}$ represents the part of $\overline{\mathbf{G}}_i$ corresponding to resultant forces at the object. Similarly, the matrix $\overline{\mathbf{G}}_{int,i}$ represents the part of $\overline{\mathbf{G}}_i$ that contributes towards the object's internal forces.

The grasp description matrix contains a model of the internal force representation as well as the relationship between applied grasp forces and object resultant forces, and as such it is central to the control scheme employed by the *virtual linkage* model.

Compared to other methods employed to characterize internal forces, the virtual linkage has the advantage of providing a physical representation of internal forces and moments.

2.1 *Centralized Control Structure*

For fixed base (non-mobile) manipulation, the *augmented object* and *virtual linkage* have been tested experimentally in a system of three Puma manipulators and implemented in a multiprocessor system using a *centralized* control structure. However, this type of control is not suited to the more autonomous nature inherent in mobile manipulation systems, where a *decentralized* control scheme is more appropriate. Before presenting the decentralized implementation we propose, we begin with a brief summary of our previous centralized control structure. The overall structure of the centralized implementation is shown in Figure 3. The force sensed at the

grasp point of each robot, $\mathbf{f}_{s,i}$, is transformed, via $\mathbf{G}$, to sensed resultant forces, $\mathbf{F}_{res,s}$, and sensed internal forces, $\mathbf{F}_{int,s}$, at the operational point, using equation (5)

$$\begin{bmatrix} \mathbf{F}_{res,s} \\ \mathbf{F}_{int,s} \end{bmatrix} = \mathbf{G} \begin{bmatrix} \mathbf{f}_{s,1} \\ \vdots \\ \mathbf{f}_{s,N} \end{bmatrix}.$$

The centralized control strategy consists of (i) a unified motion and contact force control structure for the *augmented object*, $\mathbf{F}_{res}$; and (ii) $\mathbf{F}_{int}$, corresponding to the control of internal forces acting on the *virtual linkage*. The first part of the controller, associated with motion and contact force control, is

$$\mathbf{F}_{res} = \mathbf{F}_{motion} + \mathbf{F}_{contact}; \qquad (10)$$

where

$$\mathbf{F}_{motion} = \hat{\Lambda}_\oplus \Omega \mathbf{F}^*_{motion} + \hat{\mu}_\oplus + \hat{p}_\oplus \qquad (11)$$

and

$$\mathbf{F}_{contact} = \hat{\Lambda}_\oplus \overline{\Omega} \mathbf{F}^*_{contact} + \mathbf{F}_{contact,s}. \qquad (12)$$

$\hat{\Lambda}_\oplus$, $\hat{\mu}_\oplus$ and $\hat{p}_\oplus$ represent the estimates of $\hat{\Lambda}_\oplus$, $\mu_\oplus$, and $p_\oplus$. The vector $\mathbf{F}^*_{motion}$ and $\mathbf{F}^*_{contact}$ represent the inputs to the unit mass, decoupled system. Ω is the *generalized selection matrix* associated with motion control and $\overline{\Omega}$, its complement, is associated with contact force control. The control structure for internal forces is

$$\mathbf{F}_{int} = (\hat{\Lambda}_1 + \hat{\Lambda}_2)\mathbf{F}^*_{int} + \mathbf{F}_{int,s}; \qquad (13)$$

where $\hat{\Lambda}_i$ are the inertia matrices at the ith grasp point and the vector $\mathbf{F}^*_{int}$ represents the inputs to the decoupled, unit mass system. A suitable control law can be selected to obtain $\mathbf{F}^*_{motion}$, $\mathbf{F}^*_{contact}$ and $\mathbf{F}^*_{int}$. The con-

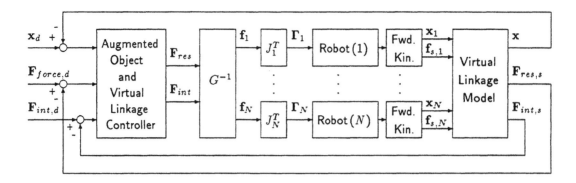

Fig. 3. Centralized Control Structure

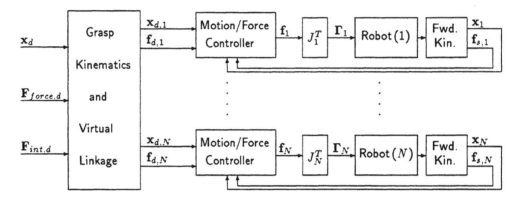

Fig. 4. Decentralized Control Structure

trol forces of the individual mobile manipulator at its grasp point, $\mathbf{f}_i$, are given by using equation (9),

$$\begin{bmatrix} \mathbf{f}_1 \\ \vdots \\ \mathbf{f}_N \end{bmatrix} = \mathbf{G}^{-1} \begin{bmatrix} \mathbf{F}_{res} \\ \mathbf{F}_{int} \end{bmatrix}.$$

2.2 Decentralized Control Structure

The *virtual linkage* and *augmented object* models have been successfully used in the cooperative control of two and three fixed PUMA arms, with a *centralized* control scheme. In a centralized control setup, each arm sends its sensory information to an off-board central controller which then commands the motion of each arm based on information from all the arms in the system.

However, for systems of a mobile nature a *decentralized*, on-board control structure is more suitable, to account for the greater autonomy that such systems entail. In such a decentralized control scheme, each robot makes decisions and closes its own control loop based on its local sensory information and that received from other platforms.

In the proposed decentralized control structure (Khatib *et al.*, 1997), the object level specifications of the task are transformed into individual tasks for each of the cooperative robots. Local feedback control loops are then developed at each grasp point. The task transformation and the design of the local controllers are accomplished in consistency with the *augmented object* and *virtual linkage* models, (Khatib, 1987a) and (Williams and Khatib, 1993). The overall structure is shown in Figure 4. The local control structure at the ith grasp point is

$$\mathbf{f}_i = \mathbf{f}_{motion,i} + \mathbf{f}_{force,i}. \tag{14}$$

The control vectors, $\mathbf{f}_{motion,i}$, are designed so that the combined motion of the various grasp points results in the desired motion at the object operational point. On the other hand, the vectors $\mathbf{f}_{force,i}$ create forces at the grasp points, whose combined action produces the desired internal forces on the object. The motion control at the ith grasp point is

$$\mathbf{f}_{motion,i} = \hat{\Lambda}_{\mathcal{Q},i}\Omega\mathbf{f}^*_{motion,i} + \hat{\mu}_{\mathcal{Q},i} + \hat{p}_{\mathcal{Q},i}; \quad (15)$$

with

$$\hat{\Lambda}_{\mathcal{Q},i} = \hat{\Lambda}_{g,i} + \overline{\mathbf{G}}_{res,i}\hat{\Lambda}_{\mathcal{L}}\overline{\mathbf{G}}^T_{res,i}; \quad (16)$$

where $\hat{\Lambda}_{g,i}$ is the kinetic energy matrix associated with the ith effector at the grasp point. The second term of equation (16) represents the part of $\hat{\Lambda}_{\mathcal{L}}$ "assigned" to the ith robot, described at its grasp point. The vector, $\hat{\mu}_{\mathcal{Q},i}$, of centrifugal and Coriolis forces associated with the ith effector is

$$\hat{\mu}_{\mathcal{Q},i} = \hat{\mu}_{g,i} + \overline{\mathbf{G}}_{res,i}\hat{\mu}_{\mathcal{L}}; \quad (17)$$

where $\hat{\mu}_{g,i}$ is the centrifugal and Coriolis vector of the ith robot alone at the grasp point. Similarly, the gravity vector is

$$\hat{p}_{\mathcal{Q},i} = \hat{p}_{g,i} + \overline{\mathbf{G}}_{res,i}\hat{p}_{\mathcal{L}}; \quad (18)$$

where $\hat{p}_{g,i}$ is the gravity vector associated with the ith end effector at the grasp point. The sensed forces at the ith grasp point, $\mathbf{f}_{s,i}$, combine the contact and internal forces felt at the ith grasp point, together with the acceleration force acting at the object. The sensed forces associated with the contact and internal forces alone, $\mathbf{f}_{\bar{s},i}$, are therefore obtained by subtracting the acceleration effect from the total sensed forces

$$\mathbf{f}_{\bar{s},i} = \mathbf{f}_{s,i} - \overline{\mathbf{G}}_{res,i}\left(\hat{\Lambda}_{\mathcal{L}}\ddot{\mathbf{x}}_d + \hat{\mu}_{\mathcal{L}} + \hat{p}_{\mathcal{L}}\right). \quad (19)$$

Here, the object desired acceleration has been used instead of the actual acceleration, which would be difficult to evaluate. The force control part of equation (14) is

$$\mathbf{f}_{force,i} = \hat{\Lambda}_i\mathbf{f}^*_{force,i} + \mathbf{f}_{des,i}; \quad (20)$$

The vector $\mathbf{f}_{des,i}$ is the desired force assigned to the ith mobile manipulator. This vector is

$$\mathbf{f}_{des,i} = \overline{\mathbf{G}}_{int,i}\mathbf{F}_{int,des} \quad (21)$$

where $\mathbf{F}_{int,des}$ is the desired object internal force. $\mathbf{f}^*_{force,i}$ represents the input to the decoupled, unit mass system associated with the internal forces. It can be achieved by selecting

$$\mathbf{f}^*_{force,i} = -\mathbf{K}_f\left(\mathbf{f}_{\bar{s},i} - \mathbf{f}_{d,i}\right) - \mathbf{K}_{vf}\dot{\mathbf{f}}_{\bar{s},i}. \quad (22)$$

The assumptions in the above control structure is that the object and the grasps are rigid, and that there is no slippage at the grasp points. Significant flexibilities and gripper slip in the real system will result in errors in the grasp kinematic computation and inconsistencies

Item	Length	Mass
Link 1	1	1
Link 2	1	1
Link 3	1	0.5
Object	0.2	0.1

Table 1. Kinematic and Inertial Simulation Parameters

Grasp	K_x	K_y	K_θ
1	2000	2000	200
2	2000	2000	200

Table 2. Grasp Simulation Parameters

with the *virtual linkage* model. One way to compensate for these effects would be to use grasp and force information from other robots. This requires some level of communication between the different platforms for updating the robot state and modifying the task specifications. The rate at which this communication is required is much slower than the local servo control rate and can be achieved over the radio Ethernet link (at 10-20 Hz) connecting our experimental SAMM mobile manipulator platforms.

Communication between robots is also specially important in tasks requiring external sensory information. For example, during certain cooperative force control operations the robots need to "talk" to each other to decide what to do next. We plan to add in this extra slower loop for data updating, task modification and robot communication in our future work.

Below we present simulation results obtained using the decentralized control structure. As a next step, we plan to implement and test it on the two SAMM platforms.

2.3 Results

In this section we present simulation results for a planar two-manipulator system implementing the decentralized internal force and motion controller described above.

Each manipulator has three degrees of freedom, and is connected to the object with a three degree-of-freedom spring that resists relative displacements between the gripper and the object, as well as relative planar rotations. The spring is intended to model the dynamics and forces sensed by the force sensor attached to the gripper in the real system. The kinematic and inertial parameters of the simulation are given in Table 1 and the grasp parameters in Table 2. The spring stiffnesses used are much lower than those of a typical force sensor, which are in the order of tens of thousands of Newtons per meter, to facilitate numerical computations.

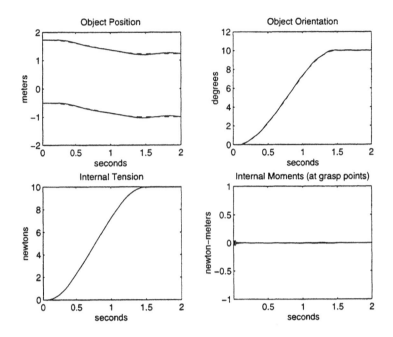

Fig. 5. Results with Decentralized Motion-Force controller

When we apply the control structure described by Equation 14 to this planar two-arm system, we find that we can accurately control internal forces as well as obtain the desired motion. This decentralized controller is in accordance with the *augmented object* and the *virtual linkage* models. The kinematic structure of the *virtual linkage* consists of a linear actuated joint joining the two grasp points to model internal tension stresses, and actuated spherical joints at each grasp point to resist internal moments, as described in (Williams and Khatib, 1993).

Figure 5 shows the response of the system when asked to track a cubic spline for both motion and internal forces for each manipulator (both internal moments are controlled to remain at zero). The broken lines show the desired trajectory and the solid lines the actual response. In this idealized scenario control is precise with little deviation from the desired trajectory and no steady-state error.

3. CONCLUSION

Multiple vehicle/arm compliant operations have been successfully implemented on two mobile manipulators, SAMM platforms, built at Stanford University. We have developed a new decentralized control structure based on the *augmented object* and *virtual linkage* models that is better suited for mobile manipulator systems and proved its efficacy in simulation. As a next step we plan to implement and test the decentralized cooperation strategy on the SAMM platforms, introducing an extra layer of control based on inter-robot communication and sharing of sensor information.

4. REFERENCES

Adams, J. A., R. Bajcsy, J. Kosecka, V. Kumar, R. Mandelbaum, M. Mintz, R. Paul, C. Wang, Y. Yamamoto and X. Yun (1995). Cooperative material handling by human and robotic agents: Module development and system synthesis. In: *Proc. IROS.* pp. 200–205.

Khatib, O., K. Yokoi, K. Chang and A. Casal (1997). The stanford robotic platforms. In: *IEEE Int. Conf. Rob. and Auto. Video Proc.*

Khatib, Oussama (1987a). Object manipulation in a multi-effector robot system. In: *The International Symposium of Robotics Research.* pp. 137–144.

Khatib, Oussama (1987b). A unified approach to motion and force control of robot manipulators: The operational space formulation. *IEEE Journal on Robotics and Automation* 3(1), 43–53.

Pin, F.G. and Killough S.M. (1994). A new family of omnidirectional and holonomic wheeled platforms for mobile robots. *IEEE Trans. Robotics and Automation* 10(4), 480–489.

Uchiyama, M. and P. Dauchez (1988). A symmetric hybrid position/force control scheme for the coordination of two robots. In: *Proc. ICRA.* pp. 350–356.

Williams, David and Oussama Khatib (1993). The virtual linkage: A model for internal forces in multigrasp manipulation. In: *Proceedings of the 1993 IEEE International Conference on Robotics and Automation.* pp. 1025–1030.

TASK OF OPENING AND PASSING THROUGH A DOOR BY A MOBILE MANIPULATOR BASED ON RECS CONCEPT

Hiroaki SEKI* Atsushi OMOI Masaharu TAKANO***

* Department of Mechanical Systems Engineering, Kanazawa University
2-40-20 Kodatsuno, Kanazawa 920, JAPAN
** Fujitec Co. Ltd. 1-28-10 Syou, Ibaraki, Osaka 567, JAPAN
*** Department of Industrial Engineering, Kansai University
3-3-35 Yamate-cho, Suita, Osaka 564, JAPAN

Abstract: RECS (Robot - Environment Compromise System) means the technology to compromise robots and environment in order to increase robot performance. A mobile manipulator system has been developed as a home robot based on this concept. The task of opening a door with spring and passing through a doorway, which is one of the standard tasks for a home robot, has been realized by the minimum modification of a door, the detection of door position utilizing the LED - reflecting marks - CCD camera system, and the cooperative motion of a mobile base and a manipulator. *Copyright © 1998 IFAC*

Keywords: Mobile Robots, Manipulation Tasks, Navigation, Cooperative Control, Visual Pattern Recognition

1. INTRODUCTION

Universal robots will be needed in near future used in the human-robot coexisting environment such as office robots, home robots, and welfare robots(Hashino, 1989; Robertson, 1991; Nakano et al., 1981; Saitoh et al., 1995). The difficulty of these human-support robots is that environment and required tasks have much diversity and are easily changeable, unlike industrial robots in factories. Two approaches are considered to practice the human-supporting technology. One is to use exclusive devices for each need, the other is to develop universal robots. The former should prepare many devices to satisfy all needs and it takes much cost and large space. The latter has the great problem whether universal robots can be practical in near future even if the most advanced techniques are applied, because it requires high intelligence and reliability. But the development of welfare robots is too urgent. In order to solve these dilemma, the new concept RECS is proposed, which will enable to develop universal robot utilizing the currently used technologies.

In this study, a mobile manipulator system have been developed as a home robot based on RECS concept. We deal with the task of opening a door and passing through a doorway as one of the standard tasks for home robots(Nagatani and Yuta, 1995). Although automatic doors can be considered, it takes high cost to equip them at all doorway in home. Therefore, a door is applied minimum modification such as marking for robots to recognize and operate easily and reliably. The detection of door position utilizing reflecting marks and the cooperative motion of a mobile base and a manipulator are also key issue to realize this task based on RECS.

2. RECS CONCEPT

RECS (Robot - Environment Compromise System) (Takano et al., 1996) means to arrange the environment in order to make the robot task possible or easy. This concept aims to share the technical difficulties with robot and environment. When we consider the robot-environment system, the environment has almost 100% technical burden in

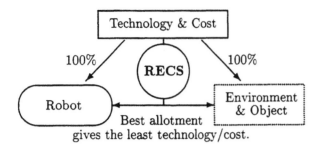

Fig. 1. Total amount of technological effort/cost in relation to allotment to robot and environment/object

factories. While in welfare robot researches, many people try to give 100% burden to robots without any environment modification, the practicality of which would not be attained in near future. The sharing of technical problems to robot and environment will give minimum difficulty, minimum cost, and maximum practicality (**Fig.1**).

When the robot tasks in human-robot society is analyzed, most part of the difficulties consists of recognition of environment/objects and end effectors to handle various kinds of workpiece. RECS insists arrangement on environment/objects in order to make easy their recognition and handling. It will be better and give more practicality if environment (marking) has any intelligence to co-operate with robot. But the arrangement should not prevent men from free daily life and should not be conspicuous. Therefore, the modification of environment should be minimized for users (societies) to accept. How to arrange devices to be convenient for robot and not conspicuous for humans is one of the problem peculiar to RECS. These technologies should be developed for early robot utilization though RECS is the midway technology to ultimate robot in far future.

3. MOBILE MANIPULATOR SYSTEM BASED ON RECS

An indoor mobile robot system has been developed according to the RECS concept. The task of opening a door and passing through it was realized as one of the basic tasks needed for a home robot.

3.1 *Target Door with Spring and Lever*

A target door is shown in **Fig.2**. This door has some modifications for a robot to operate easily according to RECS concept.

- *Door with spring;* The door has spring which closes it automatically. Though a robot must hold the door while passing, it does not need to close the door after passing.

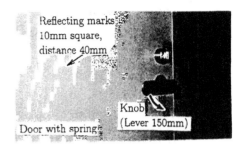

Fig. 2. Target door and marking by RECS

- *Lever-type door knob;* The door knob is lever with spring. Though a manipulator should be applied force control because of positioning error in case of a cylindrical knob, a lever knob is only pulled down without any complicated control. It also requires not a multi-fingered hand but only a simple gripper.
- *Marking;* In order to detect the door (knob) position easily, two pieces of reflecting tape are pasted on the door near the knob and they are in vertical line to the floor. The relative positions between the marks and the rotation axes of the door and the knob are given previously. Reflecting tapes are normally used for safety goods for example and they have property to reflect light strongly in the direction where it is thrown.

These are minimum modification base on RECS.

3.2 *Task of Opening and Passing through a Door*

The sequence to open and pass through the door is as follows.

(1) *Navigation;* We assume that the robot arrives near the door (within about 1m) by a certain navigation system.
(2) *Approach and position detection;* The robot approaches the door measuring the door position by detecting the reflecting marks on it.
(3) *Release of door lock;* The robot rotates the door lever after grasping it to release the door lock. The end of rotation should be found.
(4) *Opening and passing;* The robot base goes forward and the manipulator on it opens the door simultaneously. The robot must hold the door until it passes through the doorway because of the spring of the door.
(5) *Closing;* The robot releases the door by pulling the manipulator at high speed and then the door is closed automatically.

3.3 *Developed Mobile Manipulator System*

The structure of the developed mobile manipulator system for door opening task is shown in **Fig.3** and **Fig.4**. This system consists of 4 subsystems.

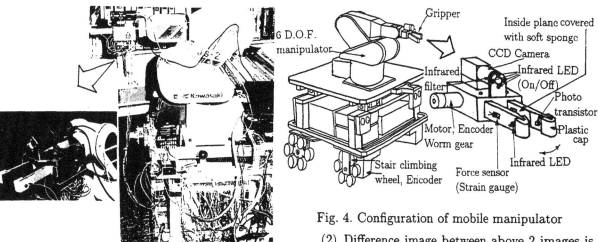

Fig. 3. Mobile manipulator system

- *Robot base;* It has 2-drive 2-caster wheels and driven wheels has encoders for ranging from dead reckoning.
- *6 D.O.F. manipulator;* It pushes the door to open in cooperation with the robot base It is applied only position control.
- *Gripper;* It grasps the door knob (lever) and it has photo sensor to find the lever between 2 fingers. Fingers are driven by a DC motor through a worm gear in order to keep their positions even if they receive strong forces and each finger equips force sensor in order to detect the contact to the lever and the end of its rotation. The range and resolution of this sensor are $0 \sim 20$N, 0.05N respectively. The fingertips are covered with plastic caps so as not to damage the door while pushing.
- *Mark detection system;* The robot detects reflecting marks and its position from the images taken by a CCD camera with 512×512 dots and 8 bits gray scale. Infrared LEDs of high intensity and a infrared filter are attached in front of the camera to shine the reflecting marks and to detect them reliably. Infrared LEDs are not obstructive to human.

This system utilizes not high but reliable technology and the modification of environment is small.

4. DOOR OPENING TASK BY RECS

4.1 Detection of Reflecting Marks

Assuming that the CCD camera and the door are parallel, the door and the knob positions can be easily obtained by detecting the position of 2 reflecting marks on the door surface as follows.

(1) 2 images are taken by CCD camera on the manipulator when the infrared LEDs are on and off. Only when LED is on, marks reflect its light in the camera direction.

Fig. 4. Configuration of mobile manipulator

(2) Difference image between above 2 images is obtained in order to avoid disturbance light.

(3) The positions of the centers of mark images are obtained after binarization and noise removal processing.

(4) The distance (pixel) of 2 marks on the images is inversely proportional to the distance between the camera and the door and the displacement (pixel) from the image frame center is proportional to the displacement from the optical axis of the camera. Therefore, the relative position of the marks from the robot can be calculated. If the orientation of the door should be measured, one more mark must be pasted.

(5) The robot approaches the door by measuring the relative position of the marks repeatedly. The manipulator moves the camera position during this approach so that the marks may not go out of the image frame.

(6) The robot stops at the specified distance from the door and measures the mark position finally.

4.2 Knob Rotation Task to Release Lock

In order to release the door lock, the robot grasps the door knob according to the mark position and rotates it as follows.

(1) When the robot grasps the knob (lever), the lever entering between 2 fingers can be found by the photo sensor.

(2) If the robot grasps the lever rigidly, excessive force is caused by the positioning error while rotating. Therefore, it grasps the lever loosely to pull it down by the upper finger (**Fig.5**).

(3) Though force control is not necessary to rotate the lever, the end of its rotation must be detected by the force sensor of the finger. When the force increases more than the effect of the spring in the lever, the robot stops to rotate the lever by monitoring differential coefficient of the force.

(4) The robot pushes the door $2 \sim 4$cm forward to release the lock and rotates back the lever.

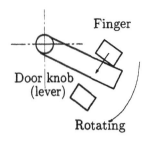

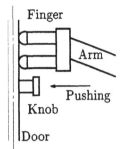

Rotating a lever with grasping loosely

Pushing a door without grasping a knob

Fig. 5. Manipulation considering position error

(5) After the fingers are moved to the upper position of the lever sliding on the door surface (**Fig.5**), the robot pushes the door again not grasping the lever rigidly but using its fingertips because of the positioning error.

RECS need no complicated and no intelligent control.

4.3 *Opening Door Task by Cooperation of Robot Base and Manipulator*

The robot opens a door and passes through it in cooperation with the robot base and the manipulator. It must also hold the door while passing because the door has spring which closes it automatically. The collision between the door and the robot body should be taken into account.

The manipulator rotates the door so as to fix the hand to the knob position d at an angle of ψ (normal:$\pi/2$). The base coordinate system Σ_D is located at the door axis as shown in **Fig.6**. Let Θ_d ($-\pi/2 \le \Theta_d \le \pi/2$) be the door angle and $R(X_r, Y_r, \Theta_r)_{\Sigma_D}$ be the position and orientation of the robot base. The hand $H(x_h, y_h, \theta_h)_{\Sigma_R}$ of the manipulator on the mobile robot (coordinate system Σ_R) should be moved as satisfying the following equation.

$$\begin{bmatrix} d\cos\Theta_d \\ d\sin\Theta_d \\ \Theta_d + \psi \end{bmatrix} = \begin{bmatrix} X_r \\ Y_r \\ \Theta_r \end{bmatrix} + \begin{bmatrix} \cos\Theta_r & -\sin\Theta_r & 0 \\ \sin\Theta_r & \cos\Theta_r & 0 \\ 0 & 0 & 1 \end{bmatrix} \begin{bmatrix} x_h \\ y_h \\ \theta_h \end{bmatrix} \quad (1)$$

In order to open and pass through a door, 2 stages are necessary as follows.

(1) The robot base goes forward and the manipulator pushes a door to open simultaneously. The door angle Θ_d is increased until Θ_{dmax} (normal:$\pi/2$).

(2) A robot base goes forward with the door holding at an angle of $\Theta_d = \Theta_{dmax}$.

The next condition is necessary to avoid the collision between the door and the nearest corner $C(x_c, y_c)_{\Sigma_R}$ of the robot body.

$$F = X_c \sin\Theta_d - Y_c \cos\Theta_d \ge F_{min} > 0, \quad X_c > 0 \quad (2)$$

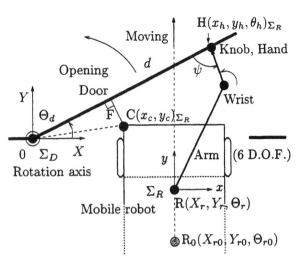

Fig. 6. Position of robot hand and door

$$\begin{bmatrix} X_c \\ Y_c \end{bmatrix} = \begin{bmatrix} X_r \\ Y_r \end{bmatrix} + \begin{bmatrix} \cos\Theta_r & -\sin\Theta_r \\ \sin\Theta_r & \cos\Theta_r \end{bmatrix} \begin{bmatrix} x_c \\ y_c \end{bmatrix} \quad (3)$$

F is the perpendicular position to the door and F_{min} is the closest distance for safety.

Because the number of variables is redundant against the constraint equations (1), the door angle Θ_d and the robot position $R(X_r, Y_r, \Theta_r)$ can be specified. The robot goes straight from the position in front of the door and it moves in proportion to the door angle Θ_d (rate : $a (> 0)$) until the end of door rotation so as not to increase the opening speed of the door suddenly. In case of the developed system, the moved distance of the robot base is given by the dead reckoning.

$$\Theta_d = \begin{cases} \Theta_d' & (\Theta_d' < \Theta_{dmax}) \\ \Theta_{dmax} & (\Theta_d' \ge \Theta_{dmax}) \end{cases} \quad (4)$$

$$\begin{bmatrix} X_r \\ Y_r \\ \Theta_r \end{bmatrix} = \begin{bmatrix} X_{r0} \\ a\Theta_d' + Y_{r0} \\ 0 \end{bmatrix}, \quad \begin{bmatrix} X_c \\ Y_c \end{bmatrix} = \begin{bmatrix} X_r + x_c \\ Y_r + y_c \end{bmatrix} \quad (5)$$

X_{r0}, Y_{r0} is the robot's position when the door angle $\Theta_d = 0$. When the door is held at a certain angle ($\Theta_d = \Theta_{dmax}$), Y_r should be increased independently. If the robot position $R(X_r, Y_r, \Theta_r)$ and the door angle Θ_d are given by these equations, The hand position $H(x_h, y_h, \theta_h)_{\Sigma_R}$ can be obtained from the Eq. (1). **Fig.7** shows the example of the cooperative motion of the robot base and the manipulator when the robot opens the door until $\Theta_d = \pi/2$ and then hold it.

We should determine the motion rate a when the corner of the robot body C does not collide with the door. This condition (Eq.(2)) becomes

$$f = \frac{F}{X_c} = \sin\Theta_d - (\alpha\Theta_d' + \beta)\cos\Theta_d > \frac{F_{min}}{X_c} > 0 \quad (6)$$

$$\alpha = \frac{a}{X_c} > 0, \qquad \beta = \frac{Y_{r0} + y_c}{X_c} < 0 \quad (7)$$

$$Y_c = X_c(\alpha\Theta_d' + \beta) \quad (8)$$

254

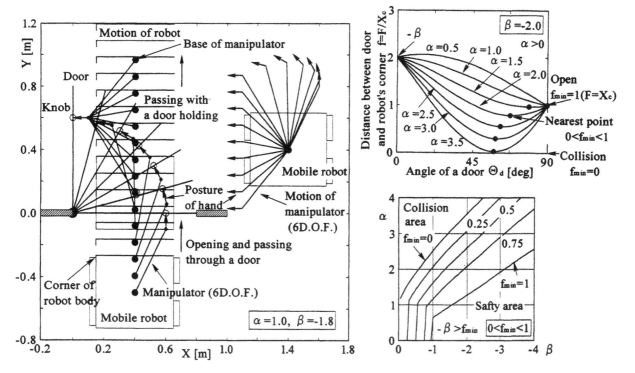

Fig. 7. Cooperative motion of robot base and manipulator to open and pass through a door

Fig. 8. Parameters of robot - door motion α, β to avoid collision

If the perpendicular position to the door F satisfies this condition, the robot body does not enter within the distance F_{min}. Namely, the minimum f_{min} of f in the region of $0 \leq \Theta_d \leq \Theta_{dmax}$ must be more than F_{min}/X_c and the maximum of the motion rate α exists because the larger α becomes, the closer the robot goes to the door (the smaller f_{min} becomes) as shown in **Fig.8**. This maximum of α can not be solved analytically but can be obtained by convergence calculation. Moreover, the minimum of α is determined by the motion space of the manipulator and so on. **Fig.8** shows the region of α, β to avoid collision. These were calculated using the parameters in **Fig.7**.

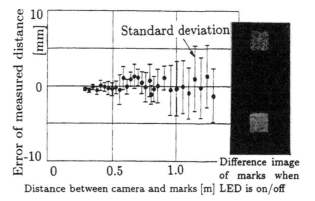

Fig. 9. Detection of a camera - marks distance

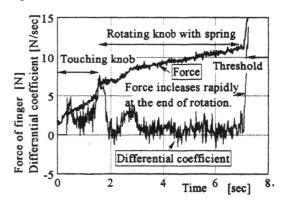

Fig. 10. Force during the rotation of the door lever with spring

5. EXPERIMENTAL RESULTS

5.1 *Detection of Mark Position*

Fig.9 shows the experimental result of measuring the distance by 2 reflecting marks and a CCD camera and it plots the error of measured distance. The difference image is also shown in this figure. This image is robust against disturbance lights. Since the distance between 2 marks' images is inversely proportional to the distance between the camera and the door, the closer the camera gets to the door, the smaller the error becomes. It can be seen that the accuracy of the measured distance is ±5mm within 1m. The displacement of the marks from the optical axis of the camera also could be measured with about the same accuracy.

5.2 *Rotation of Door Lever*

Fig.10 shows the force of the upper finger during the rotation of the door lever. It can be seen that the end of the rotation can be detected by monitoring differential coefficient of the force.

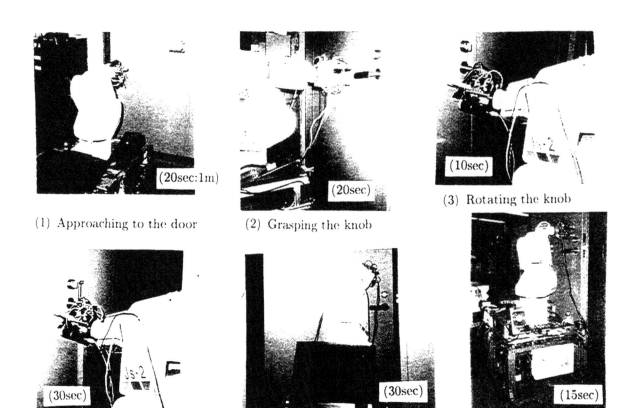

(1) Approaching to the door (20sec:1m)

(2) Grasping the knob (20sec)

(3) Rotating the knob (10sec)

(4) Releasing the lock (30sec)

(5) Passing/opening the door (30sec)

(6) Passing/holding the door (15sec)

Fig. 11. Mobile manipulator in opening and passing through a door

5.3 Realization of Door Opening Task

The total task of opening and passing through a door was implemented on the mobile manipulator. The sequence of this task and the time of each step are shown in **Fig.11**. The robot almost succeeded to open the door in about 2 minutes. But sometimes the wheels slid and the path of the robot base curved because of the reacting force from the door spring. The outside wheel should be driven more than the inside wheel for the robot base to go straight. Another problem was that the manipulator sometimes stopped owing to its motion space. Wide motion space is required for the manipulator on a mobile robot in home.

6. CONCLUSION

An mobile manipulator system has been developed according to the RECS concept and the method for opening and passing through a door with spring by this robot has been proposed. RECS requires the minimum modification of environment for a robot to recognize and operate easily and reliably. The opening door task has been realized by the detection of position utilizing reflecting marks and cooperative motion of the mobile base and the manipulator.

REFERENCES

Hashino, S. (1989). Control of patient care robot "melkong". *Proc. 20th Int. Symp on Industrial Robots* pp. 461–466.

Nagatani, K. and S. Yuta (1995). An experiment on opening-door-behavior by an autonomous mobile robot with a manipulator. *Proc. IEEE/RSJ Int. Conf. on Intelligent Robots and Systems* **2**, 45–50.

Nakano, E., T. Arai, K. Yamaba, S. hashino, T. Ohno and S. Ozaki (1981). First approaches to the development of the patient care robot. *Proc. 11th Int. Symp. on Industrial Robots* pp. 87–94.

Robertson, G. I. (1991). Helpmate delivery robot operaties safely amongst the general public. *Proc. 22nd Int. Symp. on Industrial Robots* pp. 12–17.

Saitoh, M., Y. Takahashi and A. Sankaranarayanan (1995). A mobile robot testbed with manipulator for security guard application. *Proc. IEEE Int. Conf. on Robotics and Automation* pp. 2518–2523.

Takano, M., T. Yoshimi, K. Sasaki and H. Seki (1996). Development of indoor mobile robot system based on recs concept. *Proc. 4th Int. Conf. on Automation, Robotics and Computer Vision* **2**, 868–872.

SELECTION AND PLACEMENT OF A MANIPULATOR
FOR THE MOBILE ROBOT RAM-2

M. A. Martínez and J. L. Martínez

*Dep. de Ingeniería de Sistemas y Automática. E.T.S. Ingenieros Industriales.
Universidad de Málaga. Plaza El Ejido s/n, 29013 Málaga (Spain).
Fax: (+34) 5 213-14-13; Tel: (+34) 5 213-14-08;
E-mail:mams@ctima.uma.es*

Abstract: Recently, robot arms mounted on mobile platforms, often called mobile manipulators, have been studied in relation to several different subjects. The great amount of mobility provided by the vehicle combined with the manipulation abilities allow the global system to perform complex tasks. This paper deals with the study of several small size robot arms in order to select one for being installed on the mobile robot RAM-2. The special characteristics that mobile manipulators must fulfil are discussed. The manipulator selected, the Performer MK2, is presented and the reachable workspace over the vehicle for carrying out several tasks is shown. *Copyright © 1998 IFAC*

Keywords: mobile robots, robot arms, manipulation tasks, implementation, autonomous vehicles.

1. INTRODUCTION

Among the research developed for mobile manipulators, it is possible to find systems that work in teleoperated mode (Fogle and Heckendorn, 1992), and others autonomously (Pin, et al., 1994). Also, it is possible to distinguish between when the vehicle and manipulator motions proceed in parallel, moving the arm while the base is moving (Carriker, et al., 1990), or sequentially in two phases: navigation to a suitable area, and manipulation (Seraji, 1995). In the latter case, it is possible to consider the base as a mere positioner for the arm, rather than an active degree of freedom. This strategy leads to a simpler control scheme, but it is less efficient.

Most of the proposed methods move the base and the arm simultaneously in order to accomplish user defined tasks. For proper and effective coordination of the motions of the mobile platform and the manipulator, it is necessary to look at the new issues introduced by the combined system that are not present in every individual component, such as redundancy and dynamic interactions.

Kinematic redundancy is created by the addition of the degrees of freedom of the mobile vehicle to those of the manipulator. A particular point in the workspace may be reached by moving the manipulator, by moving the base, or by the combined motion of both. In spite of being quite desirable since it allows mobile manipulators to operate under many modes of motion and to perform a wide variety of tasks, the coupling of the mobility and manipulation functions introduces a new order of complexity in the planning problem.

The second new issue is that the mobile vehicle and the manipulator dynamically interact with each other. Examples of coordinated control of mobile manipulators that do not consider the effect of the dynamic interaction are found in studies like (Carriker, et al., 1990) where path planning is treated as an optimization problem in which the decision variables for mobility (base position) are separated from the manipulator joint angles in the cost function. In other cases, the redundancy introduced by the mobile base is used to accomplish a set of user-defined tasks in addition to end-effector motion (Lim, *et al.*, 1994). The discussion about usefulness of minimax criteria for generally resolving the redundancy of mobile manipulator systems, and in particular for calculating commutation configurations on the basis of various task objectives is presented in (Pin, *et al.*, 1994).

There are also papers that consider the dynamic interaction between the manipulator and the mobile platform. In (Liu and Lewis, 1990) a description of a decentralized robust controller for a mobile robot is presented by considering the platform and the manipulator as two separate systems. In others, the study is directed to obtain the stability criteria that ensures that the vehicle does not tip over (Wiens, 1989). A system in which a manipulator is mounted on a platform with soft suspension and an algorithm that enables the manipulator to follow a desired trajectory in the presence of dynamic disturbances is considered in (Hootsmans, *et al.*, 1992). In other works, the objective is to actively compensate for the dynamic interaction through a nonlinear feedback to improve the performance of the overall system (Yamamoto and Yun, 1994).

The main subject of this paper is the selection of a robot manipulator to be fixed on the RAM-2 mobile robot. In Sect. 2 a summary of mobile manipulators built to accomplish tasks in different fields is presented. In Sect. 3 the restrictions for choosing manipulators for mobile platforms are discussed. Section 4 describes RAM-2 and concludes with the selection of the best robot to be mounted on the vehicle. Finally sections 5, 6 and 7 are devoted to conclusions, acknowledgements and references respectively.

2. EXISTING MOBILE MANIPULATORS

Many mobile manipulators have been built to work in different environments and to carry out several tasks. Examples of the wide range of uses given to composite systems and the experiments accomplished with them are given in the following paragraphs.

1.- The use of mobile manipulators for increasing efficiency in material-handling tasks such as warehouse management, maintenance, or repair justifies the research and development performed in this area. HERMIES-III is an autonomous battery-powered robot comprised of a seven degrees of freedom manipulator designed for human scale tasks and an omnidirectional wheel-driven vehicle. The research points include performance of human scale tasks like valve manipulation, use of tools, etc., that are usual to find in an industrial environment, and the study of motion planning and control of mobile manipulators (Pin, *et al.*, 1994, Weisbin; *et al.*, 1990).

2.- The use of robotics in agricultural tasks requires a higher sensory capability than in industrial tasks because the agricultural environment is less structured. It is possible to distinguish between greenhouses and open field applications because the cultivation environment inside a greenhouse is more constrained, and the position and shape of the plants are more regular. AGROBOT is a mobile manipulator developed for agricultural operations in greenhouses. The robot architecture is based on an autonomous vehicle with a six degrees of freedom arm and is able to navigate between rows of plants, stop near each plant, identify the relevant objects, and pick ripe fruits or locally spray substances on leaves or flowers (Buemi, *et al.*, 1995). In open field applications, a prototype robot for picking citrus fruit has been developed at Florida University (Harrel, *et al.*, 1990) where the picking manipulator has three degrees of freedom with the two first joints rotatory and the last sliding. The manipulator is mounted on a portable grove-lab which provides power units for stand-alone operation and computational power for robot control.

3.- The field of rehabilitation robotics, mainly the area of disabled and elderly people, is becoming more and more important due to the growing number of severely or moderately impaired people in industrialised countries. Several prototypes of mobile manipulators capable of performing a limited number of home assistance functions and selected tasks in medical facilities have been developed.

The MOVAR was the first mobile manipulator platform developed for this purpose. The URMAD project consists of a vehicle equipped with an eight degrees of freedom robotic arm and a bedside workstation that allows the user to operate the robot via a man-machine interface with communication between the workstation and the mobile unit achieved by a radio link. The MOVAID project is composed of a Electric Pedestrian vehicle and a manipulator derived from the prototype built for the URMAD system. The MOVAID concept is extended by making the whole system modular, which will allow users to choose the support which best suits their individual requirements (Dario, *et al.*, 1995). In (White, *et al.*, 1987) the HCR for assistance to the elderly and the handicapped is described. The system is composed of a battery powered mobile platform and a manipulator of six degrees of freedom that extends approximately one meter for reaching objects on the floor and a low cost control station for operating the mobile manipulator.

4.- The use of mobile manipulators in hazardous environments like nuclear plants has been studied for a long time. The mobile surveillance robot SURBOT was developed to perform visual, sound, and radiation surveillance within radiologically hazardous rooms at nuclear power plants. The robot arm with seven degrees of freedom is provided mainly for obtaining contamination smears from the floor and equipment, and for other tasks such as opening doors, actuating valves, etc. The manipulator can be teleoperated using the master arm on the control console or programmed (White, *et al.*, 1987).

The OAO-150 has been built for removing the cesium source used for radiation detection equipment at the site where some problem has occurred, before maintenance work by operators can be initiated. It is composed of a small, tracked, skid-steered vehicle that can be operated by either radio control or tether cable, and the arm with five degrees of freedom with a parallel jaw gripper. The TSR-700 WASP was developed to remotely decontaminate process rooms and is composed of a hydraulically driven skid-steered mobile platform with a six axis manipulator arm. For video surveillance work in sand filters for radioactive particles like plutonium, uranium, etc., located in the separations areas, the HORNET robot was developed with a simple skid-steered vehicle powered by motorcycle batteries and a two degrees of freedom arm with a parallel jaw-gripper end effector (Fogle and Heckendorn, 1992).

5.- In other cases, the mobile manipulator has been constructed for researching tasks, like HERBERT, composed of a two degrees of freedom arm and an omni-directional vehicle. The mobile manipulator is used for testing the multiple partial representations performance when collecting soda cans in an unstructured office environment (Connell, 1991). The mobile manipulator built by the IKERLAN centre is another example of construction for research purposes (Ezquerra *et al.*, 1997).   .

3. CHARACTERISTICS OF MANIPULATORS FOR MOBILE ROBOTS

As can be inferred from the previous section, the characteristics to be studied in order to select a manipulator for a mobile platform mainly depend on the platform where the robot arm is going to be installed and the tasks that the overall system will carry out. There are also some important characteristics regarding manipulators that must be taken into account:

Speed/repeatability: for a manipulator over a mobile platform, speed and repeatability are less important, mainly because the manipulator usually does not have to do the same operation under the same conditions, and also it is not necessary to work at a high speed and accuracy. Moreover, the manipulators supplied by manufacturers are very restrictive as regards these points.

Price level: for achieving the aims of the project, it is necessary to incur many expenses, so only a limited amount of money is available for the manipulator purchase.

Arm and controller weight: if the arm and the controller are less weighted, the power consumption supplied by the vehicle decreases and it is possible to work autonomously for a longer time. Also, the available space on the platform requires a small manipulator.

Workspace: the arm robot has to work mainly outside the mobile vehicle and sometimes with objects placed over the platform. The manipulator workspace is not as essential as in industrial manipulators because of the mobility of the vehicle, but it is still very important.

Programming language/Operating system: as the manipulator will be integrated in the overall robot system, it is necessary to communicate with other systems in order to receive orders and to supply information about its state. Thus, the controller language and operative system should be open enough to carry out these tasks.

Degrees of freedom: generally the applications carried out by the mobile manipulator do not require a high number of joints, so five is usually enough. Nevertheless, a higher number has the advantage of a more dextrous workspace.

Load capacity: this is the parameter that fixes the type of object to be manipulated by the arm robot and the type of end effector. The higher the value, the bigger the flexibility for the task that the system can carry out.

Power supply: the manipulator has to be powered by some system. If the manipulator is powered by alternating current from the electrical network, a special adapter has to be designed.

The optimization of some parameters has a negative repercussion on others, for instance, if the price decreases the features decrease too, or if the robot arm is bigger, the workspace will be bigger too. So, the selection process must balance out all the critical characteristics.

4. APPLICATION TO RAM-2

RAM-2 is an autonomous mobile robot designed and built at the University of Málaga for research applications in indoor environments. It is based on the RAM-1 mobile robot (Ollero, *et al.*, 1993), but in contrast to the latter, its control system is supported by PCs.

The dimensions of RAM-2 are constrained by its ability to navigate passing between objects and through corridors and doors. On the other hand, the robot needs space to accommodate the power system, standard electronic and computer enclosures, and a variety of sensors for navigation and operation: cameras, laser sensors, and a ring of sonars.

In Fig. 1 it is possible to observe the space available in front of the vehicle for the manipulator's placement.

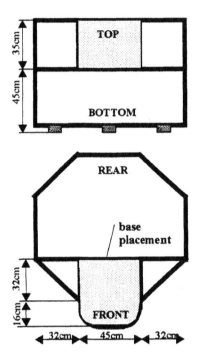

Fig. 1. Mobile robot RAM-2 scheme

The electrical system is composed of 12V/54AH batteries grouped in two circuits; the first, the 48V circuit, with four batteries connected serially, and the second, the 24V circuit, with two batteries also connected serially.

In addition, it is possible to work directly with the power supplied by the electrical network because the corresponding commuters have been installed. Two emergency buttons for stopping the vehicle quickly have been installed on the mobile robot.

The vehicle has four wheels located in the vertices of a rhombus with a diagonal in the longitudinal axis. The two parallel wheels are each driven by a DC motor. The front and rear wheels are steered by a single DC motor with a kinematic rigid link. The locomotion system can provide a zero turning radius. For manoeuvring, the directional wheels are steered and differential steering applied to the parallel wheels. The top speed, which is 1.6 m/sec, can only be reached when the vehicle moves along a straight line, and must decrease as the vehicle's curvature increases.

The low-level control of the vehicle is carried out by a specialised board. Navigation is controlled by a

Pentium running at 120 MHz under the real time operating system Lynx. The tasks that the mobile manipulator will perform are:

- To pick up, carry, and place low weight objects such as books, video tapes, etc.
- To manipulate the objects while the vehicle is moving.
- To press or turn little components such as machine controls, door-handles, etc.

The building of a manipulator specifically for RAM-2 has not been considered because of the associated financial and time costs. Although this choice provides the best solution, the aim of the project is the study of the possibilities of a manipulator on board a vehicle.

In Table 1 it is possible to compare the parameters applied to several commercial manipulators in order to select one of them for RAM-2. The robot arms studied have been catalogued by low (L), medium (M), or high (H) price, the operating system by close (C) and open (O) and the symbol ? means information unknown.

It is necessary to point out that industrial manipulators of medium or high size have been directly eliminated because of the impossibility of being powered by the vehicle or not having enough space for placing them. The study of small-size robot arm characteristics obtained from advertisements in specialist magazines is summed up in Table 1.

The first option was the Zebra-Zero but availability problems led to the selection of a Performer MK2. In Fig. 2 it is possible to see the structure of this manipulator of five degrees of freedom.

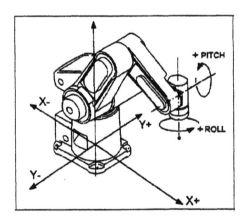

Fig. 2. The Performer MK2 manipulator

This manipulator can easily fit into the available space at the front of the vehicle. Also, it is possible to install the controller on the vehicle without making any change to the rest of the vehicle's electronics, except for the commuter introduced for powering the

<u>Table 1: Analysis results</u>

ROBOT ARM	PRICE LEVEL	WEIGHT (Kg.)	WRIST REACH (mm.)	OPERATING SYSTEM	d. o. f.	LOAD CAPACITY (Kg.)	POWER SUPPLY (V)
SCORBOT ER-VII	L	66	600	C	5	2	110/220 AC
ZEBRA-ZERO	L	11.6	510	O	6	1	110 AC/24 DC
CRS A-255	M	48	508	C	5	2	110/220 AC
RV-M1	L	42	410	C	5	1. 2	220 AC
PUMA 260B	M	50.2	406	C	6	1	110/220 AC
WAM7-100	H	35+?	860	O	7	?	110/220 AC
ROBOSOFT	M	35+?	700	O	6	2	Battery
PERFORMER MK2	L	62	600	C	5	2	110/220 AC

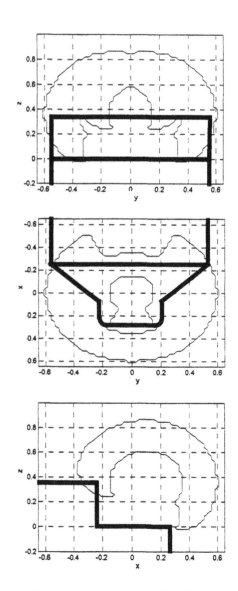

Fig. 3. Workspace sections over RAM-2

manipulator by batteries instead of the electrical network. The maximum power consumption of the robot arm is 1500W, less than the maximum supplied by the 48V battery circuit.

To better understand the space shape, the intersection with the planes $x=0$, $y=0$ and $z=0$ has been represented in Fig. 3. In the intersections, the proportional space occupied by the vehicle has been represented.

5. CONCLUSIONS

The commercial manipulator Performer MK-2 has been selected for being installed over the mobile vehicle RAM-2. The manipulator placed over the vehicle can be observed in Fig. 4.

The points taken into account have been presented as well as the restrictions that the solution assumes; for instance, it is impossible for the manipulator mounted at 45cm from the floor to collect small objects placed there. Nevertheless, other tasks like collecting books from a table or shelf, carrying them to another place, opening a door, picking up a bar reader for recognizing coded video-tapes, etc., can be carried out.

Future work includes the integration of the robot arm into the mobile robot control architecture and the experimental study of dynamic interactions.

Fig. 4. Performer MK2 on mobile robot RAM-2

6. ACKNOWLEDGEMENTS

This research has been partially supported by the C.I.C.Y.T project TAP-96-0763.

7. REFERENCES

Buemi F., M. Massa and G. Sandini (1995). AGRO-BOT: A Robotic System for Greenhouse Operations. *Proc. IARP '95*, pp 172-184.

Carriker W. F., P. K. Khosla and B. H. Krogh (1990). The use of Simulated Annealing to Solve the Mobile Manipulator Path Planning Problem. *Proc. IEEE Int. Conf. on Robotics and Automation*, pp. 204-209.

Connell J. (1991). Controlling a Mobile Robot Using Partial Representations. *SPIE Vol. 1613 Mobile Robots VI*, pp. 34-45.

Dario P., E. Guglielmelli, C. Laschi, C. Guadagnini, G. Pasquarelli and G. Morana (1995). MOVAID: A New European Joint Project in the Field of Rehabilitation Robotics. *ICAR '95*, pp. 51-59.

Ezkerra J. M., J. Mujika, J. Uribetxebarria and J. Basurko (1997). Vehículo dotado de brazo manipulador capaz de operar de forma autónoma y teleoperada. *Ponencia del 5° congreso de la AER*, pp. 311-326.

Fiorini P., K. Ali and H. Seraji (1997). Health Care Robotics: A Progress Report. *Proc. IEEE Int. Conf. on Robotics and Automation*, pp. 1271-1276.

Fogle R. F. and F. M. Heckendorn (1992). Teleoperated Equipment for Emergency Response Applications at the Savannah River Site. *Journal of Robotic Systems 9(2)*, pp. 169-185.

Harrell R. C., P. D. Adsit, R. D. Munilla and D. C. Slaughter (1990). Robotic Picking of Citrus. *Robotica, Vol. 8*, pp. 269-278.

Hootsmans N. A. M., S. Dubowsky and P. Z. Mo (1992). The Experimental Performance of a Mobile Manipulator Control Algorithm. *Proc. IEEE Int. Conf. on Robotics and Automation*, pp. 1948-1954.

Lim D., T. S. Lee and H. Seraji (1994). A Real-time Control System for a Mobile Dexterous 7 DOF Arm. *Proc. IEEE Int. Conf. on Robotics and Automation*, pp. 1188-1193.

Liu K. and F. L. Lewis (1990). Decentralized Continuous Robust Controller for Mobile Robots. *Proc. IEEE Int. Conf. on Robotics and Automation*, pp. 1822-1827.

Ollero A., A. Simón, F. García and V. E. Torres (1993). Integrated Mechanical Design of a New Mobile Robot. *Proc. IFAC Symposium International Conferences on Advanced Robotics*. Pergamon Press.

Pin F. G., J. C. Culioli and D. B. Reister (1994). Using Minimax Approaches to Plan Optimal Task Commutation Configurations for Combined Mobile Platform-Manipulator Systems. *IEEE Transactions on Robotics and Automation, Vol. 10, No. 1*, pp. 44-54.

Seraji H. (1995). Reachability Analysis for Base Placement in Mobile Manipulators. *Journal of Robotic Systems 12(1)*, pp. 29-43.

Weisbin C. R., B. L. Burks, J. R. Einstein, R. R. Feezell, W. W. Manges and D. H. Thompson (1990). HERMIES-III: A Step Toward Autonomous Mobility, Manipulation and Perception. *Robotica, Vol. 8*, pp. 7-12.

White J. R., H. W. Harvey and K. A. Farnstrom (1987). Testing of Mobile Surveillance Robot at a Nuclear Power Plant. *Proc. IEEE Int. Conf. on Robotics and Automation*, pp. 714-719.

Wiens J. G. (1989). Effects of Dynamic Coupling in Mobile Robotic Systems. *Proc. of SME Robotics Research World Conference*, pp 43-57.

Yamamoto Y. and X. Yun (1994). Modelling and Compensation of the Dynamic Interaction of a Mobile Manipulator. *Proc. IEEE Int. Conf. on Robotics and Automation*, pp. 2187-2191.

AN ELECTRO-OPTICAL SURVEILLANCE SYSTEM APPLICATION FOR SHIP COLLISION AVOIDANCE

Antonio Criado Garcia-Legaz, Alfonso Cardona Peral

Empresa Nacional Bazán de C.N.M.,S.A. Fábrica de Artillería
Carretera de la Carraca. San Fernando. Cádiz 11100 (Spain)
phone: +34-56-599613, fax: +34-56-599587, Email: acriado@faba.es

Abstract: This paper summarises the concept of an electro-optical navigation aid complementary to the radar, SERVIOLA naval surveillance system, based on the exploration of the visible and infrared spectral environment surrounding a naval platform by video cameras. The paper covers the system design methodology from operational requirements analysis, according the anticollision and surveillance missions, and the characteristics of the ship, through the required system performance and functions identification, down to the system components definition. As specific technical aspects, the paper discusses more in deep the proper selection of the infrared camera sensor for naval surveillance. *Copyright © 1998 IFAC*

Keywords: Marine systems, Obstacle detection, Infrared detectors, Image sensors, Cameras, Visual pattern recognition

1. INTRODUCTION

Safety at sea navigation under poor visibility conditions either in the night or under heavy weather is a main issue specially in critical situations such as navigation in high density traffic areas or confined waters. Under these conditions, navigation aids complementary to the radar based on EO (electro-optical) sensors have been demonstrated as a reliable mean of reducing the risk of collision preventing human injury, loss of life and material damages.

SERVIOLA NSS (Naval Surveillance System), see fig. 1, designed and manufactured by BAZAN FABA is an EO stabilised system based on the exploration by video cameras of the visible and IR (infrared) spectral environment surrounding a naval platform. The visible or IR video images are monitored to the helmsman in the ship bridge and, simultaneously, real time processed searching objects that enter the field of view covered by the exploring cameras and generating alarms automatically when an intrusion is detected (Criado, 1997). The NSS is also able to perform automatic video tracking of objects or references. The system can be completed, moreover, with a laser range finder, as an auxiliary device for target mapping.

Similar systems based on electro-optical sensors are being extensively used on board of military naval platforms. Typical applications cover IRST (Infrared Search and Track) systems for surveillance, target automatic detection and tracking, and EO pedestals comprising daylight, IR cameras, and laser range finder acting as main director in fire control systems. Main advantages of this kind of sensors compared with radar are discretion doubt to its passive operation and counter measures resistance. Latest advances in technologies like IR FPA (focal plane arrays), micromechanical silicon inertial sensors and high speed digital signal processors manufactured in serial production lines, have made these components become affordable for industrial and civilian applications and consequently the range of applications of EO systems is being extended to new fields.

Fig. 1 SERVIOLA EO sensor head

The use of EO sensors as a navigation aid provides several advantages compared with radar based instruments: continuous operation even during the night and fog, good performance at short ranges and, as being passive, lack of multi-path reflections when pointed in low elevation angles close to the horizon. Moreover, a system based on vision has the advantage of providing defined video pictures enabling target visual discrimination and identification. In general, EO systems are considered complementary to the radar and the highest degree of navigation safety is obtained by means of a combination of both systems.

2. SYSTEM DESIGN METHODOLOGY

System design process of a NSS for a particular application of ship collision avoidance comprises several steps that could be summarised on the scheme of fig. 2. The process starts from the top level *operational requirements* defined by the user for its specific vessel and operational scenario, that during the *analysis* process are converted into the system *performance requirements* and where basic *functions* are identified and allocated. These system *performance functions and data* are the inputs for system *synthesis* phase that produces the individual components *technical requirements* as the basis for the appropriate selection of components and system detailed design.

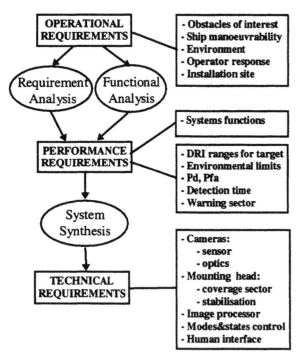

Fig. 2 NSS design process

2.1 OPERATIONAL REQUIREMENTS

The need of navigation safety requires to observe the navigation environment and detect any possible obstacle with enough time to avoid collision. This capability becomes critical in the case of fast vessels navigating in areas frequented by small ships, or in any type of vessels under night-light or low visibility conditions, specially near the coast, when approaching a harbour or an area of high density of traffic. In all mentioned cases, it is not possible to reach the required detection level only by means of visual observation or by radar. EO instruments to aid the crew to obtain a safe navigation in such situations should be defined according with user required effectiveness, comprising following *operational requirements*:

- *Obstacles of interest.* NSS should be able to detect those typical marine obstacles that appear undetected by navigation radar such as small ships, sailing boats, floating objects, cetaceans and wreckage rests. Apparent target dimensions range from 0.5 m. width by 0.5 m height for a floating barrel, to 10 m by 3 m. for a fishing boat. In a first approach, obstacles can be considered stationary relative to own ship, provided that is out of the scope of current version of the NSS to obtain historic trails of detected contacts.

- *Ship manoeuvrability.* Determines the minimum ranges at which obstacles of interest must be detected. Manoeuvrability limits for turn angles, speed, acceleration and ship dynamic performance are characterised for each new built ship at shipyard during sea acceptance trials respecting passenger safety. Results of these trial are recorded and permanently displayed to the helmsman on the bridge board. Allowed limits include maximum ship turn angles at different speeds, crash stop travelled range and duration, and maximum deceleration.
For fast ferries reaching speeds up to 40knots, 20 m/s, at open sea and assuming a 20 s. required time for evasive manoeuvres, leads to a minimum detection range of 200 m to avoid the obstacle at a safe distance.

- *Operational environment.* The conditions under which the ship must operate should be clearly defined provided that are strong constraints for the EO system performance doubt to the atmosphere attenuation effects in the energy received from the target at the sensor front end. These conditions include weather conditions, visibility, time of day, sea state and operational zone of the world.

- *Operator response.* The lag time since the operator reacts to an obstacle detection alarm, evaluating the danger and giving the evasive command to the ship steering systems, is translated in an additional travelled range that causes an increase in the minimum required detection range. Operator reaction time depends mainly on fatigue, motivation and training level. Assuming a 10s. delay, it results in 200 m. additional travelled range towards the obstacle. This figure can be reduced by integration ship steering system with automatic obstacle detection instruments.

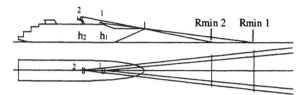

Fig. 3 Sensor installation site

- *Sensor installation site.* EO platform location in the ship superstructure is chosen considering minimum required range, as shown in fig. 3, blind sectors caused by own ship obstructions and structural stiffness. The place is usually on amidships line at an appropriate height so that sensors FOV covering sector keeps close to the ship bow, and also to reduce ranging errors of triangulation for low elevation angles. Practical limitations on installation height, doubt to lack of accessibility and ship structural vibrations, cause blind sectors in the surveyed sector like own ship beacons, antennas or masts, that can act as false targets.

2.2 SYSTEM FUCTIONS

To cover the above *operational requirements*, the EO system should perform the following basic *functions*:

- Automatic scanning of a fixed sector ahead of the ship or any other sector programmed by the operator.
- IR and visible images monitoring of ship course on a screen located at ship bridge and cameras control.
- Line of sight stabilisation to compensate own ship movements.
- Unmanned operation, under proper background conditions, providing automatic detection of obstacles with potential collision risk and generating alarms.
- Manual sensors pointing enabling the observation and automatic video tracking of any object of interest like references, obstacles or buoys.
- Visual recognition and identification capabilities of observed objects by the operator.

2.3 PERFORMANCE REQUIREMENTS

Operational requirements analysis defines the *system performance requirements*:

- *Target signature.* Signal energy received by the EO sensor is caused by target self emission and reflections from external sources. Main features of targets of interest are: apparent surface section, reflectivity and emissibity coefficients at IR and visible wavelengths, that lead to target to background thermal contrast and visual contrast. Target to background thermal contrast for sea floating objects (target air dried surface to sea or sky) is in the order of 2°C. Target detection performed by a human operator looking at a monitor that displays EO sensor output requires a **minimum apparent scene contrast** of 2%, which corresponds to a ratio of target to background brightness of 0.02. This figure is generally accepted and defines the minimum contrast limit for the international code of visibility ranges (Jursa, 1985).

- *Detection, recognition and identification, DRI, ranges.* Minimum ranges at which different obstacles types must be detected to avoid collision can be determined from both ship manoeuvring limits and operator response time. Needed recognition and identification ranges for different targets of interest should be defined from the ship surveillance missions.

For a particular EO system and target type, DRI ranges are a direct function of both sensor sensitivity and resolution performance, that are jointly expressed in terms of **MRTD (minimum resolvable temperature difference)** as a curve which gives the minimum ΔT, target to background thermal contrast, that the sensor is able to detect for vertical bars targets shapes at different spatial frequencies (cycles/mrads). Experimental studies (Johnson, 1958) have established the generally accepted **DRI criteria** for a human observer, based in the number of resolvable l-p (line pairs) over the target critical dimension that the EO system provides at the corresponding environmental conditions. This criteria states that one resolvable l-p is needed for detection, 4 l-p for recognition and 6 l-p for identification, all referred at 0.5 detection probability.

- *Probabilities of detection and false alarm.* Target signal detection in a noisy environment is a statistical problem that can be expressed as the P_d, **(probability of detection)**, probability to detect a target when it really exists, and the P_{fa} **(probability of false alarm)**, probability to declare a target when it does not exists. This last parameter is also expressed as **FAR (false alarm rate)** as the number of false alarms per time period.

NETD (noise equivalent temperature difference), is another sensor performance parameter that express the minimum ΔT that the sensor is able to detect, or the target ΔT that results in a *SNR (signal to noise ratio)* at the sensor output equal to the unit. From radar signal theory, theoretical models have been constructed for sinusoidal signals detection in random Gaussian noise (Skolnik, 1981), that allow calculating P_d for a given P_{fa} and SNR. SNR can roughly be estimated as the ratio of the target received ΔT to the EO sensor NETD performance parameter. This models assume that target detection is based in signal pre-filtering and later comparison with a threshold level whose value is either fixed, from a priori knowledge, or adaptive, being a function of measured scene noise at the detector.

- *Environmental operational limits.* All the above performance parameters are affected by environmental conditions under which the NSS must operate. Atmospherics attenuation effect is the most limiting factor in EO detection performance since energy radiated or reflected from the target is attenuated and blended with external radiation sources from the atmosphere along the optical path towards the sensor. The optical transmittance τ_a of the atmosphere at a certain wavelength λ over a uniform horizontal path of length R at sea level is expressed by the Beer- Lambert law as:

$$\tau_{a\lambda} = e^{-\sigma_\lambda R} \qquad (1)$$

where σ_λ is the *extinction coefficient* that includes the effects of molecular and aerosol absorption and scattering.

For marine environment, aerosols scattering by haze, consisting mainly of salt particles from sea water, is the dominant optical attenuation factor. Atmosphere extinction doubt to haze is also strongly dependent on air RH (relative humidity) and temperature. As RH increases, water vapour condenses on particles which grow in size resulting in an increased optical attenuation. *Maritime aerosol* models have been constructed that are valid for the 2-3 Km atmosphere boundary layer covering the ocean. For the inner layer, 10-20 m over the sea, a different model called *fresh sea-spray* is applied, which is strongly dependent on wind speed. When RH comes to 100% other phenomena appear such as clouds, fog, snow and rain whose effects are treated independently of haze aerosols since the particles size is greater. In this situations no exact theory exist to predict the extinction and experimental models are used.

All above atmospheric effects can be predicted through simulation tools based on experimental models providing accuracy better than 10% compared with results obtained from field radiometric measures.

- *Warning sector.* Safety area forward to ship course where potential obstacles must be early detected to avoid collision. This area must be covered by EO sensors field of regard, defined by sensor head bearing limits centred on midships line, and range limits defined by sensor head elevation limits. Warning sector might be covered with static wide FOV (field of view) cameras looking forward to the ship head, only when required DRI ranges are short enough to allow the use of short focal lengths optics.

- *Detection time.* Maximum elapsed time from a potential target entering the defined warning sector till the automatic generation of an alarm by the image processor. This is affected by the sensor head scanning angular rate, the sensors FOV and the time period consumed by the automatic detection process for analysing each video frame.

2.4 TECHNICAL REQUIREMENTS

The above performance requirements lead to the proper selection of system components:

- *Cameras:* Sensor type and optics MTF (modulation transfer function) are selected to meet the system required MRTD, NETD and FOV. IR sensor spectral response, medium or long wave IR band, will be further discussed in paragraph 4 of this paper.

- *Pedestal performance*: Field of regard of the sensors in azimuth and elevation are selected to cover the warning sector in bearing and range. Angular scanning rate should be compatible with image processor performance, processing rate, at the required P_d. LOS stabilisation maintains sensor elevation angle constant relative to the horizon, required to remove scene changes due to own ship movements and enables both human or automatic detection processes. Provided that sensor is normally forward looking towards the ship course sector, reliable operation can be achieved compensating only ship pitch in the mounting head elevation axis.

- *Image processing*: Basic functions assigned to the image processor are: graphics generation superimposed to the scene, image enhancement, target automatic detection and target tracking. Image enhancement increases target to background contrast and removes scene noise to improve human detection performance. Applying time and spatial average filtering techniques, target SNR is increased, producing a subsequent improvement of P_d. Automatic detection is only feasible with uniform backgrounds, like open sea, otherwise the FAR would not be acceptable.

3. SYSTEM DESCRIPTION

The block diagram of SERVIOLA NSS as shown in fig 4, is divided in four main subsystems: EO sensors, stabilised pedestal, image processor and control and human interface.

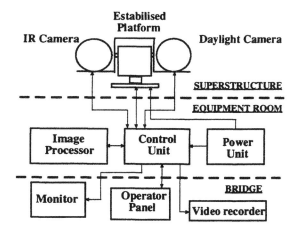

Fig 4. SERVIOLA NSS block diagram

Electro-optics sensors. The sensor suite comprises visible and infrared cameras both aligned in a common boresight and installed in a two axis mounting head. Depending on the visibility condition the operator selects one of the cameras to be processed and displayed at the ship bridge.

Stabilised pedestal. The stabilised pedestal is specifically designed and custom manufactured for this naval application . It points the sensors LOS towards the required direction of the inertial space, maintaining that direction regardless of the own ship movements by compensating the deck pitch and roll tilt angles by inertial rate sensors closing a digital servo loop.

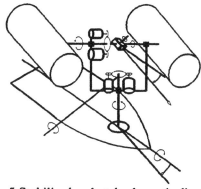

Fig. 5 Stabilised pedestal schematic diagram

The electromechanical part of the pedestal, as shown in fig 5, is a two axis mounting head, elevation over azimuth style, driven by DC brushless gear motors. Angular position sensors are optical absolute encoders 12 bits resolution, providing elevation and azimuth LOS angles readout referenced to the ship deck. Inertial angular motion sensors are two rate gyros strapped on the cameras mounting plate in order to sense angular rates related to the LOS.

Image processor. The image processing of the camera video outputs takes place in a dedicated video processor with a custom designed frame grabber whose main characteristics are:

- Real-time video analog-to-digital conversion and frame storage
- Dual video memory in ping-pong architecture
- Real time grey level histogram calculation
- Graphics and symbols overlaid to video.
- Processed video output synthesis for monitoring
- Video test pattern generation for built in test

Control and human interface. Comprises the following units: Control unit, Power unit, Operator panel, Video monitor and Video recorder. Main functions of the Control Unit are:

- Control of system states and modes.
- Digital servocontrol of the sensors platform.
- Interface with external systems such as radar.
- Operator panel commands decoding and signalling
- Cameras settings and video recorder control.
- Video source selection: visible or IR

Functions under operator control are:

- Exploration control : fixed ahead, sector automatic scan or manual (joystick).
- Limits of scan sector programming
- Image processing: automatic detection and tracking
- Video selection : visible or IR
- IR camera settings : FOV, focus, gain and pedestal
- Visible camera settings : zoom and focus
- Display of LOS angles, system modes and states.

4. MWIR VERSUS LWIR SENSORS IN MARINE APPLICATIONS

Infrared cameras selection is in the practice almost reduced to two options: 3 to 5 µm MWIR, (medium wave infrared) or 8 to 12 µm LWIR, (long wave infrared). Atmosphere spectral transmission along the IR region has two maximum transmission windows that correspond also to these bands. Many trade-off theoretical studies and field trial efforts have been devoted to determine which of them is the most convenient for different applications. Depending on the waveband there are available different sensors technologies. Typically, MWIR cameras are based on FPAs of PtSi, HgCdTe or InSb staring sensors with a great number of pixels, while LWIR cameras are based on linear arrays of HgCdTe sensors with mechanical scanning.

When evaluating target detection by EO IR systems, the key aspects that affect the final system performance are target and background temperatures, sensitivity of IR sensor and atmosphere attenuation including range. Leaving apart the effects of atmosphere and range, one of the EO sensor systems performance indicator used to predict sensor sensitivity for detection tasks, is NETD[-1], that can be expressed as (Lloyd, 1978):

$$NETD^{-1} = \frac{\alpha\beta\, A_o\, \tau_o\, D^*(\lambda_p)\, \dfrac{\Delta W}{\Delta T}}{\pi\left(ab\,\Delta f_R\right)^{1/2}} \qquad (2)$$

α,β = detector instantaneous FOV (IFOV) in rads.
A_o = entrance pupil area of the optics in cm^2.
τ_o = optical transmission at corresponding IR band.
$D^*(\lambda_p)$ = peak spectral detectivity in cm Hz$^{1/2}$ / watt.
$\Delta W/\Delta T$ = change in target emittance with temperature in watts/cm^2 °K.
a, b = individual detector dimensions in cm.
Δf_R = reference filter equivalent noise bandwidth in Hz, also expressed as $\Delta f_R = \pi/4\,\tau_d^{-1}$, where τ_d is the dwell or integration time of the detector.

For the purpose of comparing sensor sensitivities for both IR bands, it can be assumed equal optics transmission and detector size, then from (2) results:

$$\frac{NETD^{-1}{}_{LW}}{NETD^{-1}{}_{MW}} = \frac{\tau_{dLW}{}^{1/2}}{\tau_{dMW}{}^{1/2}}\; \frac{D^*(\lambda_p)_{LW}}{D^*(\lambda_p)_{MW}}\; \frac{\dfrac{\Delta W}{\Delta T}_{LW}}{\dfrac{\Delta W}{\Delta T}_{MW}} \qquad (3)$$

These three ratios, can be independently analysed for both sensors and then consider the joint product.

For linear parallel scanning sensors of N elements at a frame rate F_f, typical in LWIR, it results $\tau_d =$ N IFOV / FOV F_f, while for staring arrays, typical for MWIR, $\tau_d = 1 / F_f$. The ratio results:

$$\frac{\tau_{dLW}{}^{1/2}}{\tau_{dMW}{}^{1/2}} = \left(\frac{20\,\mu s}{10\,ms}\right)^{1/2} \cong \frac{1}{22} \qquad (4)$$

Typical sensor detectivities in MWIR band, correspond to 6 x 10^{10} (PtSi) and 4 x 10^{11} cm Hz$^{1/2}$ / watt (InSb). In MWIR band, HgCdTe detectivity corresponds to 6 x 10^{10} cm Hz$^{1/2}$ / watt. Hence, for the available sensor technologies MWIR InSb sensors have higher detectivities than LWIR HgCdTe in the order of 7 times, while PtSi MWIR sensor have the same.

In marine scenarios target to background thermal contrast ΔT varies from a few °K for obstacles like wind surfers, sailing boats or floating objects to tens of °K for ships engine exhaust gases. Background temperatures are obviously dependent on the weather state and climatic zone of the world. In conditions of targets and backgrounds at a relative medium temperature around 300 °K, the $\Delta W/\Delta T$ emitted energy per ΔT has a spectral distribution where the fraction corresponding to the LWIR band is about 30 times larger than the corresponding to the MWIR band. More precisely, emitted energy depends also of target emissibity, reflectivity and the presence of

external radiators, which might not be neglected in a complete analysis.

Independently of atmospherics effects and range, above figures jointly give a total better predicted sensitivity performance of MWIR InSb versus LWIR HgCdTe sensors of 5 times. And in the opposite, LWIR HgCdTe sensors are superior to MWIR PtSi in the order of 1.4 times. These figures must be considered only as an approach to the final performance that can only be known after field tests.

For a constant length of the optical path, atmosphere attenuation value strongly depends on the weather conditions, in particular RH and T_a. Marine environments are characterised by high RH, haze and fog or rain Experiments and simulation models results (Lloyd, 1975) show that in mid-latitudes regions at low humidity conditions, transmission in LWIR is superior than in the MWIR doubt to the negative absorption effect of CO_2 between 4.2 and 4.4 μm. The attenuation doubt to haze aerosols scattering, which are very common in long path marine scenarios, affect more negatively the MWIR windows than the LWIR. On the other hand, in high humidity conditions, fog or rain, LWIR transmission window is more adversely affected by H_2O molecules absorption than the MWIR window. In summary, MWIR sensors are more degraded by aerosols while LWIR sensors are more affected by molecular absorption.

Current situation is that most IR cameras in operation for long range naval surveillance in mid-north latitudes zones belong to the LWIR range. However, considering all the above factors, there is not a big difference of final performance between MWIR and LWIR cameras, even MWIR sensors will usually give a better performance and are the recommended option when operated in a marine scenario with warm temperatures and high humidity conditions like tropical zones. Experimental trials detecting targets at long ranges using different sensor types in both bands at different environment also confirm the above criteria.

REFERENCES

Skolnik, M. (1981). *Introduction to radar systems*. Mc Graw-Hill. International Editions.

Lloyd, J. M. (1975). *Thermal Imaging systems, Honeywell Inc.*, Plenum Press.

Criado, A. (1997). *Sistema de Vigilancia Naval*. Revista de Ingeniería Naval.n° 736. Madrid.

Johnson, J. (1958). *Analysis of image forming systems*. Image Intensifier Symposium. Fort Belvoir. Virginia

Jursa, A. (1985). *Handbook of geophysics and the space environment*. AFGL

NAVIGATIONAL DATA DISTRIBUTION AND POSITION FIXING AUTOMATION IN MOBILE MARINE APPLICATIONS

Manuel Ariza Toledo, Francisco Paños García

Empresa Nacional BAZAN de C.N.M., S.A. Fábrica de Artillería
Carretera de la Carraca S/N -P.O. Box 18-, 11100-San Fernando, Cádiz (Spain)
Phone : 34-65-599600, Fax : 34-56-599587, Email: pagnos@faba.es

Abstract : This paper provides a general overview of a real-time practical development of the Spanish industry, applicable to current and future platforms, for the optimisation and automation of the real time accurate position fixing of mobile platforms, and for the distribution of navigational data to remote end-users on board marine vehicles. The paper describes the major functional capabilities, configuration elements and interfaces of a Spanish Navigation System (*SP-NAVSYS*), as an integral part of the navigation system of a generic mobile marine vehicle (*mmv*). *Copyright © 1998 IFAC*

Keywords : Ethernet, Kalman filters, Least-squares algorithm, Multiprocessing systems, Multiprocessor systems, Probability density function, Quality control, Reliability, Statistics.

1. INTRODUCTION

SP-NAVSYS is a hard/soft system capable of acquiring, processing and distributing navigational data, operating in conjunction with the *mmv* movement, attitude, meteorological and positioning sensors, for the acquisition, conditioning and processing of data coming from the *mmv* integrated navigation sensors, and for the computation and distribution, to different types of *mmv* end-users, of the best position and movement vector estimates.

Major SP-NAVSYS performance characteristics are:

- *Mmv* position estimate in geographical and XY formats.

- Distribution of time to external systems, synchronised with GPS.

- Geodetic Systems transformation capability.

- UTM, Mercator, Transverse Mercator, Lambert and azimuthal transformation capability.

- Statistical analysis of positioning errors.

- Capable to be interfaced with a wide range of navigation sensors.

- Dual redundant Ethernet IEEE 802.3 interconnection capability.

- Capable to be integrated in a wide range of platform types.

SP-NAVSYS is a multiprocessor and multiprocess based architecture system, basically composed of three CPU cards, two memory cards, a time and frequency processor card with external GPS synchronisation input, two input/output cards to interface with the navigation sensors, and a control and monitoring panel to interact with the operator.

The basic functions of the system are provided by the position fixing module, which is the major software component of the system. Main tasks developed by this module include :

- Computation and distribution of the statistically best (*most likely*) position of the *mmv* and its associated error (*position fixing quality control*). The computation takes into account the data filtering of the interfacing sensors, the application of corrections (due to the movement of the platform, the antennae layback and the time deskew of the data), and finally the Kalman filtering of the computed position.

- Dead reckoning estimation based on the last data received from the navigation sensors and/or computed by the system.

- Computation of the true wind direction and speed, set and drift, and leeway.

- Geodetic Systems transformations and projections computations.

2. SYSTEM FUNCTIONAL CAPABILITIES

Next, the major functional capabilities of the system are described in detail. Mainly, the position fixing process, the fix quality control determination and the Kalman filtering.

2.1 Position Fixing

The key feature of the system is the position fixing of the *mmv*, using a weighted least squares method that combines (or equivalently integrates) the available positioning measurements (lines of positions or LOPs) provided by the platform sensors. The method yields the simultaneous maximum likelihood position for uncorrelated LOP errors.

The system computes the *most likely* position by weighted least squares combination of XY coordinates, LAT/LON coordinates, ranges, and bearings, provided by the different interfacing sensors; once converted to an internal datum, checked against errors, and corrected to a common reference point.

The algorithm combines the available measurements, applying weighting factors that are proportionally inverse to the measurement variances. It can be mathematically showed, in terms of the optimisation of the measurements residual errors, that the computed position is statistically the most likely position that can be achieved. It can be also showed that the residual error of the combined position is lower than any of the residual errors of the participating sensors.

The objective of LOP combination is to estimate a position, **r**, which best fits the observed LOP values taking into account the relative variances of the individual LOPs. The basic approach is to linearise the LOP equations about a trial position, r(i), to create a linear least squares problem and make the solution of this problem the next estimate, r(i+1), in an iterative scheme.

If $\hat{L}_n$ is the observed value for the n'th LOP and $L_n(\mathbf{r})$ the LOP value that would result if the true position were **r**, the least squares objective function is :

$$\sum_n \frac{1}{\sigma_n^2}\left(\hat{L}_n - L_n(r)\right)^2$$

where the actual functions L_n depend on the type of LOP and the precise geometry of the system. Linearising L_n about r(i), taking partial derivatives with respect to **r** and equating the resultant expression to zero gives a set of two simultaneous linear equations which can be expressed as :

$$A\left(r(i+1) - r(i)\right) = b \qquad (1)$$

where A is a 2 by 2 matrix with elements :

$$A_{11} = \sum_n \frac{1}{\sigma_n^2}\left(\frac{\partial L_n}{\partial x}\right) \qquad (2)$$

$$A_{12} = A_{21} = \sum_n \frac{1}{\sigma_n^2}\left(\frac{\partial L_n}{\partial x}\right)\left(\frac{\partial L_n}{\partial y}\right) \qquad (3)$$

$$A_{22} = \sum_n \frac{1}{\sigma_n^2}\left(\frac{\partial L_n}{\partial y}\right)^2 \qquad (4)$$

and b is a 2 by 1 vector with elements :

$$b_1 = \sum_n \frac{1}{\sigma_n^2}\left(\frac{\partial L_n}{\partial x}\right)\left(\hat{L}_n - L_n\right) \qquad (5)$$

$$b_2 = \sum_n \frac{1}{\sigma_n^2}\left(\frac{\partial L_n}{\partial y}\right)\left(\hat{L}_n - L_n\right) \qquad (6)$$

Rearranging the equation (1), the solution is given by :

$$r(i+1) = r(i) + A^{-1}b \qquad (7)$$

Applying equations (2) to (7), the *mmv* position fix is computed from the LOP values, the LOP variances and a trial position, using the following equations :

$$x = x_0 + \frac{\left(A_{22}b_1 - A_{12}b_2\right)}{\left(A_{22}A_{11} - A_{12}A_{21}\right)}$$

$$y = y_0 + \frac{\left(A_{11}b_2 - A_{21}b_1\right)}{\left(A_{22}A_{11} - A_{12}A_{21}\right)}$$

$$R_{xx} = \frac{A_{22}}{\left(A_{22}A_{11} - A_{12}A_{21}\right)}$$

$$R_{yy} = \frac{A_{11}}{\left(A_{22}A_{11} - A_{12}A_{21}\right)}$$

where (x_0, y_0) is the trial position, (x,y) is the fix position, R_{xx} is the variance of the fix x coordinate, and R_{yy} is the variance of the fix y coordinate.

The summations are over all valid LOPs where n refers to the LOP number. The functions L_n, $\left(\frac{\partial L_n}{\partial x}\right)$ and $\left(\frac{\partial L_n}{\partial y}\right)$ depend on the particular LOP type being used and are always evaluated at the trial position. The quantities $\hat{L}_n$ are the supplied LOP values and σ_n^2 the supplied LOP variances.

For instance, the LOP function and derivatives for circular (range) LOPs are :

$$L = r_0$$

$$\frac{\partial L}{\partial x} = \sin\theta_0$$

$$\frac{\partial L}{\partial y} = \cos\theta_0$$

where r_0 is the range and θ_0 is the bearing of the trial position from the tracked object.

A trial position is required in order to approximate LOP's by straight lines. The closer the trial position is to the actual position the better the approximation.

If the computed position fix is not sufficiently close to the trial position, then the procedure is repeated with the new estimate in place of the trial position. The process is repeated until a closeness criterion (based on the difference between the trial and the estimate) is met.

When a position fix is obtained, the residual sum of squares is subjected to a chi-squared test to guard against a bad fit.

The test statistic is :

$$c - b_1\left(x - x_0\right) - b_2\left(y - y_0\right)$$

where c is defined by :

$$c = \sum_n \frac{1}{\sigma_n^2}\left(\hat{L}_n - L_n\right)^2$$

2.2 Position Fixing Quality Control

Errors will remain in all observations of any type of measurement process, such as in the weighted least squares combination of LOPs. Thus, an assessment of the quality of the position fixing process is required in order to ensure the quality of the computed positions. Usually, navaid systems (like SP-NAVSYS) provide quality assessment expressed in terms of the precision and the reliability of the position fix *(fix)*.

SP-NAVSYS automatically computes in each processing cycle two position fixing quality control parameters :

1. The 95% *a posteriori* horizontal error ellipse of the fix, as the primary source to assess the precision of the fix. This parameter describes the quality of the fix with respect to random errors. The smaller these errors are, the higher the precision of the fix shall be.

2. The marginally detectable error *(MDE)*, to assess the reliability of the fix. This parameter describes the quality of the fix with respect to biases, errors that are not caused at random but systematically (systematic errors) or due to changes in prevailing physical circumstances. *MDE* is measured by stating the size of error that might remain undetected with a specified probability. The smaller the size of the undetected error is, the higher the reliability of the fix shall be.

Error Elipse

Assuming the random error of the computed fix is a bivariate normal random variable $f_{XY}(x,y)$, with statistically independent x and y coordinates, then the probability density function of the error is the joint probability density function :

$$f_{xy}(x,y) = \frac{1}{2\pi\sigma_x\sigma_y\sqrt{1-\rho^2}} e^{-\frac{1}{2(1-\rho^2)}\left[\frac{(x-x_c)^2}{\sigma_x^2} - \frac{2\rho(x-x_c)(y-y_c)}{\sigma_x\sigma_y} + \frac{(y-y_c)^2}{\sigma_y^2}\right]} \quad (8)$$

where x_e is the mean of the distribution of the random error in the X axis, y_e is the mean of the distribution of the random error in the Y axis, σ_x is the standard deviation of the mean in the X axis, σ_y is the standard deviation of the mean in the Y axis, and ρ is the correlation coefficient between the x and y components of the random error.

The corresponding marginal distributions derived from the joint distribution have the density:

$$f_x(x) = \frac{1}{\sqrt{2\pi}\sigma_x} e^{-\frac{1}{2}\left[\frac{(x-x_e)^2}{\sigma_x^2}\right]}$$

$$f_y(y) = \frac{1}{\sqrt{2\pi}\sigma_y} e^{-\frac{1}{2}\left[\frac{(y-y_e)^2}{\sigma_y^2}\right]}$$

Equating *(8)* to a constant probability C gives the equiprobability surface

$$f_{xy}(x,y) = C$$

the contour of which is an equal-height curve that projects in the XY plane into the ellipse :

$$\frac{(x-x_e)^2}{\sigma_x^2} - \frac{2(x-x_e)(y-y_e)\rho}{\sigma_x\sigma_y} + \frac{(y-y_e)^2}{\sigma_y^2} = K^2 \quad (9)$$

Points on the ellipse may be thought of as equally likely combinations of the x and y error components. If ρ is zero, the x and y components are uncorrelated, and the ellipse have their semimajor and semiminor axes parallel to the X and Y axes. If σ_x is equal to σ_y (and $\rho=0$), the ellipse degenerates to a circle. In the other extreme, as ρ approaches unity, the ellipse becomes more and more eccentric.

The size of the error ellipse, and therefore the corresponding probability, is a function of K in the equation *(9)*. When this constant takes the value 2.447, the corresponding ellipse is the 95% error ellipse.

MDE

MDE is a term in close relationship with the detection of biases (or outliers). Due to the random nature of the observations, biases can never be detected with certainty but only with a certain probability. The smaller the bias, the smaller the probability of finding it.

Consequently, the detection of outliers is a topic related with the statistical testing and to understand what the MDE is, first an understanding of the basis of statistical testing is

required. It is not within the scope of this paper to go into details in the field of the mathematical testing of hypotheses, and only a brief review of the major concepts will be showed.

A statistical hypothesis is an assumption about the distribution of a random variable. In the present case, the distribution of the error of the LOPs used to compute the fix is assumed to be known by the system *a priori* for every LOP, and outliers are detected against each particular distribution. Although according to statistical theory very high or very small values may occur with a very small probability, in practice such values are flagged as outliers. The probability beyond which observations are marked as outliers, is called in statistics the *significance level of the test* and has to be chosen sufficiently small to leave a small chance of rejecting good data *(Type I error)*. It is recommended to standardise on a value between 0.1% and 5% as the significance level of the test.

On the other hand, a statistical test of a hypothesis is a procedure used in general to find out whether the hypothesis may be accepted or rejected, or more specifically to estimate some parameter about the nature of the biases of a random variable, such as the size or the probability of detection of the bias. In the present case, based on the result of a specific statistical test, the system computes the size of the bias, associated with the fix, that can be found with a preselected probability. This probability is called in statistics the *power of the test*, and has to be chosen sufficiently large to leave a small chance of not discovering biases *(Type II error)*. It is recommended to standardise on 80% as the power of the test.

According to this, assuming that the significance level of the test *(α)* and the power of the test *($1-\beta$)* are fixed, the number of measurements used in the position fixing process *(n)* is known and the standard deviation of the fix *(σ)* is also known, the system computes the *MDE* as a function of α, β, n and σ as follows :

$$MDE = \frac{\sigma}{\sqrt{n}}\left(\Phi\left(\frac{1-\beta}{2}\right) - \Phi\left(\frac{\alpha}{2}\right)\right)$$

where Φ is the distribution function of the normal distribution with mean 0 and variance 1.

2.3 Kalman Filtering

Each navaid interfaced with SP-NAVSYS is smoothed by its own Kalman filter. The *most likely* position computed by combination of the available LOPs is also smoothed by a Kalman Filter.

These Kalman filters use position inputs and are uncoupled in the x and y coordinates.

The Kalman filters are implemented using the standard set of recursive equations that can be found in the specialized literature.

It is assumed that the target dynamic process can be modelled in the discrete Markov form :

$$x_{k+1} = \Phi_k x_k + w_k$$

where x_k is the n-dimensional target state vector that includes the quantities to be estimated, Φ_k is the assumed known transition matrix, relating x_k to x_{k+1} in the absence of a forcing function, and W_k is the zero-mean, white, Gaussian process noise with assumed known covariance Q_k.

Additionally, the measurements are assumed to occur at discrete points in time in the form of linear combinations of the system state variables, corrupted by uncorrelated noise, according with :

$$z_k = H_k x_k + v_k$$

where z_k is the vector measurement at time t_k, H_k is a matrix giving the ideal (noisless) connection between the measurement and the state vector at time t_k, and v_k is the zero-mean, white, Gaussian measurement noise with known covariance R_k.

Initial estimates of the target state vector and of the error covariance matrix are also assumed.

Taking into account all these assumptions, the Kalman filter is run performing the following standard iterative algorithm :

1- Kalman filter gain (*Kalman gain*) calculation, according with the following matrix equation :

$$K_k = P_k^- H_k^T \left(H_k P_k^- H_k^T + R_k \right)^{-1}$$

2- Target state estimate update, based on the last measurement z_k, according with the following matrix equation :

$$\hat{x}_k = \hat{x}_k^- + K_k \left(z_k - H_k \hat{x}_k^- \right)$$

3- Kalman filter covariance matrix update, according with the following matrix equation :

$$P_k = \left(I - K_k H_k \right) P_k^-$$

4- Target state estimate prediction for the next cycle *(k+1)*, according with the following matrix equation :

$$\hat{x}_{k+1}^- = \Phi_k \hat{x}_k$$

5- Kalman filter covariance matrix prediction for the next cycle *(k+1)*, according with the following matrix equation :

$$P_{k+1}^- = \Phi_k P_k \Phi_k^T + Q_k$$

2.4 Other Major System Functions

Next, the remaining functional capabilities of the system are briefly summarised.

System Commanding Interface

SP-NAVSYS is capable to interface with other systems throughout a standard Ethernet IEEE 802.3 commanding interface, in order to carry out the distribution of navigational data to external users located throughout the platform and to receive commanding orders.

Electronic Chart Display Console Interface

The system is capable to transfer standard NMEA messages containing navigational data (eg.. position, speed, heading, wind data, depth. etc...) to an Electronic Chart Display Console.

Peripherals interfacing

The system is capable to be interconnected, through standard RS-422 interfaces, with a specific set of peripherals (such as printer, plotter, remote display, etc...), for the transfer of navigational information to be remotely displayed or registered.

Geodetic Transformations and Projections

Calculations related with position fixing are referred to the selected spheroid and datum. However, the system is capable to transform geographic coordinates between different spheroids and datums.

Once the position is calculated, the system is able to project the computed value in a XY coordinate system selected by the operator : Transverse Mercator (TM), Universal Transverse Mercator (UTM), Mercator, Lambert Conical and Azimuthal.

3. SYSTEM ARCHITECTURE

Figure 1 is a block diagram showing the major SP-NAVSYS hardware components : the *System Processor*, the *Navigation Sensors Interface*, the *MMI*, the *Data Bus Interface*, the *Internal Communications Bus*, and the *System Enclosure*.

The *System Processor* comprises a set of independent general purpose CPU boards for system control, external interfaces control, and position fixing and navigation calculations.

The *Navigation Sensors Interface* comprises a set of independent serial communications boards of microcontroller based design with standard RS232C/RS422A I/O ports.

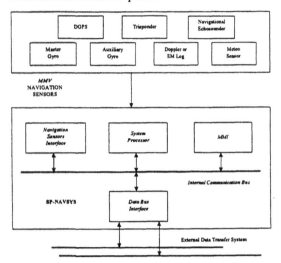

Fig. 1. SP-NAVSYS System Block Diagram showing the major hardware components.

SP-NAVSYS has a high degree of growth potential, being capable to include additional interfacing boards; both serial and/or of different types (Synchro/digital, analog/digital, etc...).

The *Data Bus Interface* consists of an interfacing PCB for the connection of SP-NAVSYS to a redundant local area network with bus topology and CSMA/CD network access protocol, according to IEEE 802.3.

The PCB has two intelligent network access modules, which are connected to the network transmission medium (network cables) through the corresponding transceivers.

The *MMI* consists of a control and monitoring panel based on a software programmable keyboard and an electroluminiscent display.

SP-NAVSYS PCBs and power supply units are internally connected through the *Internal Communications Bus*, which is in accordance with the standard "VME Bus Specification".

Finally, the *System Enclosure* comprises : the mechanical enclosure, a standard 19" and 6U VME rack to support the set of PCBs, a standard extended VME backplane, an air ventilation unit, and the power supply unit.

4. SYSTEM EXTERNAL INTERFACES

SP-NAVSYS is capable of interfacing with the following external elements:

1- External commanding system through the Data Bus Interface specified in section 3 of this paper. By means of this interface the system is capable to stablish a bi-directional data link with the external commanding system to obtain and transmit, among other types of data, the following information:

- Mobile platform heading, pitch and roll and their rates of change.

- Mobile platform relative and absolute speed.

- Mobile platform geographical and XY position and the corresponding accuracy and time of fix.

- Date and GMT.

- Drift and set.

- Depth.

- Atmospheric pressure and air temperature.

- True and relative wind direction and speed.

- Status indications and time validity of the data.

2- Navigation sensors, through the Navigation Sensors Interface specified in section 3 of this paper. The system is capable to interface, through standard RS232C and/or RS422A ports, with the following navigation sensors :

- Mobile Platform Positioning Sensors: DGPS and Trisponder (or equivalent position fixing system).

- Own Ship Movement/Attitude Sensors: Inertial Navigator System (INS) and Doppler or Electromagnetic (EM) Log.

- Other Sensors: Navigational Echosounder and Meteorological Sensors.

The data received from these sensors (movement data, attitude data, position measurements, time, etc...) are time stamped and stored in the system data base for future processing as required.

SP-NAVSYS supports standard NMEA 0183 and specific (particular) sensors interface protocols.

3- Electronic Chart Display Console, through one of the standard System Processor serial interfaces (RS 232C I/O port). Through this interface, SP-NAVSYS is capable to distribute navigational data (position, speed, heading etc...) to the Electronic Chart Display Console.

5. CONCLUSION

This paper has introduced the Spanish Navigation System (SP-NAVSYS). A system developed and built by the Spanish industry to provide real time accurate position fixing and navigational data distribution on board current and future marine platforms.

SP-NAVSYS is a multiprocessor and multiprocess based architecture system. The position fixing module, which is the major software component of the system, mainly provides for computation and distribution of the statistically best (*most likely*) position of the platform and its associated error (*position fixing quality control*). The computation takes into account the data filtering of the interfacing sensors, the application of corrections, and the Kalman filtering of the computed position. Other involved computations are geodetic systems transformations and projections computations.

The key feature of the system is the position fixing of the platform, using a weighted least squares method that combines the available lines of positions provided by the platform sensors, onced converted to an internal datum, checked against errors, and corrected to a common reference point..

Another processes involved in the computation of the most likely position are the statistical analysis of the LOPs combination (W-Test and unit variance test) and the osition fixing quality control parameters determination (95% error ellipse and marginally detectable error).

6. REFERENCES

Barry, B. A. (1991): *Errors in Practical Measurement in Surveying, Engineering, and Technology*. M. D. Morris, P. E.

Basker, G (1991): "GPS Observables & Algorithms". In *4th International Seminar on GPS*. Nottingham University, April 1991.

Bomford, G. (1980): *Geodesy*. Clarenden Press London. (4th Edition).

Cross, P. A. (1990): *Advanced Least Squares applied to position fixing*. Working Paper No. 6, Dept. Of Land Surveying, University of East London.

Cross, P. A., Hawksbee, D. J. and Nicolai, R. (1994): "Quality Measures for Differential GPS Positioning", *Hydrographic Journal-April 1994*.

Grover, R. B. & Hwang, P.Y.C (1985): *Introduction to random signals and applied Kalman filtering*. John Wiley & Sons Inc. Second edition.

Hofmann-Wellenhof, B., Lichtenegger, H. and Collins, J. (1994): *GPS Theory and Practice*. Springer-*Verlag* Wien New York. Third, revised edition.

Kreyszig, E. (1970): *Introductory Mathematical Statistics. Principles and Methods*. John Wiley & Sons, Inc.

Moore, T. (1991): *Coordinate Systems and Datums*. Nottingham University.

Taylor, J. R. (1982): *An Introduction to Error Analysis. The Study of Uncertainties in Physical Measurements*. University Science Books. Oxford University Press.

THE VARIABLE STRUCTURE SYSTEM SYNTHESIS FOR AUTONOMOUS UNDERWATER ROBOT

Filaretov, V.F., Lebedev, A.V.

Far-Eastern State Technical University, Vladivostok, Russia.

Abstract: The adaptive control system of autonomous underwater robot (AUR) spatial motion has been developed. The method of decomposition has been applied in order to simplify the synthesis procedure. The sliding mode properties have been used for compensation of AUR model parametrical uncertainties and viscous environment influence. The adaptive algorithm allowing to improve the system processes has been applied. The synthesized control system was simulated under the various AUR's work regimes. *Copyright © 1998 IFAC*

Keywords: autonomous robots, variable structure systems, sliding mode, adaptive algorithms, position control, velocity control, robot dynamics.

1. INTRODUCTION

Resently a great number of autonomous underwater robots are used to perform scientific investigations and different technological operations. It is often required to provide the accurate moving of these robots along complex spatial trajectories and/or their fast and precise approach to work object.

The conditions AUR functionate are differ from ordinary robots ones, because of viscous environment and underwater streams, in particular. Therefore, both well studied forces and moments dued by degrees of freedom interaction and hydrostatic and hydrodynamic forces (moments), which are often unknown or too difficult for identification, act to AUR. Hence, it is necessary to design an adaptive AUR control systems, which are invariant to changing of the control object parameters and viscous environment influence.

The synthesis problem of such system is discussed in the paper. Here it is shown how the usage of sliding mode properties rejects the influence of variable and often unknown control object parameters and viscous environment on whole robot functionate quality.

The algorithm of sliding surface parameters adjustment allowing to supply the high speed of

motion and to use the thrusters power more effectively is considered.

An efficiency of developed control system has been confirmed by the results of mathimatical simulation.

2. THE AUR DYNAMICS DESCRIPTION

It is known (Fjellstad, Fossen and Egeland, 1992), that the AUR dynamics in general case can be described by system of two-order nonlinear differential equations:

$$M\ddot{q} + D\dot{q} + C\dot{q} + g = t, \qquad (1)$$
$$\dot{x} = J(x)\dot{q}, \qquad (2)$$

where x is a vector of AUR position and orientation in the earth-fixed reference frame, $\dot{q}$ and $\ddot{q}$ are vectors of AUR linear and angular velocities and accelerations in the AUR-fixed reference frame, M is an inertia matrix, C is a matrix containing Coriolis and centripetal terms, D is a matrix containing viscous damping terms, g is a vector of hydrostatic forces and moments, t is a vector of thrusters forces and torques, J is a transformation matrix depending on the choice of coordinates. M and C matrices also contain the added masses of viscous environment.

The thrusters dynamics is considered to be described by the first order differential equations (for each degree of freedom):

$$T_{di} \, \dot{t}_i + t_i = K_{di} \, u_i, \quad i = 1, 6, \qquad (3)$$

where K_{di} and T_{di} are gain coefficient and time constant of i-th thruster, u_i is an input control signal of i-th thruster.

The method of decomposition is used to simplify the AUR control system synthesis. Under this, the complete system being described by the equations (1)-(3) can be represented as a set of separate subsystems corresponding to different AUR degrees of freedom. The cross-connections being in whole system will be observed as external disturbances for each subsystem. In future it s convenient to analyse the position control loop and velocity control loop in the received subsystems separately. The general structural scheme of i-th AUR's subsystem is given in figure 1.

Taking into account (1)-(3), dynamic equation of i-th subsystem velocity control loop takes the form:

$$m_{ii} \, \dot{v}_i + d_{ii} \, v_i \, |v_i| + f_i = t_i, \quad i = 1, 6, \qquad (4)$$

where $v_i = \dot{q}_i$, f_i is an external disturbance acting on i-th subsystem from other subsystems and including the elements of equation (1) which are not depending from q_i; m_{ii} and d_{ii} are the diagonal terms of M and D matrices.

This terms satisfy with follow inequalities:

$$m_{ii \, min} < m_{ii} < m_{ii \, max}$$
$$d_{ii \, min} < d_{ii} < d_{ii \, max} \quad i = 1, 6, \qquad (5)$$

where $m_{ii \, min}$, $m_{ii \, max}$, $d_{ii \, min}$, $d_{ii \, max}$ are known boundary values of AUR parameters.

So, the expressions (3) and (4) allow to describe the velocity control loop of each subsystem by nonlinear second-order differential equations with uncertain parameters and uncontrolled external disturbances.

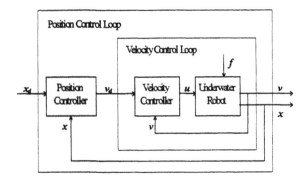

Fig. 1. Structural scheme of i-th AUR subsystem

The i-th subsystem synthesis can be devided into two stages. At the beginning the control algorithm for internal loop (velocity control loop) is developed in order to stabilize the subsystem parameters and to eliminate the disturbance f_i influence. Then the position control law is synthesized providing the required dynamic properties and subsystem work quality criteria.

3. VARIABLE STRUCTURE SYSTEM SYNTHESIS FOR AUR POSITION CONTROL

The solution of AUR velocity control loop synthesis task will be searched in the class of variable structure systems (Dyda and Filaretov, 1989; Yoerger and Slotine, 1991). As well known, such system are robust in conditions of parameters variability and uncertainty. The velocity error is equal to:

$$e_i = v_{di} - v_i, \quad i = 1, 6, \qquad (6)$$

where v_{di} is an internal control loop input signal. The law of thrusters control for nonlinear system (3), (4) are proposed to form as follows:

$$u_i = |e_i| \, K_{u1i} \, \text{sign}(s_i) + e_i^2 \, K_{u2i} \, \text{sign}(s_i), \quad i = 1, 6, \qquad (7)$$

where K_{u1i}, K_{u2i} = const > 0. Values of s_i are determined by expressions:

$$s_i = \dot{e}_i + C_i \, e_i, \quad i = 1, 6, \qquad (8)$$

where C_i = const > 0. The sliding mode existence condition for each subsystem has the form:

$$s_i \, \dot{s}_i < 0, \quad i = 1, 6. \qquad (9)$$

The inequalities for the choice of coefficients K_{u1i}, K_{u2i} and C_i were received by using the condition (9) and relations (3), (4), (7) and (8):

$$K_{u1i} > \max \, (|1 - T_{di} \, C_i| \, C_i \, m_{ii}/K_{di}), \quad i = 1, 6.$$
$$K_{u2i} > \max \, (|2T_{di} \, C_i - 1| \, d_{ii}/ \, K_{di}), \quad i = 1, 6. \qquad (10)$$

This inequalities must be satisfied in order to stabilize sliding mode in velocity control loop when parameters of control object take any values in given range (5). In this case the error e_i of each subsystem will be changing in accordance with the solution of first-order differential equation:

$$\dot{e}_i + C_i \, e_i, = 0, \quad i = 1, 6, \qquad (11)$$

which is received from expression (8) if $s_i = 0$.

This means that the velocity control loop behaviours always depends on values of parameters C_i only and does not depend on object parameters, control channels interaction effects and viscous environment influence.

The signal v_{di} changing law which generates in the position control loop is taken as follows:

$$v_{di} = J^{-1}(x)\,(\dot{x}_{di} + K_{pi}\,(x_{di} - x_i)), \quad i = 1, 6, \quad (12)$$

where x_i is an i-th component of AUR position and orientation vector; x_{di} is a desirable law of x_i variation; $J^{-1}(x)$ is an inverse matrix for $J(x)$ transformation matrix; $K_{pi} > 0$ is a certain constant coefficient determined by designing.

Using the expressions (2) and (12) and taking into account that the $v_i = v_{di}(t)$ mode is a stable mode of velocity control loop, a position error equation is received:

$$\dot{e}_{xi} = -K_{pi}\,e_{xi}, \quad i = 1, 6, \quad (13)$$

Obviously, in this case $e_{xi} \rightarrow 0$, $x_i \rightarrow x_{di}(t)$ and the AUR motion is carried out in accordance with a given AUR's coordinates changing law.

Now it is possible to give required dynamic properties (including monotonous character of the transition process) to the system described by equations (1), (2), (3), (7) and (12) by the choosing of parameters C_i and K_{pi}.

Two approaches can be used in order to implement the control law (7). Using the first approach the constant value of C_i is determined for the "worst" possible values of AUR's parameters m_{ii} and d_{ii} in accordance with inequality (10).

In this case the coefficient C_i will have a minimal value and transition process will be slow, because the solution of equation (11) has a form:

$$e_i = e_i(0)\,e^{-C_i t}, \quad i = 1, 6. \quad (14)$$

However, it is obvious that the control object parameters can be differ from "worst" values during the AUR operation. Therefore, it is possible to accelerate the system transition processes by parameter C_i increasing till certain $C_{i\,max}$ value, of cause, without violation of relation (10), so that the sliding mode was kept for current object parameters values and external perturbances.

4. ADAPTIVE ALGORITHM SYNTHESIS

In practice the control law (7) takes the form:

$$u_i = (|e_i|\,K_{u1i} + e_i^2\,K_{u2i})\,f(s_i), \quad i = 1, 6, \quad (15)$$

where $f(s_i)$ is a special function (the view of this function and its time structure are given in figure 2).

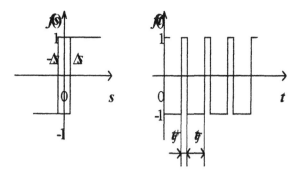

Fig. 2. Function $f(s)$ and $f(t)$

In this case the sliding mode parameter m_i used to investigate the variable structure system has a view:

$$m_i = t_f^+/(t_f^- + t_f^+), \quad i = 1, 6, \quad (16)$$

where t_f^+ is a time interval, when $f(t) > 0$; t_f^- is a time interval, when $f(t) < 0$.

As shown in (Dyda and Filaretov, 1993), $m_i \rightarrow 1$, when the coefficient C_i increase. Therefore, parameter m_i is the undirect indicator of vicinity of C_i and C_{imax}. Sliding surface parameter adjustment can be based on the property mentioned.

Different adaptive algorithms of C_i adjustment are developed. In particular, the following law was used:

$$\dot{D}_i = K_{ci}\,(m_0 - m_i),$$
$$C_i = C_{i\,min} + D_i, \quad i = 1, 6, \quad (17)$$

where $C_{i\,min}$ is a constant calculated for "worst" control object parameters by the inequality (10); D_i is an additional adjustment signal; $K_{ci} = const > 0$; $m_0 = 1 - d$; $d > 0$ is a small constant (the sliding mode stability is provided by value d choice).

The structural scheme of adaptive controller realising the control algorithm (8), (15), (17) is showm in the figure 3. Here the value of m_i is determined by special logical block.

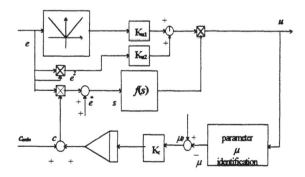

Fig. 3. The structural scheme of adaptive controller

Here it is necessary to provide fast variation of D_i value in comparison with AUR parameters variation.

It is important that the parameter C_i value is determined and changed without control object parameters identification in accordance with this adaptive control algorithm.

5. MATHEMATICAL SIMULATION

The mathematical simulation of the developed control algorithm was carried out under the following values of AUR and controllers parameters: m_{iimin} = 330 kg, m_{iimax} = 600 kg, $d_{ii\,min}$ = 500 N sec^2 m^{-2}, $d_{ii\,max}$ = 1000 N sec^2 m^{-2}, K_{di} = 50 N V^{-1}, T_{di} = 0.3 sec, K_{u1i} = K_{u2i} = 50 V sec m^{-1}, K_{pi} = 25 sec^{-1}, $C_{i\,min}$ = 2. The longitudinally-vertical motion and AUR rotation was considered. The results of simulation is shown in the figure 4, 5, 6, 7, 8.

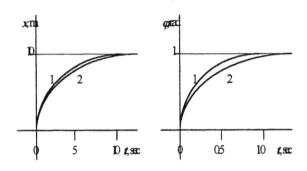

a) The longitudinal motion (x_d = const)
b) The rotation (j_d = const)

Fig. 4. The control processes in the "conventional" system (C = const)
curve 1 - transition process, when m_i=$m_{i\,min}$, d_i = d_{imin};
curve 2 - transition process, when m_i=$m_{i\,max}$, d_i = d_{imax}.

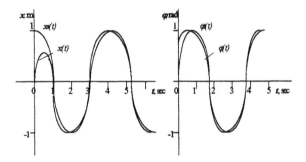

a) The longitudinal motion (x_d = cos(wt))
b) The rotation (j_d = cos(wt+y))

Fig. 5. The control processes in the "conventional" system (C = const)
curve 1 - transition process, when m_i=$m_{i\,min}$, d_i = d_{imin};
curve 2 - transition process, when m_i=$m_{i\,max}$, d_i = d_{imax}.

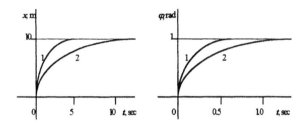

a) The longitudinal motion (x_d = const)
b) The rotation (j_d = const)

Fig. 6. The control processes in the adaptive system (C = var)
curve 1 - transition process, when m_i=$m_{i\,min}$, d_i = d_{imin};
curve 2 - transition process, when m_i=$m_{i\,max}$, d_i = d_{imax}.

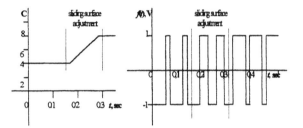

Fig. 7. Parameter C variation

Fig. 8. The signal f(t)time structure

6. CONCLUSIONS

As a results of reserch has shown, the variable structure system usage makes possible to provide high quality AUR motion to the given point of space independently on current control object parameters values and viscous environment influence. The adaptive adjustment of switch surface parameters allows to accelerate considerably (more than 2 times) the system fast-action.

REFERENCES

Dyda, A.A. and Filaretov, V.F. (1993) "Algorithm of Time-Sub-Optimal Control for Robot Manipulator Drives", *Proc. of the 12th World IFAC Congress*, Sydney, Australia.

Dyda, A.A. and.Filaretov, V.F. (1989) "Self-Adjusting System with Variable Structure for Manipulator Actuators Control", *Electromechanics (Russia)*, **N2**, pp.102-106.

Fjellstad, O.-E., Fossen, T.I. and Egeland, O. (1992) "Adaptive Control of ROVs with Actuator Dynamics and Saturation", *Proc. of the Second International Offshore and Polar Engineering Conference*, San Francisco, USA..

Yoerger, D.R. and Slotine, J.-J.E. (1991) "Adaptive Sliding Control of an Experimental Underwater Vechicle", *Proc. of the IEEE Conference on Robotics and Automation*, Sacramento, USA.

Course-keeping and course-changing H_∞ controller for a Ferry

M.J. López*, J. Terrón*, A. Consegliere*, C. Mascareñas**

* Dpto. Ingeniería de Sistemas y Automática, ** Dpto. Ciencias y Técnicas de la Navegación.
Facultad de Ciencias Náuticas. Universidad de Cádiz. 11510 Puerto Real, Cádiz, Spain.
E-mail: manueljesus.lopez@uca.es

Abstract: In this work we present a study carried out for a ship of Ferry type. An H_∞ regulator is developed to keep a fixed course of the ship, and another H_∞ controller is designed for course changing operation. Both designs have into account the physical limitations of the rudder servosystem, and uncertainties in ship dynamics. The objective of the control system is to achieve robust performance. *Copyright © 1998 IFAC*

Keywords: Robust performance, H_∞ regulator, uncertainty level, ship control.

1 Introduction

In most ships a simple PID regulator is used by the autopilot for course keeping, and if regulator parameters are tuned in a suitable way, the performance may be good. Nevertheless, ship dynamics depends strongly on factors like cruise speed and load conditions. Needless to say that environmental characteristics are also important for autopilot tuning.

If a model-based control technique is employed, it is necessary to obtain an adequate mathematical model of the ship, in order to be useful for design, and to take into account the essential characteristics of the ship for the operation conditions. In this work, a mathematical model of the Ferry Ciudad de Zaragoza (C/Z) has been obtained based on experimental data (and navigation simulator data) for different navigation conditions. Since ship dynamics depends on many factors, we have analyzed the robustness of the designed control system with respect to uncertainty in the nominal model of the ship.

In ship operation there are two different conditions: 1) course keeping, and 2) course changing; and in general, regulator must be adjusted for each condition in a different way. In this work two H_∞ controllers are designed for C/Z, so that robust control philosophy is employed in the setting out of the problem, and standard algorithms are used for solving H_∞ optimization problem. Both regulators achieve robust performance, for the uncertainty level which is taken into account in the mathematical model of the ship.

This work is structured as follows: in paragraph two the mathematical model of the ship and en-

vironmental disturbancies used in this work are presented, in paragraph three the control problem and objectives are set, in the fourth paragrph simulations results are described, and finally conclusions are summarized.

2 Ship mathematical model

It is known that ship dynamics can be described in a precise way by complex mathematical models (Lewis, 1989; Lloyd, 1989), which take into account the high nonlinear nature of ship behaviour. The following equations describe this nature for yawing $(r = \dot{\psi})$: $-N_{\dot{v}}\dot{v} + (I_z - N_{\dot{r}})\dot{r} = f(u,v,r,\delta)$, where: $f = f_1 + f_2 + f_3 + f_4 + f_5 + f_6$.

$$
\begin{aligned}
f_1 &= N^0 + N_u^0 du + N_{uu}^0 du^2 + N_v v + \frac{1}{6} N_{vvv} v^3 \\
f_2 &= \frac{1}{2} N_{vrr} vr^2 + \frac{1}{2} N_{v\delta\delta} v\delta^2 + N_{vu} vdu \\
f_3 &= N_r r + \frac{1}{6} N_{rrr} r^3 + \frac{1}{2} N_{rvv} rv^2 + \frac{1}{2} N_{r\delta\delta} r\delta^2 \\
f_4 &= \frac{1}{2} N_{ruu} rdu^2 + N_\delta \delta + \frac{1}{6} N_{\delta\delta\delta} \delta^3 \\
f_5 &= \frac{1}{2} N_{\delta rr} \delta r^2 + N_{\delta u} \delta du \frac{1}{2} N_{\delta uu} \delta du^2 + N_{vr\delta} vr\delta \\
f_6 &= \frac{1}{2} N_{vuu} vdu^2 + N_{ru} rdu + \frac{1}{2} N_{\delta vv} \delta v^2
\end{aligned}
$$

and similar equations are used for sway and longitudinal motions. For using this model, hydrodynamics derivatives $(N_r, N_v, N_{vvv}, N_{vrr}, \dots)$ of the ship must be obtained.

Nevertheless, if some hypothesis are considered (a ship of large dimensions, constant speed, and open sea operation, mainly), a simpler model may be obtained for controllers design. A Ferry ship (C/Z) has been used for this work, and some of the main characteristics of the ship are the following: length between perpendiculars 92 m, beam 16.8 m, draught 4.9 m, cruise speed 15 knots, displacement 4239 Tn. It is known that a useful model of a ship with these characteristics can be describeb by Norrbin's model (Astrom and Wittenmark, 1989; Lewis, 1989):

$$\alpha \ddot{\psi} + H(\dot{\psi}) = \beta \delta$$

where $H(\dot{\psi})$ is a non-linear function that takes into account the nonlinear relation between rudder angle (δ) and yaw ($\dot{\psi}$) when large values of rudder angle are applied to the ship. Parameters α and β depend mainly on ship speed and load conditions.

The C/Z has been studied and a Norrbin type mathemathical model has been adopted:

$$\tau \ddot{\psi} + \dot{\psi} + \gamma \dot{\psi}^3 = K\delta$$

so that, for small rudder angles and constant speed τ represents a time constant, and K is a stationary gain. To take in account unmodelled high frequency dynamics, a time delay (τ_d) is added to this model.

Steering gear is modelled by a first order system, with unit gain and a time constant (τ_m). The rudder actuator hydraulics has a non-linear behaviour, which can be characterized by limitations in rudder angle ($\pm\delta_{max}$) and in its derivative ($\pm\dot{\delta}_{max}$). In the C/Z steering gear can be operated by one or two hydraulic pumps, so that derivative limitation and time constant depend on that too.

In course-keeping operation, the severity of the yawing motions depends on the wave height and encounter frequency. To take into account wave disturbances, the Bretschneider spectrum (Lewis, 1989) is used, where H is the significant wave amplitude and T the wave period.

$$S(\omega) = \frac{691H^2}{4T^4\omega^2} \exp\left(\frac{-691}{T^4\omega^4}\right) \ \text{m}^2\text{s}$$

A stochastic process with this spectral density is obtained in simulation as output from a filter with white noise as input.

Wind is an important disturbance too, it depends on the largeness and distribution of the wind areas of the ship, ship speed, direction and velocity of the wind. In computer simulations, wind effect is taken into account by a constant signal, where its amplitude depends on the foregoing factors.

3 Controller design

Figure 1 shows the feedback arrangement of the course control system. There are two different basic operations for controlling a ship: course-changing and course-keeping. For course-changing the objective is to make the manoeuvre in a clear

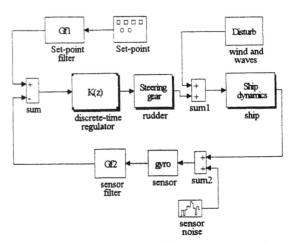

Figure 1: Block diagram of the course control system

way for other ships (without overshoot), and in its realization a significative speed loss can happen for large rudder angles (in large magnitude and length). In course-keeping operation the objective is to keep the heading angle of the ship, and therefore to reject disturbances due to waves and wind.

Ship dynamics depends mainly on its load conditions and speed, so that for a controller design, uncertainties in the mathematical model must be taken into account to achive a robust controller. Robust in the sense that the control system must be stable even though uncertainties, and a suitable performance must keep for all possible plants due to a determined uncertainty level. In other workds, the control system must be characterized by a robust performance (RP).

Linear H_∞ control technique deals with the control system robustness in an explicit way, and a frequency domain approach is carried out to take into account design specifications. Performace specifications can be expressed by two weighting functions, one W_S for sensitivity function $S = 1/(1 + GK)$, and other W_{SK} for the product of this one by controller transfer function SK (control sensitivity function). This is due to the fact that disturbance attenuation, command errors and bandwidth depend on S; and control demand depends on SK. Ship dynamics, mechanical limitations of the steering gear, and disturbance characteristics must be taken into account in the selection of these weighting transfer functions.

Ship unmodelled dynamics is taken into account like an equivalent multiplicative uncertainty, so that a weighting transfer function W_T for the complementary transfer function ($T = 1 - S$) is used. The magnitude of this weighting function represents the uncertainty level of the ship nominal model in the frequency domain. In this way, the following cost function is used for H_∞ controller synthesis:

$$\|T_{zw}\|_\infty$$

where T_{zw} is given by:

$$T_{zw}^T = [W_S S \ \ W_{SK} SK \ \ W_T T]$$

For a multiplicative uncertainty $E(\omega)$, which is bounded by W_T,

$$G_{actual} = G_{nominal}(1 + E), \quad | E | \leq | W_T | \quad \forall \omega$$

where $G_{nominal} = G$ is the nominal transfer function of the ship (linearized dynamics), and G_{actual} is its actual transfer function. In this case, robust performance (RP) is achieved provided that

$$J_{RP} < 1, \quad J_{RP} = \| \ | W_S S | + | W_T T | \ \|_\infty$$

Two less demanding requeriments are included in th robust performance condition: 1) nominal performance specification (NP) for the ship nominal model:

$$J_{NP} < 1, \quad J_{NP} = \|W_S S\|_\infty$$

and 2) robust stability (RS) for the system control:

$$J_{RS} < 1, \quad J_{RS} = \|W_T T\|_\infty$$

In order to solve H_∞ optimization problem, algorithms given in *Robust Control Toolbox* for Matlab have been used (Chiang and Safonov, 1992). Selection of the three weighting transfer function $\{W_S, W_{SK}, W_T\}$ depends on the type of ship operation and therefore, will be different for course-keeping and course-changing regulators.

4 C/Z Ferry control studies

The main points of the proposed approach are:

1. Ship model is obtained by system identification techniques. Nominal model of the plant is obtained for a ship speed of 15 knots, load conditions given in section two, and for two operating pumps in the steering gear, for which: $\tau = 38.4 \ s, K = 0.0411 \ (deg/s)/deg, \gamma = 0.43 \ (deg/s)^{-2}, \tau_d = 2 \ s, \tau_m = 1 \ s, | \dot{\delta}_{max} | = 3.8 \ deg/s, | \delta_{max} | = 35 \ deg$.

2. Regulator design is made using a first order Pade approximation for the delay term.

$$e^{-\tau_d s} \approx \frac{-\tau_d/2 \ s + 1}{\tau_d/2 \ s + 1}$$

3. Nominal specifications consist of load constant disturbances attenuation, and stationary error to step changes in the set-point, smaller than a value $(1/\alpha_1)$. The system band-width is made depending on: a) the effective time

constant of the ship (τ), the equivalent delay time (τ_d), and c) an independent parameter μ_a. The following weighting function are employed:

For course-changing:

$$W_S(s) = \frac{(s + pS)}{(s + pS1)} \alpha_2$$

$$pS1 = \left(\frac{\alpha_2}{\alpha_1}\right) pS$$

where: $\alpha_1 = 1e5, \alpha_2 = 0.333$

$$W_S(s) = \frac{0.333s + 0.0145}{s + 1.45 \ 10^{-7}}$$

For course-keeping:

$$W_S(s) = \frac{(s + pS)(s + pS2)}{(s + pS1)} \alpha_2$$

$$pS1 = \left(\frac{\alpha_2}{\alpha_1 K_6}\right) pS$$

where: $\alpha_1 = 1e5, \alpha_2 = 0.333$, and the physical meaning of these parameters are the following: $1/\alpha_1 \equiv$ permitted maximum value of $| S(j\omega) |$ (low frequency); $1/\alpha_2 \equiv$ permitted maximum value of $| S(j\omega) |$ (high frequency) $(pS2 = pS/K_6, K_6 = 3)$, where:

$$pS = \frac{1}{\tau} \frac{1}{1 + 2(\tau_d/\tau)} \mu_a$$

and the parameter μ_a is used to modify the band-width of the system ($\mu_a = 1.85$ for course-changing, $\mu_a = 1.1$ for course-keeping).

4. With respect to the operation point, ship dynamics equivalent uncertainty is bounded by the following transfer function:

$$W_T(s) = \frac{3.1623s + 0.0474}{s + 0.15}$$

It is considered an uncertainty level up to K_T 100% (32%) at low frequency (where the ship dynamics is better known), and a level up to β_2 100% (316%) at high frequency (variations in ship speed and delay time 30 % and 100 % respectively). For that, it is used:

$$W_T(s) = \frac{s + pT}{s + pT_1} \frac{pT1}{pT} K_T$$

where, $pT1 = pT\beta_2/K_T, pT = pS/0.6$

5. To take into account limitations in actuator (rudder engine), the following W_{SK} is used:

$$W_{SK}(s) = \frac{s + 26.013}{s + 260.13}$$

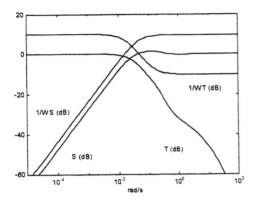

Figure 2: $|S|,|1/W_S|,|T|,|1/W_T|$ for course-changing controller

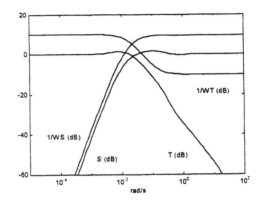

Figure 3: $|S|,|1/W_S|,|T|,|1/W_T|$ for course-keeping controller

It is used a first order weighting transfer function W_{SK}:

$$W_{SK} = \frac{s + \rho_1}{s + \rho_2}\left(\frac{\rho_2}{\rho_1}\right)K_\rho, \quad (\rho_1 < \rho_2)$$

where ρ_1 and ρ_2 are selected depending on effective time constant of the ship:

$$K_\rho = 0.1$$

$$\rho_1 = (1/\tau)1e3, \quad \rho_2 = (1/(\tau)1e4$$

6. Final implementation uses a discrete time regulator, and the design procedure is summarized here:[1] [2]

$$G(s) \xrightarrow{ZOH} G(z) \xrightarrow{bilin^{-1}} G(w)$$

$$G(w) \xrightarrow{H_\infty} K(w) \xrightarrow{bilin} K(z)$$

A sample time of 5 seconds is used for both controllers (5th order), and the final implementation of the control system is represented in figure 1, which shows the structure used in computer simulation. In this figure a filter

$$G_{f1}(s) = \frac{1}{(\tau_{f1}s + 1)^2}$$

for set-point changes can be seen, which is used to carry out large changes in the set-point of the ship (reference heading angle) without getting the rudder saturation (τ_f depends on set-point magnitud, for 90 degrees $\tau_{f1} = 27$ seconds). Another filter (G_{f2}) is used for heading sensor noise.

$$G_{f2}(s) = \frac{1}{(\tau_{f2}s + 1)^2}, \quad \tau_{f2} = 0.5\ s$$

To analyze performance and robustness of each controller, some parameters are calculated: gain margin (MG), phase margin (MF), delay margin

[1]ZOH: Zero Order Hold approximation.
[2]bilin: Bilinear transformation.

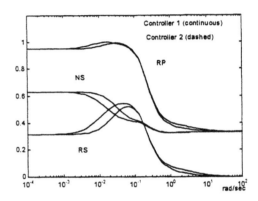

Figure 4: NP, RS and RP tests for controllers 1 and 2

(MR), multiplicative stability margin (MSM), nomimal performance indicator (J_{NP}), robust stability indicator (J_{RS}), and robust performance indicator (J_{RP}). These measurements of performance and robustness are given in tables 1 and 2, where type 1 belongs to course-changing controller and type 2 to the course-keeping regulator.

Nominal performance (NP), robust stability (RS) and robust performance (RP) tests are shown in figure 4 for both controllers. Time responses to set-point changes of 90 degrees and 180 degrees are given in figure 5. In figure 6, time responses of the system are shown and compared for each controller for course-keeping operation. Course-keeping regulator has been designed in such a way that it does not try to reject disturbances effects for a frequency interval belongs to waves (yawing motions of the ship due to waves), since that would generate an unnecessary stress in the steering gear. Figure 7 shows the maximum singular value of T_{zw} ($\overline{\sigma}T_{zw}$) for both controllers.

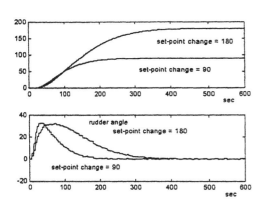

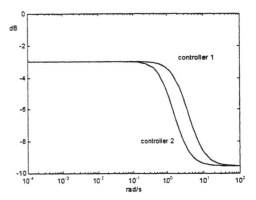

Figure 5: Time responses for course-changing controller

Figure 7: $\overline{\sigma}(T_{zw})$, course-changing controller (1), course-changing controller (2)

Controller type	MG (dB)	MF (deg)	MR (sec)	MSM (%)
1	16.1	71.5	6.9	100
1	18.0	67.7	6.0	91

Table 1: Robustness indicators

Controller type	J_{NP}	J_{RS}	J_{RP}
2	0.51	0.50	0.97
2	0.63	0.54	0.99

Table 2: Performance and robustness indicators

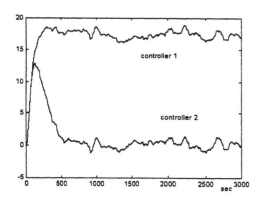

Figure 6: Time responses for controllers 1 and 2, course-keeping operation

5 Conclusions

A mathematical model of a Ferry ship has been obtained from empirical data, as well as an equivalent uncertainty for using in autopilot design based on robust control techniques. Two regulators are designed based on H_∞ control technique, one for course-changing and another one for course-keeping. Based on nominal model and on estimated uncertainty, both controllers achieve robust performance, and therefore they are indicated for real operations of the ship.

References

[1] K.J. Astrom and B. Wittenmark (1989), *Adaptive Control*, Addison-Wesley.

[2] T. Chen and B. Francis (1995), *Optimal Sampled-Data Control Systems*, Springer-Verlarg.

[3] R. Y. Chiang and M.G. Safonov (1992), *Robust Control Toolbox*. The MathWorks Inc.

[4] M.J. Grimble (1994), *Robust Industrial Control: Optimal Design Aproach for Polynomial Systems*. Prentice Hall.

[5] E.V. Lewis (1989), *Principles of Naval Architecture*, SNAME.

[6] A.R.J.M. Lloyd (1989), *Seakeeping: Ship Behaviour in Rough Weather*, Ellis Horwood Limited.

[7] M. Morari and E. Zafiriou (1989), *Robust Process Control*, Prentice Hall.

[8] A. Stoorvogel (1992), *The H_∞ Control Problem*, Prentice Hall.

[9] K. Zhou, J.C. Doyle and K. Glover (1996), *Robust and Optimal Control*, Prentice Hall.

SHORT DISTANCE ULTRASONIC VISION FOR MOBILE ROBOTS

J.R. Llata, E.G. Sarabia & J.P. Oria

Electronic Technology and Automatic Systems Department
E.T.S.I. Industriales y Telecomunicación, University of Cantabria. SPAIN

Abstract: Using ultrasonic sensors it is possible to detect the presence of an object, to localise its position and to realise a superficial scanning in order to give information about the three dimensional shape of the piece. Consequently, it permits to endow the robot with a certain capacity of interaction with the environment. Besides, this could be used to provide a visual information to the robot operator or to supply this data to an artificial intelligent system. Important aspects such as radiation angle, edge detection and edge tracking are presented. *Copyright © 1998 IFAC*

Keywords: Robot vision, Ultrasonic transducers, Multisensor Integration, Recognition.

1. INTRODUCTION

This paper deals with the utilisation of ultrasonic transducers as a vision system for robotic applications.

It is well know that ultrasonic sensors are widely used in robotic applications, see (Llata, et al.,1997a; Llata, et al., 1997b; Calderon, 1989). This is because they have several properties whose can complement the information obtained by other type of sensors.

For example, the more important ones are the followings:

They are able to give information both digital and analogue. Then, it is possible to detect the presence of an object (It is or It is not) and furthermore to know how much is the distance from the sensor to the object by means of echo shape, flight time, etc.

There are models with a much-reduced weigh and size. That is very interesting in order to build arrays

of ultrasonic sensors whose do not modify the dynamic characteristics and do not reduce the workspace or movements of the robot.

They can be used to detect and even to see objects in conditions such as absence of light and presence of smokes. Their price is reduced compare to other types of sensors, and then it is not a problem to replace

Fig. 1. Array of ultrasonic sensors on the robot grip.

them in a crash case.

Besides, these types of sensors are robust enough to bear the requirements of acceleration and deceleration of standard industrial robot, and they are able to deal with the industrial ambient.

The object of the present paper is to use these characteristics in order to carry out a vision system based on ultrasonic transducers. The idea is to place an array of this type of sensors on the grip of the arm (fix or mobile robot). Figure 1 shows the real system.

Later on, it is possible to carry out some type of learning system, see (Llata, et al.,1997a; Llata, et al., 1997b), such as neural nets, expert systems, etc., which will be able to decide and provide information about what sort of objet it is.

In the present work a rule-based expert system has been use in order to control the scanning process.

2. OPERATION PRINCIPLE

The whole system is base on travel time measurement for ultrasonic signals when they are used with the echo-signal method.

The pressure field generated, by a flat and circular emitter, is given by:

$$\mathbf{p}(r,\theta,t) = j \cdot \frac{\rho_0 \cdot c \cdot k \cdot U_0}{2 \cdot \pi} \cdot \int_s \frac{e^{j(wt-k\cdot r')}}{r'} \cdot ds \qquad (1)$$

It can be simplify to obtain the pressure at any point on the axis, in this way:

$$P = 2 \cdot P_0 \cdot \left| Sen\left[\frac{k \cdot r}{2}\left(\sqrt{1+\left(\frac{a}{r}\right)^2} - 1\right)\right]\right| \qquad (2)$$

Expression (2) permits the definition of two well-known zones: The near field (o Fresnel zone) and the far field (or Fraunhofer zone), see (Krautkramer and Krautkramer, 1990). In the far field zone, the wave front tends to take a curved shape inside a cone, defined by the directional term:

$$\left[\frac{J_1(kasin(\theta))}{kasin(\theta)}\right] \qquad (3)$$

Drawing expression (3) on a polar graph it is possible to observe the form of the emitted beam, see figure 2.

The visual information provided by an ultrasonic sensor is a function of both sensor characteristics (beam angle, emitter-receiver distance, etc.) and focus

object characteristics (object-sensor distance, shape, rough, etc.).

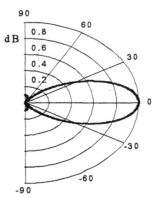

Fig. 2. Beam shape for a circular ultrasonic emitter.

The reception has been carried out by means of a data acquisition board. The signal coming from the receiver, see figure 3, has been digitally filtered with a third order Butterworth low pass cut off frequency of 1 kHz in order to obtain the envelope of the incoming signal. See figure 4.

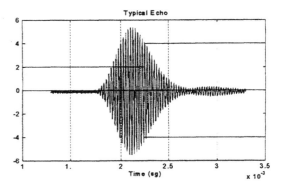

Fig. 3. Ultrasonic echo signal.

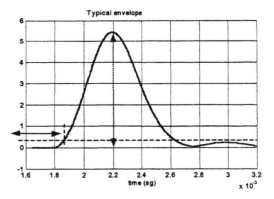

Fig. 4. Echo signal envelope.

In the following sections, the effect of slopes and edges will be analysed in an approximate way.

Besides, It will be presented methods for edge detection and edge tracking based on the previous operation principle.

288

3. ECHO DEFLECTION

3.1 Planar Surface with no Inclination

In (Llata, et al. 1998) there is an expression for the size of the rebounded echo in case of planar surface with no inclination. In this case, a symmetrical echo does exist and its size is defined by:

$$b = h \cdot tg(\gamma_{-6dB}) \tag{4}$$
$$L = 2b \tag{5}$$

Figure 5 shows a scheme of the echo signals when an emitter-receiver pair is used.

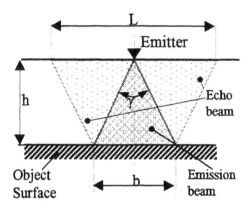

Fig. 5. Emitted and reflected signals on a planar surface with no inclination.

3.2 Planar Surface with Inclination

Also in (Llata, et al. 1997c) there is an expression to obtain the inclination angle (θ) of a planar surface based on the ultrasonic signal travel times.

In Figure 6 it is possible to observe the schematic description of the ultrasonic echo signal when it is reflected by a planar surface with $\theta°$ inclination. Here it is clear that there is a deformation of the echo and the symmetric shape is lost.

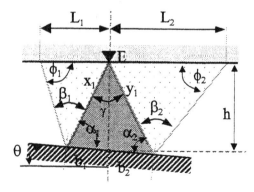

Fig. 6. Emitted and reflected signals on a planar surface with inclination.

The incidence angles of the –6 dBs lines are:

$$\alpha_1 = \frac{\pi}{2} + \theta - \frac{\lambda}{2}; \quad \alpha_2 = \frac{\pi}{2} - \theta - \frac{\lambda}{2} \tag{6}$$

The b1 and b2 are:

$$b_1 = \frac{sin(\gamma/2)}{sin(\alpha_1)} \cdot h; \quad b_2 = \frac{sin(\gamma/2)}{sin(\alpha_2)} \cdot h \tag{7}$$

Besides, it is very easy to obtain:

$$\beta_1 = 2\left(\frac{\pi}{2} - \alpha_1\right); \quad \beta_2 = 2\left(\frac{\pi}{2} - \alpha_2\right) \tag{8}$$

$$\phi_1 = \pi - \left(\frac{\pi}{2} - \frac{\gamma}{2}\right) - \beta_1; \quad \phi_2 = \pi - \left(\frac{\pi}{2} - \frac{\gamma}{2}\right) - \beta_2 \tag{9}$$

$$x_1 = b_1^2 + h^2 - 2b_1 h \cdot cos\left(\frac{\pi}{2} - \theta\right) \tag{10}$$

$$y_1 = b_2^2 + h^2 - 2b_2 h \cdot cos\left(\frac{\pi}{2} + \theta\right) \tag{11}$$

And:

$$L_1 = \frac{sin(\beta_1)}{sin(\phi_1)} \cdot x_1; \quad L_2 = \frac{sin(\beta_2)}{sin(\phi_2)} \cdot y_1 \tag{12}$$

Therefore, it is clear, by the (12) expression and by the figure 6 that as bigger the inclination it is a lower L1 distance and a bigger L2 distance.

It is important to headlight that although in the L_1 zone the amplitude remains in a similar value, in the L_2 zone the amplitude decrease due to the dispersion effect.

In this way, as bigger it is the inclination angle (θ), bigger it is the echo deflection. That is clear in figure 6.

On the limit, there is an inclination angle (θ_L) which provide a direction of reflection, for the –6dB line, parallel to the emitter plane. This is:

$$\theta_L = \frac{\gamma}{2} - \frac{\pi}{2} \tag{13}$$

It must be clear that this angle is for the limit and that in practical cases this angle is much more reduce. That is due to several factors such as the difraction and attenuation of the beam, etc.

4. EDGE DETECTION

It is possible to use the behaviour of the ultrasonic signal showed in the previous section, to detect edges. In fact, there are several useful methods. Here two methods are showed.

A graphical description of the first method used is showed in figure 7. As it is clear in this picture this method uses the same configuration of emitters-receivers as the system presented in (Llata, et al. 1998c).

As it's clear in this figure, if there are a abrupt slope change in the object surface, a non linear echo signal is obtained. Therefore, a gap appears which size and boundary angle are defined by the variation of the inclination angle (ϕ).

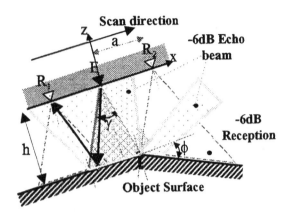

Fig. 7. Emitted and reflected signals on a planar surface with an edge. First method.

If the variation of the inclination angle (ϕ) is near to $\pi/2$, in the direction showed in figure 7, this angle will be close to the maximum. If it is near to zero, this angle will take the same value.

In this case is easy to observe that the position of the edge, at the moment in which the –6dB signal is lost by R_2 and related to the emitter cordinates system, is:

$$P_{edge} = [a/2; \ 0; \ -h]^T \qquad (14)$$

The second method proposed, based on one emitter and two receivers, is presented. The difference with the previous one is that, in this case, the receivers used are in the transversal direction to the scanning movement.

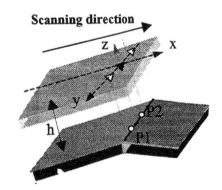

Fig. 8. Second Method for edge detection.

In this case, each one emitter-receiver pair works similar as the previous case.

At the moment of losing the signal for each receiver, the edge will be at the points:

$$P1_{edge} = [0; \ a/2; \ -h]^T \qquad (15)$$

$$P2_{edge} = [0; \ -a/2; \ -h]^T \qquad (16)$$

6. EDGE TRACKING

Now that methods has been developed for scanning smooth surfaces and for detecting edges, it could be very interesting to carry out a system to follow these edges. This is an important point because solid object boundaries are recognised as essential elements in the extraction of useful features from these objects.

In the previous section, an especially interesting method for edge tracking has been presented. This method use an emitter and two receivers, all of them placed on a common line perpendicular to the scanning direction.

When these sensors are scanning a surface and a emitter-receiver pair detect a point of an edge (in fact do not detect echo signal), the sensor rotate until the other emitter-receiver pair detect the edge. This behaviour can be seen in figure 9.

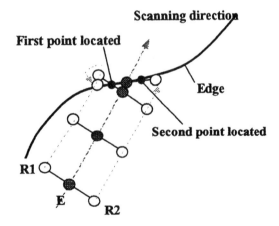

Fig. 9. Edge tracking based on the second method of the edge detection.

It is important to observe that the cordinates of these two points are known by means of the expression of the section four and by means of (Llata, et al. 1998).

In this way, the edge has been detected and a approximation to the tangent in the middle of both previous points ($P_1=[x_1, y_1, z_1]$ and $P_2=[x_2, y_2, z_2]$) obtained as:

290

$$\frac{x - x_1}{x_2 - x_1} = \frac{y - y_1}{y_2 - y_1} = \frac{z - z_1}{z_2 - z_1} \qquad (17)$$

This equation provides the direction of the edge and therefore the information necessary to continue the movement along the edge.

However, it has been tested that with this information alone the sensor loses the edge in a short number of iterations. The reason is the sensor loses the reference of the initial surface.

A method to overcome this problem is simply taking a Z-shaped trajectory. It means, to force the sensor to keep the reference of the initial surface using a zig-zag path. This is showed in figure 10.

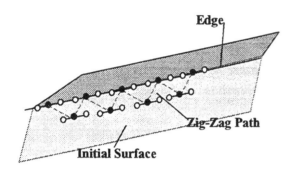

Fig. 10. Zigzag Path. Used to keep references during the edge tracking.

7. SCANNING

This stage take charge of cover the whole object surface (only for its superior side) under study in such a way that the vertical distance between the object and array will be in the established range for a correct operation.

In order to fix the range of suitable vertical distances is necessary to take into a count several aspects as, for example, the frequency and angle of radiation of the transducers and the behaviour of the data acquisition system.

It is preferable to use a data acquisition system able to work in a parallel way. It means, to be able of "listening" with all receivers at the same time, waiting for the signal coming from its emitter. However, that would produce hard requirements about the maximum separation. For example, let us see the requirements when this method of data acquisition is used with the array used for this experimental study.

The distance between emitters and receivers centres is 13 mm. The distance from the centre of one emitter to the nearest receiver, for which it is not desirable, the reception (from this emitter) is about 29 mm. If the same previous expressions are applied, we obtain that the working distance to achieve that there are no direct echoes in second rows of receivers is:

$$r = h \cdot k_{(\gamma - 6dB)} \cdot \frac{\lambda}{2 \cdot a} \qquad (18)$$

$$h = \frac{2 \cdot a \cdot r}{\lambda \cdot k_{(\gamma - 6dB)}} \qquad (19)$$

This is the maximum distance at which the array has to be places to avoid receiving first echoes from an emitter in second rows of receivers. In this case, it was about 56.8 mm.

The reason for being this distance so short is the low directionality of these sensors. A possible solution would be to work with higher frequency sensors so that their ultrasonic beam was narrower.

At this point we must to remark it is not necessary that the hardware employed into the system has a very high performance. That is in this way because the obtained information from each receiver is only two values: Flying time of the ultrasonic beam and the highest amplitude value of the signal.

8. SURFACE VISION

It is clear that in this moment, the developed system is able to detect presence, measure distances and surface inclinations, to detect and follow edges, etc.

All of this is obtain for a set of four receivers and one emitter, see (Llata, et al. 1998), but if it is necessary, it is possible to add other emitter-receiver pairs to improve the system. Them, scanning the surface and measuring distances and slopes, It is obtained a three-dimensional map. It permits us to make a rough representation of the surface of the object.

Fig. 11: This figure shows the effect of the typical noise. In this case the measure was done over the floor with sensors of 55° of beam in different conditions and with out o used average measures.

However, the accuracy of the vision system developed is function of the scanning resolution, of the angle beam, of the variation of the humidity, temperature, etc. For example, figure 11, shows the effect of the noise in the measures.

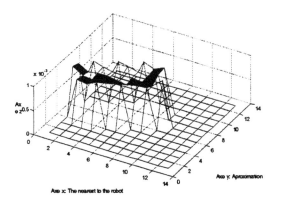

Fig. 12: Three-Dimensional Ultrasonic Vision of a prismatic piece with low scanning resolution.

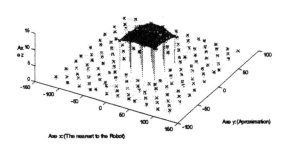

Fig. 13: Three-Dimensional Ultrasonic Vision of a prismatic piece with higher scanning resolution.

9. CONCLUSIONS

In this paper, several methods to improve the artificial vision for robotics applications have been presented. Besides, it had tried to remark the importance of the size of the angle beam, the noise, the scanning accuracy, etc, on the results.

During the developing of this work it has been clear that ultrasonic sensors arrays provide a very interesting method for objects detection, objects localisation, surface tracking and surface reconstruction, edge detection and edge tracking, etc.

It is important to remark here that this system never will be finalised without an artificial intelligent system, which takes the decisions about direction, rotations, etc, based on the ultrasonic information.

10. ACKNOWLEDGEMENT:

This work has been carried out under the C.I.C.Y.T sponsorship in the Spanish project: TAP-95-0361.

11. REFERENCES:

Brown M. K, (1985). Feature extraction techniques for recognizing solid objects with an ultrasonic range sensor., *IEEE Journal of Robotics and Automation*, Vol. RA-1, **4**, 191-205.

Brule M., Soucy L. (1990). Modular algorithms for depth characterization of objects surfaces with ultrasonic sensors. *SPIE Vol 1388*. 432-441.

Caicedo E., (1995). Sistema de Identificación de Objetos Mediante un Modelo Paramétrico. PhD. Thesis. ETSII Madrid.

Calderon L., (1989). Sensor ultrasonico adaptativo de medidas de distancias. PhD Thesis. Ed. complutense de Madrid. Madrid.

Krautkrämer J., Krautkrämer H. (1990). Ultrasonic Testing of Materials", Springer-Verlag.

Llata J.R., Sarabia E.G., Arce J., Oria J.P. (1997). Probabilistic expert systems for shape recognition applied to ultrasonic techniques. *CIRA97 Proceedings*, 158-163.

Llata J.R., Sarabia E.G., Oria J.P. (1997). Shape recognition by ultrasonic sensor arrays in robotic applications. *ACRA97 Proceedings*, 183-188

Llata J.R., Sarabia E.G., Oria J.P. (1997). Three dimensional ultrasonic vision of surfaces for robotic applications. *AVI98. Madrid*

Oria J.P., González A. G., (1993). Object recognition using ultrasonic sensors in robotic applications. *IECON'93 Proceedings*. Hawaii. USA

LOW-COST IMPROVEMENT OF AN ULTRASONIC SENSOR AND ITS CHARACTERIZATION FOR MAP-BUILDING

Jesús Ureña, Manuel Mazo, J. Jesús García, Emilio Bueno, Álvaro Hernández, Daniel Hernanz

Electronic Department. University of Alcalá. Escuela Politécnica. Campus Universitario s/n. 28871 Alcalá de Henares (MADRID).
Tfno: (+34 1) 8854810-18   Fax: (+34 1) 8854804-99
E-mail: {urena, mazo, jesus, emilio, alvaro}@depeca.alcala.es

Abstract: Ultrasonic sensors have often been used as rangefinders in mobile robots. In many cases their use has been reduced to the measurement of distances to the closest object in front of each transducer, in isolation of the rest of the transducers. This paper describes the design, implementation and application of an ultrasonic sensor made up of four transducers suitably set up. Each transducer is equipped with a low cost electronic system, improved as compared with basic versions, so it may be used in conjunction with the rest and provide accurate Time of Flight (TOF) readings. This makes it possible to use triangulation techniques. The new sensor has been characterised for use in map building by means of certainty grids. *Copyright © 1998 IFAC*

Keywords: intelligent instrumentation, ultrasonic transducer, enhancement, rangefinders, mobile robots.

1. INTRODUCTION

When ultrasonic transducers are used in mobile robots (MRs), these are distributed evenly around the robot's periphery. The distance readings are obtained by determining the times of flight (TOFs) of the ultrasounds from the moment they are emitted to when they are reflected back by surrounding objects. The readings obtained are processed in the light of one or several of the following objectives:

a) Direct implementation of an algorithm for guidance of the MR. The readings obtained are sometimes introduced directly and processed into high level algorithms, whose outputs directly provide the robot's guidance speeds (reactive control). This research work normally lays greater stress on the techniques and algorithms used than on the use and problems of the ultrasounds *per se*. They therefore deal with the use of fuzzy techniques, neural network learning (Song & Sheen, 1995), etc.

b) Map building of the MR's environment: Although some attempts at map building have been made using

"adaptive" models (visibility graphs, free regions, etc), errors and lack of precision in the ultrasound readings are more suited to "rigid" models (for example the use of certainty grids and probability functions for post-reading updating (Elfes, 1987)). Along the same lines, some authors have proposed simplified updating models following heuristic rules for their implementation in real time (Borenstein & Koren, 1991a) (Song & Chen, 1996). The algorithms for the MR guidance are built up from the maps in environments with fixed or mobile objects.

c) Locating the MR. Determining the position of the MR in relation to its environment involves identifying some of the environment's objects, whose position is known a priori. Normally the objects detected are classified into three types: planes, corners and edges. For the identification of this type of reflectors, using TOF readings, an active perception of the environment is necessary (i.e., different "points of view"). This is done either by taking advantage of successive readings as the robot moves along (Leonard & Durrant-White, 1991) or by using more than one ultrasound receiver for each emitter. Kleeman and Kuc (1995) showed that the classification of the three reflector types (planes, corners, edges) calls for at least two emitters, their pulses also being received by at least two receivers. For classification algorithms it is essential to have transducers capable of measuring distances with sub-millimetre accuracy. This involves the use of echo signal processing techniques such as optimum filtering and the compression of pulses (Audenhart *et al.*, 1992, Hamadene & Colle, 1997).

This paper will show, firstly, the improvements made in the basic transducer, such as that of Polaroid (POLAROID, 1991), geared towards achieving the following objectives with a low-cost electronic system:

1. Ease of combining one transducer with another, so they may make up more complex sensory modules.
2. High sensitivity, enabling the detection of echoes from small objects or those apart from transducer axis.
3. Ability to detect multiple echoes in a single reading.
4. High accuracy in the determination of the TOFs, (distances determined with sub-millimetre accuracy).

Four improved transducers can be arranged to form a sensor module. The study of a sensor of these characteristics, its possibilities of measuring bearing angles and classifying edges or walls/corners, plus the arrangement of several of them on an industrial vehicle, may be seen in (Ureña *et al.*, 1998).

Lastly, in light of the possibilities of a sensor module of these characteristics, a description is given of an algorithm for map building, using a gridded environment. The translation of the readings obtained into certainty values on the corresponding cells is a costly, computer-intensive process (especially if the number of cells is high). To solve this problem a previous, *off-line* processing is made of the different solutions that may arise, once the parameters to be used in the map building have been reduced to some discrete values. After each pre-processing, templates have been generated (*predetermined templates*) with the certainty values to be updated on the cells remaining in the influence zone of the sensor. These templates are first stored in memory and thereafter used in updating the map as readings are taken (*on-line* processing).

2. IMPROVEMENT OF A TRANSDUCER ELECTRONIC SYSTEM

A basic transducer, such as that of Polaroid [1], employing integration and thresholding techniques of the echo signal, gives an accuracy (see figure 1) that may suffice for many MR applications requiring only

[1] "Probably the single most significant sensor development from the stanpoint of its catalytic influence on the robotics research community" H.R. Everett *"Sensors for mobile robots"*. Ed. A K Peters. Pag. 144 .(1995)

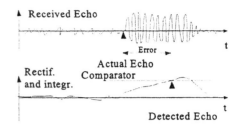

Figure 1.- Echo detected after rectification, integration and thresholding.

distance measurement but not for the determination of reception angles nor for a correct classification of the detected reflectors.

2.1. Proposed electronic system

To improve the transducer's capacity of measuring TOFs accurately, the echo signal is processed along similar lines to those used in radar. The great drawback is that each transducer must necessarily be equipped with a processing system for carrying out all necessary operations. This may significantly increase the complexity of the total system, making its assembly more difficult and cutting down some of the great advantages of ultrasonic sensory systems: their relatively low cost and simplicity. To avoid this, the algorithms have been implemented into a specific digital system, based on an FPGA 4005E of Xilinx, the block diagram of which is shown in figure 2. From an external point of view the working of the transducer may be controlled by a micro-controller system, through an interface circuit, using only four digital lines, namely: *a) mode signal (E-R/R)*; this digital line indicates whether the transducer has to function in EMITTER-RECEIVER mode (E-R) or as RECEIVER only (R), b) *Initiating signal (INIT)*; this signal indicates the moment when the measuring process starts, c) *CLOCK signal*: the functioning of the whole system, in its digital processing facet, calls for a single

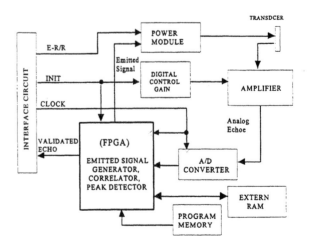

Figure 2.- Electronic system for every transducer.

clock synchronising signal and d) *Echo reception signal* (VALIDATED ECHO): as the echo signal is received it is processed to determine the arrival times of the various echoes detected. Each time one of the echoes is validated the ECHO VALIDATED signal is turned on for 32 µs. The system resolution has been established with a minimum TOF difference of 64 µs (a distance resolution of about 1 cm).

2.2. Implemented algorithm

A practical approach to an optimum filtering (Hamadane et al, 1997) involves taking the emitted signal as reference and making a cross-correlation between this and the signal received (see figure 3). A peak search is made at the correlator output, the peaks coinciding with the arrival of the echoes.

From a digital point of view, if N is the sample number of the emitted signal, y(n) will be:

$$y[n] = \sum_{k=0}^{N} x_e[k] \cdot x_r[n+k] \qquad (1)$$

The ultrasonic transducer used requires binary excitation, so the signal emitted has to be of this type. A binary sequence whose autocorrelation function has a straight lobe is the 13-bit Barker code (Hovanessian, 1984). The main lobe has a width of one bit and an amplitude 13 times higher than the side lobes.

The signal emitted by the transducer must be 50 Khz and last long enough for the energy emitted to allow the detection of echoes from reflectors at a considerable distance. For each bit of the Barker code a symbol has been emitted composed of two periods of the carrier signal (at 50 Khz). In total 13 x 40 µs = 520 µs. A two-phase modulation was used, emitting with phase zero if the code bit is 1, and complemented phase if it is -1. A separation time equal to the sampling time (2 µs) is set between consecutive samples, and the signal emitted can be obtained as the correlation between the symbol used (two periods) and the Barker code sequence (but in this case with consecutive samples separated by a symbol - 20 periods), as may be seen in figure 4, from which it follows :

$$x_e[n] = c[n] * b[n] \qquad (2)$$

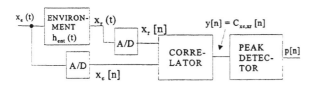

Figure 3.- Block diagam for the prccessing of the received signal.

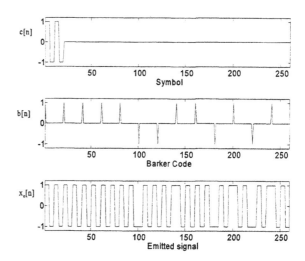

Figure 4.- Generation of the emitted signal.

2.3. Implementation of the correlation and detection of peaks

The sampling period used in the digitalisation of the signal received is 2 µs (sampling at 500 Khz). In other words, for each sequence emitted there will be 520/2 260 samples. The reception of the echo signal is maintained for 32 ms (to detect distant targets up to about 6m.) But the correlation is effected while the signal is being digitalised, maintaining only a reception memory capacity of 260 samples (equal to the duration of the signal emitted). The total correlation between the signal emitted and received may be broken down into two successive correlations:

$$y[n] = x_r[n] * x_e[n] = x_r[n] * (c[n] * b[n]) =$$
$$= (x_r[n] * c[n]) * b[n] \qquad (3)$$

* *First correlation*: Between the signal received, once digitialised, and the sequence corresponding to a symbol (20 samples). As the latter only takes on the values +1 and -1, the correlation is simplified since only additions and subtractions are necessary (not products). The process will involve taking twenty samples of the signal received and, in units of five, adding or subtracting them according to the followingsequence: {+++++-----+++++-----}. Figure 5 shows a block diagram of the circuit for this process.

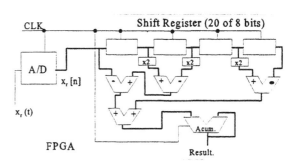

Figure 5. Block diagram for the first correlation.

Second correlation: In each sampling period the result of the previous correlation is taken to an external RAM memory whose addresses are accessed by a pointer - counter- in cyclical mode. The capacity of this memory is 13x20 = 260 words. The block diagram is shown in figure 6. During the sampling period (2 μs) a 16-state sequential process is synchronised by a frequency clock 16 times higher (8 MHz)-. In these states the following operations are carried out:

State 1: While the value of the previous correlation is stored in the relevant memory address (picked out by the counter acting as pointer) , it is taken to an accumulator where it is added to the initial value.
States 2-13: It is added twenty to the pointer value of the previous state and the contents of the new direction are introduced by the relevant sign to the Barker code sequence in the accumulator.
States 14 to 16. These are used for peak detector operations with the last correlation result.

The combinational logic block performs the function shown in table 1:

Table 1.- Combinational function (in figure 6).

STATES	A	B	C	D	E
0	0	1	1	0	0
1,2,3,4,5,8,9,11	0	0	1	0	0
6,7,10	0	0	0	0	0
12	0	0	0	0	1
13	1	0	x	0	x
14,15	0	0	x	1	x

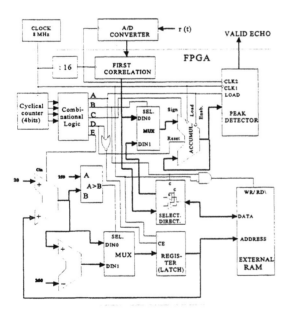

Figure 6.- Circuit for the second correlation.

Peak detector: A comparator is used so that only output values of the second correlation higher than a certain threshold are taken into account. When this obtains, this instant is stored and is definitely validated as an echo reception moment if there is no higher value in the 16 following samples (and at the end of same). In other words two echoes separated (in flight time) by less than 32 x 2 μs = 64 μs (tantamount to about 10 mm of reflector separation) will never be distinguished.

When a valid peak is detected the VALID ECHO output is activated (at a high level) during 32 μs (guaranteeing that during at least 32 μs it will not again be activated). It should be pointed out that from the moment an echo is received until it is actually validated with the corresponding signal there is a systematic processing time of 260 periods for both correlations and 32 periods for the echo validation, i.e., 292 x 2 μs = 584 μs. The external system must subtracting this time from the value actually measured. Figure 7 shows the block diagram of the peak detector.

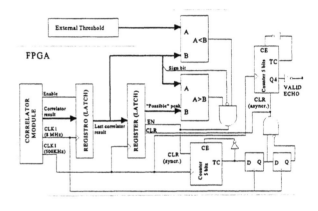

Figura 7.- Block diagram for the peak detector.

2.4 Results

The experimental results have proved the enhacement of the system obtaining a time-of-fly precission of 2μs. Figure 8 shows the analog signal recived and the pulse of validation of the echo for a isolated reflector (note that this pulse is 584 μs delayed since the begining of the echo). In figure 9 can be observed the case of multiple echoes (even with matching beetwen them).

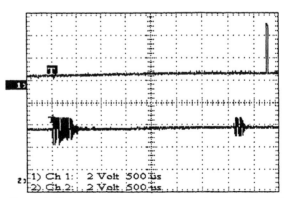

Figure 8. Detection of a echo (one isolated reflector).

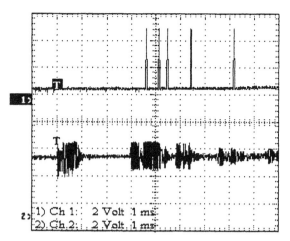

Figure 9. Detection of several echoes (five reflectors)

3. SENSOR MODULE

A single sensor (see figure 10) has been assembled from four transducers with the electronic system indicated in the above section, plus one microcontroller Intel 87C51FB to configure, synchronise and collect the readings of the whole system. The possibilities of measuring distances, reception angles and even discrimination between reflector types (edge or plane) has been described in Ureña et al., 1998. Five sensors of these characteristics were fitted on an industrial vehicle, constituting one of the sensory nodes thereof, together with a CCD colour vision camera and a laser rangefinder (Mazo et al., 1997).

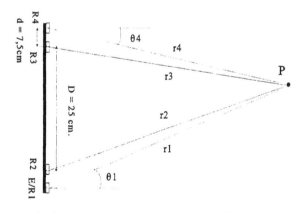

Figure 10. Four-transducer sensor (diagram and prototype).

4. CHARACTERISTICS OF THE SENSOR FOR MAP BUILDING

When one of the sensor's end transducers is emitting, the possible echo signals are received separated by the four transducer, obtaining four TOFs (t_1, t_2, t_3, and t_4). A general rule such as that shown in table 2 is derived from the analysis of the correspondence of the echoes between themselves (verification of whether the transducers have detected the same reflector or another nearby one (Ureña et al., 1998). The sensor has been simplified, each end transducer pair being considered sufficiently close to take in the same "viewing" angle (see figure 11).

Table 2.- Possible data obtained after one reading.

Correspondence: R1-R2 R3-R4 R1-R4			It can be obtained:
No	No	Indiferent	r_A y r_B
No	Yes	Indiferent	r_A, r_B y θ_B
Yes	No	Indiferent	r_A, θ_A y r_B
Yes	Yes	No	r_A, θ_A, r_B y θ_B
Yes	Yes	Yes	r_A, θ_A, r_B, θ_B, type

Once points A and B have been located and the orientation of the transducers is known, the values resulting from a reading determine which zones in front of the sensor can be considered "empty" of obstacles and which "full". A function (variable between 0 and 1) will also be assigned to indicate the degree of certainty involved in both cases. Several algorithms have been proposed for the case of a transducer. Basically, a zone is considered to be "full" when it comes within the opening angle of the transducer at a distance equal to the reading, more or less the degree of accuracy with which said reading is taken. The "empty" zone is that existing between the transducer and the aforementioned "full" zone. Algorithms for a single transducer, such as the HPF (Elfes, 1987), the HIMM (Borestein and Koren, 1991b) or the HAM

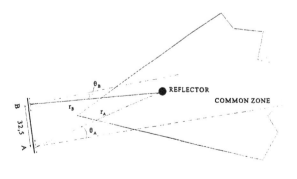

Figure 11.- Simplified model of the sensor.

297

(Song & Chen, 1996) assign maximum certainty (for both zones) along the whole axis, decreasing as the angles grow therefrom. Greater confidence is also granted to short than long readings.

The algorithm herein proposed takes into account not only the distance measured but also the following: whether the distance reading is confirmed by two transducers on one end (double confirmation) - in the event of any discrepancy the lower distance is chosen - whether the reception angle has been obtained (maximum certainty is assigned there to) and whether the reflector type has been determined (if it is a wall its inclination angle is taken into account to determine the "full" area) For both zones the following updating function is used (f):

$$ f = \left[1 - \frac{r - d_{min}}{2 \cdot (d_{max} - d_{min})} \right] \cdot e^{-\left[\frac{(\theta - \theta_c)^2}{2 \cdot \sigma^2} \right]} \quad (4) $$

where (r, θ) is the point of the "empty" or "full" zone considered (with respect to A or B), d_{min} and d_{max} are the maximum and minimum distances processed, Q is the reception angle detected and σ is a value that varies as the function decreases from its maximum (fixed at 5° of the full zone if it is not a wall and 30° if it is a wall; for the empty zone it will in any case be 10°). This function is applied to the central point of each cell of the gridded environment, and the value obtained is used to update the certainly value (CV) thereof. If CV(k-1) is the previous value, after updating the new one, CV(k), will take the value:

$$ CV(k) = m \cdot f + (1 - m) \cdot CV(k-1) \quad (5) $$

where m is a factor between 0 and 1 that weighs up the influence of past history of the map updating. Following the empirical considerations of the HAM method, 0.4 has been fixed for the "full" zone and 0.2 for the "empty" zone.

The evaluation of the *f* function, mentioned above, has been carried out *off-line* for all existing cells in an octant of the co-ordinate system centred on the position of the transducer. For remaining octants an easy extrapolation can be made thereafter in terms of symmetry. A first template for each measurement has therefore been generated that assumes an orientation - according to the axis - and distance coinciding with the cell in question. For each cell additional templates have also been considered, taking into account the angles measured (for the values -12, -9, -6, -3, 0, +3, +6, +9, +12). Figure 12 shows a template generated for one side of the sensor with an orientation of 60°, after a reading giving a value of r_A = 500cm and θ_C = 5° with a cell width of 20 cm, d_{min} = 30 cm and d_{max} = 600 cm for the case of an edge. Negatives values are for the "empty" zone and positives for the "full" one.

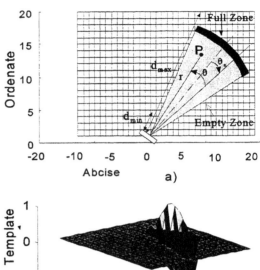

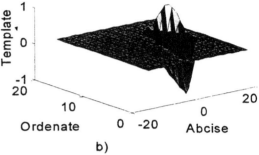

Figure 12.- a) Full and empty zones, b) Template.

5. ACKNOWLEDGEMENTS

This work has been carried out thanks to the grants received from the CICYT (Interministerial Science and Technology Committee) projects TAP94-0656-003-01 and IER96-1957-C03-01

6. REFERENCES

Audeanert K., Peremans H., Kawahara Y., Van Campenhout J. "Accurate Ranging of Multiple Objects using Ultrasonic Sensors". *Proc. of IEEE International Conference on Robotics and Automation* pp. 1773-1738. Nice, May 1992.

Borenstein J., Koren Y. "The Vector Field Histogram - Fast Obstacle Avoidance for Mobile Robots". *IEEE Transact. on Robotics and Automation.* vol. 7, n° 3, pp. 278-288. June 1991 (a).

Borenstein J., Koren Y. "Histogramic In-Motion Mapping for Mobile Robot Obstacle Avoidance". *IEEE Trans. Robotics and Automation.* Vol 7, no. 4, pp. 535-539. August 1991(b).

Elfes A. "Sonar Based Real-World Mapping and Navigation" *IEEE J.of Rob. and Autom..* Vol. RA-3, n° 3, pp. 249-264, June 1987.

Hamadene H., Colle E. "Optimal estimation of the range for MRs using ultrasonic sensors" *SICICA'97*, pp. 141-146, June 1997.

Kleeman L., Kuc R. "Mobile Robot Sonar for Target Localization and Classification" *The International Journal of Robotics Research.* Vol. 14, No. 4, pp. 295-318, August 1995.

Leonard J., Durrant-White H. "Mobile Robot Localization by Tracking Geometric Beacons". *IEEE Transactions on Robotics and Automation.* Vol. 7, n° 3. pp. 376-382. June 1991.

Mazo M., Rodríguez F.J., Sotelo M.A., Ureña J., García J.C. Lázaro J.L., Espinosa F. "Automation of an industrial fork lift truck, guided by artificial vision, ultrasonic and infrared sensors in open environments". *Int. Conf. On Field and Service Robots FSR'97.* Camberra. December 1997.

Song K.T. y Chen Ch. Ch. "Application of Heuristic Asymmetric Mapping for Mobile Robot Navigation Using Ultrasonic Sensors". *J.of Intell. and Rob. Syst.* Vol. 17, pp. 243-264, 1996.

Song K.T. y Sheen L.H. "Fuzzy-neuro control design for obstacle avoidance of a MR". *IFES'95.* Vol. 1, pp. 71-76, July 1995.

Ureña J., Mazo M., García J.J. y Bueno E. "Ultrasonic Sensor Module for an Autonomous Industrial Vehicle". *3rd IFAC Symp. on Intell.Auton.Veh.(IAV'98) .* Madrid, March 1998.

MAP BUILDING INTELLIGENT SENSOR FOR AN AUTONOMOUS VEHICLE.

G. Benet, F. Blanes, M.Martínez., J. Simó

Dep. Ingeniería de Sistemas, Computadores y Automática. Fax: 34 (9) 6 387 75 79.
Universidad Politécnica de Valencia. Cno. Vera 14. E-460022. Valencia.
{gbenet, pblanes, mimar, jsimo@aii.upv.es}

Abstract: In this paper the results in the development of a sonar array sensor for the use in mobile robot platforms are presented. The work involves hardware development and signal processing techniques for the reconstruction of the scene. The developed sensor uses an array of four transducers to improve the distance measurement and angular resolution. Its angular response has some sidelobes around the main lobe. The main advantage of this approach is a sharper main lobe, compared with the main lobe of the only one pair emitter-receiver device. However, the lateral sidelobes have significant amplitude and can produce false peaks detection. This problem can be addressed using adequate deconvolution methods. This approach decreases the amplitude of the sidelobes and produces a high quality data with low angular ambiguity. This allows map building with better data, obtaining good results in real environments.
The tests performed show good enough obstacle detection results in a normal office environment. The signal processing treatment is adequate for the DSP implementation allowing the use of the sensor on a real-time mobile vehicle. The use of the measurements in sensor fusion levels for the map building, reduces the complexity and uncertainty of this process, allowing a simple signal-fusion level and increasing the quality of the final map.
Copyright © 1998 IFAC

Keywords: Ultrasonic Transducers, Sensor Fusion, Obstacle Detection, Signal Processing.

1. INTRODUCTION. [1]

In recent years, the use of ultrasonic sensors for distance measuring by robots and the construction of environmental maps has become increasingly common. Some reasons why ultrasonic sensors have been chosen over other devices like laser or vision are:

1. Low cost: for some applications the final cost of a robot is important
2. Low energy consumption: the autonomy of the robot depends on the installed sensors and actuators

3. Simple signal processing: interesting from the real-time point of view.

On the other hand, some factors make ultrasonic sensors less efficient.

1. Not-enough directional measuring: the beam of an ultrasonic sensor is wider (30°-45°) than other distance measuring sensors like laser.
2. Non-constant sound speed: depends mainly of the air temperature.
3. Specular effects of the signal.
4. Low bandwidth: it is difficult to resolve closely spaced objects.

Therefore, the measures obtained by ultrasonic sensor are uncertain, because it cannot detect the exact position of the object at the far edge of the cone. Consequently the information provided by these sensors is poor and requires the conformation through the repetition of measures with the sensor in different

[1] This work has been supported by the projects TAP97-1164-C03-03 of the Spanish Comisión Interministerial de Ciencia y Tecnología (CICYT) and GV-C-CN-05-058-96 under the program Projectes d'Investigació Científica i Desplegament Tecnològic "Generalitat Valenciana".

positions (robot exploration). The most common device used on robots is the POLAROID (Borenstein and Koren, 1991; Lee, 1996; Pagac *et al.*, 1996; Leonard and Durrant-Whyte, 1992), which gives a measure of the closest object in the front of the sensor into a cone around 30°. In this work, we propose a sonar sensor device with an accurate obstacle distance measurement, based on signal processing technique. This system is under development for the integration into the YAIR architecture (Gil *et al.*, 1997) (Simo *et al.*, 1997).

2. SONAR SYSTEM.

2.1 Ultrasonic sonar module

This intelligent module uses the 80C592 microcontroller and includes a 10-bit A/D converter. The microcontroller is used to:

1. Control the angular position of the rotating transducer by means of a stepper motor with 1.8 degrees per step.
2. Generate the ultrasonic waveforms to be supplied to the transducer.
3. Process the echoes received from surrounding objects.

The sonar systems used in robotics are usually based on a single transducer that acts as emitter and as a receiver, as is the case for Polaroid devices. Other approaches use two separate transducers, one acting as emitter and the others as receiver. In the present paper, a linear array of four transducers is presented (two emitters and two receivers). The layout of the transducers is the following: E-R-R-E (E means Emitter, and R means Receiver) and they are placed each 17.78mm on a little piece of mounting printed circuit board. They are piezo-ceramic devices adjusted to operate at 40kHz. This sensor has an effective measuring range from about 15cm to 5m. The prototype of the sonar module can be observed in the Figure 1.

The four transducers are equally distributed on the array, with the receivers situated at both ends. Therefore, the signals of the two receivers are in-phase only in several angular positions, making the angular response to have some sidelobes around the main lobe. The main advantage of this approach is the better angular resolution of the main lobe (± 5 degrees approx.), compared with the main lobe of the only one pair emitter/receiver device (± 15 degrees approx.).

In the figure 2 are compared the normalized angular magnitude response of each sensory device. These plots have been obtained from the echo of a vertical stick located at 1m of distance. Note that the lateral sidelobes have significant amplitude. This can produce false peak detection in the map-building algorithms.

Fig. 1: Prototype of the rotary sonar head with four-transducer array.

The module sends a short train of ultrasonic pulses in a given direction. The resulting echo is received, digitized and processed to obtain an estimate of the robot environment. This process can be repeated for each one of the 200 angular positions that the stepper motor can be settled. Using this data, a map of the world surrounding the robot can be made.

Depending on the requirements fixed by the main controller, the module can produce different quality of data. These can vary from simple rough detection of objects in one single direction, to more sophisticated and time-consuming measurements such as: mapping with information on the instantaneous speed of each point in the scenario; or assessing about the relative hardness of objects. However, the module is normally set to measure distances to objects, using fast algorithms as the one described in (Benet *et al.*,1992).

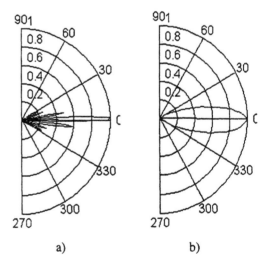

Fig.2. Normalized angular amplitude response of the four-transducer array (a) compared with the two-transducer array (b) (Angles are in degrees).

2.2 Ultrasonic signal processing:

When the ultrasonic array of transducers has been correctly positioned, a short train of 40kHz pulses is applied to the emitters using a full bridge power driver. It produces a train of ultrasonic waves that is propagated through the air at 343 m/s approx. When a solid object is found, an echo returns to the receivers. The amplitude of this echo depends on the surface properties of the object. The time elapsed between the emission and the reception of an echo is called Time of Flight (ToF), and is used to determine the distance d between the object and the transducers. If one denotes the speed of sound as c, d can be obtained as follows:

$$d = \frac{c \cdot ToF}{2} \qquad (1)$$

Accurate measurement of the time taken for the echo to reach the receivers (ToF) is essential for achieving sufficient precision. Unfortunately, this is not an easy task, because of the shape of the signal supplied by the transducers.

Usually, most of ultrasonic systems employ specific integrated circuits, designed to convert the pulsed input signal into an edge that is used to find the ToF. These IC's do not detect the true ToF. Instead, they generate the edge of the signal transition over a fixed threshold, thus loosing some initial cycles, which do not reach the threshold amplitude. This method is very popular and modules based on this approach are widely used in robotics. This is the case for the ubiquitous Polaroid-based modules.

Unfortunately, the pulse amplitude is very dependent on the distance and surface reflection characteristics, and cannot be easily modeled Thus; those ToF measurements made with a fixed threshold, will be affected by a non-constant offset and will produce large errors. In practice, the error suffered by these non-compensated ultrasonic systems totals some tens of millimeters when using low-cost piezoelectric transducers at frequencies of around 40-50 kHz.

Alternatively, digital signal processing methods can be used for a more accurate determination of d, but they require high computational loads and large amounts of data. Moreover, the frequency spectrum of the echo is centered around 40kHz, with a narrow bandwidth of few kilohertz's. This implies sampling frequencies above 100kHz if we intend to digitize this signal directly from the receiver.

Fortunately, it is not necessary to sample the signal at high frequencies to capture all the information present in the echo. Instead, the signal received can be previously demodulated, eliminating the 40kHz component and reducing the bandwidth of the signal to be digitized to a few kilohertz. Consequently the sampling frequency required and the total number of points digitized will be drastically reduced. To retain all the information present in the echo, a coherent demodulation must be used. This means to obtain both in-phase and in-quadrature components of the echo envelope. These two components can be digitized at low sampling frequencies and processed digitally.

In our approach, the signal supplied by the receivers is first delivered to a programmable gain amplifier. The range of gains is enough to compensate the attenuation of the signals due to the propagation losses. After this amplification, the signal is passed into a coherent demodulator, that uses the same 40kHz clock signal used in the emission of the pulse train. The demodulation uses also a 90 degrees shifted clock to obtain the quadrature component. These two components of the demodulated signal are passed into a fifth order Butterworth filter stage to remove all the frequencies above 4kHz (the effective signal bandwidth). This filtering also serves as an anti-aliasing filter before the A/D conversion, which samples these signals at 10kHz before storing them before the digital signal processing.

3. WORLD MAP BUILDING.

3.1 The scene model.

The solution to the scene reconstruction after the sensor readings is based on a inverse formulation (Soumekh, 1994). The target model in this inverse problem is useful in order to identify the properties (distance) of the objects.

The model used in this work (it is not the only one) assumes that the target region is homogeneous (water, air) and composed of a M non-sized reflector objects. Each object has its own reflection coefficient f_n that depends on its physical properties, and its coordinates x_n in the spatial domain. The model is shown in the figure 3.

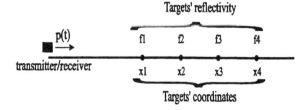

Fig. 3. Target model.

This model is correct for our work because it could be assumed that every object is a combination of infinite point reflectors on its surface.

3.2. Problem formulation.

In order to simplify the following results, we assume that the signal has no propagation losses. The transmitter, located at $x = 0$, sends a time-dependent signal $p(t)$. This is a short duration signal (pulse) with a finite duration in the time domain. The signal reaches the first target at $t_1 = x_1 / c$ (c is the speed of the wave propagation). Then the time-dependent signal that arrives is $p(t - t_1)$ and the echo is $f_1 p(t - t_1)$ and arrives after t_1 seconds. The receiver records the signal. Assuming no losses, the same $p(t)$ arrives at the second target at $t_2 = x_2 / c$. As in the previous target, the echo could be expressed as $s_2 = f_2 p(t - 2t_2)$. Finally, the total echo signal that arrives to the receiver is

$$s(t) = \sum_{n=1}^{M} s_n(t) = \sum_{n=1}^{M} f_n p(t - 2t_n) \quad (2)$$

3.3 Inversion: The deconvolution problem.

The goal is to identify the targets from the signal $s(t)$, then the last equation can be expressed as:

$$s(t) = p(t) * f_0(x) \quad (3)$$

where (*) denotes time-domain convolution, with $x = ct / 2$ being a linear transformation from x to t, and $f_0 = \sum_{n=1}^{M} f_n \delta(x - x_n)$, a spatial domain signal sum of delta functions at the coordinates of the targets. Then, we can solve the problem retrieving $f_0(x)$ from $s(t)$, this involves the deconvolution of the model, as represented in (3). This can be better done in the frequency domain, using the Fourier transform, obtaining

$$S(\omega) = P(\omega) F_0(k_x) \quad (4)$$

where k_x is a lineal function of ω. Thus, we can obtain

$$F_0(k_x) = \frac{S(\omega)}{P(\omega)} \quad (5)$$

This last formula is valid only if $P(\omega) \neq 0$ for all ω. This is not always true, because sometimes the noise makes this restriction false. The problems of deconvolution have been addressed by numerous authors (Crilly, 1991), and several approaches have been proposed. In (Anaya *et al.* 1992) a method is proposed that is compatible with real time operation,

and in this paper, we have applied it to our measurements obtaining good results, as will be described later.

The deconvolution method used consists of replace $1 / P(\omega)$ in (5) with a compensating filter $C(\omega)$ that can be defined as:

$$C(\omega) = \frac{1 - \left[1 - k \mid P_n(\omega) \mid^2 \right]^m}{P(\omega)} \quad (6)$$

where k must be comprised between 0 and 2, and m is called the expansion coefficient, and $P_n(\omega)$ denotes $P(\omega)$ normalized to its maximum. The optimum k and m must be determined in order to minimize deconvolution errors. The scenario can be now obtained using:

$$F_0(k_x) = S(\omega).C(\omega) \quad (7)$$

in addition, after inverse transform of $F_0(k_x)$, one can obtain an estimate of the scenario, $f_0(x)$.

3.4. Lateral resolution enhancement.

Given the angular response of the sensors, a punctual obstacle, like a vertical thin stick will produce echoes that will vary in amplitude depending on the angular position of the rotary sonar. This implies that an obstacle, even if it is punctual, will be represented in the map with a considerable angular uncertainty due to the transducers angular response.

This ambiguity in the angular location of the objects produces poor quality maps. However, this ambiguity can be reduced by means of data processing at various levels. The first one is to reduce the width of the lobe of the sonar array. As it has been already commented, the four-transducer array has a main lobe of ± 5 degrees approx., but has also lateral sidelobes that will produce large errors in the map. To reduce the importance of the sidelobes, a second deconvolution process has been adopted. In this deconvolution, the lateral response showed in the figure 2 is used as $p(\alpha)$ in the previous formulation. Thus, assuming that the lateral response obtained from an horizontal scan $s(\alpha)$ is the result of the convolution of $p(\alpha)$ with the scenario objects $f(\alpha)$; the previous discussion is applicable to the case of the lateral response of the sonar system, and can be used to enhance these angular readings.

The second processing level is carried out in the subsequent data processing, that is, the data fusion and integration of different scans taken from different locations, to obtain the final map of the scenario. In the construction of this maps, several approaches can be used. The bayesian approach (Abidi, 1992) is one of the most employed with good results. As the sensor described in the present work presents good lateral response, this data fusion and integration stage can be simplified and simpler data fusion algorithms can be used to obtain satisfactory results. In this work, a grid map is obtained from the different scans with a simplified algorithm that consists of:

1. After a scan has been performed, two consecutive deconvolutions are applied to the data: one axial and one lateral. In the figure 4 are represented the results of the lateral deconvolution of a scan taken at 50cm of distance of two vertical sticks placed at 90° and 270°. The figures 4a and 4b correspond with a scan from the two-transducer sensor array, and the figures 4c and 4d correspond with the four-transducer array. In these figures, the 4a and 4c are the plot of the data before the lateral deconvolution, and the 4b and 4d show the same data after the deconvolution. As can be observed from these figures, although the lateral deconvolution enhances both cases, the corresponding to the four-transducer array has considerably better angular resolution and the lateral sidelobes have decreased its amplitude.
2. The amplitude values of each scan corresponding to a particular grid cell are added and the mean is obtained as representative value for this cell. Therefore, the grids with higher amplitude values are associated with higher occupancy probability.

This approach is extremely simple and easy to implement. The preliminary results obtained show good results and are satisfactory.

4. EXPERIMENTAL RESULTS.

The four-transducer sonar head has been placed into a room with brick walls and with 6 vertical wood sticks (diameter approx.: 1cm) distributed on it. Ten scans have been obtained from different sonar positions and its data have been processed using the described method. The data processing after each scan give us a high quality data about the position and the distance of the surrounding objects (sticks), and also the walls and the corners. In the figure 5 can be observed the grid map obtained after a single scan position(cell size: 2x2cm). As it can be observed, the quality of the map is good, although not all the objects are detected in this single scan. To build a high level map of the scenario, the data obtained from the ten scans have been fused, yielding a more accurate map with reduced uncertainty. So, in the figure 6, is depicted the grid map corresponding to the fusion of the 10 scans. Notice how all the sticks are perfectly located and detected.

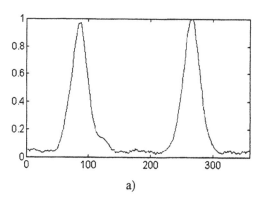

a)

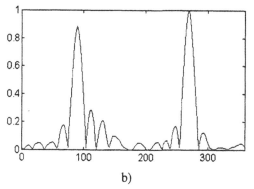

b)

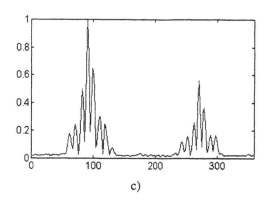

c)

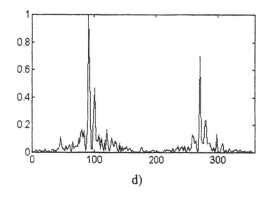

d)

Figure 4.- Results of the lateral deconvolution for the two different sonar heads. a) and c) are the amplitude scan plots at 50cm, showing two objects(sticks), located at 90° and 270°. b) and d) show the results after lateral deconvolution.

5. CONCLUSIONS.

In this paper a new ultrasonic array sensor is presented, for its use in mobile autonomous vehicles. The features of the sensor include accurate distance measurement, low cost, and reduced angular ambiguity. The preliminary tests show good enough obstacle detection results in a real environment.

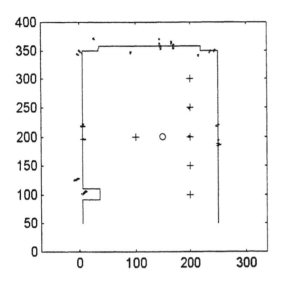

Fig. 5.- Contour plot of the scenario. It has been obtained from only one scan, taken from location marked with(O). Solid lines stand for room walls. Symbols (+) stand for vertical sticks. Coordinates are in centimeters.

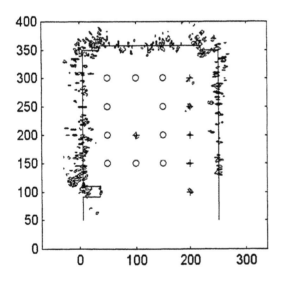

Fig. 6.- Contour plot of the scenario. It has been obtained from 10 different scans, taken from locations marked with(O). Solid lines stand for room walls. Symbols (+) stand for vertical sticks. Coordinates are in centimeters.

The signal processing treatment is adequate for a DSP implementation, allowing the use of the sensor in a real time mobile platform, like the one described in (Gil *et al.*,1997).

The use of the measurements in higher levels like sensor fusion for the map building, reduces the complexity and uncertainty of this process, allowing a signal fusion level and increasing the quality of the final map.

6. REFERENCES.

Abidi, M.A. and R.C. González (1992). *"Data fusion in robotics and machine intelligence".* Academic Press Ed.

Anaya, J.J. *et al.* (1992)."A method for real time deconvolution". *IEEE Trans. Instrum. Meas.,* **41**(3), 413-419.

Benet, G. *et al.* (1992)."An intelligent ultrasonic sensor for ranging in an industrial distributed control system". In *proceedings of the SICICA-92.* pp. 97-51. Málaga.

Borenstein, J. And Y. Koren. (1991). "Histogramic in-motion mapping for mobile robot obstacle avoidance". *IEEE Journal of Robotics and Automation,* **7**(4).

Crilly, P.B. (1991)."A quantitative evaluation of various iterative deconvolution algorithms". *IEEE Trans. Instrum. Meas.,* **40**, 558-562.

Gil, J.A. *et al.* (1997). "A CAN architecture for an intelligent mobile robot". *In Proceedings of the SICICA-97.* pp.65-70. Annecy.

Lee, D. (1996) *"The Map-Building and Exploration Strategies of a Simple Sonar-Equipped Mobile Robot".* Cambridge University Press.

Leonard, J.J. and H.F. Durrant-Whyte (1992) *"Directed sonar sensing for mobile robot navigation"* Kluwer Academic Press.

Pagac, D. *et al*(1996). "An evidential approach to probabilistic map-building". *In Proceedings of ICRA-96.*

Simó, J. *et al.* (1997). "Behaviour Selection in the YAIR Architecture". *In proceedings of IFAC Conference on Algorithms and Architectures for Real Time Control. AARTC'97* Vilamoura, Portugal.

Soumekh, M. (1994).*"Fourier Array Imaging".* Prentice Hall Ed. State University of New York at Buffalo.

DEVELOPMENT OF INDOOR MOBILE ROBOT NAVIGATION SYSTEM USING ULTRASONIC SENSORS

Seiji AOYAGI, Masaharu TAKANO and Hajime NOTO

Dept. of Industrial Engineering
Faculty of Engineering
Kansai University

Abstract: A simple localization system for indoor mobile robot navigation using ultrasonic sensors is proposed according to the "RECS"(Robot Environment Compromise System) concept. In this system some ultrasonic receivers are located on the ceiling of the room and one ultrasonic transmitter is attached to the mobile robot. The time-of-flights of the ultrasonic pulses from the transmitter to the receivers are measured and the position of the transmitter, namely the position of the robot, is obtained using the triangulation principle. In this paper we describe the method calculating 3-D coordinates and correcting the delay time of transmitting an ultrasonic pulse. *Copyright © 1998 IFAC*

Keywords: Mobile robots, Navigation systems, Localization systems, Ultrasonic transducers, Electric spark, Triangulation, Radio communication, Accuracy

1. INTRODUCTION

In order to develop intelligent robots which do welfare or home works in near future, authors have already proposed the "RECS"(Robot Environment Compromise System) concept (Takano, et, al., 1996). This concept means the technology to compromise the robot and environment in order to increase the robot performance. This concept is based on the evident fact that the development of a wholly automated intelligent robot in the following ten years is very difficult considering the present technology level of robotics. So this concept suggests that in stead of imposing all responsibility to the robot, arranging the environment so that it assists the robot to move and work smoothly and skillfully is important. This arrangement should be minimized to the extent that the human beings who live and work in the space such as hospitals or houses do not feel uncomfortable and are not obstructed. As one of the practical systems based on this RECS concept, a localization system for indoor mobile robot navigation using ultrasonic sensors is proposed in this paper. In this case the arrangement of environment is only setting small ultrasonic receivers

on the ceiling but this arrangement is very effective for the mobile robot to obtain the accurate position and orientation in the room.

2. OUTLINE AND CHARACTERISTICS OF NAVIGATION SYSTEM

2.1 Outline of the System

In this system, several ultrasonic receivers are located on a ceiling at positions of which coordinates are known in advance and one ultrasonic transmitter is attached to a mobile robot. The time-of-flights of the ultrasonic pulses from the transmitter to the receivers are measured and the position of the transmitter, namely the position of the robot, is obtained using the triangulation principle. If two transmitters of which relative distance is known are employed, the orientation of the robot also can be obtained. While all receivers are connected to each other by wire communication, it is impossible to connect the robot with the receivers by wire communication since the robot moves around in the room. So we adopted radio frequency wireless communication in the case that the robot declare to the receivers that the position measurement procedure starts from now and in the case that the receivers send detected arrival time of the ultrasonic pulses to the robot. The outline of the system is shown in **Fig.1**. A processing unit which is loaded on the robot carries out all procedures of calculating the position and orientation of the robot.

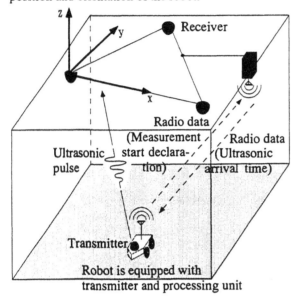

Fig.1 System composition

2.2 Transmitter Using Electric Spark Discharge

Some position measuring systems using ultrasound have been already proposed (Takamasu, et, al.,1984;Seki,et,al.,1996). These systems generally use the piezoelectric bimorph type ultrasonic transducers as transmitters. But the directivity of the transducer of this type is so narrow that the receivers can not catch the ultrasounds reliably. Moreover the

waveform of ultrasound transmitted by the transducer of this type is a burst wave of which leading edge is so dull that detecting the precise arrival time of the wave is very difficult. On the other hand, the system proposed in this paper adopts an electric spark discharge as a transmitter. The directivity of this transmitter is approximated to that of the non-directional point sound source as shown in **Fig.2**. The sound pressure of the ultrasonic pulse is 10Pa at 1m and it is enough for robot metrology. The waveform of the transmitted ultrasonic pulse is shown in **Fig.3**. It has a sharp leading edge so that the arrival time is easily detected by setting an appropriate threshold level.

Since an electric spark discharge is impulsive, the frequency band of the transmitted sound is widely distributed from DC to infinite frequency theoretically. So this transmitter produces not only necessary ultrasonic pulses but also audible noise. However this audible noise is thought to be not offensive to the ears of human beings considering that the localization procedure is not carried out so frequency because the speed of the mobile robot of this system is not so fast. According to the RECS concept as mentioned before the level of this noise is to the extent that human beings can compromise with.

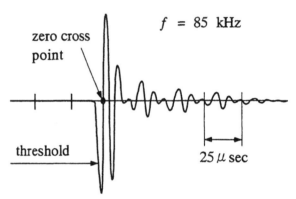

Fig.3 Received pulse waveform by electric spark discharge

Sound pressure at 1 m on middle line was 10.0 Pa and this is regarded as 0 dB. Each data is the average of 200 times measurements.

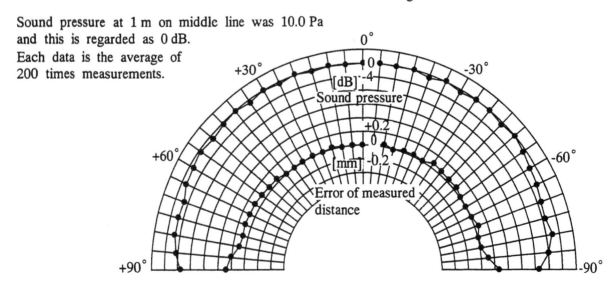

Fig.2 Directivity of electric spark discharge

2.3 Zero-Cross Method

The first zero-cross point is taken as the arrival time of the ultrasonic pulse as shown in **Fig.3**. This point is stable and gives high resolution to the ranging system even when the amplitude varies according to the change of measured distances.

2.4 Calibration of Delay Time of Radio Wireless Communication

The indoor wireless radio frequency communication is generally realized by using frequency modulation (FM) method as shown in **Fig.4**. This modulation and demodulation procedures consume delay time of about $\pm 160 \mu$ s and this time is not constant. Moreover multipass effect which is caused by the reflection at the wall or the floor varies this delay time. **Table1** shows the example of the average and the standard deviation of the delay time at carrier frequency of 260MHz. The standard deviation of 2μ s equals to 0.68mm provided that the sound velocity is 340m/s. So it is not negligible for precise position measurement of the mobile robot. In this system the method to calibrate the delay time accurately by using one redundant receiver is newly invented. The detail of this principle and procedure are explained in the next chapter 3.

3. PRINCIPLE OF POSITION MEASUREMENT

3.1 Delay Time Correcting Method

The 3-D position of a transmitter can be obtained by using at least three receives. Assume a transmitter $T(X,Y,Z)$ and three receivers $R_i(x_i, y_i, z_i)$ and let t_i denote the measured transit time of the ultrasonic pulse between T and R_i where i is 1 to 3 and let C denote the monitored sound velocity in the measured space. Then following three equations exist:

$$\left(X - x_i\right)^2 + \left(Y - y_i\right)^2 + \left(Z - z_i\right)^2 = \left(C \cdot t_i\right)^2$$
$$(i=1 \sim 3) \qquad (1)$$

However it is impossible for receivers to know the precise transmitting time as far as the robot radiates the wireless start signal to the receivers at the moment of transmitting, because the delay time exists in the wireless communication as mentioned above. So t_i includes the delay time and the precise position of $T(X,Y,Z)$ is not obtained by solving Eq.(1).

Considering these facts, the method which

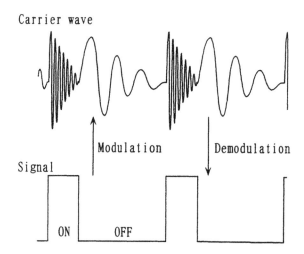

Fig.4 Frequency modulation (FM) method

employs one redundant receiver and calibrates precise delay time is newly invented in this system. The time chart of the measuring procedure is shown in **Fig.5**. First the robot declare to the receivers that the measuring procedure starts from now by using a wireless signal. This signal is called as 'start declaration' in this paper. The receivers reset their timer counters when they get the start declaration. This reset procedure can be done simultaneously by all receivers because they are connected with each other by wire communication. This time is called 'start time' in this paper. From this time, the timer counter of each receiver is counted up at intervals of 200ns. The robot transmits the ultrasonic pulse after some intervals which is set intentionally. Let the period between start time and transmitted time be t_d as shown in **Fig.5** and measured arrival time be t_i, then real transit time is obtained by $(t_i - t_d)$ where i is the number of receiver. In this procedure it is important and inevitable to estimate t_d accurately. So one redundant receiver is employed and unknown parameter t_d is taken into account. Then the equivalent formula to Eq.(1) is expressed as follows:

$$\left(X - x_i\right)^2 + \left(Y - y_i\right)^2 + \left(Z - z_i\right)^2 = C^2 \cdot \left(t_i - t_d\right)^2$$
$$(i=1 \sim 4) \qquad (2)$$

Equation (2) is able to be solved analytically because the number of unknown parameters of X, Y, Z, t_d and that of equations are both four and coincident.

3.2 Consideration on Number of Receivers

In this system typically four receivers are used. If five receivers are employed in near future, the sound velocity C is also able to be estimated precisely by

Table 1 Example of average and standard deviation of delay time at carrier frequency of 260MHz (unit; μ s)

Distance	1m	2m	3m
Delay time	158.31	165.84	164.81
Standard deviation	2.35	2.10	1.63

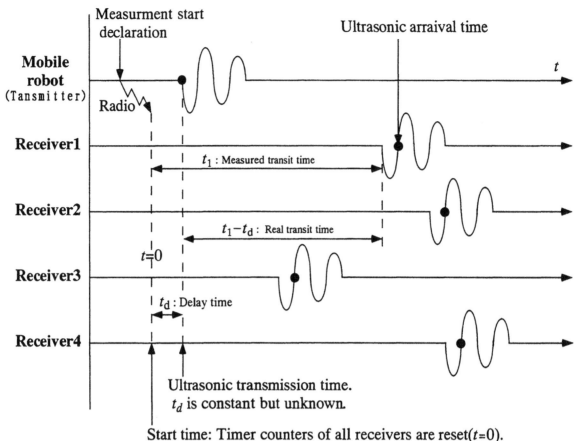

Fig.5 Timing chart of measuring procedure

dealing not only X,Y,Z,t_d but also C as variable.

When the robot moves in a wide room or corridor etc., the distances between transmitter and receivers will be too long for the ultrasonic pulse to reach the receiver and the distances will not be measured accurately. This problem will be cleared by dividing the measured area to several small areas and employing four or five receivers for each area.

4. PRELIMINARY EXPERIMENTS ON 2-D PLANE

In order to estimate the accuracy of position measurement of this localization system, preliminary experiments on 2-D plane were carried out. Three receivers are enough for obtaining the position of $T(X,Y)$. The experimental condition is shown in **Fig.6**. Nine lattice points were set as a calibration standard of which intervals are just 100mm. One transmitter was positioned at these lattice points and 2-D coordinate of $T(X,Y)$ was measured for 100 times at each point by using the developed localization system. Sound velocity C was always monitored directly by measuring the time-of-flight of an ultrasonic pulse which goes and backs for the known distance of 668mm. In this experiment the transmitter and the receivers were connected by wire communication and radio frequency wireless communication was not used in order to simplify the

experimental setup. Under this condition, the delay time of t_d was intentionally given by setting the interval between start time and transmission time to just 160 μ s by using a precise function generator.

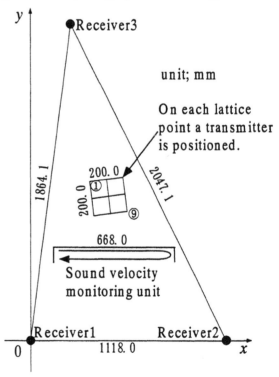

Fig.6 Experimental setup on 2-D plane

Experimental results are shown in **Table2**. The error of distance measurement, which is equivalent to the error of position measurement, was about $\pm$ 0.6mm. The standard deviation of the measured distances was less than 0.1mm (the data of the standard deviation were omitted for want of space).

The average of estimated t_d was 159.66μs and the deviation of it from 160μs was 0.34μs. The standard deviation of the estimated t_d was less than 0.5μs. These 0.34μs and 0.5μs are equivalent to 0.11mm and 0.17mm respectively provided that the sound velocity is 340m/s.

Considering that the accuracy of 2-D position measurement is better than 1mm, this measuring system has good potential to measure 3-D position with accuracy of submillimeter order and has good potential for the practical mobile robot navigation.

Table 2 Detected distance among the lattice points

Lattice point number	Distance (mm)	Deviation (mm)
①⇔②	99.47	0.53
①⇔④	99.96	0.04
②⇔③	99.59	0.41
②⇔⑤	99.24	0.76
③⇔⑥	99.02	0.98
④⇔⑤	100.26	0.26
④⇔⑦	99.65	0.35
⑤⇔⑥	100.83	0.83
⑤⇔⑧	99.46	0.54
⑥⇔⑨	99.90	0.10
⑦⇔⑧	100.40	0.40
⑧⇔⑨	99.57	0.43
Average	99.78	0.62

※Each data was measured for 100 times.

※Deviation means absolute deviation from 100mm.

※Distance and deviation for twelve distances between each lattice point were averaged and showed in the average column.

5. CONCLUSION

The localization system for a mobile robot navigation using ultrasonic sensors is proposed according to the RECS(Robot Environment Compromise System) concept. The concept and the advantages of this system were explained. This system has high accuracy of position measurement because of adopting an electric spark discharge as the transmitter, taking the first zero-cross point of the pulse waveform as the arrival time and estimating the delay time of radio wireless communication accurately. Preliminary experiments on 2-D plane were carried out. It was proved that the accuracy of this system is about $\pm$ 0.6mm for 2-D position measurement and that this system has good potential for the practical mobile robot navigation.

REFERENCES

Takano, M., Yoshimi, T., Sasaki K., and Seki, H. (1996). Development of Indoor Mobile Robot System Based on RECS Concept. *J. Japan Soc. for Precision Eng., Vol.62, No.6* , pp.815-819.

Takamasu, K., Furutani, R. and Ozono, S. (1984). Development of the Three Dimensional Coordinate Measuring System with Ultra-Sonic Sensors. *Proc. Int. Symp. Metrology for Quality Control in Production*, pp.300-305.

Seki, H., Tanaka, Y., Takano, M. and Sasaki, K. (1996). Position and Orientation Measuring System for Indoor Mobile Robot Using Ultrasonic Lighthouse. *Proc. Japan Soc. Precision Eng. Fall meeting*, pp.13-14.

A DYNAMIC MODEL FOR IMAGE SEGMENTATION. APPLICATION TO THE PROBLEM OF WEED DETECTION IN CEREALS

A. Rodas and J.V. Benlloch

Departamento de Ingeniería de Sistemas, Computadores y Automática
Universidad Politécnica de Valencia
P.O. Box 22012, 46071-Valencia, Spain
{arodas, jbenlloc}@disca.upv.es

Abstract: Starting from the study of embedded snakes applied to the case 2D, image segmentation problem has been tackled. In order to adapt these techniques, typically used in visualization of noisy data, to further interpretation tasks, a dynamic model that evolves according to certain morphological transformations is proposed.
Finally, the application of the method to a real problem, weed detection in cereals under actual field conditions, is studied. Taking into account the feasibility of a hardware implementation, future work will aim at developing a prototype of autonomous vehicle able to perform farming tasks in a precision agriculture framework. *Copyright © 1998 IFAC*

Keywords: Computer vision, Binary images, Scene segmentation, Modelling, Agriculture.

1 INTRODUCTION

Segmentation plays an important role in image analysis as well as in high-level image interpretation and understanding. Both in 2D (Haralick and Shapiro, 1985; Kasturi and Jain, 1991) and 3D images (Ayache, et. al., 1995), authors distinguish between:

- Edge detection: includes filtering stage (pre-process + gradient or laplacian), edge tracking and closing, and high-level representation of edges with three approaches: *linking, polygonal approximation* and *optimal boundaries*.

- Region extraction either through *classification methods* (partition of the data space: luminance, colour, texture...) or *region growing algorithms*.

Related to *optimal boundary methods* are *active contours* (also known as *snakes* or *deformable models*), which represent the major interest of this paper. These methods can also be divided in:

- Energy-minimizing methods. A *deformable model* that evolves under the action of internal (elasticity properties of the model) and external forces, by attracting the model towards some detected edges, is used.
- *Snake Splines*. Considers only extern energy provided that internal energy is embedded in the border representation.

Here, several modified algorithms have been proposed for improving the original method introduced by Kass, et al. (1988) and/or to adjust the contour energies in order to achieve a certain purpose (Lin, et al., 1996; Park, et al., 1996; Yu and Yla-Jaaski, 1995;...). In this paper, starting from the studies of Terzopoulos et al. (1987) and Terzopoulos and Metaxas (1991), the work is focus on Whitaker and Chen approximation (1994) applied to case 2D for segmentation tasks.

This paper is structured as follows: in section 2 the background of such models is given. Section 3 is devoted to the description of a dynamic model that found inspiration in the ideas stated previously. In section 4 a real application of the proposed method is studied and finally, in section 5, the conclusions and further work are presented.

2. SNAKES AND EMBEDDED SNAKES

2.1 Snakes

Kass, et al., (1988) describe an energy-minimizing curve C (*snake*), parameterized along its length s, which combines internal and external energy constraints according to (1).

$$\frac{\partial \bar{C}}{\partial t} = \alpha \frac{\partial^2 \bar{C}}{\partial s^2} + \beta \frac{\partial^4 \bar{C}}{\partial s^4} - \gamma \bar{N} g(\bar{C}) \quad (1)$$

The first two right terms correspond to internal forces and the last one corresponds to external or image force. The first term, the "membrane" effect, causes the contour to shrink and the second one, the "rigidity" term, causes the contour to become straight. In absence of external forces, the *snake* evolves over time under elastic properties. The last term pushes the curve towards salient image features like lines or edges, which correspond, to the desired image attributes[1].

The relative influence of the above terms is controlled by a set of parameters: α and β are elastic parameters of the model so that α determines resistance to stretching and discontinuity whereas β determines resistance to bending. The parameter γ controls the influence of external forces.

The evolution process is achieved through the numerical solution of equation (1) that corresponds to two independent equations (for x and y coordinates).

2.2 Embedded Snakes

Whitaker and Chen (1994) use a scalar function, represented by a greyscale image, where the minimization problem between internal and external forces is tackled. The method proposes an implicit representation of a snake as a level curve of the scalar function, what is known as *embedded snake*. The advantages are: independence of any particular parameterization, possibility of "splitting" the model into pieces to form multiple objects and versatility in creating the initial model.

---

[1] Here, edge points have been considered:
$\bar{\nabla} g(\bar{C}) = \bar{\nabla} \left(\sqrt{\bar{\nabla} I(\bar{C}) \cdot \bar{\nabla} I(\bar{C})} \right)$ where **g** is a function indicating presence of interest points and **I** denotes the image.

It is expected obtaining an equation in the form:

$$\frac{\partial F(\bar{C}, t)}{\partial t} = f_{ext} - f_{int} \quad (2)$$

In such a way that $F(x, y, t) : \Re^2 \to \Re$ is the scalar function and the terms f_{int} y f_{ext} represent internal and external forces equivalent to those of equation (1). In this case, the forces have to be referred to F instead of C.

The function F, which evolves over time, satisfies $F(\bar{C}, t) = K$ and the time derivative is:

$$\frac{dF(\bar{C}, t)}{dt} = \frac{\partial F(\bar{C}, t)}{\partial t} + \bar{\nabla} F(\bar{C}, t) \cdot \frac{\partial \bar{C}}{\partial t} = 0 \quad (3)$$

Thus,

$$\frac{\partial F(\bar{C}, t)}{\partial t} = -\bar{\nabla} F(\bar{C}, t) \cdot \frac{\partial \bar{C}}{\partial t} \quad (4)$$

Combining equations (1) and (4) and considering, for simplicity, only second-order derivatives, results the following evolution equation:

$$\frac{\partial F(\bar{C}, t)}{\partial t} = -\bar{\nabla} F(\bar{C}, t) \cdot \left(\alpha \frac{\partial^2 \bar{C}}{\partial s^2} - \gamma \bar{N} \sqrt{\bar{\nabla} I(\bar{C}) \cdot \bar{\nabla} I(\bar{C})} \right) \quad (5)$$

It is shown from (2) that the external force is:

$$f_{ext} = \gamma \bar{N} F(\bar{C}, t) \cdot \bar{\nabla} \sqrt{\bar{\nabla} I(\bar{C}) \cdot \bar{\nabla} I(\bar{C})}$$

The goal is also to express f_{int} as a function of F, in order to work on the image space (model) instead of doing it in the parameterized space of the curve. The method described proposes to express derivatives of F in *Gauge coordinates* (v,w), using subscripts v, w to indicate derivatives in direction tangent and perpendicular to the level curves, respectively (Fig. 1).

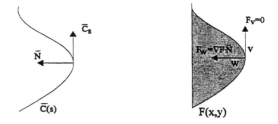

Fig. 1. Equivalences between a *Snake* and a level curve

Thus, it can be seen on Whitaker and Chen (1994).

$$f_{int} = \alpha \frac{\partial^2 F(\bar{C}, t)}{\partial v^2} \quad (6)$$

In order to overcome the complexity of solving (6), a simplified version of f_{int} depending only on the function gradient of F, has been proposed. Below, Fig. 2 shows a practical example of both a *snake* and *embedded snake* evolution, following this approach. It can be noticed that the *embedded snake* evolution is not a single contour, but a contour density in the image space.

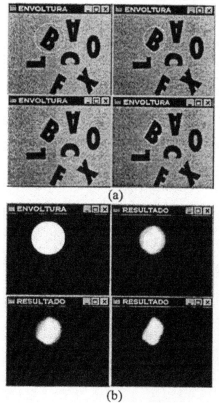

(a)

(b)

Fig. 2. *Snake (a)* and *embedded snake (b) evolution*

3. DYNAMIC MODEL WITH MORPHOLOGICAL OPERATIONS

In the former section a model that evolves over time has been introduced. This technique is typically used in the visualization of sparse, fuzzy or noisy data, because the resulting model is in fact a greyscale image from which a rendering can be carried out.

Nevertheless, in the case of segmentation tasks, where a final binary model is required, it is necessary to include further steps, as thresholding, which is not always a simple process.

Another idea consists of building a binary model, called *dynamic model*, in such a way that evolves according to certain morphological transformations and stops in the presence of interest points of the image to be segmented. It is convenient to point out that this model is not a true *deformable model*, but inspired on it.

The algorithm, which is outlined after, decides the evolution according to the local comparison between internal and external forces in equation (2)

$$f_{int} > f_{ext} \qquad (7)$$

If the inequality is accomplished, the model shrinks (through a modified erosion), otherwise, the model stops in the point under analysis. Another morphological operation (closing) is performed in each iteration in order to avoid noise effects own to the local analysis. Fig. 3 depicts the process.

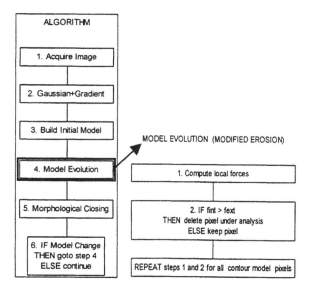

Fig. 3. Dynamic model algorithm

It can be noticed that the dynamic model described above is related to hybrid-linkage region growing algorithms, based on edge operators.

Moreover, this approximation has the advantages reported in Section 2.2 for embedded models as well as simplifies computing complexity of the evolution equations. Next, Fig. 4 shows an example of the model evolution.

Fig. 4. Example of model evolution.

It is necessary to bear in mind that the time to achieve the model convergence is strongly influenced by the chosen initial model: as close the initial model is to the targeted objects as fast the algorithm runs. Some work leading to automatic generation of optimal initial models has been carried out.

In any case, application to real-time problems requires a hardware implementation of the proposed method. In this line, FPGA based circuits have been used to implement morphological operations, showing a significant reduction in time processing (Albaladejo, et al., 1997).

4. APPLICATION: WEED DETECTION IN CEREALS

The method presented above can be applied to a good number of problems including segmentation tasks. It results particularly interesting the cases of image analysis developed for agricultural applications (Tillett, 1991), where the variability of both natural objects (in size, shape, ripeness...) and lighting conditions (agricultural operations mostly occur outdoors where controls are difficult), forces to more complex solutions.

As recent investigations have demonstrated that weeds are distributed non-uniformly across arable fields, weed control based on conventional practice of broadcast or banded applications of herbicide is therefore undesirable, in both economic and environmental terms. In order to implement site-specific weed management, information on weed location is required. Since manual surveying is a labour-intensive method, automatic techniques for determination of weeds have been proposed.

Even though monitoring weeds a few days before the herbicide application ("weed mapping") may seem cumbersome compared to real-time patch spraying, it has several advantages. Christensen, et al. (1996) states that the optimum time for weed detection may not be the optimum time to spray. Moreover, other sources of information (species, nutrient maps, available herbicides or even historic data about weed patches and/or crop yield maps in previous years) may be very useful to support the "dose map".

This approach supposes to survey a set of field samples and to build up a complete weed map by using geostatistical interpolation procedures.

In this paper, the application to the problem of automatic weed mapping in cereals under actual field conditions is undertaken.

The idea is to acquire images while an autonomous vehicle borne sensor is moving forward through the tramlines at a typical speed of 0.25m/s. To reduce segmentation errors due to lighting changes, a diffuser is placed over the scene. The vision system includes a 768*574 pixel array CCD focusing a 0.67*0.5m area, giving a resolution of about 1 pixel per mm.

In order to validate the dynamic model, a typical assortment of colour images representing 40 locations of spring-barley (including different row crop and weed densities, various weed species, diverse soil background ...) has been tested. Fig. 5 (left) shows a typical sample.

Segmentation steps have as final objective to distinguish between three classes: soil (background), crop and weeds. Fig. 5 (right) depicts a graylevel representation of an index image (using Red and Green channels) introduced to make easier the discrimination between plants and background (Benlloch, et al., 1995).

In spite the enhanced contrast of the index image, it results difficult to seek an automatic threshold that permits the discrimination of vegetal material to be fulfilled. Some approaches tend to consider as plants many pixels that are in fact background, whereas other classifies as soil, points belonging to vegetation. On the other hand, the noise present into the images is responsible of many misclassifications, considering little soil spots as weeds (Benlloch, et al., 1997; Pérez, et al., 1997).

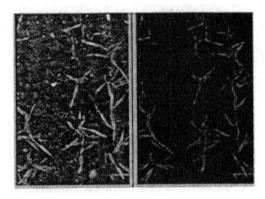

Fig. 5. Initial and NDI images.

In this paper the discrimination between soil and plants by using the former *dynamic model* is undertaken. The approach is based on the definition of an initial model for vegetal material (starting from a safe threshold in the index image histogram); that evolves according to the algorithm described in previous sections (Fig. 3). Once the model is stabilised, this result is combined with the original index image in such a way that segmented objects (crop and weeds) are clearly outlined.

Fig. 6. Initial model and segmentation results.

Fig. 6 shows an example of how the approach works. On the left, the initial model; at the centre, the final state evolution of the model and on the right, the segmented image, after combining the former result with the index image. This last step has been included to achieve a more realistic solution.

Results show that the presented method is able to reduce one of the main error sources on global thresholding approaches ("twisted leaves"). On the other hand, contour definition improvement will reduce misclassifications in further shape analysis steps (López, et al., 1997; Pérez, et al., 1997).

Considering a 512x512 image, total computing time is about 50 seconds (using a Pentium 200 MHz). This time allows the weed mapping approach to be fulfilled but results inadequate to process image sequences where a hardware implementation has to be used. In this case, a more accurate definition of weed spatial distributions would be reached, making unnecessary geostatistical interpolation procedures.

5. CONCLUSIONS AND FUTURE WORK

In this work a review of the literature on segmentation has lead to focus on the generally known as *deformable models*. After testing classical *snake* approaches, an implicit representation of a *snake* as a level curve of the scalar function, *embedded snake,* has been selected. This variant has many advantages compared to classical models and is typically applied to visualization problems of sparse, fuzzy or noisy data.

In order to tackle segmentation tasks, where a final binary model is required, a *dynamic model* that evolves according to certain morphological transformations has been proposed.

The application of this approach to the difficult problem of weed detection in cereals, under actual field conditions, has demonstrated a very good performance.

Present work is devoted to find alternative external force functions as well as improve initial model creation. Generalization to the 3D case is also another challenging objective.

Taking into account the feasibility of a hardware implementation, further work will aim at developing a prototype of autonomous vehicle able to perform farming tasks in a *precision agriculture* framework.

REFERENCES

Albaladejo, J., J. Gracia, L. Lemus, D. Gil, J.C. Baraza and P.Gil (1997). Diseño de Operadores Morfológicos en VHDL. In: Proceedings of the Seminario Anual de Automática, Electrónica Industrial e Instrumentación (SAAEI'97), Valencia, Spain, pp 745-750.

Ayache N., F. Cinquin, I. Cohen, L. Cohen, F. Leitner and O. Monga (1995). Segmentation of Complex 3D Medical Objects. *Data Acquisition and Segmentation.* Cap 4.

Benlloch J.V., M. Agusti, A. Sanchez and A. Rodas (1995). Colour Segmentation Techniques for Detecting Weed Patches in Cereal Crops. In: Proceedings of *4th Int. Workshop on Robotics in Agriculture & the Food Industry,*. Toulouse. France, pp. 71-81.

Benlloch J.V., T. Heisel, S. Christensen and A. Rodas (1997). Image analysis techniques for determination of weeds in cereal. In: *International Workshop on Robotics and Automated Machinery for Bio-productions. (Bio-Robotics 97)*, Valencia (Spain), pp. 195-200.

Christensen, S., T. Heisel, T., and A.M. Walter (1996) Patch spraying in cereals. In: Proceedings of the Second International Weed Control Congress, Copenhagen, Denmark, pp 963-968.

Haralick R.M. and L.G. Shapiro (1985). Image Segmentation Techniques. *Computer Vision, Graphics and Image Processing,* **Vol. 29,** pp. 100-133.

Kasturi R. and R.C. Jain (1991). *"Computer Vision. Principes".* IEEE Computer Society Press. pp. 65-75.

Kass M., A. Witkin and D. Terzopoulos. (1988). Snakes: Active Contour Models. *International Journal of Computer Vision,* **Vol. 1,** pp. 321-331.

Lin F., H.S. Seah and Y.T.Lee (1996). Deformable Volumetric Model and Isosuperface: Exploring a New Approach for Surface Boundary Construction. *Computers and Graphics,* **Vol. 20,** pp. 33-40.

López F., A.J. Pérez and .M. Agusti. (1997). A Comparative Study of Pattern Recognition Methods on Weed Detection Problem. In: *Preprints of the VII National Symposium on Pattern Recognition and Image Analysis,* Barcelona, Spain, pp. 44-45.

Park J., D. Metaxas, A.A: Young and L. Axel (1996). Deformable Models with Parameter Functions for Cardiac Motion Analysis from Tagged MRI Data. *IEEE Trans. on Medical Imaging.* Vol. 1, pp. 378-289.

Pérez, A.J., F. López, J.V. Benlloch and S. Christensen, (1997). Colour and Shape Analysis Techniques for Weed Detection in

Cereal Fields. In: *Proceedings of the First Europen Conference for Information Technology in Agriculture*, Copenhagen, Denmark (H. Kure, I. Thysen, A.R. Kristensen (Ed.)), pp. 45-50.

Terzopoulos D.and D. Metaxas (1991). Dynamic 3D Models with Local and Global Deformations: Deformable Superquadrics. *IEEE Trans. on Pattern Analysis and Machine Intelligence*, **Vol. 13**, pp. 703-714.

Terzopoulos D., J. Platt, A. Barr and K. Fleischer (1987), Elastically Deformable Models. *Computer Graphics*, **Vol. 21**, pp. 205-214.

Tillett, R.D. (1991). Image Analysis for Agricultural Processes: a Review of Potential Opportunities. *Journal of Agricultural Engineering Research* **Vol. 50**, pp. 247-258.

Whitaker R.T. and D.T. Chen (1994). Embedded Active Surfaces for Volume Visualization. *SPIE*, **Vol. 2167**, pp. 340-352.

Yu, X. And J. Yla-Jaaski (1995). Interactive Surface Segmentation for Medical Images. *SPIE*, **Vol. 2564**, pp. 519-527.

ON LINE MEASUREMENT OF GRASS FLOW IN THE FIELD

C.Werkhoven

IMAG-DLO, P.O.Box 43
6700 AA Wageningen, The Netherlands
C.Werkhoven@imag.dlo.nl

Abstract

At IMAG-DLO, a project is carried out for precision farming in grassland. A Mobile
Weighting Unit (MWU) is used for the on-line weighting of the grass crop after mowing of
the crop (Fig.1). Under the MWU (a modified Heston swath mower) a conveyer belt with a
width of 1.2m is mounted. On the IMAG experimental farm in Duiven measurements are
done to get information about the yield and distribution of the grass along the field. 8 test
rows are measured and the results are stored on the computer on the MWU.
A ground radar system is used for the exact positioning of the MWU in the row. In the future
a D-GPS system will be tested for the accurate positioning in the field.
Copyright © 1998 IFAC

Keywords: precision farming, weight measurement on line

Fig. 1 Mobile Weighing Unit

1. Introduction

Precision farming (Yield mapping) for crop production
is useful for the reduction of fertilizers by applying only
on plant level if necessary. Reducing loss of fertilizer is
then possible because the field is divided in small maps
and the yield of every small field is known.

For the registration of the yield of the crop it is
necessary to know the position of the equipment in the
field. Currently, the position of the MWU (mobile
weighting unit) is very accurate detected in one
direction by radar. To measure the flow with sufficient
accuracy, a weight belt conveyer is needed with a small
mass. The movements caused by the driving over the
field has then little influence on the accuracy of the
measurements.

For the measurements with the MWU, a conveyor belt is built under a Heston swath mower; the mower part is replaced by a pickup. So the mowed grass swath can easily picked up on the conveyer belt, weighted and put back on the ground. On the back of the MWU a platform is built where the computer and the box with the interfacing is placed.

2. Material and methods

2.1 Measuring system mechanical

A conveyer is used with a length of 250 cm and a width of 120 cm and a Ramsey weight unit. The conveyor is attached to a Heston mower, the mowing part being replaced by a pick-up. The conveyor and the pick-up are driven by hydro-motors and controlled by a computer on the Heston by electric driven valves.

The speed of belt is adjustable from 1 to 2 m/s. the speed of the MWU is 1 to 5 m/s.

The Ramsey unit calculates the load (kg/m), rate (kg/h), and Total (kg). The data is recorded on the hard disk of the computer.

A 220V alternator supplies the power for the computer and the weighting unit.

2.2 Measuring system electronic

For the registration of the yield of the crop production it is necessary to know exactly the position of the MWU in the field. The position of the MWU is now detected by two-ground radars mounted on each side of the MWU.

The Ramsey weight integrator unit MIDI 44-101 is microprocessor-based equipment designed for industrial application.

The load cell is connected to a weight idler under the conveyor belt, see fig 2.

Functions integrated into the instrument are:

- Load cell supply.
- Speed signal measurement.
- Rate calculation.
- Automatic calibration
- Auto Zero Tracking
- Indication of: Rate, speed, belt loading, totalized.
- Weight signal measurement.

The Field mounting box is waterproof (height 370 mm, width 255 mm, depth 138 mm).
The speed of the belt is measured with a pulse sensor

on the shaft of the belt. In the hydraulic motors for the pickup and the belt also pulse sensors are available.

They are used to control the speed of the belt and the pickup.

The box is connected to the computer over a serial interface. The unit is so programmed in such a way that every 10 s the data is available in the PC.

Output Ramsey every 10 s:

Date	dd/mm/yy
Time	hr/mm/ss
Rate	kg/h
Speed	m/s
Load	kg/m
Tot	kg
Rtot	kg

In the PC an interface card Microlink 551 is used with analog and digital I/O to actuate the electrical valves for the conveyor belt and the pickup and for counting pulses.

2.3 Software

Under Windows 95, Windmill software is used for the communication and controlling the MPU. In the background, the input lines are monitored and available via DDE-links (direct data exchange). The program for controlling the MPU is written in Delphi. (Fig. 4).
The measured data is logged every 10 seconds during

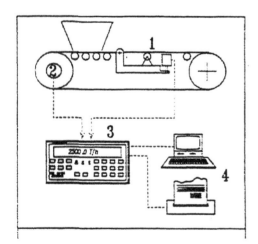

Fig. 2 weight integrator

Fig.3 Ground radar

one session and stored in a separate file.

Software written in Delphi is used for the processing of the data in the files.

The speed of the belt and pickup can be chosen with the cursor on a scrollbar. The speed is held constant by processing the pulses from the sensors. If necessary the speed is corrected via electric/hydraulic valves for the hydro motors. Another off-line program is used to process the data files and achieve as result the yield of the field in kg/m2.

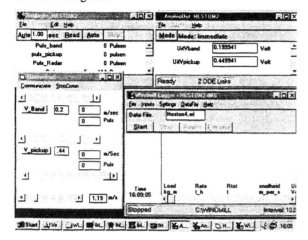

Fig. 4 screen layout

2.4 Experimental setup

At the experimental farm of IMAG-DLO at Duiven, a test field (110 x 110m) is used for the weight measurements. The field is divided into 8 paths for experiments with the MWU. Before the measurements the 8 paths are mowed out. Each path has a width of 6 m and a length of 85 m. The mower divides the path in three rows width a width of 3 m. One row is used for the measurements. Every row in the 8 paths is measured separately. At the begin of the row the datalogging is started and at the end the logging is stopped, the result is 8 data-files of the test field. After the measurement the grass is picked up from the field and the weight of 8 rows is registrated on the weighbridge on the farm.

3. Result of the measurements

3.1 Grass flow

The test field of 1ha is first mowed and then divided in 8 rows with a width of 1m and a length of 85m. Every row is recorded in a separate file. The mowing of the grass started at 9.30 a.m. The results are used to make yield maps of the field and a graphic representation of the measurements (load, rate, total and speed). Fig. 5 gives the results of the row 4. Every 10 s the data is recorded. The load is expressed in kg/m, the rate in kg/h and the speed of the MWU in m/s. Table 1 gives the recorded data of one row.

The mowing of the grass started 10-7-1997 at 9.30.

Every 10 s the flow is recorded.

After the field measurements the grass is weighted on the farm, the results are:

Weight path 1,2,3,4 1340 kg

Path 5,6,7,8 1256 kg

Total 8 rows 2596 kg

The results of the MWU are:

Ramsey	1	340	kg
	2	330	kg
	3	390	kg
	4	310	kg
	5	390	kg
	6	310	kg
	7	340	kg
	8	210	kg
Total:		2620	kg

A delay of about 15-min occurs between the MWU measurements and the weighting afterwards on the farm weighting bridge.

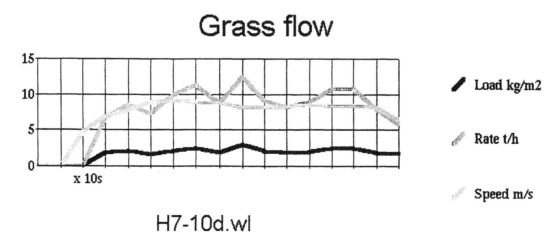

Grass flow

/ Load kg/m2

/ Rate t/h

Speed m/s

x 10s

H7-10d.wl

Fig. 5 result row 4 load in kg/m2

319

3.2 Positioning

The two signals of the radars are recorded every 10 seconds while driving over the field, fig.6 gives the result while driving a rectangle and fig. 7 the result of driving in a circle.

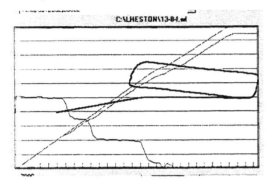

Fig.6 Two radars: rectangular

4. Future developments

Up to now the measurements of the speed and distance were first done with one radar sensor, but a second sensor is placed on the MWU and experiments has been carried out to see if it is possible to position the MWU in the field at a higher accuracy. A Differential Global Position system (D-GPS) will be placed on the MWU to test its accuracy; the MWU will then be used as an AGV (Autonomous Guided Vehicle). The mechanical gearbox of the MWU will be replaced by 2 hydraulic motors, which can be controlled by an electric signal from the computer. The sampling rate for the radar will be 1 second or higher if needed.

5. Conclusions

With the MWU accurate in-line measuring of the crop flow during the grass harvest is possible. The yield is now calculated in intervals of 10s, with a speed of 0.56 m/s a map of 2.1 x 0.56 = 12 m2 is measured. Now the

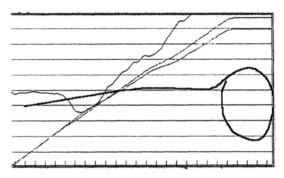

Fig.7 Two radars: circular

position of the MWU can only be recorded in the drive direction of the MWU with one ground-radar. With D-GPS and an extra ground-radar sufficient accurate positioning in the field will be possible with a higher reliability. When obstacles like trees and buildings block the GPS satellite signals, positioning with only GPS can lead to erratic results. With the characteristics of the of the two systems, combined, it is possible to construct a reliable positioning system with long term accuracy.

References

Bergeijk, J. van, Goense D.,Keesman K.J. (1996) Enhancement of global positioning system with dead reckoning. AgEng 96 Madrid paper 96G-013

Larsen, W.E., Tyler, D.A., Nielsen, G.A. (1991) Using the GPS satellites for precision navigation. Automated Agriculture for the 21st century. ASAE publication No. 11-91,201-208.

Zuydam, R. van, Werkhoven, C. (1997) High-Accuracy remote position fix and guidance of moving implements in the open field. Precision Agriculture 1997 1997 Bios Scientific Publishers Ltd

MODELLING AND OPTIMIZATION OF THE DYNAMIC BEHAVIOUR OF SPRAYER BOOMS.

Patrik Kennes [*], Jan Anthonis, Herman Ramon

K.U.Leuven, Faculty of Agricultural and Applied Biological Sciences
Department of Agro-Engineering and -Economics
Kardinaal Mercierlaan 92, B-3001 Heverlee, BELGIUM
(*)Tel: +32-16-32.14.78; fax: +32-16-32.19.94
email: patrik.kennes@agr.kuleuven.ac.be

Abstract: Finite element simulations show that existing suspensions approximately halve vertical sprayer boom movements compared to a fixed boom. The reduction of horizontal boom vibrations however is negligible. This shows the need for a performant horizontal suspension. Because of the high mass and inertia of sprayer booms active suspensions would require too much engine power. Therefore the possibilities of a passive suspension are investigated. Optimal suspension settings (spring stiffness, damping constant) are determined by optimization of the singular value plot derived from a parameterized linear multi-body model of the boom and suspension. Validation experiments confirm that the used modelling techniques are reliable. *Copyright © 1998 IFAC*

Keywords: passive suspension, modelling, optimization, finite elements, dynamics, validation

1. INTRODUCTION

Chemical products for crop protection are usually distributed as liquids by a field sprayer. Unwanted movements of the sprayer boom, effected by soil unevenness, create local under- and overdoses of spray liquid due to a varying velocity and position of the nozzles with respect to the ground. Vertical vibrations give rise to variations in spray deposit between 0 % and 1000 % and horizontal vibrations between 20 % and 600 %, while 100 % is ideal (Langenakens et al, 1995). To obtain the desired response on the total field, the advised dose is in excess of that required to make sure that there is enough product in each area. However, environmental and economical reasons demand that farmers use less chemicals which must be better distributed to maintain their effectiveness. This requires more performant boom suspensions because

they reduce the unwanted boom motions and thus increase the homogeneity of the spray pattern, just as the suspension of a vehicle influences its driving comfort and road-holding. Finite element models for different existing (mainly vertical) boom suspensions reveal their effectiveness for reducing vertical boom movements, while they are ineffective for the horizontal boom movements. Horizontal movements are much more difficult to reduce by a passive suspension than vertical vibrations because there is no reference level (as the soil is for vertical movements), and there is no gravity force that defines the static horizontal equilibrium position of the boom. Just as for a vehicle suspension, a sprayer boom suspension should pass the low frequent tractor motions (e.g. turning at the headland of the field), while high-frequent tractor movements, which cause boom deformations, have to be filtered. Active suspensions would require much engine power because of the large weight of the sprayer boom. Therefore the possibilities of a passive boom suspension are investigated.

2. MODELLING BOOM VIBRATIONS

A non-linear Finite Element Modelling software package (Samcef®) is used to simulate boom movements of a sprayer during field operation. This requires knowledge of the soil profile under the wheels and data of the geometric and physical properties of the tractor and the sprayer (stiffness of the tires, friction in hinges, ...). In the first stage, the sprayer and tractor body are represented as a rigid bodies with corresponding mass and inertia characteristics. Numerical values of tractor mass and inertia, the tire stiffness and damping are based on data given in the article of Crolla *et al* (1990). Contact between the wheels and the ground is modelled by flexible wheel elements. The tires can have radial deformation, while the ground is assumed to be rigid. The front axle is connected to the tractor body by a hinge that allows it to pendulate. Movements at the hitchpoint of the boom suspension are calculated for a tractor driving across a standardized bumpy road (according to the ISO norm 5008) at a speed of 5 km/h . In a next stage, these disturbances are applied as input excitations for the sprayer models to simulate boom movements for several types of existing vertical suspensions (pendulum, single and double trapezium). The modelled sprayer boom has a working width of 24 meters and consists of five parts, which allow the boom to be fold for transportation (figure 1).

Figures 2 and 3 show the vertical movements of the boom tip for a 24 meters width spray boom respectively fixed to the sprayer frame and with a pendulum suspension. It is clear that the suspension filters out high frequency boom movements, so boom deformation and stress in the beams are reduced. The low-frequent sinusoidal signal that can be found in the boom tip motion is due to the rigid body rolling motion of the boom with respect to the pivot point of the pendulum. The amplitude of the vertical motion of the boom tip with suspension is almost halve of the motion of the boom tip without suspension. The reduction of horizontal boom vibrations by existing suspensions however is negligible (figures 4 and 5), while very small horizontal vibrations with an amplitude of 30 cm can cause large fluctuations in spray distribution (Ramon, 1993). From this it can be concluded that the development of a performant horizontal suspension is necessary. Because of the high mass and inertia of sprayer booms (up to 36 meters wide) active suspensions would require too much engine power. Therefore the possibilities of a passive boom suspension are investigated.

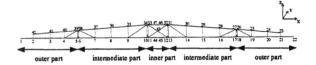

Fig. 1. Finite element model of 24-m sprayer boom.

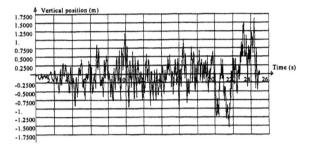

Fig. 2. Simulated vertical movements of the boom tip (fixed boom).

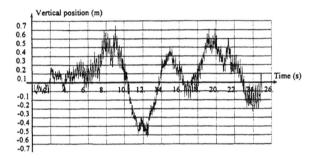

Fig. 3. Simulated vertical movements of the boom tip (pendulum suspension).

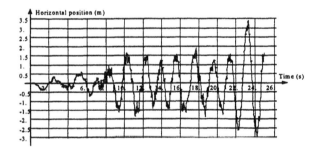

Fig. 4. Simulated horizontal movements of the boom tip (fixed boom).

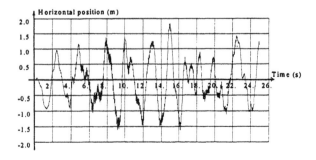

Fig. 5. Simulated horizontal movements of the boom tip (pendulum suspension).

3. OPTIMIZATION OF THE BOOM SUSPENSION

3.1. Description of the experimental suspension

An experimental improvement of the performance of the boom suspension would involve costly and time-consuming prototyping and measurement efforts. Therefore a parametric linear multi-body model of a boom with a passive horizontal suspension is built. This implies that the important characteristics of the boom and its suspension, such as the mass and moments of inertia of the boom and the properties of applied springs and dampers are represented by parameters in the equations of motion, so that their value can be easily adapted. The suspension has two degrees of freedom, one translational and one rotational, to isolate the boom from respectively jolting and yawing tractor vibrations. Practically the translation of the boom is realized by mounting it on a sledge that can slide on two linear bearings. On this sledge there is a vertical axle around which the boom can rotate. The modelled suspension is worked out on a test rig as showed schematically in figure 6. The springs are necessary to define the static equilibrium position of the boom because there is no returning force as gravity for a pendulum suspension. Just as for a vehicle suspension, a sprayer boom suspension should pass the low frequent tractor motions (e.g. turning at the headland of the field), while high frequent tractor movements, which cause boom deformations, have to be filtered. An optimization routine points out which are the optimal numerical values of the suspension parameters to reach this aim as good as possible.

3.2. Linear multi-body models

Based on the modelling technique of Ramon (1993), the linearized equations of motion of a rigid spray boom with the passive horizontal suspension as described earlier are determined:

$$\mathbf{M}.\ddot{\mathbf{q}} + \mathbf{C}.\dot{\mathbf{q}} + \mathbf{K}.\mathbf{q} = \mathbf{W}.\mathbf{w} \qquad (1)$$

with: $\mathbf{M}$ = mass matrix
$\mathbf{C}$ = damping matrix
$\mathbf{K}$ = stiffness matrix
$\mathbf{W}$ = input distribution matrix
$\mathbf{q}$ = vector of generalized co-ordinates
$\mathbf{w}$ = input forces

The mass matrix $\mathbf{M}$ is function of the mass and moment of inertia of the boom for a rotation about the pivot point and of the mass of the sledge. The matrices $\mathbf{C}$ and $\mathbf{K}$ depend on the values (k_1, k_2) and (c_1, c_2) respectively for translational and rotational spring stiffness and damping.

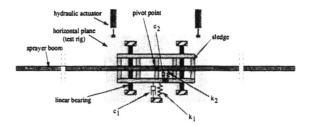

Fig. 6. Scheme of the spray boom with passive horizontal suspension.

The equations of motion with force input (eq. 1) can be transformed to position input according to the method described by Anthonis (1998). This results in

$$\mathbf{M}^u.\ddot{\mathbf{u}}_p + \mathbf{M}^q.\ddot{\mathbf{q}}_2 + \mathbf{C}^u.\dot{\mathbf{u}}_p + \mathbf{C}^q.\dot{\mathbf{q}}_2 + \mathbf{K}^q.\mathbf{q}_2 = \mathbf{K}^u.\mathbf{u}_p \qquad (2)$$

with: $\mathbf{u}_p$ = vector of imposed positions
$\mathbf{q}_2$ = vector of remaining generalized coordinates

The advantage to have position-input is clear because the hydraulic actuators of the test rig to reproduce horizontal tractor movements are position controlled. The delivered force contrarely is unknown, unless a force cell is mounted.

3.3. Optimization procedure

The advantage to have linear equations of motion (eq. 2) is that they can be written in state space form as:

$$\begin{cases} \dot{\mathbf{X}} = \mathbf{A}.\mathbf{X} + \mathbf{B}.\mathbf{u}_p \\ \mathbf{Y} = \mathbf{C}.\mathbf{X} + \mathbf{D}.\mathbf{u}_p \end{cases} \qquad (3)$$

The vector $\mathbf{u}_p$ represents the input translation and rotation of the tractor, while the output vector y consists of the boom translation and rotation (absolute with respect to the ground). From this, a singular value (SV) plot of the system is calculated. Besides this, it is also possible to define a desired SV plot. For low frequencies, the boom should follow the tractor motions, which requires SV's of 1. This happens for example when the tractor turns at the headland or accelerates slowly. For high frequencies contrarily the boom should be isolated from the tractor to reduce boom deformation and fatigue. This corresponds with low SV's. Based upon field measurements (CRAFT-project, 1997) a breaking frequency can be defined beyond which tractor vibrations must be filtered. Another important thing to be considered is the resonance peaks which must be avoided because they induce large boom movements. The difference between the desired and modelled SV's is the goal function that has to be minimized by adapting the parameters k_1, k_2, c_1 and c_2 (figure 7). The used optimization algorithm is a quasi-Newton method with the simplex search method of Nelder and Mead. The initial SV's and the SV's after optimization are plotted in figure 7, from which it can be seen that the proposed dynamic characteristics of the suspension are reached.

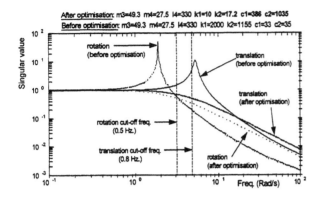

Fig. 7. Initial and optimized singular values plot.

3.4 Validation of the linear multi-body models

The multi-body models are validated by experiments on a hydraulic test rig that can reproduce jolting and yawing tractor motions. Hydraulic actuators impose a swept sine translation excitation from 0.1 to 5 Hz. Meanwhile, movements of the centre of mass of the boom with the passive (not yet optimized) suspension are registered. From this measurements the frequency response function (FRF) is calculated. The simulated SV plot and the measured FRF can be compared in this way (figure 8). The slight difference just before resonance is due to stick-slip effects of the linear bearings, which is not included in the linear model.

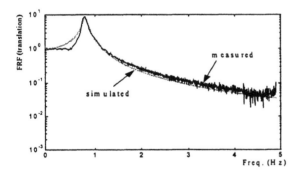

Fig. 8. Measured FRF and simulated SV plot for translation of the boom.

3.5 Results

After validation, the models are used to simulate the effect of adaptations to the suspension set-up. Figures 9 and 10 show respectively the simulated translations and rotations of the boom centre for a 12-m spray boom with an optimized passive horizontal suspension. The input signal is calculated from the non-linear finite element model of a tractor with mounted sprayer that is driven across a field at a speed of 5 km/h. The suspension filters high frequency translation and yawing, while the global tractor movement is followed.

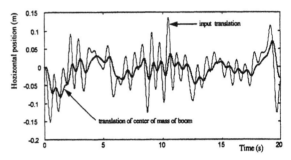

Fig. 9. Input and output translation with optimized suspension.

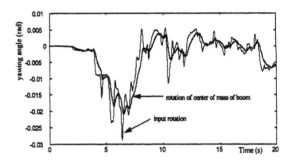

Fig. 10. Input and output rotation with optimized suspension.

4. PRACTICAL IMPLEMENTATION OF THE OPTIMIZATION RESULTS

The optimization procedure described above determines the optimal stiffness and damping constant for the used springs and dampers. In reality a normal spring behaves like in the model (F=k.q). A practical problem occurs when we have to choose the right damper, because most (oil)dampers have a non-negligible friction and don't correspond with the modelled behaviour $F = c.\dot{q}$. Friction and corresponding stick-slip effects cause deformations of the boom at its eigenfrequencies and should be avoided as much as possible. Therefore a low friction adjustable air damper is developed. The working principle is like an air cylinder. Linear bearings at both ends supply a low friction motion of the damper. Changing the orifice area for the incoming/outgoing air (figure 11) alters the damping constant. The same test procedure as for the model validation is executed, but now with the air damper mounted on the suspension. FRF's for translation of the boom with mounted air damper are given in figure 12 for two different orifice areas. The resonance peak disappeared when the correct orifice opening is applied (critically damped system).

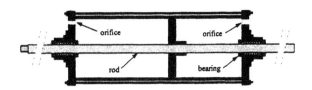

Fig. 11. Scheme of the air damper

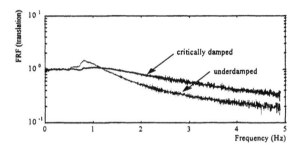

Fig. 12. FRF's with air damper

5. CONCLUSIONS

Finite element models are a useful method to simulate the dynamic behaviour of sprayers. Simulations show that a pendulum or trapezoidal suspension of the boom decrease the rolling motion significantly. The sprayer models can also be used to simulate the effect of structural adaptations on the sprayer, without the need for time-consuming experiments. A lot of work and material costs can be saved in this way.

The singular value decomposition of the state space representation of the linearized equations of motion for a boom with suspension is a valuable tool for design optimization. Simulations show that an optimized passive horizontal suspension of the boom · acts as a low pass filter, decreasing the yawing and jolting motions significantly for the high frequencies, while the low frequencies are passed. This enables the boom to follow field undulations and turning of the tractor at the headland, while boom vibrations and deformations are strongly reduced.

First practical tests with a low friction damper prototype are promising for building a prototype suspension.

ACKNOWLEDGEMENT

This paper was elaborated within the Belgian IWONL Project D 1/4-6115/5702A.

REFERENCES

Anthonis, J., Ramon, H. (1998) Generalized procedure to derive the linearized equations of motion of position controlled mechanical systems. Paper in preparation

CRAFT project FA-S2-9009 (1997), Progression Report.

Crolla, D. A., Horton, D. N., Stayner, R. M. (1990). Effect of tire modelling on tractor ride vibration predictions. *Journal of Agricultural Engineering Research*, 47, 55-77.

Langenakens, J., Ramon, H., De Baerdemaeker J. (1995). A model for measuring the effect of tire pressure and driving speed on the horizontal sprayer boom movements and spray patterns. *Transactions of the ASAE*, 65-72.

Nelder, J. A., Mead, R. A simplex method for function minimization. *Computer Journal*, 7, 308-313.

Norme Internationale, *ISO 5008* (1979) (F). Organisation Internationale de Normalisation.

Ramon, H. (1993). *A design procedure for modern control algorithms on agricultural machinery applied to active vibration control of a spray boom*, Ph.D. thesis NR. 231, Faculteit der Landbouwwetenschappen, K.U.Leuven.

OBSTACLE AVOIDANCE FOR TELEOPERATED ROBOTS FOR LIVE POWER LINES MAINTENANCE, USING ARTIFICIAL VISION.

R. Aracil*, F. M. Sánchez*, D. García*, J.M. González*,L.M. Jiménez**, J.M. Sebastián*.

* DISAM, Universidad Politécnica de Madrid, MADRID, Spain (www.disam.upm.es)
** DISC, Universidad Politécnica de Alicante. ALICANTE, Spain

Abstract: Nowadays uninterrupted power supply has become a must for electrical companies, due to the increasing technological advances, present in our society, that lead to higher demands of electricity. This strong need is obliging the electrical companies to develop processes and techniques for outage-free maintenance of electrical power supply, whose major application lies in aerial distribution lines. In these conventional techniques workers have to do their job on a live electrical power line indirectly, with various kinds of hot-sticks or directly touching the line with rubber-insulated gloves. Therefore, work is performed in a hazardous environment with both, the risk of electric shock and the danger of falling from a high place. In addition, workers have to be very skilled and work cooperatively under very demanding tasks. For all these reasons, the European Community foresees the suppression of this operation in a short term. This paper presents a teleoperated system, developed at DISAM*, capable of performing electrical power lines maintenance and inspection, tasks. The general features of the project are explained, paying special attention to the artificial vision system, which is an obstacle avoidance system capable of preventing collisions between robots and the electric cable. Such a system is the result of several years of research on computer vision. Copyright © 1998 IFAC

Keywords: Obstacle Avoidance, Teleoperation, Artificial Vision, Stereo Vision, Calibration, Path Planning.

1. INTRODUCTION.

This work has been developed at the Departamento de Ingeniería de Sistemas y Automática (DISAM*), which belongs to the Escuela Técnica Superior de Ingenieros Industriales (ETSIIM) at the Universidad Politécnica de Madrid (UPM).

The system consists of two hydraulic driven manipulators with six degrees of freedom, placed on a rotating platform on top of an insulated boom, which substitute the workers in performing the hazardous work on the hot line. The operator, instead, is located on the ground, from where he

more information fsanchez@disam.upm.es

teleoperates the slave manipulators via two force-reflecting masters arm. Since many maintenance operations may be performed automatically, both, the safety of the workers and the overall efficiency of the task are increased. In the other hand, this system includes a vision module, which consists of two different subsystems:

Visual inspection system for the operator. Due to the fact that the operator does not have a clear view of the operation that is being performed, two cameras have been installed to help him supervise his work. The first camera is attached to the left manipulator and allows the operator to visualise the operations performed by the other manipulator. The other camera includes a motorised lens with zoom, focus and iris control and it is set on a pan-tilt unit on top of the system. This camera provides a panoramic view of the working area to the operator, and therefore, a feeling of being performing the job directly. Since the operator has both arms engaged, he can command this camera by means of a multimedia interface with voice synthesis and recognition.

Collision detection system. It basically consists of a binocular stereo mount and a laser illuminating system. This system is responsible for avoiding collisions between the manipulators and the lines that would lead to a short-circuit situation. The main problems related to the use of a vision system are found in the need for reducing the noise introduced by sun's light and the quick detection of the position of the cable. This issue is of special importance, as the true spatial position of the cables must be known to avoid any action, whether deliberate or unintentional, that may cause a disaster.

The system may be considered as semiautomatic with the operator on the ground. In a first stage, a supervisory control in some degree has been implemented, although the final objective is to reach supervisory control next to fully automatic control. Some advantages of this way of operating, in contrast to a manual operation, are a higher level of protection, safety for the worker and quickness. As disadvantages, may be considered those related to the higher complexity and cost.

2. SYSTEM DESCRIPTION AND ARCHITECTURE

2.1. General structure

The system has been designed to develop maintenance and repairing tasks on overhead distribution lines of up to 49 kv. Table 1 briefly describes its principal elements. The architecture is oriented to allow the operator to perform tasks in an optimal manner and in the fewest time, and to achieve the highest degree in telepresence and

Fig 1: System set up.

therefore, increasing the performance of the system.

Table 1: Main system components.

Truck	5.5 ton, 8 m long
Boom	15 m long, telescopic and with up to 69 kV of dielectric strength.
Cabin	Mounted on truck chassis, next to truck cabin. Includes the operator post with the boom, jib and manipulation control stations
Robots	Hydraulic; 7-function (6 axes + grip); articulated. Masters with force feedback; max. Payload 45 kg/arm; net weight 60 kg/arm
Jib	Hydraulic; 3 dof ; telescopic; lifting capacity: 200 kg; with winch
Rotating platform	Mounted on top of the boom. Holds the slave manipulators, the jib and the vision system
Vision System	Stereo head, laser scanner, and real time frame grabber.

The system consist of three different modules: HIC, (Human Interactive Computer), TIC (Task Interactive Computer), SPC (Sensor Processing Computer).

The HIC is responsible for the interaction of the system with the operator, so that the control loop of the system is closed. The TIC receives commands from the HIC and takes the proper actions on the manipulators and other controlled devices. Finally, the SPC processes the information that the remote sensors provide. A more detailed information in (Santamaría, et al., 1996; Aracil, et al., 1995,).

Communication between the different modules is performed via ethernet, whereas the exchange established between the TIC and SPC is made on a VME bus.

2.2 Structure of the vision system.

In this section, the different elements that constitute the SPC system are described in detail.

TRC motorized lenses stereo mount. It is a binocular stereo head developed by the American Company TRC (HelMate Robotics Inc., at present). It is controlled by an 8-axis PMAC controlling card set on a VME bus. The head has four degrees of freedom, that is: pan, tilt and vergence on each camera. Besides this, both cameras include motorized lenses with three degrees of freedom: zoom, focus and iris.

Datacube Maxvideo 200 video processing board. It is a real time acquisition and image processing VME board with a pipeline architecture. It is mounted on a

VME chassis with a Motorola 68.000 as a host, running Lynx 2.0 Operative System.

Floating point processor Max860. It is a high performance floating point RISC processor that complements the existing real-time Datacube hardware by providing speed and flexibility for implementing virtually any image-processing algorithms. It allows general purpose floating point operations, processing data at 40 MHz for a peak performance of 160 Mflops. It may be programmed using an i860 cross-compiler, assembler and linker and includes different libraries for controlling data transfers, inter-processor communication and event management. This processor, together with the Maxvideo 200 processing card and the host processor, make up the SPC.

810 nm laser scanner. This equipment has been developed at the Universidad Carlos III de Madrid and consists of a 500 mW semiconductor laser and a cylindrical lens that generate a vertical laser beam. This beam may be oriented by means of a mirror attached to a galvanometer. The engine that positions the laser beam is controlled through the host serial port. Besides, it allows a sweeping operation in which the maximum and minimum sweeping range may be selected. The laser switching on and off may be performed automatically or synchronized with the integration frequency of the cameras.

It is worth mentioning that radiation safety of laser products equipment rules, UNE (Aenor, 1993) have been taken into account to ensure a safe use of the system.

3. DEVELOPMENT OF THE PROJECT.

3.1. Technological problems and solutions presented.

The main problem of detecting a cable is that there is not any significant point of the cable available to solve a typical correspondence problem. In a laboratory, under controlled illumination conditions (as shown in figure 3), it is possible to calibrate the stereo unit with accuracy and to perform a segmentation operation on the cable in order to obtain its spatial position. Nevertheless, when attempting to perform the same set of operations under sun's light, being the stereo mount on top of a boom over a truck, the results obtained may not be as good as expected. Another factor to take into account is the fact that the cable is approximately parallel to the epipolar line, so that any lack of precision may cause very large errors.

In order to solve the first problem, one may make use of active vision resources, such as illuminating

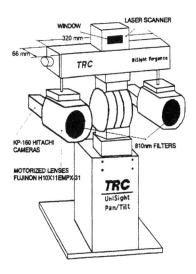

Fig 2: Stereo head mount.

the cable with a vertical laser beam so that the intersection may be within the field of view of both cameras. Thereafter, this point of intersection may be detected in both images and the problem of correspondence may be solved. If is possible to obtain a set of 3D points that belong to different portions of the cable, it is easy to create, by means of interpolating, a security area around the true position of the cable. This way, the manipulator will not be allowed to cut across this area and touch the cable.

The next problem to be solved is the amount of noise that sun's light involves. The working conditions of the system are very often subjected to a strong solar illumination, a very common situation in our country. The system though, should be able to work against the light. This fact implies that the employment of structured light in this environment may be useless.

The solution to overcome this second problem involves acting on the power and wavelength of the artificial light supply. The sun, even though it behaves as a black body, stands a fall in energy density in the infrared range, being this descent larger as the wavelength increases.

The problem of selecting a larger working wavelength is that the cost of both, cameras and laser equipment, increases radically. Furthermore, for wavelengths higher than 900 nm there are not any commercial CCD cameras and solid state lasers available nowadays. As a result, it would be necessary to consider other technologies, which besides the higher cost may occupy an excessive volume, making this solution non-viable.

Finally, the solution selected consist in a 810 nm laser with an output power of 200 mW and two Hitachi KP-160 cameras, each one provided with a 810 nm filter of ±10 nm passing band. By means of this equipment, we are able to visualize and detect the intersection of the laser with the cable very sharply, even under strong solar illumination conditions.

3.2 Functioning of the vision system.

When the platform approaches the cable, the system starts being able to detect the location of the line. As soon as the cable is spatially positioned, its 3D coordinates are added to the database, which stores the wired model of the scene in front of it. Whenever the operator commands a movement on the manipulator, the system checks if the trajectory is compatible with the model and that no collision is possible. The operator may enable or disable certain space areas in order to prevent the manipulator from crossing them accidentally. The worker performs this operation using a voice driven interface, since he has both arms engaged in controlling the master arms.

The laser is turned on and off synchronously with the camera in such a way that, it is on during the even frames and off during the odd frames. This way, the Datacube unit may process two images and detect the differences between them. The illuminated spot is detected in both images, corresponding to the left and right cameras, and the coordinates are then passed to the max860 equipment, which solves the correspondence problem.

Two different detection strategies have been implemented. In the first place the area of the cable is swept around. This is followed by a more selective sweeping on the area within the field of view of both cameras.

4. ALGORITHM FOR COLLISION DETECTION

This chapter explains in detail the basis of the vision algorithm. The analysis is divided into different stages.

4.1 Strategy of image acquisition

Two different detection strategies have been implemented in the system, as was pointed out in the previous section. According to this, the system always follows the first strategy in the first place. Once the position of the cable has been correctly detected, the second strategy plays its role. If there is a chance that the collision detection algorithm is failing, an alarm message will be displayed on the operator screen indicating that the operations are not completely safe and that every precaution must be carried into extremes.

First strategy. The stereo head moves steadily form left to right sweeping the cable area around horizontally. The intersection of the laser beam with the cable remains inside the field of view of both cameras at every instant of time. The cameras

Fig 3: Image correspondences made at the laboratory

are fixed at a constant position with respect to the eyes coordinate system although they present a small vergence angle. During this motion, the vision system will be identifying points that belong to the cable. The spatial position of these points is calculated in real time by the algorithm explained in this section.

Second strategy. The mount remains still while the vertical laser beam is sweeping the cable around within the cameras field of view. Just like in the previous case, the points are calculated in real time and the point coordinates are added to the TIC database.

4.2. Calibration of the vision system

The calibration of any vision system is crucial in order to obtain a degree of accuracy. In the case of this binocular stereo mount with motorized lenses, a calibration table has been calculated for a wide range of values of the zoom and focus.

Due to the fact that the working distance is approximately constant, being this distance limited by the length of the manipulators, the cameras are thus set almost at a constant distance with respect to the cable. Thereby, it is only necessary to make certain adjustments on the focus by means of an automatic focusing system, (Jiménez, 1995b). Since the stereo mount is rotating horizontally from left to right, it is necessary to perform a kinematic calibration (González, 1996). In the following paragraph, the method employed for the kinematic and camera calibration is analyzed.

Methods: Calibration of motorized lenses involves a higher complexity than that of a fixed lens. It can be considered that a lens behaves according to the mathematical pinhole model for a set of specific zoom, focus and iris values. Nevertheless, once any of these values is modified, the intrinsic and extrinsic parameters of the method are affected, mainly because of the displacement that the center of the image suffers. (Gonzalez, 1996; Wilson, 1994). The mathematical model employed is the well-known pinhole model which is explained in detail in a vast number of publications (Tsai, 1989; Li and Lavest, 1994; Wilson, 1994; Gonzalez, 1994).

Intrinsic parameters. The projection of a point P on the CCD plane comply with equation (1) in which (x_w, y_w, z_w) are the coordinates of P with respect to the world coordinate system. (x_f, y_f) represent the coordinates of a point in the image measured in pixels. The rotation matrix [R] defines the spatial transformation of the coordinate system, being its 9 parameters a function of the angles rotated. The other 3 parameters of the translation matrix [T] correspond to the displacement of the origin of the coordinate system from the optical center. The parameter f represents the focal length of the camera. D_x and D_y represent the tangential and radial distortions in each direction. In order to change from coordinates in millimeters to coordinates in pixels, It is necessary to multiply such values by a scale factor K_x and K_y and thereafter, also to displace the principal point C_x, C_y

$$X_f = C_x - K_x D_x + K_x f \frac{r_{11}x_w + r_{12}y_w + r_{13}z_w + t_x}{r_{31}x_w + r_{32}y_w + r_{33}z_w + t_z} \quad (1)$$

$$Y_f = C_y - K_y D_y + K_y f \frac{r_{21}x_w + r_{22}y_w + r_{23}z_w + t_y}{r_{31}x_w + r_{32}y_w + r_{33}z_w + t_z}$$

Finally, the coordinates of the image or frame can be expressed in terms of the intrinsic and extrinsic parameters of the system (1). These are referred to by photogrammetrists *as collinearity equations.*

Calibration process. Several algorithms have been tested in order to obtain the intrinsic and extrinsic parameters. Among them, some deserve special attention: Tsai's method (Tsai, 1987), DLT method (Fan and Yuan, 1993) and Vanishing Point method (Wang and Tsai, 1991). The algorithm that finally has been employed in the system is Tsai's method, both Reg Willson's (Carnegie Mellow University) and Janne Heikkiläs (Heikkilä and Silven, 1996) (University of Oulu, Finland) implementations. In both cases, the results obtained are much better than when other methods are used.

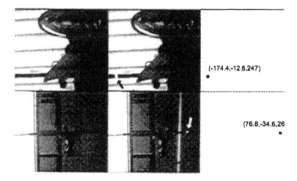

(-174.4,-12.6.247)

(76.8,-34.6,26

Fig 4: Example of cable coordinates detection

Kinematic calibration. In any active vision system, the intrinsic and extrinsic parameters change when the position or the camera state vary. Therefore, it is necessary to develop certain calibration techniques such that may maintain the system calibrated under any conditions. The intrinsic parameters will be modified according to the position of the stereo mount. That is, whenever the tilt, vergence or pan values change, the system will read the new parameters from a set of tables (Tsai,1987). Other authors propose the development of mathematical formulas such that the intrinsic parameters may be obtained in terms of the position of the zoom, focus and iris (Wilson, 1994).

The problem of continuously knowing the values of the extrinsic parameters can be solved by means of a previous calculation of a set of transformation matrixes. Such matrixes relate the camera coordinate system, located at the center of projection of the image, and the coordinate system that defines the axis of rotation of the stereo mount.

By knowing the extrinsic parameters at an initial position in which we have calibrated the system, and by obtaining previously every transformation relationship between the different coordinate systems, it is possible to calculate the extrinsic parameters of the camera at any time. (Li *et al.*, 1994)

4.3. Description and on-line implementation of the algorithm

The process of acquiring and processing the images taken by both cameras may be summarised in the following steps: first an acquisition of both images, being the laser turned off. Second a filtering and noise reduction of every image. Third a subtraction of the laser-on-image minus the laser-off-image in each camera. Forth a binarization on each result images. Fifth, Morphological operations to eliminate false responses. Sixth a detection of possible points and their coordinates. Seventh shape analysis and suppression of false hypothesis. Eighth stereo correspondence. Ninth 3D coordinate calculations according to the intrinsic and extrinsic calibration Tenth, update of the spatial coordinates points database

Steps 1 through 8 are performed in real time by the image acquisition and processing board Datacube Maxvideo 200. This board is a part of the SPC, as was previously mentioned.

Once the cable position has been detected in both images, both coordinates in pixels are passed to a Datacube Max860 processor, which stores the calibration parameters and the position of the stereo mount. Finally, if the operation has been successful, the 3D coordinates of the point are transferred from the SPC to the TIC via ethernet, and the obstacles database is updated. In step 9, the correspondence between two candidate points or primitives takes into account the epipolar constraints and the continuity of each figure (Jiménez, 1995a). On the other hand, in step 10, a real epipolar geometry of the system is developed in order to minimize the mean quadratic error of the false intersections between the projective lines that link the focal center and the primitives of the images.

5. RESULTS

The set of tests that were carried out at the laboratory was completed satisfactorily. It has been possible to measure accurately the localization error introduced by the motion of the stereo mount.

In this case, the manipulators were fixed on a table at a constant distance of the cable and, therefore, the cable spatial position was easily estimated. The only factor that may influence for the correct localization of the cable was the movement of the stereo mount. This error has always been lower than 1 cm., which makes the system accurate enough to work in the field. If we consider the fact that the cable and the platform on which the manipulators and stereo unit are set are going to be swinging, a cylindrical security area of about 30-mm of diameter will be created around the cable. In this way, the manipulators will not be able to cut across this security area.

At present, the system has being tested in the field with a total success. The vibrations introduced to the whole system, by the movements of the robots, and the swing of the platform is corrected by the vision system. The shake added to the stereo mount has been minimized by a shock absorber system. The total collition-free operation has been completely achieved.

6. CONCLUSIONS

The system presented constitutes an important contribution to solve a real problem, as it is the maintenance of electrical power lines. It provides a great amount of information to the operator in order to help him in maintenance and repairing tasks without the risk of an accident. The main technical difficulties encountered during the development of this project are those inherent to any 3D vision system (lack of precision, correspondence problem...), together with the aggressive environment (sun's light) in which the system must operate.

REFERENCES.

AENOR (1993). Seguridad de radiación de productos láser, clasificación de equipos, requisitos y guía del usuario. UNE EN 60825,

Aracil, R., L.F. Peñín, M. Ferre, L.M. Jiménez. (1995). Teleoperated System for Live Electrical Power Lines Maintenance. *SPIE*.

Chen, W. and B. C. Jiang (1991). 3D Camera Calibration Using Vanishing Point Concept. *In Pattern Recognition*. Vol. 24, No 1 pp.57-67, 1991

Fan, H. and B. Yuan (1993). High Performance Camera Calibration Algorithm. *SPIE* Vol. 2067 Videometrics II (1993) pag 2-13

Gonzalez J.M.(1996). Calibración de cabezal estereoscopico con óptica motorizada, *DISAM Technical reports.*

Heikkilä, J. and O. Silvén (1996). Calibration Procedure for Short Focal Length Off-the-shelf CCD cameras. *Proc. 13th International Conference on Pattern Recognition, Vienna*, Austria p.166-170

Jimenez, L.M (1995a). Modelado de entornos 3D para planificación de Trayectorias de Robots Mediante Visión Estereoscópica y Control Activo de Parámetros Ópticos. *DISAM Technical report.*

Jiménez L.M.(1995b). Enfoque a partir del zoom. *DISAM Technical reports*.

Li, M. and J.M. Lavest (1994). Camera Calibration of a Head-Eye System for Active Vision, Proc. *Third European Conf. Computer Vision, J.O. Eklundh*, ed., pp. I:543-554, Stockholm.

Li, M., D Betsis and J.M. Lavest, (1994) Kinematic Calibration of the KTH Head-Eye System. Report from Computational Vision and Active Perception Laboratory (CVAP).

Santamaría, A. P.Mª Martinez Cid, R. Aracil, M. Ferre, L.F. Peñín, L.M. Jiménez, F. Sánchez, E. Pinto y A. Barrientos, A. Tuduri, F. Val (1996). Aplicación de la Robótica al Mantenimiento en Tensión de Instalaciones de Distribución" MATELEC.

Tsai, R.Y. (1987). A Versatile Camera Calibration Technique for High-Accuracy 3D Machine Vision Metrology Using Off-the-Shelf TV Cameras an Lenses. IEEE J. Robotics and Automation, vol. 3 no.4, pp.323-331.

Wang, L. and R. Tsai (1991). Camera Calibration by Vanishing Lines for 3-D Computer Vision. IEEE Transactions on Pattern Analysis an Machine Intelligence Vol. 13.NO4 ,

Wilson R.G. (1994). Modeling and Calibration of Automated Zoom Lenses, PhD thesis, CMU-IR-TR-94-03, Robotics Inst., Carnegie Mellon Univ,

EXPERIMENTAL RESULTS FROM STEREOSCOPIC VIEWING AND AUTOMATIC CAMERA ORIENTATION FOR VEHICLE TELEOPERATION

J. M. Gómez-de-Gabriel, J. J. Fernández-Lozano, A. Ollero and A. García-Cerezo.

Dept. Ingeniería de Sistemas y Automática. E.T.S.I.Industriales.
Universidad de Málaga. Plaza de El Ejido s/n, 29013, Málaga, SPAIN
degabriel@ctima.uma.es

Abstract: This paper presents and analyzes experimental results from the teleoperated guidance of a mobile robot. Different modes have been tested, from mono to stereo viewing and from static to automatically oriented cameras, in order to determine the best configuration for teleoperated guidance by means of elemental curvature and speed commands. This system also features an original low-cost stereo viewing system that allows easy wireless transmission combined with a camera pan and til unit. In addition an automatic camera positioning method wich helps avoiding control unstability due to the vehicle dynamics. *Copyright © 1998 IFAC*

1. INTRODUCTION

Remote operation of a mobile robot include many different tasks from simple navigation to manipulation, with different visual requirements. In order to determine the best configuration, a set of experiments have been done for manually teleoperated guidance. The configuration parameters considered for the experiments are mono vs. stereo viewing and static vs. mobile cameras as well as the operator skill from medium to experienced. Other researchers studied similar problems like skill acquisition (Drascic, 1991) for similar parameters.

The robotic testbed is the AURORA robot, an autonomous wheeled mobile robot designed for automatic greenhouse operations without the physical presence of a human operator or supervisor. Its aim is to perform purposeful operations in its particular working environment. Although typical operation for this system is chemical spraying, its flexibility allows its adaptation to other operations and applications, such as greenhouse monitoring, transportation or production inspection. Its control architecture integrates sensors, controllers and actuators as well as

software and communication components.

Next sections include an overview of the robotic system, then teleoperator station, the design of the low-cost stereo viewing system, the automatic camera positioning method, and finally the experiment results and analysis for a teleoperated navigation task. The last two sections are for conclusions and references.

2. THE TELE-ROBOTIC SYSTEM

The vehicle mechatronic system consists of an octagonal mobile platform that accommodates a spraying device, the power system, standard electronic and computer enclosures, and a variety of sensors. Its dimensions, constrained for the ability to navigate in narrow greenhouse corridors, are 80 cm in width and 140 cm in length.

The locomotion system is dual configuration that renders high maneuverability (zero turning radius), which is essential in constrained environments. It has four wheels located in the vertices of a rhomb. This a redundant locomotion system with two steering wheels (front and back) and two differential wheels (left and right). A more detailed description of AURORA can be found in Ollero *et al* (1995).

An special configuration of ultrasonic sensors has been chosen considering the characteristics of constrained environments. Thus, a combination of three different types of ultrasonic sensors has been placed in the front half of the robot, allowing unidirectional autonomous navigation.

Besides odometric encoders, additional proprioceptive digital sensors are included in order to obtain the vehicle's power status, the spraying device status, and the detection of malfunction conditions.

The on-board communication system includes a video transmitter that sends analog TV images to the teleoperator station and a 1200 bauds semi-duplex radiomodem for remote commands and status reporting.

Details about the teleoperation original architecture, mixed hierarchical and behavior based, can be found in Gómez-de-Gabriel (1996).

Another important task implemented in the vehicle control module is the responsible for wheels coordination in this locomotion redundant system, because of the difference between the driving motors acceleration, and the steering motor speed. Bad coordination causes wheel's slippage that increases the vehicle position uncertainty. This coordination tasks periodically sets the speed of the driving motors (fast response) as a function of the steering motor actual position (slower response). This causes a significant delay for manual control.

3. THE TELEOPERATOR STATION

The operator is located away from the robot working environment. The operator communicates with the robot through the Teleoperator station (TOS), which is a real-time computer system, with man/machine

Fig. 1. The AURORA robot in an indoor environment.

interfaces attached, and a video monitor.

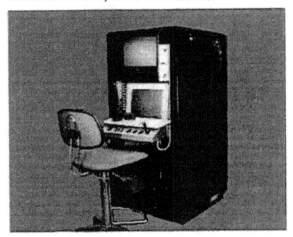

Fig. 2. The teleoperator station rack.

This station allows the operator the performance of several tasks such as monitoring, task programming, task intervention, and shared or traded vehicle control, which means communication at different levels of control: user, supervision, reference generation and executive. This makes possible the achievement of many different operations from the operator station, allowing complete robot control.

The TOS scheme is shown in Fig. 3. This is a PC based workstation, with a radio modem, a TV receiver and an stereo viewing decoder. Several man/machine interfaces have been experimented during development: SpaceBall (6 dof force/torque sensor), mouse, digital (incremental) joystick, analog (absolute) joystick and keyboard in various operating modes. For this study incremental keyboard control has been successful used.

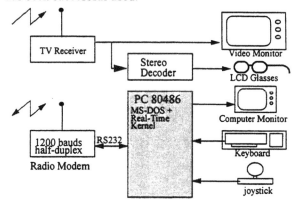

Fig. 3. Teleoperator Station.

The computer screen layout gives graphical representation of the remote vehicle's status and sensor's readings (see Fig. 4.). This provides actual information of the commanded speed (necessary with incremental interfaces) and obstacle proximity (not in the field of view) while visually driving the vehicle.

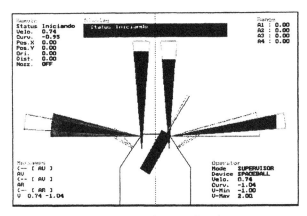

Fig. 4. Status, sensors and control representation on computer screen

5. The stereo image multiplexer and the vi transmitter.

The TOS is the responsible of initiating the message transfer, and AURORA of returning the appropriate response. The TOS sends commands, and AURORA returns status messages. Every message consist of a single command or status message.

4. THE STEREO REMOTE VIEWING SYSTEM

A low-cost stereo vision system has been developed. This is an active time-multiplexed system, with a video multiplexer module and a sync detector for the control of a pair of LCD glasses. The video multiplexer unit is mounted on the robotic vehicle and the glasses control unit is located in the teleoperation station as seen in Fig. 5. Video merged images are sent to the station through a commercial video-sender. Finally, stereo vision was obtained by watching the monitor wearing LCD glasses.

The main aspects involved to get an appropriated performance in stereo vision are camera separation and camera convergence. A strong camera separation can result in better performances and resolution. But it can cause strong image disparity, making the stereo fusion hard. Likewise, camera convergence can bring good depth resolution and image quality, but also depth distortion (Brooks, 1991). Both items determine how strong the stereo effect is, and both can cause image distortions. This way, choosing a camera configuration is a key aspect in stereo vision. The pair of cameras on the pan & til unit can be seen in Fig. 6.

In previous works with the stereo system mentioned above, we got some conclusions about the best attachments for our system. One of the most useful configurations (this meaning stereo vision results for almost every users, and depth perception as near as possible to that of natural vision) was the disposition adopted for our trials. Camera separation was 45 mm (between camera axes at the basis of the lenses), and the convergence angle was 1° 20' degrees.

5. AUTOMATIC CAMERA ORIENTATION

The presence of time delays in the control loop can cause instability for direct teleoperation tasks. Many teleoperation systems include a transportation delay due to the long distance (spacial robots) or the nature of communication media (acoustic transmission for underwater robots). In this case there is no significative communication delay, but there is a important delay due to the vehicle dynamics.

The manual control mode for this experiment is based on low level commands of desired speed and desired curvature. The user interface for vehicle guidance consists of incremental changes of speed and curvature by pressing the teleoperator station cursor keys. So there is not kinesthetic feedback for the last commanded desired curvature.

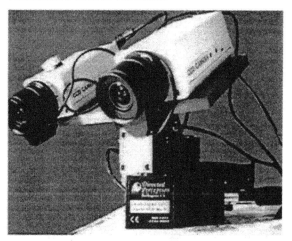

Fig. 6. The stereoscopic camera arrangement mounted on the pan & tilt unit.

The above considerations can be avoided by using an automatic camera orientation method. The camera pan & tilt unit features faster response time and gives to the teleoperator absolute information about the last desired curvature commanded. Furthermore camera elevation can be implemented as a function of the last desired speed. The automatic positioning method can be so simple as:

$$Pan = (desired\_curvature - actual\_curvature) * Kp$$
$$Tilt = offset\_tilt + desired\_speed * Kt$$

Whit this method, the user, located away from the robot, sees an undelayed image for the robots set points. As will be seen later this method improves the task performance.

6. EXPERIMENTS

Our main purpose was to evaluate operator performance by using stereo and mono viewing, and active and passive camera motion. To reach this objective, we planned an experiment in which a human operator used all four possible configurations for our vision system, trying to complete a course along a fixed path, along over 20 meters. To consider the possible different effects of four configurations depending of operator experience, we tested two human operators: one with many hours controlling Aurora Robot ("Expert operator"), and one with only a few hours in such operations ("Medium operator"). Every one of them tried three times every vision system configuration. In every trial we recorded robot data such as speed, curvature, time from the start, range, and number of commands. Trials with a failure, such as collision, were discarded in order not to lengthen experiment time, giving outlayered results. Anyway, only two of these failures happened all over the 24 experiments.

A picture of the course can be seen in Fig. 7.Top view of the experiment. Along the path, Aurora had to turn four times, two times left and two times right. Total length of the course was about 22 m. The human operator controlled the robot from a teleoperation station a few meters away, but disposed in such a way that the human operator could see the robot only by watching his monitor, and not by directly viewing.

Fig. 7. Top view of the experiment

For every experiment, the above mentioned data (speed, curvature, etc.) were recorded, and then represented on graphics. About 24 different executions have been done, and in order to illustrate this paper 8 of them have been selected.

Figures 8 to 11 correspond to the medium operator experiments with all the parameter combinations. Same experiments for the experienced operator can be found in figures 12 to 15.

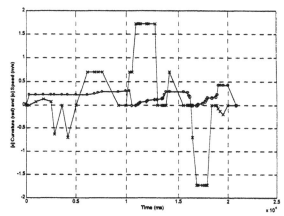

Fig. 8. Single experiment curvature and speed data for medium operator with stereo viewing and static camera conditions.

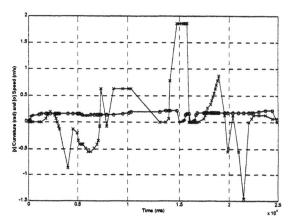

Fig. 9. Single experiment curvature and speed data fo medium operator with mono viewing and stati camera conditions.

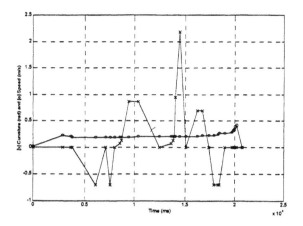

Fig. 10. Single experiment curvature and speed data fo medium operator with stereo viewing an automatic camera orientation.

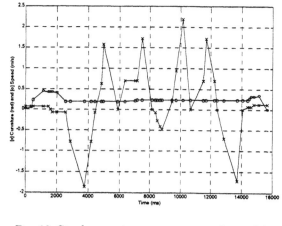

Fig. 13. Single experiment curvature and speed data fo experienced operator with mono viewing an static camera conditions.

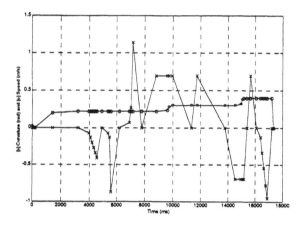

Fig. 11. Single experiment curvature and speed data for medium operator with mono viewing and automatic camera orientation.

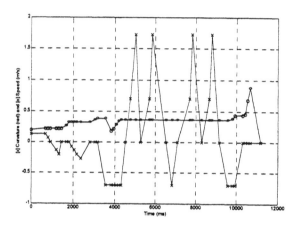

Fig. 14. Single experiment curvature and speed data for experienced operator with stereo viewing and automatic camera orientation

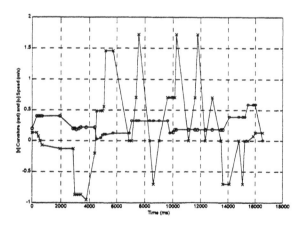

Fig. 12. Single experiment curvature and speed data fo experienced operator with stereo viewing an static camera conditions.

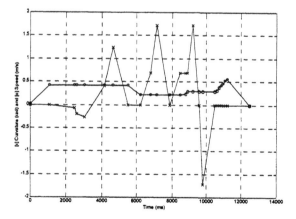

Fig. 15. Single experiment curvature and speed data for experienced operator with mono viewing and automatic camera orientation.

7. DATA ANALYSIS

To analyse the recorded data, we considered the trials with mono vision and static cameras as a reference.

Likewise, we fixed our attention on some aspects:

1) Number of commands.

Number of commands diminished by using stereo vision or camera motion, but not too much with the expert operator. However, a light difference can be seen, but it would not justify the use of such a scheme for a trained operator. The strongest reduction was identified between static and mobile cameras for medium skill operator.

2) Average speed.

The highest medium speed was got by the expert operator using camera motion and stereo vision. The medium operator got his best performances by using mono vision and camera motion. The best improvement was also detected under this configuration, for medium operator. Expert operator did his best by passing from static cameras to active ones, both in mono and stereo.

3) Maximum speed.

Every experiment got a point with a maximum speed, which was very different depending of operator experience, camera motion, vision system and, sometimes, number of trial. Since no collision is allowed, a higher speed means better control. Thus, expert operator did his best under stereo vision and camera motion, though records with static cameras are not too far. Medium operator got his better results with stereo vision, no difference between active or static cameras.

4) Zero speed points.

A zero speed point means a stop, a place where the operator must think what to do or where he must arrange so much robot trajectory that it can not be made in motion. Expert operator had rarely to stop. Medium had to do that in almost every trial, but important reductions were identified when using camera motion, stereo vision, or both systems. As medium speed was not too fast, main problem for this operator was curvature control, so the faster camera dynamics (when using camera motion) meant an important advantage to predict robot course. Expert operator also reported verbally this profit.

As a conclusion from the full set of experiments the next summary has been obtained and can be seen in tables 1 and 2.

Table 1: Experimental results for medium skill operator

	Static Cameras	Camera Motion
Mono vision	reference	< number of commands > medium speed only one zero speed point > maximum speed
Stereo Vision	< number of commands > medium speed < zero speed points > maximum speed	< number of commands = medium speed < zero speed points = maximum speed

Table 2: Experimental results for experienced operator

	Static Cameras	Camera Motion
Mono vision	reference	< number of commands (light improvement) > medium speed No zero speed points > maximum speed
Stereo Vision	= number of commands > medium speed (light improvement) < zero speed points > maximum speed	= number of commands > medium speed No zero speed points > maximum speed (highest of all trials)

8. CONCLUSIONS

From the obtained data we can extract some conclusions:

1) Both mobile camera and stereo vision gives a sensible improvement in teleoperated driving tasks for medium trained operator and smaller for the expert one.

2) Stereo vision makes medium operators get driving results near those of expert drivers, though difference is still important.

3) However, camera motion brings a considerable profit on operations, for all kind of operators.

4) Skill improvement is based on a major knowledge of the environment, acquired in previous executions. By using stereo vision, a medium skill operator can estimate depth more accurately, getting near expert skill. This way, remaining disadvantage for medium operator is using the teleoperator station.

9. REFERENCES

Gómez-de-Gabriel, J.M., Martínez, J.L., Ollero, A., Mandow, A. and Muñoz, V.F., "Autonomous and Teleoperated Control of The AURORA Mobile Robot", (IFAC'96) San Francisco, June 1996.

Ollero, A., J.L. Martínez, A. Simon, J.M. Gómez-de-Gabriel, V.F. Muñoz, A. Mandow, A. García-Cerezo, F. Garcia, and M.A. Martínez, (1995). The Autonomous Robot for Spraying AURORA, *IARP 95, Robotics in Agriculture and the Food Industry*, Toulouse, France.

Brooks, T, I. Ince, and G. Lee, "Vision Issues for Space Teleoperation Assembly and Servicing (VISTAS)", STX Robotics, 1991.

Dascic, D., "Skill acquisition and Task Performance in Teleoperation using Monoscopic and Stereoscopic Video Remote Viewing", Poc. of the Human Factors Society 35th Annual Meeting, 1367-1371, San Francisco, September 1991.

ESTIMATION OF MOTION PARAMETERS FOR AUTONOMOUS LAND VEHICLE FROM A MONOCULAR IMAGE SEQUENCE

H. Hassan * , B. A. White**

*Ph.D. Research, ** Head of DAPS
RMCS (Cranfield University), Shrivenham, Swindon, Wilts, SN6 8LA, UK
Aerospace, Power and Sensor Department (DAPS).

Abstract: This paper presents an approach to the estimation of the motion of a vehicle using a fixed order Extended Kalman Filter. The estimation assumes a fixed environment and a moving camera, mounted on the vehicle. The estimation of yaw rate and sideslip angle are shown to be possible using this technique. A feature correspondence based approach is used and the motion estimation algorithm employs the dynamics of the vehicle to increase the accuracy of the estimation. A simple algorithm to avoid the correspondence and occlusion problems is also developed. The algorithm depends on creating a sequence of virtual images from the real image sequence. This sequence has a fixed number of virtual feature points but retains the true motion information. *Copyright © 1998 IFAC*

Keywords: Estimation, Motion , Autonomous, Navigation, Kalman filters

1. INTRODUCTION

Estimating the relative camera motion from a sequence of images is an important research topic in the areas of autonomous land vehicles. There are three basic approaches for recovering camera motion parameters from image motion information: (1) motion from optical flow fields; these methods represent motion in the image plane as sampled ,Adive (1985), or continous velocity fields (Bruss and Horn 1982). (2) motion from image displacements (Feature based approach) which depends on the recognition of the same set of correspondence points in two or more images (Borida and Chellappa, 1990; Peng and Sanjeev, 1995). (3) direct motion; this approach depends on the image brightness information to estimate the motion parameters (Tall , 1992).]. Most of these techniques assume that the camera is not moving through the environment and the camera position is the origin of the reference coordinates. But in the case of navigation of an autonomous vehicle, the

camera position is changing, which results in more complexity in the algorithms for extracting motion parameters. Also, none of the perivious methods employ any information of the dynamics of the camera motion, which is neccessary when studying the motion estimation problem. In this work we assumed that the environment is rigid and stationary. We use the feature based approach and formulate the motion estimation problem as a recursive problem which uses the dynamics of the vehicle. We use the Extended Kalman Filter (EKF) to estimate the vehicle motion parameters and introduce a simple algorithm to avoid the main problem for feature based approach; that of feature correspondence and occlusion. The new algorithm depends on creating a sequence of virtual images from the real image to provide a fixed number of virtual features that do not suffer from occlusion. The EKF can then continue to estimate the motion if real feature points in the image sequence move out of the image frame or become occluded, due to the motion of the vehicle.

2. MODELS

The model in this paper is based on the assumption that the camera is stationary w.r.t the vehicle and both of them are moving through a stationary environment with the Z axis pointing along the line of sight (figure 1).

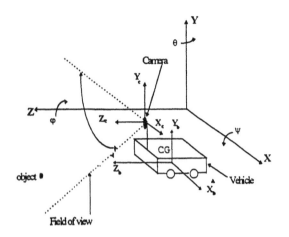

Fig. 1. Coordinate System.

2.1 Vehicle model.

The autonomous land vehicle is assumed to have two driven wheels (figure 2), and is represented by the following linearized model (Ackermann, 1993).

$$\begin{bmatrix} \dot{\beta} \\ \dot{q} \end{bmatrix} = \begin{bmatrix} a_1 & a_2 \\ a_3 & a_4 \end{bmatrix} \begin{bmatrix} \beta \\ q \end{bmatrix} + \begin{bmatrix} b_1 \\ b_2 \end{bmatrix} \delta_F \quad (1)$$

$$a_1 = \frac{-(C_F + C_R)}{\tilde{m} V} \quad, \quad a_2 = -1 + \frac{(C_R L_R - C_F L_F)}{\tilde{m} V^2}$$

$$a_3 = \frac{C_R L_R - C_F L_F}{\tilde{J}} \quad, \quad a_4 = \frac{-(C_R L_R^2 + C_F L_F^2)}{\tilde{J} V}$$

$$b_1 = \frac{C_F}{\tilde{m} V} \quad, \quad b_2 = \frac{C_F L_F}{\tilde{J}}$$

where:

β : the side-slip angle.

q : the yaw rate around the Y axis.

δ_f : the steering angle.

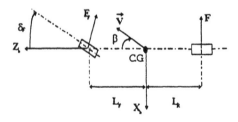

Fig. 2. Land vehicle with two front driven wheels

C_f, C_r are the cornering stiffnesses for the front and the rear wheels. The vehicle mass m is normalized by the road adhesion factor μ, i.e. $\tilde{m} = m / \mu$ and the moment of inertia J is normalised as $\tilde{J} = J / \mu$.

2.2 Imaging Model.

A central projection imaging model is assumed, defined by :

$$h : S \rightarrow s$$

where :

$$S = (X, Y, Z)^T \in S = \left\{ (X, Y, Z)^T \in R^3, Z > 0 \right\}$$

is a spatial point coordinate, and:

$$s = (x, y)^T \in P \subset R^2$$

is an image plane point coordinate. The space P is normally a finite rectangle, corresponding to the image plane of camera, with the focal length of the camera f set to 1 then:

$$x = \frac{X}{Z} \quad ; \quad y = \frac{Y}{Z} \quad (2)$$

2.3 Motion Model.

Our aim is estimation of the motion parameters (U,V,W,p,q,r) and we assumed that the motion of a rigid body can be represented by a small rotation and translation. Let the translation vector $T = (U \ V \ W)^T$, the rotation vector $\Omega = (p \ q \ r)^T$. The position vector $A = (X \ Y \ Z)^T$ represents the point S in world coordinates. The velocity V of the point S is then given by:

$$\mathbf{V} = -\mathbf{T} - \Omega \times A \quad (3)$$

$$\mathbf{V} = (\dot{X}, \dot{Y}, \dot{Z})^T \quad (4)$$

$$\dot{X} = -U - qZ + rY$$

$$\dot{Y} = -V - rX + pZ \quad (5)$$

$$\dot{Z} = -W - pY + qX$$

2.3 Noise Model.

The images are assumed to be collected from a digital picture frame. A such the features will be contaminated by noise that results from pixel quantisation and by the software process that defines the centre of the feature. It is thus assumed that the measured image coordinates (x,y) are corrupted by white zero mean Gaussian noise.

3. THE EXTENDED KALMAN FILTER

The recursive Extended Kalman Filter (EKF) algorithm is used to estimate the motion parameters of the autonomous vehicle. The motivation behind using this type of approach is that recursive methods require much less computation time for each new set of data is introduced (each new image). The filter continually computes the state at the current time, based on all past data, and can easily extrapolate the state at the next time step. The general problem can be formed for solution by recursive (EKF) algorithm as follows:

The plant model is:

$$\dot{S}_t = f(x_t, t) + G(t)w_t \qquad (6)$$

and the measurement model is

$$y(t_k) = h(x_{tk}, t_k) + v_k \qquad (7)$$

where w_t and v_t are zero mean, white Gaussian noise processes, and h is the measurements function.

In our case the plant and the measurement processes are nonlinear, and so approximate filters must be used. The Extended Kalman Filter (EKF) is the well-known Kalman filter form which handles nonlinear models. It models the unknown parameters as additional state variables to be estimated. Linearisation of the nonlinear functions is essential to obtain a real time implementation. The state vector S and its covariance matrix P in the prediction state are :

$$\hat{S}_{K/K-1} = \Phi_{K-1}\hat{S}_{K-1} + \int_{tk}^{tk+1} f(\hat{S}_{K/K-1}, t)\, dt \qquad (8)$$

$$P_K = \Phi_{K-1}P_{K-1}\Phi_{K-1}^T + Q_{K-1} \qquad (9)$$

where f is the function from (6). The states are propagated by numerically integrating equation (8). The state transition matrix (Φ) is simply the exponential of $F(S_t)$, where

$$F(S_t) = \frac{\partial f(S_t)}{\partial S_t}. \qquad (10)$$

The innovation sequence is :

$$v_K = y_K - h(\hat{S}_{K/K-1}) \qquad (11)$$

The correction state is given by;

$$\hat{S}_{K/K} = \hat{S}_{K/K-1} + K_K v_K \qquad (12)$$

$$P_{K/K} = [I - K_K H_K]P_{K/K-1} \qquad (13)$$

The kalman gain K is :

$$K_K = P_K H_K^T (H_K P_K H_K^T + R_K)^{-1} \qquad (14)$$

where: $H_K(\hat{S}_{K/K-1}) = \dfrac{\partial h_K(\hat{S}_{K/K-1})}{\partial \hat{S}_{K/K-1}}$

It is not possible to determine all the motion parameters because of the depth ambiguity associated with recovering 3-D motion from the image plane. This results in an unknown scale factor that can not be determine without a - prior knowledge of any one of the translation rates. To overcome this problem, one of the translation parameters is used to normalise the other parameters. We choose the forward velocity (W). The motivation for this choice is that the normalised parameters (U/W) and (V/W) can be approximated as side-slip and incidence angles respectively. With this assumption, and assuming n feature points, the state vector S and a measurements vector h are :

$$S = \left[\frac{U}{W} \frac{V}{W} \frac{\dot{U}}{W} \frac{\dot{V}}{W} \frac{\dot{W}}{W} p\, q\, r\, \dot{p}\, \dot{q}\, \dot{r}\, \frac{z_1}{W} x_1\, y_1 \ldots \frac{z_n}{W} x_n\, y_n\right]^T$$

$$h = [x_1\, y_1\, x_2\, y_2 \ldots\ldots\ldots\ldots\ldots x_n\, y_n]^T$$

from equations 2 and 5 we can get :

$$\dot{x}_i = \frac{-U + x_i W}{Z_i} + px_i y_i - q(x_i^2 + 1) + ry_i$$

$$\qquad (15)$$

$$\dot{y}_i = \frac{-V + y_i W}{Z_i} + p(y_i^2 + 1) - qx_i y_i - rx_i$$

$$\frac{\dot{z}_i}{w} = -1 - py_i \frac{z_i}{w} + qx \frac{z_i}{w}$$

343

then by using equations 1 and 15 we can compute

$$F(S_t) = \frac{\partial f(S_t)}{\partial S_t} \; .$$

4. CORRESPONDENCE PROBLEM.

The estimation of autonomous vehicle motion parameters based on the feature point approach has two main problems, the correspondence problem and the occlusion problem. Most of the recursive techniques assume that the correspondence between sequence of images is available, and some introduced different approaches to solve the correspondence and occlusion problems by using the optical flow computation, Weng and Huang (1993), In this section we introduce a simple approach to avoid these problems. This approach creates a sequence of virtual images from the real sequence of images. The virtual images have a fixed number of virtual points (n) which can be used to estimate the vehicle motion using the EKF. To create this sequence of virtual images, assume that each real image has (N) feature points. These are sectioned into (n) groups by regions in the image(figure3). The mean position (Mi , i=1:n) for each group (figure4) is then computed and this defines the virtual point for that group. This results in a virtual image containing (n) virtual points which retains the original motion information. If a feature point from the real image moves out the image frame or is occluded due to the vehicle motion, the number of virtual points (n) will not changed and the filter algorithm not effected. The second step in the algorithm is to periodically re-select the groupings for the n virtual points, as the original grouping will become invalid as the real image group moves out of the image plane.

5. SIMULATION RESULTS.

The vehicle was simulated to perform the trajectory shown in figure 5, which involves a simple turn. The vehicle is moving at 1 m/sec and the simulation consists of 100 frames: each frame itself consisting of 25 images. For each frame a set of 4 virtual points was selected from 24 random features (a typical image frame of 24 points with the 4 groupings in shown in figure 3, with the virtual image in figure 4.

Figures (6) and (7) show the total image point trajectories over the simulated trajectory, together with the trajectory of the virtual points. Each feature has noise of variance 1% or 10% added. The percentage error is an independent position error on each of the 24 feature points.

Figure (8) shows that the side-slip angle (β) and yaw rate (q) converge rapidly to their true value with different levels of noise and the initialise of the virtual points at the end of each frame has not effected the motion estimation process.

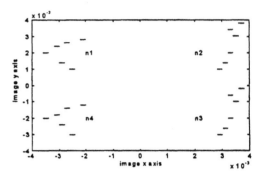

Fig. 3. :An image from a real sequence after classify the objects to n groups

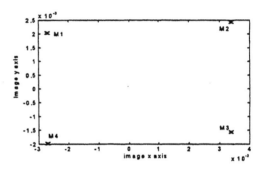

Fig. 4. A virtual image with n virtual points

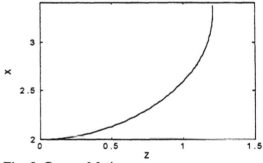

Fig. 5. Camera Motion

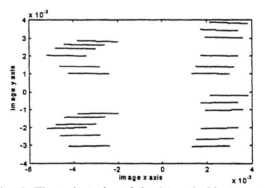

Fig. 6. The trajectories of the 24 real objects on the image plane without noise

344

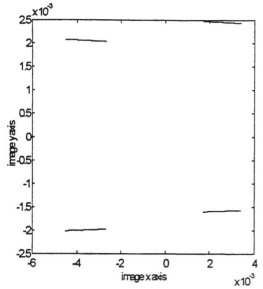

Fig. 7.The trajectories of the 4 virtual objects on the image plane without noise

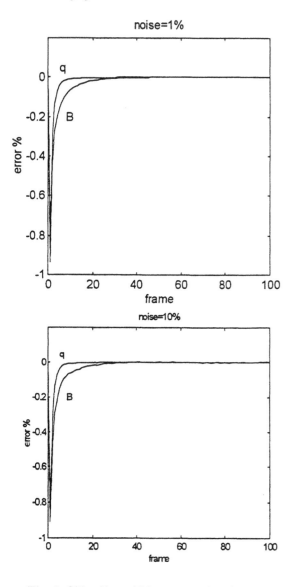

Fig. 8. Side-slip and Yaw rate estimation error

6. CONCLUSION.

The purpose of this paper is to produce a recursive method to estimate the motion parameters for an autonomous land vehicle from a sequence of noisy monocular images. The recursive estimation is performed using an EKF which uses a dynamic model of the dynamics of the vehicle, which allows for quicker convergence of the states of motion parameters. A virtual feature generation technique is introduced to avoid the main problems for feature based approach: that of correspondence and occlusion problem. The technique generates an image sequence that has a fixed number of virtual feature points while retaining the original motion information. This approach has flexibility in that the number of virtual points can be increased to increase fidelity, and means that a fixed order EKF can be used.

REFERENCES

Adive, A. (1985) Determining Three- Dimensional Motion and Structure from Optical Flow Generated by Several Moving Objects. *IEEE trans. on Patt. Anal. and Mach. Intell., PAMI-7, July.*

Bruss, R. and Horn K.P (1982). Passive Navigation . In: *Image Understanding Proceedings of a Workshop held at Palo Alto, CA , September.*

Borida, T. and Chellappa, R. (July 1990). Recursive 3-D Motion Estimation from a Monocular Image Sequence. *IEEE Trans. on Aerospace and Elect. Sys*

Jurgen Ackermann. (1993). *Robust Control systems with Uncertain physical parameters.* Springer-Verlag London Limited

Peng, T. and Sanjeev, R. (1995). A New Method for Camera Motion Parameter Estimation. In: *International Conference on Image Processing.* pp 406-416. **Vol. 1.**

Tall.M.A. " Towards Autonomous Motion Vision" Memorandum rep. M.I.T April 1992.

Weng and Huang (1993). *Motion and Structure from Image Sequences.* Springer-Verlag Berlin Heidelberg

SIMULATION OF A RADIAL LASER SCANNER FOR DEVELOPING MOBILE ROBOTIC ALGORITHMS[*]

Antonio Reina and Javier Gonzalez

Dpto. Ingenieria de Sistemas y Automatica. Universidad de Malaga.
Plaza El Ejido s/n. 29013 Malaga. FAX: +34- 5-2131413.
E-MAIL: reina@ctima.uma.es; jgonzalez@ctima.uma.es

Abstract: A Radial Laser Scanner is a device that provides distances to the surrounding objects by scanning the environment in a plane (usually parallel to the ground). This paper is concerned with the precise modeling and simulation of one of such as sensor, called the Explorer. Based on the calibration of this sensor, we propose a model where the sensor readings are a function of three independent parameters: a gaussian error noise (standard deviation), the true distance to the target and its orientation. By using this model we have developed a software that accurately simulates the data provided by the Explorer as it was working on a mobile robot navigating in a specified environment. Others parameters such as the tilt of the laser beam, the specular behavior of the surface and the mixed point effect are also taking into account when simulating. This simulator, that incorporates a powerful graphical interface, permits to develop and test new robotic navigation algorithms (basically for position estimation and map building) and has demonstrated to accurately reproduce the real sensor measurements. *Copyright © 1998 IFAC*

Keywords: *Calibration, Modeling, Simulation, Laser Range-finders, Mobile Robots*

1. INTRODUCTION

Radial laser scanners are becoming increasingly popular in the mobile robotic community. The reasons are twofold: first, the information they provide is of great interest to mobile robot applications since it can be directly used to estimate the robot position (Cox, 1991; Gonzalez et al.,1993; Lu, 1995; Castellanos and Tardos, 1996; Reina and Gonzalez, 1998) and to build or update the map of the environment (Gonzalez et al., 1994; Shaffer, 1995); second, they are becoming cheaper.

Another advantage that we would like to add to these two main features is the feasibility to be simulated. Simulation of intensity images of video cameras, for example, are quite difficult to perform since many parameters are involved (light sources, reflectance surface properties, camera model, etc.) and the relations between them are certainly complex (Horn, 1975; Lee et al., 1990). On the contrary, simulating

radial laser scanners is much more simple and reliable.

The model needed for the simulation is obtained through experimental calibration. Although this model may also be utilized to develop a virtual sensor that corrects by software the known deviations in the real sensor measures (Reina and Gonzalez, 1997), in this paper we use it to construct the simulator. This simulation software facilitates both the development of new laser-based algorithms without physically using the scanner and testing the algorithms in any imaginable synthetic environment.

This paper is organized as follows. The Explorer sensor and its model are described in section 2 and 3, respectively. Then, the characteristics of the simulator are presented. Section 5 analyzes and compares both simulated and real data. Finally, some conclusions are outlined.

---

[*] This work has been supported by the Spanish "Comision Interministerial de Ciencia y Tecnologia", project TAP96-0763.

2. THE EXPLORER LASER RANGEFINDER

The Explorer is a time-of-fight radial laser range scanner, manufactured by Schwartz Electro-optics Inc. (SEO) under specification of the Department of System Engineering and Automation, University of Malaga. The components of the Explorer are an emitter/receiver pulsed gallium infrared laser, a rotating prism, a driving motor, and an encoder mounted on a steel housing (figure 1).

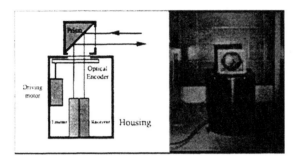

Figure 1. The Explorer laser radial scanner.

In the scanning process a laser beam is sent by the emitter and deflected horizontally by the prism. By rotating the prism, the Explorer is able to scan 360 degree field of view in a plane perpendicular to its axis of rotation (parallel to the ground) providing a two dimensional description of the environment in polar coordinates (ρ, θ).

A scan supplied by the Explorer consists of a one-dimensional array of range indices I_R, which are integer values between 0 and 1023 (ten bits of resolution). The angle θ corresponding to each range index is given by its order in the array.

The specification sheet states the following characteristics: maximum range of 35 m., range accuracy of ±10 cm., the angular resolution can be programmed to measure 128, 256, 512, 1024 and 2048 data per revolution, and the rotation speed is programmable between 0.5 and 4 revolutions per second.

3. A PRECISE MEASURING MODEL OF THE EXPLORER

According to the specification sheet the Explorer is able to measure distances from 0 to 35 meters and this interval is mapped into the range index interval. Thus, to convert a range index into a real distance (in meters) the manufacturer gives an scale factor K=3.418 cm. per index (35m. divided by 1024).

Obviously, this factor also states the resolution of the measurements. A simple model of the sensor can be accomplished using the equation:

$$I_R = Round\left(\frac{d_T}{K}\right) \qquad \text{(EQ 1)}$$

where I_R is the range index provided by the sensor model, $Round(\cdot)$ is the rounding function, d_T is the true distance and K is the scale factor above mentioned.

Through the calibration of the sensor we have tested that the range indices obtained by the above model do not reproduce the real indices supplied by the sensor. In order to get a more realistic behavior, the model should account for the following aspects:

a) **Non-linear behavior of the sensor.** To take into consideration the observed non-linearity between the actual distances and the range indices we have experimentally computed a conversion table that maps a given distance value into a range index (Reina and Gonzalez, 1997). The improvement obtained using this look-up table (LUT) can be observed in figure 2. This look-up table substitutes the linear transformation function provided by the manufacturer.

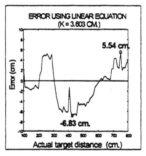 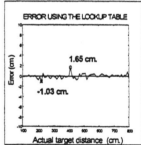

Figure 2. Measures errors using equation 1 (left) and using the look-up table (right).

b) **Additive Gaussian noise.** We have also observed that for a target at a fixed distance the sensor mostly supplies three different range indices non-equally distributed (see figure 7). To model this behavior, beside to be rounded, we assume that the actual distance to the target is affected by a white gaussian error. According to the experiments being carried out (Reina and Gonzalez, 1997) the standard deviation of this gaussian error results to be 1.2 centimeters.

c) **The incident angle.** As shown in figure 3, the incident angle between the target and the laser beam affects the measure mean. We have also checked that the variance of the measures is not affected neither by this angle nor by the range distance, thus the standard deviation above mentioned is considered to be valid for all the possible surface orientations and distances.

By considering these aspects, a more complete model of the sensor may be implemented using the equation:

$$I_R = Round(T(d_T + \eta) + A(\alpha)) \qquad \text{(EQ 2)}$$

where $T(\cdot)$ is the look-up table before mentioned, η is a gaussian error $N(0,1.2)$ and $A(\alpha)$ is another look-up table constructed from the experimental results of figure 3.

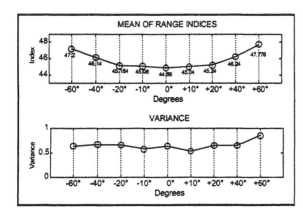

Figure 3. Mean and variance of the range indices for different target orientations.

4. THE SIMULATOR

In this section we describe both the algorithm for simulating the Explorer and the graphical interface that has been built.

4.1. The simulation process

To simulate a scan supplied by the Explorer it is first necessary to provide the simulator with a two-dimensional segment map of the environment, the angular resolution in the scanning process, and the position of the sensor (robot) in the environment.

Each line segment of the map is defined by four different parameters: the coordinates of the endpoints, its orientation, its height over the ground plane and the type of surface it represents.

The latter refers to reflectance properties of the surface. The range indices supplied by the sensor fairly vary with the color of the object intersected although they do with their specular characteristics. The calibration has demonstrates that four different categories may be considered: *normal*, where the model of equation 2 applies, *medium specular* (i.e. a glassed tile) that produces a randomly distorted image, *high specular* (i.e. a polish aluminum or a mirror) where no echo is received by the sensor, and *transparent* (a glass) that is not detected by the sensor. These types of surfaces have been easily incorporated in the simulation process, except the medium specular surface since it can not be consistently modeled.

Ideally, since the surface scanned by the sensor is a plane parallel to the ground floor at a specific height, only objects at this height can be detected. The height of the segments is used to check this fact. In practice, imprecisions in the sensor construction originates the scanned surface to be other than the expected. In our case, this surface is a cone (Chinese hat shape) with a inclination of about 1 degree, so the detection of an object does not depend only on its height but also on its distance to the Explorer.

Other issue that must be considered when simulating is the mixed footprint effect, that refers to the phenomenon that occurs when a laser beam hits the occluding edge of an object, and therefore part of the footprint lies on one segment while the other lies on a different one (see figure 4). In this case the range value is a linear combination of the range values to the two segments. Although in reality the laser beam presents a small divergence (3 mrad.), for simplifying the simulation process the cross section of the laser beam is considered to be a constant circle of 3cm. radius.

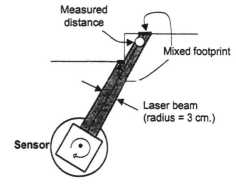

Figure 4. Mixed footprint effect

To model a mixed footprint (see figure 5) we compute the amount of laser beam that reach every surface, S_1 and S_2. The range index, I_{Rmp}, is obtained through the equation:

$$I_{Rmp} = Round\left(D_1 + (D_2 - D_1) \cdot \left(\frac{S_2}{R}\right)\right) \qquad (EQ. 3)$$

where D_1 and D_2 are the distances computed by means of the measuring model of equation 2 but without using the *Round* function and R is the radius of the laser beam.

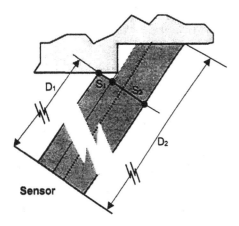

Figure 5. Parameters to compute the distance for a mixed footprint.

In the following table a pseudo-code summarizes the complete simulation algorithm.

Compute the angular resolution $A_R = 360°/n°$ points

For every laser beam direction β :

- *Determine the set of segments, **S**, in the environment intercepted by the laser beam footprint.*

- *For all the segment in **S**:*
    If not a mixed footprint:
        Compute the incident angle α
        Compute the true distance d_T
        $R[S] = Round(T(d_T + \eta) + A(\alpha))$
    Else
        Compute D_1, D_2, S_2
        $R[S] = Round\left(D_1 + (D_2 - D_1) \cdot \left(\frac{S_2}{R}\right)\right)$
- *The final measured for these laser beam β is:*
    $I_R[\beta] = \text{Minimum } (R)$

4.2. Graphic Interface

The laser sensor model has been included in a simulation package for mobile robot navigation.

For fast application development, the above simulation algorithm has been provided with a powerful graphic interface based on Xview library. This graphic interface allows to easily create 2D maps, robot navigation paths, visualize the results of the different navigation algorithms (position estimation and map building) and to directly control the real laser sensor (the Explorer) through an SCSI bus.

Figure 6 shows the main window of this graphics interface.

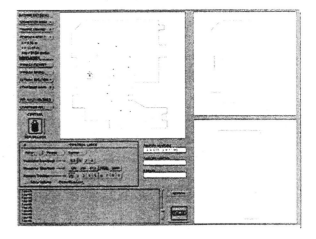

Figure 6. Main window of the graphics interface.

5. PERFORMANCE EVALUATION

The performance of the simulator has been checked by two ways: by estimating the accuracy of the measuring model and by visually comparing real scans with its simulated counterparts.

In the first case, for a target set at a fixed distance with different orientations we have measured the discrepancy between the data provided by the Explorer and the data obtained with the model of equation 2. The error between the two set of data[1] has been computed by the equation:

$$Error = \frac{1}{N}\sum_{j=1}^{1}(HistoS(j) - HistoR(j))^2 \qquad (EQ\ 4)$$

---

[1] For accuracy purposes only the central point of those hitting the target is chosen.

where HistoS and HistoR are the histograms for the real and simulated indices obtained at 1.5m. and three different orientations (0°,+40°,-40°) using N=500 samples (figure 7). The parameter "j" indexes the three most frequent values of the histograms. This experiment was repeated for the target at several distances producing in all cases an error bellow 0.3 indices.

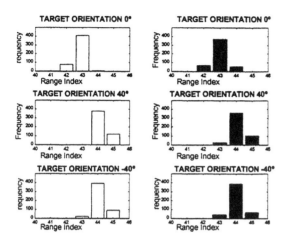

Figure 7. Real and simulated (filled) histograms for 500 samples.

On the other hand, the performance of the whole simulator can be checked by visually examining the discrepancy between synthetic and real scans.

Figure 10 shows a real scan taken in our Mobile Robot laboratory and the scan provided by the simulator using the model shown in figure 8. It includes objects (tables and bookcase) that are at a height bellow the scanning surface of the Explorer (they appear shaded) as well as a glassed partition that are transparent (dotted line).

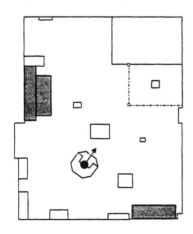

Figure 8. Polygonal model of our Mobile Robot Laboratory.

Observe that the most significant differences between the two scans (marked with circles in the figure 10) are due to objects of the environment that do not appear in the environment model.

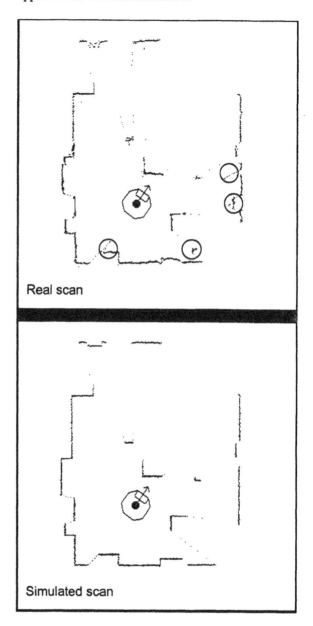

Figure 9. Real and Simulated scan (2048 points).

6. CONCLUSIONS

Based on the calibration of the Explorer radial laser scanner we have obtained a sensor model that permits to accurately simulate the data provided by it. This sensor model has been integrated into a simulation package provided with a powerful and easy-to-use interface.

This simulator has been successfully used to develop and test new robotic navigation algorithms, specially for robot position estimation (Gonzalez et al., 1995; Reina and Gonzalez, 1998) and map building (Gonzalez et al, 1994). These algorithms were finally implemented and verified in the mobile robot RAM-2 using the data supplied by the Explorer range scanner. The performance of these algorithms on the RAM-2 has been quite similar to that of the simulator, which demonstrates the validity of the simulation.

REFERENCES

Castellanos, J.A. and J.D. Tardos (1996). *"Laser-based segmentation and localization for a mobile robot"*. Second World Automation Congress. Montpellier, France.

Cox I.J. (1991). *"Blanche-An Experiment in Guidance and Navigation of an Autonomous Robot Vehicle"*. IEEE Transactions on Robotics and Automation, **Vol. 7, no. 2**.

Gonzalez, J., A. Ollero and A. Reina (1994). *"Map Building for a Mobile Robot equipped with a Laser Range Scanner"*. IEEE Int. Conference on Robotics and Automation, San Diego, CA, USA.

Gonzalez, J., A. Stentz and A. Ollero (1995). "A Mobile Robot Iconic Position Estimator using a Radial Laser Scanner". *Journal of Intelligent & Robotic Systems*.13:161-179. Kluwer Academic Press.

Horn, P. (1975). *"Obtaining Shape from Shading Information"*. Psychology of Computer Vision, P.Wiston, Ed. New York. McGraw-Hill.

Lee, H.C., E.J. Breneman, and C.P. Schulte (1990). *"Modeling Light Refection for Computer Color Vision"*. IEEE Trans. PAMI, pp. 402-409.

Lu, F. (1995). *"Shape Registration using Optimization for Mobile Robot Navigation"*. PhD. Thesis. University of Toronto.

Reina, A. and J. Gonzalez (1997). "Characterization of a Radial Laser Scanner for Mobile Robot Navigation". Proc. IEEE/RSJ Int. Conf. On Intelligent Robots and Systems, pp.579-585.

Reina A. and J. Gonzalez (1998). "Determining Mobile Robot Orientations by Aligning 2D Range Scans". *3rd IFAC Symposium on Intelligent Autonomous Vehicles*. Madrid (to appear).

Shaffer G. (1995). "Two-dimensional Mapping of Expansive Unknown Areas". *Thesis. Robotics Institute, Carnegie Mellon* University, Pittsburgh.

SKY HOOK AND CRONE SUSPENSIONS : A COMPARISON

X. Moreau, A. Oustaloup and M. Nouillant

Equipe CRONE - LAP - ENSERB - Université Bordeaux I
351, cours de la Libération - 33405 - Talence Cédex - FRANCE
Tel. (33) 56.84.24.17 - Fax. (33) 56.84.66.44
E-mail : moreau@lap.u-bordeaux.fr

Abstract :This paper deals with a comparison between sky hook and CRONE suspensions. The sky hook suspension can be defined from the Linear Quadratic (LQ) used by many authors to design state feedback laws. The CRONE suspension is defined from the second generation CRONE control (Commande Robuste d'Ordre Non Entier) and based on non integer derivation. A two degree of freedom quarter car model is used to evaluate performances. The CRONE suspension leads to a better comfort and a better road holding ability. The frequency and time responses, for various values of the vehicle load, reveal a great robustness of the degree of stability through the constancy of the resonance ratio in the frequency domain and of the damping ratio in the time domain. *Copyright © 1998 IFAC*

keywords : Two degrees of freedom quarter car model, Advanced suspension system, Linear Quadratic control, non integer derivation, second generation CRONE control, CRONE suspension, Robustness.

1. INTRODUCTION

For any vehicle, the suspension system must perform two main functions. Firstly, it provides a high degree of insulation for the vehicle body from the loads applied between the wheels and roads to ensure passenger comfort and secondly, it keeps the wheels in close contact with the road surface to ensure adequate adhesion when accelerating, braking or cornering. These functions must then be optimised within several constraints: first, the minimum value of the relative body/wheel workspace usage; second, control of vehicle attitude in manoeuvring; third minimum value of the power consumption. In fact, the conflict between these various aspects of vehicle behaviour is the main problem in suspension system design (Crolla and Abouel, 1992)

After an introduction, Part 2 of this paper presents different types of suspension and Part 3 describes the vehicle model used for comparison. Part 4 develops the synthesis method of the non integer force-displacement transmittance and describes the constrained optimisation to determine transmittance parameters. Part 5 gives the initial behaviour of the sprung mass acceleration for traditional, sky hook and CRONE suspensions. Part 6 examines the frequency and time responses which show the robustness of both the resonance and damping ratios versus load variations. Finally, in part 7 conclusions are given.

2. TYPES OF SUSPENSION

Consider the two-degree-of-freedom model of a vehicle moving with the constant forward speed V shown in Fig. 1. The two degrees of freedom $z_2(t)$ and $z_1(t)$ are associated with vehicle's sprung and unsprung masses m_2 and m_1, respectively. The model shown in Fig. 1 is often adequate in preliminary studies of vehicle ride dynamics, where the primary suspension (e.g. tires) is modeled via a linear spring with stiffness k_1 and a linear damper with damping coefficient b_1. The investigation is concentrated on the proper choice of a secondary suspension that would result in the most comfortable ride. The secondary suspension, located between the sprung and unsprung masses, develops a force $f_2(t)$ which can be generated by a traditional, active or semi-active device. (Sharp and Hassan, 1986).

2.1. Traditional suspension

The traditional suspension used in most vehicles depends on passive elements, which are various types of spring and damper. They provide forces which are linear or non-linear functions of the relative displacement and velocity between the body and wheel of the vehicle. In the linear case, the expression of $f_2(t)$ is given by:

$$f_2(t) = k_2 z_{12}(t) + b_2 \dot{z}_{12}(t) , \qquad (1)$$

in which
$$z_{12}(t) = z_1(t) - z_2(t) \qquad (2)$$

and where k_2 is the stiffness of the spring and b_2 the damping coefficient.

2.2. Active suspension

Active systems use an actuator instead of a spring and a damper of a traditional suspension. The force $f_2(t)$ is a function of state variables, namely:

$$f_2(t) = - K_c \, x \, , \qquad (3)$$

in which

$$K_c = [k_{c1} \; k_{c2} \; k_{c3} \; k_{c4}] \text{ and } x = \begin{cases} x_1 = z_1 - z_0 \\ x_2 = \dot{z}_1 \\ x_3 = z_2 - z_1 \\ x_4 = \dot{z}_2 \end{cases} \qquad (4)$$

This expression results from the use of optimal linear regulator theory studied by many reports (De Jeager, 1991; Redfield and Karnopp, 1989; Yue, *et al.*, 1989; Wilson, *et al.*, 1986) and in which k_{c1}, k_{c2}, k_{c3} and k_{c4} are linear state variable feedback gains. In the case where k_{c1} is equal to zero and by taking into account the relative velocity, the relation (3) becomes

$$f_2(t) = k_{c3} \, z_{12}(t) - k_{c2} \, \dot{z}_{12}(t) - (k_{c2} + k_{c4}) \, \dot{z}_2(t) \, . \quad (5)$$

The force developed by the actuator is the result of three terms. The first term is a force proportional to the relative displacement and can be interpreted as the force developed by a traditional spring whose stiffness k_2 is equal to k_{c3} (with constraint $k_{c3} > 0$). The second term is a force proportional to the relative velocity and can be interpreted as the force developed by a traditional damper whose damping coefficient b_2 is equal to k_{c2} (with constraint $k_{c2} < 0$). The third term is a force proportional to the absolute velocity of the sprung mass m_2. It can be interpreted as the force developed by a traditional damper, whose damping coefficient b_{sh} is equal to $(k_{c2} + k_{c4})$ (with constraint $k_{c4} > 0$), located between the sprung mass and an inertial reference. This configuration is called the "sky hook damper" scheme (Karnopp, *et al.*, 1974).

2.3. Semi-active suspension

The control requirements of a semi-active suspension are the same as those of an active system except that the actuator in the latter is replaced by a continuously variable damper which is theoretically capable of tracking a force demand signal that the actuator is assumed to be only capable of dissipating power. It needs a control law to drive it and the method adopted is that the force demand signal is given by

$$f_{sh}(t) = - b_{sh} \, \dot{z}_2(t) \quad \text{if} \quad \dot{z}_{12}(t) \, \dot{z}_2(t) < 0$$
$$f_{sh}(t) = 0 \quad \text{if} \quad \dot{z}_{12}(t) \, \dot{z}_2(t) \geq 0 \, . \qquad (6)$$

2.4. CRONE suspension

The CRONE suspension results from a traditional suspension system whose spring and damper are re-placed by a non integer order force-displacement transmittance. This device develops a force $f_2(t)$ which is a function of the relative displacement $z_{12}(t)$ and which obeys symbolically to the general relation

$$F_2(s) = C(s) \, Z_{12}(s) \, , \qquad (7)$$

in which $C(s)$ is the suspension transmittance defined by a non integer expression.

The CRONE suspension can be manufactured from passive, semi-active or active devices. The first solution, called passive CRONE suspension, uses the link between recursivity and non integer derivation (Oustaloup, 1991). This suspension has been mounted on experimental Citroën vehicles. The active and semi-active solutions are presented in this paper.

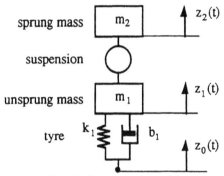

Fig. 1. Quarter car model

3. VEHICLE MODEL

If it is assumed that the tyre does not leave the ground and that $z_1(t)$ and $z_2(t)$ are measured from the static equilibrium position, then the application of the fundamental law of dynamics leads to the linearised equations of motion :

$$m_1 \, \ddot{z}_1(t) = f_1(t) - f_2(t) \qquad (8)$$
and
$$m_2 \, \ddot{z}_2(t) = f_2(t) \, , \qquad (9)$$
in which
$$f_1(t) = k_1 \left(z_0(t) - z_1(t) \right) + b_1 \left(\dot{z}_0(t) - \dot{z}_1(t) \right) , \quad (10)$$

and where $z_0(t)$ is the road deflexion. The Laplace transform of equations (8), (9) and (10), assuming zero initial conditions, are

$$m_1 \, s^2 \, Z_1(s) = k_1 \, Z_{01}(s) + b_1 \, s \, Z_{01}(s) - F_2(s) \quad (11)$$
and
$$m_2 \, s^2 \, Z_2(s) = F_2(s) \, , \qquad (12)$$
in which
$$Z_{01}(s) = Z_0(s) - Z_1(s) \, . \qquad (13)$$

To analyse the vibration insulation of the sprung mass, two transmittances are defined :

$$T_2(s) = \frac{Z_2(s)}{Z_1(s)} \quad \text{and} \quad S_2(s) = \frac{Z_{12}(s)}{Z_1(s)} \, . \qquad (14)$$

From equations (11) and (12), the expressions of $T_2(s)$ and $S_2(s)$ are given by :
- for the traditional suspension

$$T_2(s) = \frac{k_2 + b_2 \, s}{k_2 + b_2 s + m_2 s^2} \quad \text{and} \quad S_2(s) = \frac{m_2 s^2}{k_2 + b_2 s + m_2 s^2} \, ; \quad (15)$$

- for the sky hook suspension

$$T_2(s) = \frac{k_2 + b_2 s}{k_2 + (b_2 + b_{sh})s + m_2 s^2}$$

and $\qquad\qquad\qquad\qquad\qquad\qquad$ (16)

$$S_2(s) = \frac{b_{sh}s + m_2 s^2}{k_2 + (b_2 + b_{sh})s + m_2 s^2} \; ;$$

- for the CRONE suspension

$$T_2(s) = \frac{C(s)}{C(s) + m_2 s^2} \quad \text{and} \quad S_2(s) = \frac{m_2 s^2}{C(s) + m_2 s^2}.$$
(17)

To study ride comfort and road holding ability, three additional transmittances are defined :

$$H_a(s) = \frac{A_2(s)}{V_0(s)} \; , \qquad\qquad H_{12}(s) = \frac{Z_{12}(s)}{V_0(s)}$$

and $\qquad\qquad\qquad\qquad\qquad\qquad$ (18)

$$H_{01}(s) = \frac{Z_{01}(s)}{V_0(s)} \; ,$$

in which $A_2(s)$ is acceleration of the sprung mass, $Z_{12}(s)$ suspension deflection, $Z_{01}(s)$ tyre deflection and $V_0(s)$ road input velocity. A commonly used road input model is that $v_0(t)$ is white noise whose intensity is proportional to the product of the vehicle's forward speed and a road roughness parameter. In this case, the deflexion $z_0(t)$ of the road is a brownian motion (Thompson, 1973).

4. SYNTHESIS OF THE NON INTEGER FORCE-DISPLACEMENT TRANSMITTANCE

The synthesis method of the CRONE suspension is based on the interpretation of transmittances $T_2(s)$ and $S_2(s)$ which can be written as :

$$T_2(s) = \frac{\beta(s)}{1 + \beta(s)} \quad \text{and} \quad S_2(s) = \frac{1}{1 + \beta(s)} \; , \quad (19)$$

in which

$$\beta(s) = \frac{C(s)}{m_2 \, s^2} \; . \qquad\qquad (20)$$

The transmittances $T_2(s)$ and $S_2(s)$ can here be considered to be of an elementary control loop whose $\beta(s)$ is the open loop transmittance.

Given that relation (20) expresses that a variation of mass is accompanied by a variation of open loop gain, the principle of the second generation CRONE control (Oustaloup, 1991) can be used by synthesising the open loop Nichols locus which traces a vertical template for the nominal mass.

A way of synthesising the open loop Nichols locus consists in determining a transmittance $\beta(s)$ which successively presents (Fig.2) :

- an order 2 asymptotic behaviour at low frequencies to eliminate tracking error ;
- an order n asymptotic behaviour where n is between 1 and 2, around frequency ω_u and at high frequencies to ensure satisfactory filtering of vibrations.

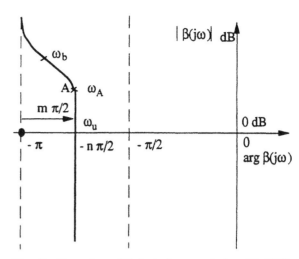

Fig. 2. Open loop Nichols locus of the CRONE suspension

Such localised behaviour (Moreau, *et al.*, 1996) can be obtained with a transmittance of the form :

$$\beta(s) = C_0 \left(1 + \left(\frac{s}{\omega_b}\right)^m\right)\left(\frac{\omega_0}{s}\right)^2 \qquad (21)$$

in which :

$$\omega_b << \omega_A \quad \text{and} \quad m = 2 - n \in \,]0,1[\; . \quad (22)$$

Identification of equations (20) and (21) gives :

$$1/\sqrt{m_2} = \omega_0 \qquad\qquad (23)$$

and

$$C(s) = \frac{F_2(s)}{Z_{12}(s)} = C_0 \left(1 + \left(\frac{s}{\omega_b}\right)^m\right) \; , \qquad (24)$$

namely, in time domain :

$$f_2(t) = C_0 \left[z_1(t) - z_2(t)\right] + \frac{C_0}{\omega_b^m} \left(\frac{d}{dt}\right)^m \left[z_1(t) - z_2(t)\right] \; . (25)$$

The force developed by the suspension is the result of two terms. The first term is a force proportional to the relative displacement and can be interpreted as the force developed by a traditional spring whose stiffness k_2 is equal to C_0. The second term is a force proportional to the non integer derivative of the relative displacement. The principle consists in replacing the order 1 damper of a traditional suspension by a damper whose non integer order is between 0 and 1. This configuration is called the CRONE damper. The force $f_c(t)$ provided by the CRONE damper, namely:

$$f_c(t) = b_c \left(\frac{d}{dt}\right)^m \left[z_1(t) - z_2(t)\right] \; , \qquad (26)$$

with

$$b_c = \frac{C_0}{\omega_b^m} \; , \qquad\qquad (27)$$

can be interpreted as the force developed by an non-stationary damper, namely :

$$f_c(t) = b_c(t) \frac{d}{dt} \left[z_1(t) - z_2(t)\right] \; . \qquad (28)$$

Identification of equation (26) and (28) leads to :

$$b_c(t) = b_c \frac{\left(\frac{d}{dt}\right)^m \left[z_1(t) - z_2(t)\right]}{\frac{d}{dt} \left[z_1(t) - z_2(t)\right]} \; . \qquad (29)$$

Two phases can be distinguished: an active phase when $b_c(t)$ is negative and a semi-active phase when $b_c(t)$ is positive.

By defining the transmittances (18) with respect to $v_0(t)$, all frequencies contribute equally to their mean square values. That is why the determination of CRONE suspension parameters is based on the minimisation of a criterion J composed of the H_2-norm of the transmittances $H_a(j\omega)$, $H_{12}(j\omega)$ and $H_{01}(j\omega)$, namely

$$J = \frac{\rho_1}{\lambda_1} \int_{\omega_b}^{\omega_h} |H_a(j\omega)|^2 d\omega \qquad (30)$$
$$+ \frac{\rho_2}{\lambda_2} \int_{\omega_b}^{\omega_h} |H_{12}(j\omega)|^2 d\omega + \frac{\rho_3}{\lambda_3} \int_{\omega_b}^{\omega_h} |H_{01}(j\omega)|^2 d\omega,$$

in which ρ_i are the weighting factors, λ_i the H_2-norm computed for the traditional suspension.

To obtain a significant comparison between traditional, sky hook and CRONE suspension performances, a constraint is fixed for the minimal sprung mass : equal unit gain frequency of open loop $\beta(j\omega)$.

5. INITIAL BEHAVIOUR

If $z_1(t)$ is a slope function of the form

$$z_1(t) = t\, u(t), \qquad (31)$$

where $u(t)$ is the unit step function, we can write, from the initial value theorem :

$$\dot{z}_2(0^+) = \lim_{s \to \infty} \frac{k_2 s + b_2 s^2}{k_2 + b_2 s + m_2 s^2} = \frac{b_2}{m_2}, \quad (32)$$

for the traditional suspension,

$$\dot{z}_2(0^+) = \lim_{s \to \infty} \frac{k_2 s + b_2 s^2}{k_2 + (b_2 + b_{sh}) s + m_2 s^2} = \frac{b_2}{m_2}, \quad (33)$$

for the sky hook suspension,

$$\dot{z}_2(0^+) = \lim_{s \to \infty} \frac{k_2 s + b_c s^{1+m}}{k_2 + b_c s^m + m_2 s^2} = 0, \quad (34)$$

for the CRONE suspension.

The comparison between the relations (32), (33) and (34) shows that the initial acceleration $\dot{z}_2(0^+)$ of the sprung mass is finite for the traditional and the sky hook suspensions and nil for the CRONE suspension. It indeed insures a better comfort for the passengers.

6. PERFORMANCES

The traditional and sky hook suspensions are rear suspensions whose parameters are given by (Moreau, 1995), namely :

- sprung mass : $150\,kg \leq m_2 \leq 300\,kg$;
- unsprung mass : $m_1 = 28.5\,kg$;
- stiffness of tyre : $k_1 = 155\,900\,N/m$;
- damping coefficient of tyre : $b_1 = 50\,Ns/m$;

for the traditional suspension
- stiffness : $k_2 = 19\,960\,N/m$;
- damping coefficient : $b_2 = 930\,Ns/m$;

for the sky hook suspension
- damping coefficient : $b_{sh} = 930\,Ns/m$;
- stiffness : $k_2 = 19\,960\,N/m$;
- damping coefficient : $b_2 = 960\,Ns/m$.

From this data, the constrained optimisation of the criterion J, computed with the optimisation toolbox of Matlab, provides the optimal parameters of the CRONE suspension, namely :

$$m = 0.75 ; \ b_c = 4500\,Ns^{0.75}/m ;$$
$$\omega_b = 0.628\,rd/s. \qquad (35)$$

6.1. Frequency responses

Fig. 3 gives the gain diagrams of $T_2(j\omega)$ for the traditional, sky hook and CRONE suspensions. For the CRONE suspension, the resonance ratio can be seen to be both weak and insensitive to variations of mass m_2. This shows a better robustness of the CRONE suspension in the frequency domain.

6.2. Step responses

Fig. 4 and 5 show the step responses of the car body and the wheel for both suspensions. For the CRONE suspension it can be seen that the first overshoot remains constant, showing a better robustness for the CRONE suspension in the time domain (fig. 4.c). The road holding ability is better for the CRONE suspension (fig. 5.c).

7. CONCLUSION

In this paper it has been shown that the CRONE suspension provides remarkable performances : better robustness of stability degree versus load variations of the vehicle. Robustness is illustrated by the frequency and time responses obtained for different values of the load.

From the concept of the CRONE suspension three technological solutions have been developed (Moreau, 1995). The first, called *passive CRONE suspension*, uses the link between recursivity and non integer derivation. This suspension has now been mounted on an experimental Citroën BX. The second solution, called *active CRONE suspension*, uses an actuator which develops a force proportional to the non integer derivative of the relative displacement. The third solution, called *passive piloted CRONE suspension*, uses a continuously variable damper. Its design permits manufacture at the traditional automobile damper cost. Bench tests on a prototype have validated theoretical expectations.

REFERENCES

Crolla D. A. and Abouel Nour A. M. A. (1992). Power losses in active and passive suspensions of off-road vehicles. *Journal of Terramechanics*, Vol.29, n°1, pp.83-93.

De Jaeger A. G. (1991). Comparison of Two Methods for the Design of Active Suspension Systems. *Optimal Control Applications and Methods*, Vol. 12, pp.173-188.

Karnopp d., Crosby M. J. and Harwood R. A. (1974). Vibration Control Using Semi-active force Generators. Journal of Engineering for Industry. Vol. 96, n°2, pp. 619-626.

Moreau X. (1995). La dérivation non entière en isolation vibratoire et son application dans le domaine de l'automobile. La suspension CRONE : du concept à la réalisation. Thèse de Doctorat, Université de Bordeaux I.

Moreau X., Oustaloup A. and Nouillant M. (1996). Comparison of LQ and CRONE methods for the design of suspension systems. *Proceedings of the 13th IFAC World Congress*, San Francisco, USA, June 30 - July 5.

Oustaloup A. (1991). *La commande CRONE*. Edition Hermes, CNRS, Paris.

Redfield R.C. and Karnopp D.C. (1989). Performance sensitivity of an actively damped vehicle suspension to feedback variation. *Journal of dynamic Systems, Measurement, and Control*, Vol.111, n°1, pp.51-60.

Sharp R. S. and Hassan S. A. (1986). The relative performance capabilities of passive, active and semi-active car suspension systems. *Proc. Inst. Mech. Engers.*, vol. 200, pp. 219-228.

Thompson A. G. (1973). Quadratic performance indices and optimum suspension design. *Proc. Inst. Engrs*, Vol. 187, pp. 129-139.

Wilson D. A., Sharp R. S. and Hassan S. A. (1986). The Application of Linear Optimal Control Theory to the Design of Active Automotive Suspensions. *Vehicle System Dynamics*, Vol. 15, pp. 105-118.

Yue C., Butsuen T. and Hedrick J. K. (1989). Alternative control laws for automotive active suspensions. *Journal of Dynamic Systems, Measurement and Control*, Vol. 111, pp. 286-291.

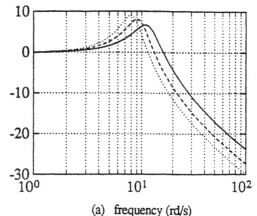

(a) frequency (rd/s)

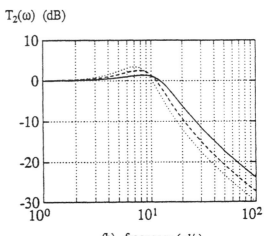

(b) frequency (rd/s)

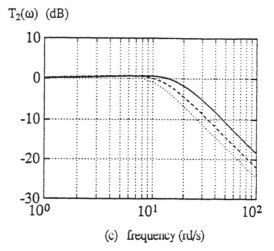

(c) frequency (rd/s)

Fig. 3. Gain diagrams of $T_2(j\omega)$ for traditional (a), sky hook (b) and CRONE (c) suspensions :

——— $m_2 = 150$ kg ; $- - -$ $m_2 = 225$ kg ;
········ $m_2 = 300$ kg

357

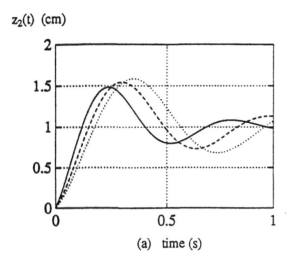

(a) time (s)

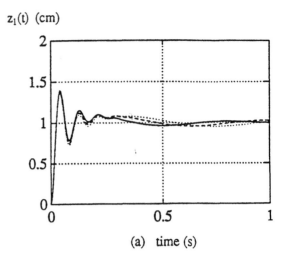

(a) time (s)

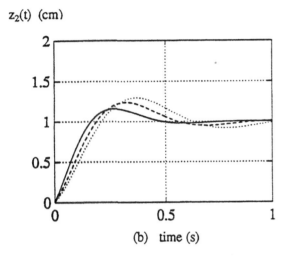

(b) time (s)

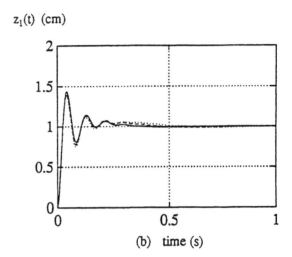

(b) time (s)

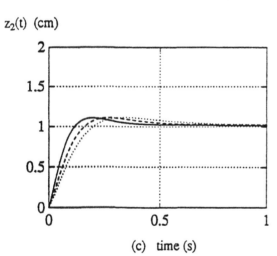

(c) time (s)

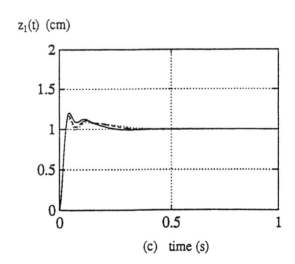

(c) time (s)

Fig. 4. Step responses of sprung mass for traditional (a), sky hook (b) and CRONE (c) suspensions :
——— m_2 = 150 kg ; – – – m_2 = 225 kg ;
······· m_2 = 300 kg

Fig. 5. Step responses of unsprung mass for traditional (a), sky hook (b) and CRONE (c) suspensions :
——— m_2 = 150 kg ; – – – m_2 = 225 kg ;
······· m_2 = 300 kg

358

AN ACTIVE SUSPENSION BASED ON SELF-ORGANIZING MAPS

Dimitrios Moshou[1], **Jan Anthonis**[2], **Herman Ramon**[3]

Department of Agro-Engineering and Economics, Laboratory for
Agro-Machinery and Processing, K.U. Leuven, Kardinaal
Mercierlaan 92, 3001 Heverlee, BELGIUM
(1) Research Engineer, tel: +32-16-321478, fax: +32-16-321994,
email: dimitrios.moshou@agr.kuleuven.ac.be
(2) Research Engineer, (3) Associate Professor

Abstract: The Self-Organizing Map Neural Network is used in a supervised way to represent a sensor-actuator mapping. The learning of the controller assumes no prior information, but only reward/failure signals that are produced by an evaluation criterion. The evaluation criterion used is based on the low-pass filtering of the gradient of a reward function and the local storing of the filtered gradient value. The control method is tested in vibration isolation of a flexible spray boom used in agriculture for pesticide application. The Neural Network learns to stabilise the boom on-line without any prior information and with a very high performance. *Copyright © 1998 IFAC*

Keywords: Neural Networks, Self-Organizing Systems, Active Vehicle Suspension, Agriculture, Machinery

1. INTRODUCTION

Many mechanical structures are subjected to vibrations that can lead to damage or to fatigue and thus shorten the operational lifetime of the structure. Passive vibration isolation gives poor results because of low selectivity. Active vibration isolation is much more performant. However, model-based techniques require persistent excitation signals. In practice persistent excitation is rarely available during the operating condition of a system. An alternative is to develop an algorithm that can discover the control actions by itself. The only source of information in this case is a "reward function" which specifies at a given moment how well the controller has performed. For this algorithm to be executed the system must now create at each learning step the control action (Ritter, *et al.*, 1992).

In the absence of any further information a stochastic search can be performed in the space of the available control values with the aim to maximize the reward received at each step. Performance based partitioning of the state-space is achieved. Current sensor, actuator and target sensor values in a vectorised form become associated with next step control actions. Through the maximisation of a certain reward function a goal directed plant inversion is performed.

For continuous state-spaces, the state-action look-up table refers to a quantization of the states of the system through the use of an adaptive algorithm. Basically two types of learning are present here:
i) the adaptation of the partitioner, and
ii) the reinforcement learning of the controller (Hermann and Der, 1995).

The determination of quantized states, which are internal states in the full control problem, represents an instance of the hidden state problem (Das and Mozer, 1994). For discrete actions, an ideal partition of the continuous state space consists of domains with each having a unique optimal action for all states belonging to that domain. In this way, an optimal partition is defined by the use of a policy function that assigns states to actions. In the current paper the partitioning is performed by using a learning vector quantizer (Ritter, *et al.*, 1992). In such a case, the cells are defined by reference vectors in the input space. The distribution of the reference vectors is usually determined by statistical properties of the inputs to the vector quantizer (Ritter and Schulten, 1986). Thus, around the stable states of the controlled system, fine-grained partitions are formed.

In this paper, the learning rule for the vector quantizer is based on Kohonen's Self-Organizing Feature Map (Kohonen, 1995) which possesses interesting noise filtering properties.

The Self-Organizing Map commonly referred to as SOM (first in Kohonen, 1982) is a neural network (NN) that converts complex, nonlinear statistical relationships between high-dimensional data into simple geometric relationships.

The determination of quantized state cannot by itself represent input-output relationships. By extending the SOM with output weights that store the output part of a mapping can provide the original algorithm with the ability to approximate continuous relationships. Such a network has been introduced earlier (Ritter, *et al.*, 1992).

In the current paper for first time a partitioner based on SOM is learned simultaneously with a reinforcement signal based learning controller. A novel training algorithm is presented for updating the parameters of this network. Then, this training algorithm is successfully applied in the on-line stabilisation of a flexible spray boom that is used in pesticide application.

2. CONTROLLER AND PARTITIONER LEARNING

The Self-Organizing Map (Kohonen, 1995) is a neural network (NN) that maps signals ($\mathbf{x}$) from a high-dimensional space to a one- or two-dimensional discrete lattice of neuron units ($\mathbf{s}$). Each neuron stores a weight ($\mathbf{w}_s$). The map preserves topological relationships between inputs in a way that neighbouring inputs in the input space are mapped to neighbouring neurons in the map space. When extended with output weights ($\mathbf{y}_s$) it can actually learn in a supervised way the mapping $\mathbf{y} = \mathbf{f}(\mathbf{x})$. This association is shown schematically in figure (1).

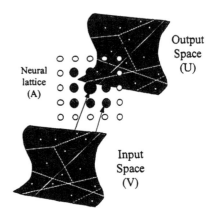

Fig. 1. Association of input-output values by using a Self-Organizing Map

The association of situation-action pairs is based on a reward/punishment scheme where a reinforcement signal is produced. A two-state reinforcement signal (+1 or 0) is produced after an evaluation of the result that a certain action has brought.

In a control situation when a system is driven by a certain control sequence:

$$\mathbf{u} = (u(k-1), u(k-2), ..., u(k-n))^T \qquad (1)$$

And the response of the system is measured:

$$\mathbf{y} = (y(k), y(k-1), ..., y(k-n))^T \qquad (2)$$

and the desired next step output is denoted as $\mathbf{y}_d$ (generally a vector, but in the example with the boom it is assumed to be a scalar), the state vector that is used as input to the state quantizer (SOM) is constructed as follows:

$$\mathbf{x} = (\mathbf{y}_d^T, \mathbf{y}^T, \mathbf{u}^T)^T \qquad (3)$$

Subsequently, these state vectors are clustered by the SOM. The control values $u(k)$ are stored as an output weight through the training procedure of Kohonen's algorithm (Kohonen, 1995). In the case of MIMO systems the data can be concatenated in the same vector. The use of SOMs to cluster concatenated sequential data has already been attempted by Kangas (1990).

A scalar reward function that determines the association of states to actions can be defined as

$$R(k) = -(y(k)-y_d)^T(y(k)-y_d) \qquad (4)$$

Where, with $y(k)$ the vector of current (at t=k) output measurements is defined, and with y_d the vector of current target output values. A policy can be based on this reward function by calculating the difference between two consecutive values of the reward function:

$$\Delta R = R(k) - R(k-1) \qquad (5)$$

In figure (2) the evaluation is performed by the look-up table block and produces a firmness signal that modifies the state quantizer. It is clear that every increase of ΔR is desirable, since the maximum target value of the reward function (R) is zero. However, a maximisation of R over a number of steps is better because temporary variations of the reward function can be due to disturbances and not caused by the control sequence. For this reason every neuron stores a moving average of the increase ΔR of the reward function R. This moving average is denoted as $\langle \Delta R \rangle$.

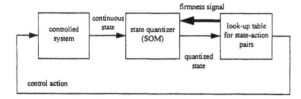

Fig. 2. Scheme of learning problem of combined controller and partitioner learning

In the presented method when a neuron is selected and a control action produces an increase over the stored moving average for this specific neuron, a positive reinforcement signal is produced. A positive reinforcement signal indicates that this neuron and its immediate neighbours are allowed to learn the state-action pair that has led to the positive reinforcement signal. But since only increases of the reward function that are greater than the stored moving average for each neuron lead to learning, a continuous improvement of the partitioner and the look-up table of control-action pairs is achieved. In the discrete time situation, the moving average of the increases of the reward function can be obtained as:

$$\langle \Delta R \rangle_k = \langle \Delta R \rangle_{k-1} + \gamma (\Delta R_k - \langle \Delta R \rangle_{k-1}) \qquad (6)$$

Where the subscript k denotes the time-step t=k and γ is a small positive constant. Note that the momentary value of the increase is denoted without brackets. After the update, the new moving average is stored in the neuron that has been activated. The whole concept has to do with supplying to each neuron a bias term to avoid overtraining. The learning algorithm for the input and output weights is derived from the original Kohonen algorithm (Kohonen, 1982) and has the form:

$$\Delta \mathbf{w}_s^{(in)} = \varepsilon h (\mathbf{x} - \mathbf{w}_s^{(in)}) \qquad (7)$$

$$\Delta \mathbf{w}_s^{(out)} = \varepsilon' h' (\mathbf{u} - \mathbf{w}_s^{(out)}) \qquad (8)$$

Where ε, ε' and h, h′ are the learning rates and the neighborhood kernels respectively. With $\mathbf{w}_s^{(out)}$ the output weight $\mathbf{y}_s$ is denoted. It must be noted that in the updating equations the winning neuron is denoted with s, i.e. the one that has the smallest distance from the input $\mathbf{x}$. However the updating equations apply to the lattice neighbours of the winning neuron at every updating step. The neighbourhood kernels used have the form of a Gaussian distribution like:

$$h = \exp(-\|\mathbf{x} - \mathbf{w}_s\|^2 / \sigma^2) \qquad (9)$$

Where $\|.\|$ denotes the Euclidean norm and σ denotes the variance of the Gaussian distribution.

In either case, the applied control action that is applied, is constructed by two components:

$$u(k) = \mathbf{w}_s^{(out)} - a_s(y(k) - y_d) \qquad (10)$$

Where a_s is a small positive value that should start from a relatively large value at the initial training phase and subsequently reduced slowly to a final small value. This allows rapid improvement at the initial period of training and allows a small margin for adaptation after the partitioner and look-up table has been learned satisfactorily.

The value of control action u(k) from equation (10) is used for updating the output weight of updating equation (8) only in case the reinforcement signal is positive. Such updating of the output weights ensures that the controller improves continuously. The updating equation for the a_s factor is:

$$\Delta a_s = \varepsilon'' h'' (a - a_s) \qquad (11)$$

In equation (11), ε'' and h'' are the learning rate and the neighborhood kernel respectively. By setting ε'' very small (in the example of the boom it is set equal to 0.005) the "exploration step" will converge to the final value denoted as (a) slowly enough to allow for satisfactory learning of control actions. If the final value (a) is set different than zero some residual plasticity will allow for continuous adaptation of the controller.

3. FLEXIBLE BOOM STABILISATION

Flexible spray booms are used in the agricultural domain for pesticide application. They usually consist of lightweight beams on which spraying nozzles are mounted. When driving a tractor over a field, the unevenness of the soil causes the flexible boom to vibrate, leading to under- and over-application of pesticides, thus resulting in environmental pollution. Stabilisation of flexible spray booms is needed in order to achieve a uniform spraying liquid distribution and avoid environmental damage.

The learning algorithm of section 2 is applied in the on-line vibration isolation of a 12th order linearised model of a flexible spray boom that has a total length of 12m tip to tip. The test set-up from which the model has been obtained is shown in figure (3). The linearised model is of the form:

$$\dot{x} = Ax + Bu + Ew$$
$$y = Cx + Du + Fv \qquad (12)$$

The E matrix is the disturbance input (w) distribution matrix and v represents sensor noise. The direct feedback (D) matrix appears because of the collocation of the sensor and the actuator and they are both accelerations.

For the simulations that are presented, only translational motion in the horizontal plane of the flexible boom are considered, thus resulting in a SISO system. In figure (3) with the accelerometer, the horizontal acceleration in translational motion is measured. The suppression of horizontal flexible deformations hasn't been tackled yet successfully, while, for the vertical vibrations a passive suspension usually suffices.

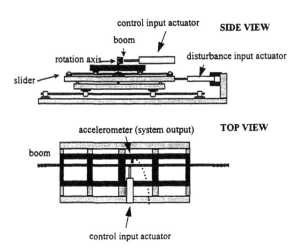

Fig. 3. Set-up for spray-boom measurements (sensor and actuator collocated)

The disturbances used are the accelerations resulting from a standardised field track (Norme Internationale ISO 5008 - 1979) fed through a model of the tractor wheels and a model of a tractor on which the spray boom has been attached. The excitation signal runs for 23 sec and is shown in figure (4). The tractor was supposed in the model to run with a constant speed of 5 km/h.

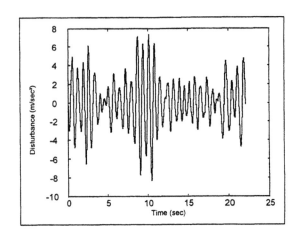

Fig. 4. The acceleration profile of the standardised track used as input to the flexible boom

The excitation signal of figure (4) is used as an input to the system. The system is discretised with a Tustin transform and integrated with Runge-Kutta of fifth order. This method of simulation is preferred because different sampling rates for the control and sensor part can be used, and so effects of latency can be assessed in a more realistic way. The uncontrolled response of the system is shown in figure (5).

The sensor (accelerometer) and electro-hydraulic actuator are supposed to be collocated at 0.25m from the connection joint of the flexible boom. The input of the network consists of vectors of previous input and output values of the system determined through a sliding time window of a certain length:

$$\mathbf{x} = (y_d, y(k),...,y(k-n),u(k-1),...,u(k-m))^T \qquad (13)$$

Two delayed values have been used for the actuator (input) and the sensor (output) which are both accelerations. The use of SOMs to cluster concatenated sequential data has already been attempted by Kangas (1990). The SOM with 100 nodes was trained for the whole period of the test signal (23 sec).

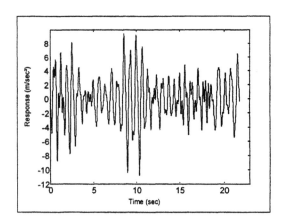

Fig. 5. The response of the system when excited by a standardised track

The entire systematic procedure of section 2 has been followed according to a certain series of steps described below:

STEP 1: Present an input to the SOM constructed from concatenated input, output data and the target output value. Here y_d is scalar and it represents the measured controlled acceleration:

$$\mathbf{x} = (\ y_d,\ \mathbf{y}^T,\ \mathbf{u}^T\ )^T \qquad (14)$$

STEP 2: The activated neuron has stored an input weight denoted $\mathbf{w}_s^{(in)}$ and an output weight denoted $\mathbf{w}_s^{(out)}$. The control action is determined partly from the output weight and partly from the current output measurement (a_s is a small positive step):

$$u(k) = \mathbf{w}_s^{(out)} - a_s(y(k) - y_d(k)) \qquad (15)$$

STEP 3: The reward (R) function for t=k is calculated based on the current output acceleration measurement and also the increase of the reward function:

$$R(k) = -(y(k) - y_d)^2 \qquad (16)$$
$$\Delta R = R(k) - R(k-1) \qquad (17)$$

STEP 4: Every neuron stores a moving average of the increases of the reward function calculated as:

$$\langle \Delta R \rangle_k = \langle \Delta R \rangle_{k-1} + \gamma(\Delta R_k - \langle \Delta R \rangle_{k-1}) \qquad (18)$$

Here a value of γ equal to 0.2 has been used. Note that this value is stored in the neuron only when it is activated.

STEP 5: A positive reinforcement signal is produced only when

$$\boxed{\Delta R_k > \langle \Delta R \rangle_{k-1}} \qquad (19)$$

STEP 6: Given the positive reinforcement the input and output weights of the neuron and its neighbours are updated according to the updating equations:

$$\Delta \mathbf{w}_s^{(in)} = \varepsilon h(\mathbf{x} - \mathbf{w}_s^{(in)}) \qquad (20)$$
$$\Delta \mathbf{w}_s^{(out)} = \varepsilon' h'(\mathbf{u} - \mathbf{w}_s^{(out)}) \qquad (21)$$

Here with u the control action at t=k is denoted. If the condition that leads to positive reinforcement doesn't hold, no updating occurs. Thus the SOM learns only from successful sensor/actuator pairs. All the weights are initialised to small random values.

STEP 7: The weighting parameter a_s is updated at all times as follows (initial value for a_s was set equal to 10^{-3} and final value equal to 10^{-4}, while ε'' was set equal to 0.005):

$$\Delta a_s = \varepsilon'' h''(a - a_s) \qquad (22)$$

STEP 8: Go back to step 1 to present a new input.

The initial and final settings of the learning parameters, namely the learning rates ε, and the widths σ of the gaussian kernels h for the updating equations were chosen to be $\varepsilon_I = \varepsilon_I' = \varepsilon_I'' = 0.9$ (initial), $\varepsilon_f = \varepsilon_f' = \varepsilon_f'' = 0.05$ (final), $\sigma_I = \sigma_I' = \sigma_I'' = 0.6 \times$(number of units in one dimension of the map) (initial), $\sigma_f = \sigma_f' = \sigma_f'' = 0.3 \times$(number of units in one dimension of the map) (final). The values of these parameters were chosen to decrease exponentially with time between the initial and the final value.

The result of following the above on-line learning of control actions is shown in figure (6). Both are accelerations at the point where the actuators are collocated with the sensors. From figure (7) it is evident that the peaks have been reduced by at least 20 dB.

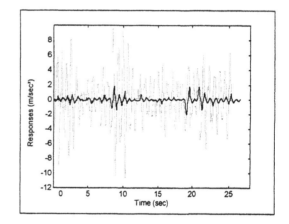

Fig. 6. The controlled vs. the uncontrolled time response of the system (the thick line represents the controlled system response)

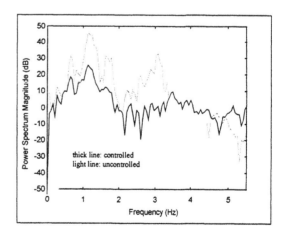

Fig. 7. The controlled vs. the uncontrolled frequency response of the system

363

The evolution of the reinforcement signal through the first sampling steps (at a sampling rate of 1 kHz) is shown in figure (8). It is clear that during the first 0.2 sec the controller learns most of the time, thus resulting in a very small acceleration from the very beginning of the training session.

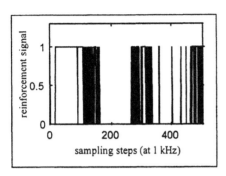

Fig. 8. The reinforcement signal during the initial half second of the training session

A way of visualising the spatial structure of the representative vectors that the Self-Organizing Map has stored is by plotting these vectors in the case that they are also two-dimensional like the SOM itself. In the case of a higher dimension of the input data the geometrical relations of the representative vectors are difficult to visualise.

As is evident from figure (9), in which the first two weights of the map are plotted, there is a very clear ordering at the end of learning. These two weights are representative values of two consecutive controlled acceleration values that have emerged through the learning process of the SOM. It has to be mentioned that during learning the SOM tends to represent the states that occur more frequently. However because of the bias that is introduced through the moving average of the reward function increase in the updating policy (STEP 5) the states that are visited tend to be equiprobable.

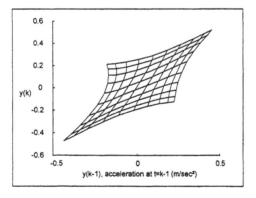

Fig. 9. The SOM at the end of the learning period

An important aspect of the final stage of the SOM is that the states around the diagonal cover a wider range. This follows from the cooling schedule of the updating equations (STEP 6); i.e. the learning rate assumes a very small final value.

4. CONCLUSIONS

A new neural network method for disturbance suppression of dynamical systems has been presented. The main advantage of this method is the local representation of the controller and state partitioner which are learned simultaneously by reward and failure signals. The whole learning scheme doesn't need any prior information but only output measurements of the controlled system. Local updating algorithms assure much faster convergence than global updating algorithms. The method is generally applicable from the point of view that it is not based on a model of the system under control. It only relies on a reward function and the moving average of locally stored rewards over time. It can be used equally well for on-line control of linear and nonlinear systems or systems with changing parameters. The new method presented can be applied in the automotive (vehicle suspensions) and the aerospace domain (flexible space structures), in the case of systems with uncertain or complex dynamic behaviour.

REFERENCES

Das, S., M.C. Mozer (1994). A unified gradient-descent/clustering architecture for finite state machine induction. In: J.D. Cowan, G. Tesauro, J. Alspector (eds.), *Advances in NIPS 6*, pp. 19-26.

Hermann, M., R. Der (1995). Efficient Q-Learning by Division of Labour. *Proceedings ICANN '95*, EC2 & Cie, Paris, **2**, pp. 129-134

Kangas, J. (1990). Time-Delayed Self-Organizing Maps. *Proceedings of the IEEE IJCNN-90 Conference*, S. Diego, CA, **2**, pp. 331-336.

Kohonen, T. (1982). Self-Organized Formation of Topologically Correct Feature Maps. *Biological Cybernetics*, **43**, pp. 59-69.

Kohonen, T. (1995). *Self-Organizing Maps*. Springer Series in Information Sciences.

Norme Internationale, ISO 5008 - (1979) (F). *Tracteurs et Matériels Agricoles à Roues-Mesurage des Vibrations Transmisses, Globalement au Conducteur*. Organisation Internationale de Normalisation.

Ritter H., T. Martinetz, and K. Schulten (1992). *Neural Computation and Self-Organizing Maps: An Introduction*. Adisson-Wesley, New York.

Ritter H., K. Schulten (1986). On the stationary state of Kohonen's self-organizing sensory mapping. *Biol. Cybernetics*, **54**, pp. 99-106.

A NEW NON-LINEAR DESIGN METHOD FOR ACTIVE VEHICLE SUSPENSION SYSTEMS

Robert F Harrison and Stephen P Banks

*Department of Automatic Control and Systems Engineering,
The University of Sheffield, Sheffield, UK*

Abstract: A novel non-linear design method based on linear quadratic optimal control theory is presented that applies both to linear and (a wide class of) non-linear systems. The method is easy to apply and results in a globally stabilising, near-optimal solution that can be implemented in real-time. The key feature of the design method is the introduction of state-dependence in the weight matrices of the usual linear quadratic cost function, leading to a *non-linear design method*, even for linear dynamics. To demonstrate the method, a simple linear suspension model is used, in conjunction with a non-linear state penalty, which better reflects the *engineering* objectives of active vehicle vibration suppression. Non-linear dynamics can equally well be accommodated. A number of simulations is conducted and compared, favourably, with a passively mounted vehicle. These preliminary results indicate the potential of the method. *Copyright © 1998 IFAC*

Keywords: Non-linear control systems; quadratic optimal regulators; active vehicle suspension; Riccati equations; global stability.

1. INTRODUCTION

Linear quadratic optimal control theory is a highly developed approach for the synthesis of linear optimal control laws and has been applied widely in studies on active vehicle suspension systems. While the approach is attractive in that it is possible to penalise different variables so as to trade-off between, say, ride comfort and handling, or comfort and suspension travel, the way these variables are treated is essentially fixed – no provision is made to allow the suspension to distinguish between a smooth road and a rough one. Evidently, while comfort might be a prime objective under normal circumstances, on rough surfaces the suspension should be stiffened to avoid hitting its limits, hence incurring damage. Although, in principle, time-varying weighting parameters are allowed in the linear quadratic approach, lack of prior knowledge of the road surface, and the anti-causal calculation for the solution makes the introduction of these difficult. The

required amplitude dependence can never, therefore, be achieved through the linear quadratic approach.

In this paper we make use of a new result that generalises the theory to non-linear systems to provide a *non-linear design* method that overcomes the shortcomings mentioned above. The method applies to systems having linear or, a broad class of, non-linear dynamics. In brief, it turns out that the infinite time-horizon linear quadratic regulator problem, when solved afresh at every state, leads to a globally stabilising, near-optimal control policy (Banks and Mhanna, 1992). Thus, for admissible system dynamics, the weighting parameters can be made to be functions of the state variables and the desired amplitude dependence obtained. Thus, the design stage allows for the introduction of non-linearity in the weighting matrices, even for linear dynamics, leading to a more "intelligent" control strategy.

In contrast to the finite-time linear quadratic optimal control problem, which must be implemented off-line, our method is causal, but has considerable computational overhead. However, by using a solution to the Riccati equation based upon the matrix sign function (Gardiner and Laub, 1986), it is possible to derive a parallel algorithm (Gardiner and Laub, 1991) suitable for real-time implementation.

In (Lin and Kanellakopoulos, 1997) a different approach to this problem is proposed using the "backstepping" design method (Kanellakopoulos *et al*, 1992) and a non-linear filter to achieve the desired behaviour. Although a direct comparison is not possible owing to the essential differences between the two methods, we illustrate our approach on the simplified (linear) model described there.

The remainder of the paper is organised as follows. In §2 the linear quadratic regulator is first set out, and the generalised results are stated. In §3 the passive and active suspension models of (Lin and Kanellakopoulos, 1997) are presented and the choice of design parameters is discussed. The results of a series of experiments are described in §4 and conclusions are drawn in §5.

2. THE DESIGN METHOD

2.1 Linear quadratic regulator

The linear quadratic optimal regulation problem is expressed as follows. Find the control policy, $\mathbf{u}$, that minimises the cost functional:

$$J = \int_0^\infty \left(\mathbf{x}'Q\mathbf{x} + \mathbf{u}'R\mathbf{u}\right)dt \qquad (1)$$

subject to the linear time invariant dynamics:

$$\dot{\mathbf{x}} = A\mathbf{x} + B\mathbf{u} \qquad (2)$$

where $\mathbf{x}$ is an n-vector of system states, $\mathbf{u}$ is an m-vector of control variables, A and B are matrices of appropriate dimension and the superscript, t, indicates transposition. The matrices Q and R are positive semi-definite and definite, respectively, and are used to penalise particular states according to the engineering objective and the control effort.

It is well known (e.g. Friedland, 1987) that the control policy which solves the above optimisation problem is a linear combination of the system states and is given by:

$$\mathbf{u} = K\mathbf{x} \qquad (3)$$

where K is in turn given by:

$$K = -R^{-1}B'P \qquad (4)$$

and P is the positive definite solution of the algebraic matrix Riccati equation:

$$0 = PA + A'P - PBR^{-1}B'P + Q \qquad (5)$$

A unique, positive definite solution to the above exists if the pair (A,B) is stabilizable and (A, Γ) is detectable, with $Q = \Gamma'\Gamma$.

2.2 Non-linear quadratic regulator

The extension of the above to non-linear systems looks identical, except that, instead of performing a single optimisation and applying the resulting gain-matrix for all time, the optimisation has to be carried out at every time step. Consider a non-linear dynamical system that can be expressed in the form:

$$\dot{\mathbf{x}} = A(\mathbf{x})\mathbf{x} + B(\mathbf{x})\mathbf{u} \qquad (6)$$

where the Jacobians of $A(\mathbf{x})$ and $B(\mathbf{x})$ are subject to some bounded growth conditions (Lipscihtz) and $A(\mathbf{0}) = 0, B(\mathbf{0}) = 0$, then at each point, $\overline{\mathbf{x}}$, on the state trajectory, a linear system is defined with fixed A and B. In (Banks and Mhanna, 1992) it is shown that solving the infinite-time linear quadratic optimal control problem, pointwise on the state trajectory, results in a globally stabilising, near-optimal quadratic control policy for systems described by equation (6). Thus, by choosing the $\mathbf{u}$ that minimises the usual quadratic cost function at every time step, we have a globally optimal control policy for a very wide class of non-linear system. Evidently, $A(\overline{\mathbf{x}})$, $B(\overline{\mathbf{x}})$ and Q are subject, pointwise, to the same conditions as for the linear case. It is clear that the proposed solution is identical to the one obtained from equations (2, 3 and 4) when the dynamics are linear.

As an aside, the dual situation follows directly from the reasoning in (Banks and Mhanna, 1992) and thus state estimation is possible via a non-linear observer although this aspect is not addressed here.

Because the control synthesis takes place pointwise, the designer is now free to select Q and R in ways that are more directly applicable to the control *engineering* objectives. In particular, these can be made functions of the instantaneous state variables, i.e.

$$J = \int_0^\infty \left(\mathbf{x}'Q(\overline{\mathbf{x}})\mathbf{x} + \mathbf{u}'R(\overline{\mathbf{x}})\mathbf{u}\right)dt \qquad (7)$$

subject to the needs for the solution of the Riccati equation and the invertibility of R. Ensuring that $A(\overline{\mathbf{x}}), B(\overline{\mathbf{x}}), R(\overline{\mathbf{x}})$ and $Q(\overline{\mathbf{x}})$ satisfy these requirements a priori is difficult in general, however, for polynomial functions which are not identically zero, the required properties will be lost only at isolated points and will not, therefore, persist.

3. SUSPENSION MODEL

Because we wish to emphasise the use of the non-linear optimal control method for *design*, i.e. how the integration of mathematical synthesis and engineering objectives can be achieved, we adopt a linear model of a vehicle suspension. The two-degree-of-freedom, quarter-car model of figure 1 has been widely studied in the literature, and it represents an active element operating in parallel with passive linear elements – a spring, k_1, and a damper, c_1.

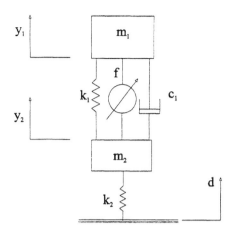

Fig. 1. Schematic of the two-degree-of-freedom, quarter-car model.

The motions of the body and wheel (sprung, m_1, and unsprung, m_2, masses, respectively) are denoted by y_1 and y_2 respectively, while the deviation of the road surface from some datum is denoted by d. The tyre is represented by a linear spring, k_2, with no damping, for simplicity. We assume that the control force, f, can be applied directly as a result of the control signal, with negligible actuator dynamics. Again this is chosen for simplicity, so as not to obscure the main point of the paper.

The equations of motion for the quarter car are given by:

$$\ddot{y}_1 = -\frac{k_1}{m_1}(y_1 - y_2) - \frac{c_1}{m_1}(\dot{y}_1 - \dot{y}_2) + \frac{1}{m_1}f \quad (8a)$$

$$\ddot{y}_2 = \frac{k_1}{m_2}(y_1 - y_2) + \frac{c_1}{m_2}(\dot{y}_1 - \dot{y}_2)$$
$$- \frac{k_2}{m_2}(y_2 - d) - \frac{1}{m_2}f \quad (8b)$$

We choose state variables thus: $x_1 = y_1, x_2 = \dot{y}_1, x_3 = y_2, x_4 = \dot{y}_2$, and identify the control signal, $\mathbf{u}$, with the force, f, leading to the form of equation (2). Note that there is a direct feed-forward path between the control force and sprung mass acceleration – one of the primary indicators of ride quality. A more realistic model would, of course, incorporate actuator dynamics.

3.1 Design objectives

For the purposes of this paper let us suppose that our primary objective is to minimise passenger discomfort. We do this by attempting to reduce the accelerations to which the passenger is subject –

vertical only, in this simple case. Thus a candidate for the cost function is $\ddot{y}_1 = C_a \mathbf{x} + D_a \mathbf{u}$, where C_a is the second row of A, and D_a is the second element of B. However, ride comfort can only take precedence when safety and integrity are not compromised. Thus it is necessary to penalise some measure which embodies these ideas, usually via the "rattlespace deflection", $y_1 - y_2 = C_r \mathbf{x}$, with $C_r = \begin{bmatrix} 1 & 0 & -1 & 0 \end{bmatrix}$. In the conventional linear quadratic approach we construct a cost function thus:

$$J = \int_0^\infty \left(q_a \ddot{y}_1^2 + q_r(y_1 - y_2)^2 + ru^2 \right) dt$$

$$= \int_0^\infty \left(\begin{array}{c} \mathbf{x}'\left(q_a C_a' C_a + q_r C_r' C_r \right)\mathbf{x} \\ + 2\mathbf{x}' q_a C_a' D_a u + \left(q_a D_a^2 + r \right)u^2 \end{array} \right) dt \quad (9)$$

Letting $N = q_a C_a' D_a$ and $R = q_a D_a^2 + r$ we accommodate the cross-term in the usual way, thus $Q \leftarrow Q - NR^{-1}N'$, $A \leftarrow A - BR^{-1}N'$ (Friedland, 1987), with the original $Q = q_a C_a C_a' + q_r C_r C_r'$. The parameters q_a, q_r are used to control the trade-off between ride and handling.

To illustrate the *non-linear design* procedure we introduce state-dependence into q_r : thus $q_r = 200\psi(y_1 - y_2, .055, .001)$ with

$$\psi(\xi, \theta, \delta) = \begin{cases} ((\xi - \theta)/\delta)^p, \xi > \theta \\ 0, \quad |\xi| \le \theta \\ ((\xi + \theta)/\delta)^p, \xi < -\theta \end{cases} \quad (10)$$

where $\theta \ge 0$ defines a deadzone, $\delta > 0$, the distance within which ψ first reaches unity, and $p=5$. The rationale for this functional form is as follows. The primary objective is to reduce body acceleration hence we choose a constant q_a (=1000). Although this could also be allowed to be state-dependent, we choose not to make it so because of technical difficulties in ensuring the necessary detectability conditions that arise from the inclusion of the cross-term in equation 9. This, in turn, arises from the specific choice of model and will be discussed later. The secondary objective, which can over-ride the first for safety reasons, is to reduce overly large excursions in the suspension strut. Thus, for a rattlespace of ±0.08m, a deadzone of ±0.055m is allowed before control action is taken. The non-linearity increasing to unity within the next 0.001m of travel and dominating the cost function very rapidly as travel approaches the limits.

We have been guided here by the function chosen in (Lin and Kanellakopoulos, 1997), however, there the non-linearity is applied to the strut closure in a very different way.

4. RESULTS

We use the passive system (equation 8 with $f(t) = 0$ for all t) as a reference and compare its behaviour with that of the *non-linearly controlled* vehicle model for a variety of road surface profiles

$$d(t) = \begin{cases} a(1 - \cos 8\pi t), & 0 \le t \le 0.25 \\ 0, & \text{otherwise} \end{cases} \quad (11)$$

as suggested in (Lin and Kanellakopoulos, 1997). Figures 2–10 display the relevant results. In each, the dashed line indicates the passive behaviour and the solid curve indicates the controlled behaviour. For $a = 0.025m$, figures 2–4 show the sprung-mass acceleration, the rattlespace deflection and the control signal, respectively, as functions of time. Likewise for figures 5–7 ($a = 0.038m$) and figures 8–10 ($a = 0.055m$). In all simulations, $r = 0.0001$ and all other parameters are as given in (Lin and Kanellakopoulos, 1997). Simulations are carried out in Matlab™ using Euler's method as the integration routine with a step length of 0.001s.

For a small bump ($a = 0.025m$) we expect little or no effect from the rattlespace weighting and that the linear situation should obtain with q_a=1000, q_r=0. This is indeed the case. Here, acceleration excursions are reduced (figure 2) and the reduction in rattlespace deflection (figure 3) arises indirectly from penalising the acceleration. The control signal that achieves these reductions is shown in figure 4.

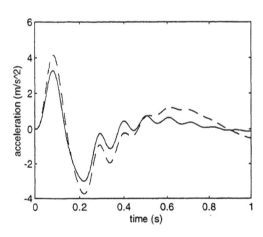

Fig. 2 Sprung-mass acceleration for $a = 0.025m$.

As the severity of the disturbance increases we expect to see the state-dependency come into play. For $a = 0.038m$ we see a sudden large reversal in the control signal (figure 7) as the relative displacement approaches the limits of travel.

Here the rattlespace penalty dominates the cost function. The control signal reverts to the linear situation for t>0.35 approximately. The control signal behaviour is manifested directly in the acceleration signal (figure 5) while seeming to have no discernible effect on rattlespace deflection (figure 6). Such

impulsive accelerations would obviously have implications for passenger comfort and for the driver's visual acuity through the mechanism of "jerk" (derivative of acceleration). We shall return to this later.

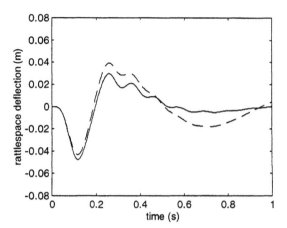

Fig. 3. Rattlespace deflection for $a = 0.025m$.

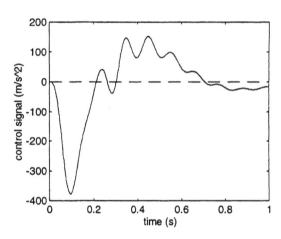

Fig. 4. Control signal for $a = 0.025m$.

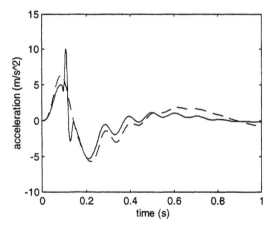

Fig. 5 Sprung-mass acceleration for $a = 0.038m$.

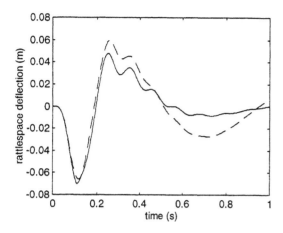

Fig. 6. Rattlespace deflection for $a = 0.038m$.

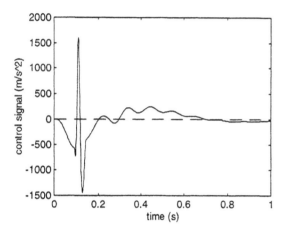

Fig. 7. Control signal for $a = 0.038m$.

For a large disturbance ($a = 0.055m$) we see highly non-linear behaviour in the closed-loop system. Now the strut deflection approaches, closely, the limits of travel. Indeed, for this size of bump, the passive system would "bottom out" with implications for safety, comfort and structural integrity. The controlled system is prevented from bottoming out by the application of large control "spikes" (figure 10). Once again these are manifested directly in the acceleration signal (figure 8) with even more severe implications for the passengers

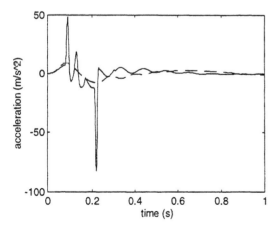

Fig. 8 Sprung-mass acceleration for $a = 0.055m$.

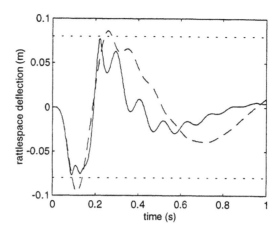

Fig. 9. Rattlespace deflection for $a = 0.055m$.

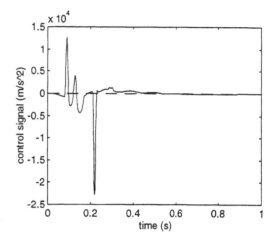

Fig. 10. Control signal for $a = 0.055m$.

Now the effect of the state-dependent weighting is clearly evident in the rattlespace deflection response, which bears little resemblance to the linear behaviour of the passive system.

	$a = 0.025m$	$a = 0.038m$	$a = 0.055m$
Accn. (passive)	2.435	5.625	11.784
Accn. (active)	1.336	4.270	79.343
Rattlespace (passive)	3.939e^{-4}	9.100e^{-4}	19.000e^{-4}
Rattlespace (active)	2.550e^{-4}	6.069e^{-4}	10.000e^{-4}
Control	1.375e^4	7.473e^4	5.181e^6

Table 1. Integrated-square values for the three bump semi-heights.

From Table 1 we can compare the integrated square of the values of the three signals of interest, over the interval zero to one second. We see that for the small and moderate disturbances the active system reduces the integrated-square of the acceleration and of the rattlespace deflection. In the case of the severe bump the rattlespace value is reduced but at the expense of seriously increased integrated square acceleration and almost two orders of magnitude additional integrated square control effort. Recall, however, that the passive system bottoms out and so the figures in the table pertaining to this situation are not reliable.

Clearly, as the state-dependent penalty dominates the cost function, the control policy becomes very aggressive, resulting in a marked deterioration in ride comfort. However, it must be recognised that the model used here is highly simplified, to the point of being unrealistic. In particular, the direct feedforward of the control signal into the acceleration response is not strictly proper and leads to a number of difficulties, both technical, as alluded to earlier, and from the point of view of interpretation of the present results; e.g. the effect of the control signal would be significantly filtered in a physical system. What is highlighted, however, is the need for careful design of the penalty function.

In (Lin and Kanellakopoulos, 1997) qualitatively similar effects are also experienced in that they are able, by virtue of the introduction of non-linearity, to prevent "bottoming" occurring. However, a direct comparison is not possible because detailed behaviour depends strongly on the individual methods and the simulations of (Lin and Kanellakopoulos, 1997) include non-linear actuator dynamics that are ignored here so as not to obscure the simplicity of the design method. It should be reiterated that non-linearities could be accommodated directly, subject to the conditions of §2.2. This is no more difficult than in the linear case, and does not require re-linearisation as might be the case with gain-scheduling.

5. CONCLUSIONS

A new method for the design and synthesis of active vehicle suspension systems is proposed, based on a generalisation of linear quadratic optimal control theory. The method is simple to apply and affords much greater design flexibility than the conventional approach. The resulting controller is non-linear, even for linear dynamics, and can be implemented in real-time.

To illustrate the applicability of the method, a simple linear two-degree-of-freedom quarter-car model has been studied using a rationale suggested in (Lin and Kanellakopoulos, 1997) to design a non-linear penalty function. Preliminary results show that the method has potential and could be tuned to provide desired closed-loop behaviour. However, the shortcomings of the simplified model prevent more concrete use being made of the results.

Some areas for future work are: the selection of more appropriate penalty functions, $Q(\mathbf{x}), R(\mathbf{x})$; the real-time implementation of such a system; the robustness of the method to modelling errors, and quantification of how near to optimal the solution is, and under what conditions.

REFERENCES

Banks, S.P. and K.J. Mhana (1992). Optimal control and stabilisation for non-linear systems, *IMA Journal of Mathematical Control and Information*, **9**, 179–196.

Gardiner, J D. and A.J. Laub (1986). A generalisation of the matrix-sign-function for algebraic Riccati equations, *International Journal of Control*, **44**, 823–832.

Gardiner, J.D. and A.J. Laub (1991). Parallel algorithms for algebraic Riccati equations, *International Journal of Control* **54**, 1317–1333.

Lin, J-S. and I. Kanellakopoulos (1997). Non-linear design of active suspensions, *IEEE Control Systems*, **17**, 45–59.

Kanellakopoulos, I., P.V. Kokotovic and A.S. Morse (1992). A toolkit for linear feedback design, *Systems and Control Letters* **18**, 83–92.

Friedland, B. (1987). *Control system design: an introduction to state-space methods*, McGraw-Hill Book Co., New York.

PATH PLANNING METHOD FOR MOBILE ROBOTS IN CHANGING ENVIRONMENTS

F. J. Blanco V. Moreno B. Curto

Dept. Informática y Automática Fac. Ciencias U. of Salamanca
Plaza de la Merced s/n Salamanca, 37008 Spain
E-mail: control@abedul.usal.es

Abstract: This paper is focused on the development of path planning techniques for mobile robots in a workspace with moving objects. More precisely, this work illustrates the use of new global methods for path planning within changing enviroments providing very good results in terms of the quality of the final solutions and the computational cost. The path planning problem has been solved in two different ways. Firstly, a time optimal path planner is presented and, secondly, an on-line version is also given to deal with the problem of adaptation when the enviroment changes. The application of these two approaches to several and suitable situations is also described, showing the reliability of both algorithms. *Copyright © 1998 IFAC*

Keywords: Mobile robots, Path planning, moving objects.

1. INTRODUCTION

Path planning is considered to be one of the main problems to be solved in Robotics in order to achieve autonomous actuation of robots. Although this problem has been addressed by many authors and some powerful tools have been developed for a wide number of applications, when a world with moving obstacles is considered, some of them either become unrealistic or present serious limitations. In this paper, an approach to deal with the particular situation of changing enviroments, is presented without the high computational load usually associated to the most common global path planning approaches found in the literature.

Collision avoidance approaches can be roughly be divided into two categories: global or local methods. The global techniques, such us the ones based on road-maps, cell decomposition and potential field methods, generate a complete trajec-

tory from the starting point to the target point (Janét *et al.*, 1997) following some optimality criterion by using a complete model to represent the world. Nevertheless, they have clear disadvantages in terms of the computational cost due to the inherent complexity of the proposed methods. In contrast, local approaches use only local representations of the world reducing drastically the computational costs.

When optimal solutions to the path planning problem is required a global approach should be necessarily used. In cases like this, additional problems, mostly associated to the complex nature of the constrained optimisation problem to be solved, have to be handled. For instance, for the particular case of global methods based on the use of potential fields, special attention should be paid to the presence and the avoidance of local minima. They also fail in the avoidance of local minima, fail to find trajectories between closed spaced obstacles and very often they produce oscillatory behaviour in narrow corridors. In this paper, a formalism to solve the collision avoidance problem with a global approach, is proposed. The

[1] The authors gratefully acknowledge the support of the Castilla y Leon Co uncil and CICYT through the projects C02/197 and TAP96-2410

computation load of the proposed technique is very low compare to other global approaches, so its real application to the path planning problem for a robot surrounded by moving obstacles can be successfully done.

The simplicity of the proposed method have been achieved, first of all, by solving the path planning problem in the configuration space (C-space) rather than in the robot workspace. Note that, in the former, the robot position and orientation are characterised by just one point (Lozano-Pérez, 1983). The planning task is carried out in two different steps: *findspace*, where the collision free robot configurations are found and *findpath* where a sequence of configurations can be found to move the robot from a given location to another one. The execution of the first stage uses a representation of the obstacles at the robot configuration space. An extra simplification is to consider the robot to have a disc shape and, consequently the configuration space is reduced to a two-dimensional one. This is not a hard restriction in terms of the applicability of the results sice in most cases the robot can be easily substituted by a circle, or a cylinder. In order to perform the second task *(findpath)*, a combination of the use of potential maps and roadmaps has been considered , (Barraquand and Latombe, 1991). In order to avoid problems associated to the complexity of previously proposed procedures (see (Latombe, 1991)), and some others, as shown in (Blanco, 1997), new techniques for the potential maps evaluation and roadmaps construction in the configuration space have been developed (Blanco, 1997). Hence, the second stage of the path planning problem can be executed in a quite fast way.

The paper presents the proposed path planning procedures together with some examples to illustrate their application and their reliability. The paper is organized as follows. In section 2 the main components of the proposed approaches for path planning are briefly described. The section is basically devoted to the description of the configuration space evaluation and the methods that allow to reduce the dimension of search space for this application. Next in section 3, the procedure to perform an off-line optimal time computation of a free-collision path is presented. The on-line version of this technique is given, in section 4. Finally, the main results and some conclusions are drawn at the end.

2. GLOBAL PATH PLANNING TECHNIQUES

As mentioned above, the paper deal with the development of procedures for path planning based on the use of global methods . In this section, those aspects that made the procedures more powerful are explained. Taking into account the task separation proposed by (Lozano-Pérez, 1983), first, the mathematical formalism applied to get a representation of the obstacles at the configuration space (*findspace* task) will be considered. Second, in order to perform the *findpath* task the procedures, used to generate the roadmap, is presented.

2.1 *Obstacles Representation in the Configuration Space*

The main goal is to obtain a structure that represents the geometrical constraints, in the C-space, imposed on the robot movements by the obstacles. This structure can be used either for path planning or for the movement robot control.

Most of the methods (Lozano-Pérez, 1983) (Guibas *et al.*, 1985) (Latombe, 1991) (Brost, 1989) that explicitly construct the representation the C-space obstacles model the robot and the obstacles as polygonal or polyhedral objects (convex or non-convex). In any case, the computational time associated to the obstacle projection from the workspace to the C-space, depends on the number of vertices of each object (both robot and obstacles) and also on the robot shape.

In (Brost, 1989) the obstacle representation in the C-space was carried out by means of an algebraic convolution of an obstacle and a mobile robot. This idea is implicitly considered in (Lozano-Pérez, 1983), where for a robot A, the mapping of a workspace obstacle B into the C-space obstacle (C-obstacle) is defined like $CO_A(B) \equiv \{x \in \text{Cspace}_A/(A)_x \cap B \neq 0\}$. In (Guibas *et al.*, 1985) is stated that this operation can be seen as the convolution of the set A with the set B. This idea is the basis of the work presented in (Curto and Moreno, 1997), which proposes a mathematical formalism to evaluate the C-obstacles by means of the convolution of two different functions: one describing the robot and the other describing the obstacles, if the proper coordinates are chosen.

The algorithms for path planning that have been developed in the last years, propose to work in discrete C-spaces (Barraquand and Latombe, 1991). More concisely, a bitmap is used to represent the obstacles in the C-space, reducing drastically the computation time to detect collisions. Kavraki (Kavraki, 1995) proposes the derivation of the C-space bitmap for a polygonal mobile robot as the convolution of the obstacle bitmap and the robot bitmap in the workspace. In order to carry out this convolution efficiently, the Fast Fourier Transform is used. This technique provides a bitmap that is independent of the complexity and shape of the robot and the obstacles. It only depends on the discretization resolution. With more general perspective, the formalism proposed in (Curto and Moreno, 1997) has been established for mobile and articulated robots.

So, the workspace and robot bitmaps are used and as algorithmic tool the FFT was chosen. The intrinsic parallelable nature of this tool at the same time as the very easy parallelation of the resulting algorithm reduces the computational time. Consistently, it is possible the application of this algorithm to changing environments.

2.2 Roadmap generation procedure

As, it was explained before, the problem of finding a path avoiding collisions usually is computed in the configuration space because there the robot can be considered as a single point. In this paper the two proposed methods to carry out the path planning comes from Artificial Intelligence as a problem of finding a solution following points of the free C- space. But, as the whole free C-space is composed by a great deal of configurations the exploration would require too computational load. For this reason, the free space must be reduced but the connectivity between the connected regions must be kept. The way to reduce the free space is by contracting it to get a roadmap of the space that states the connectivity. Other useful information to do the path planning is the proximity of a configuration to the obstacles, information that is expressed as a repulsive potential, which can be simultanously computed with the roadmap.

There are some methods to get the roadmap, one of them (Barraquand and Latombe, 1991) uses a wave front propagation from obstacles and the roadmap points are those where two wave fronts reach at the same time. But the wave front method sometimes can provide some unnecessary roads that may cause delays in the search of the free collision path, and the connectivity is not completely guaranteed. Here an alternative method, imported from the image processing field, is proposed, the thinning. The basic idea is to reduce the area surrounding obstacles, i.e., the free space, until it becomes only an unidimensional skeleton (the roadmap). This method contains an explicit way to prevent that fragments of the roadmap become disconnected each other. The thinning technique used here is an adaptation of the one that appears in (Gonzalez and Wintz, 1987).

The algorithm starts from the bitmap S, with 0s at C-obstacle positions and 1s at free C-space. The first neighbours of q are called Qs, and they are arranged, as is shown bellow, in clockwise.

$$Q(1)\ Q(2)\ Q(3)$$
$$Q(8)\quad q\quad Q(4)$$
$$Q(7)\ Q(6)\ Q(5)$$

N is the number of the first neighbours and T is the number of changes from 0 to 1 if the clockwise direction is considered.

The process is developed in two stages. First, the points q to be eliminated (they will be eliminated at the end of the stage) have to satisfy the four following conditions.

1. Q(2)*Q(4)*Q(6)=0
2. Q(8)*Q(4)*Q(6)=0
3. $2 \leq N \leq 6$
4. T=1

At this stage, south and east points, as well as northwest vertices are deleted (conditions 1st and 2nd). The 3rd one produces two consequences: the end points of the skeleton are not deleted (N=1) and those points which could cause a deformation in the potential (what in (Gonzalez and Wintz, 1987) is called an erosion in the region) are not deleted too (N=7). The 4th condition prevents disconnections of segments of the skeleton. The results can be seen at stages b) and d) of Fig. (1).

In the second stage, the points q to be deleted have to satisfy the following conditions:

1. Q(2)*Q(4)*Q(8)=0
2. Q(2)*Q(6)*Q(6)=0
3. $2 \leq N \leq 6$
4. T=1

In the same way, all these points will be deleted when the stage finishes. The only remarkable change is that the points to be deleted are those that stay at north, west or southeast vertex as it can be seen at stages c) and d) of Fig. (1).

These two stages are repeated in a cycle until no more points can be deleted. In each cycle the deleted points are valued with the potential computed as the inverse of the number of cycle done at that time.

The advantage of using the thinning is that there are special processors used in artificial vision (Matrox IMAGE-LC Image Procesor) that are able to compute the thinning of the configuration space representations in very short times (less than 500ms) with a high resolution (256x256). The roadmap can be also computed with cellular automata (Tzionas *et al.*, 1997) which also provide very short times (working at 60MHz less than 1ms).

It must be also noted that if the time t is considered as other coordinate, the space constituted by the C-space coordinates (x,y) and the time t (used in the time optimal path planner) is a three-dimensional space. If the temporal parameterization is enough to guarantee small movements of obstacles the roadmaps generated in consecutively times are so closed than in the three-dimensional space they form a general roadmap built by surfaces instead of curves (Blanco, 1997).

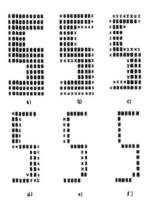

Fig. 1. Stages in the thinning process

3. TIME OPTIMAL PATH PLANNER

In this section, a brief explanation about the proposed algorithm to find a path that allows to get the desired goal with moving obstacles in the shortest period of time, is given. It is clear that, due to this temporal optimization, all the information about the environment is needed. With all the information about the environment we mean the position of the objects during the period of time when the mobile robot will have to execute the path. This is an off-line procedure, i.e., not additional information is supplied during the execution, all the information is provided before.

The starting point is the roadmap in the three-dimensional space computed by the thinning method as an array of surfaces. It must not be forgotten that the three- dimensional space (x,y,t) is formed by the two positional coordinates of the mobile robot and one coordinate related to the time. So, the computed roadmap in this way provides the connectivity of the physical space at each instant of time and how this connectivity change in the time.

The first step is to connect the starting and the goal configurations to the roadmap. The starting configuration is specified as a position (x_0, y_0) and the time when the movement starts (t_0) so it is defined an only point in the threedimensional space. The connection to the roadmap is done by following the opposite gradient of the potential (the potential was computed at the same time that the roadmap) in the following way. The opposite gradient is computed at the plane $t = t_0$. So, if the opposite gradient is followed, this one

provides the next positional coordinate (x_1, y_1) and the next time coordinate is the next instant of time $(t_1 = t_0 + \Delta t)$. The next coordinate (x_2, y_2, t_2) can be computed in the same way and so on until a point of the roadmap is reached.

The goal configuration is not completely defined, only a position is specified (x_G, y_G) but not a time instant since the time when it is reached is ignored. So, the same process done with the initial configuration must be repeated with the goal configuration (but decreasing the time) for each time instant in the known period of time starting at t0.

Once the initial and goal configurations have been connected to the roadmap the path from the initial configuration and the goal can be found using an A* graph search algorithm. This algorithm consists of the construction of a graph (a set of nodes joined by edges) from the initial configuration. As the search will be restricted to the points in the roadmap, the nodes of the graph will be the points of the configuration space in it. In order to do the searching it is necessary to pass from a node to others and this is provided be the expanding rule (it is the way in which the edges connect the nodes). In our case the expanding rule from a node is defined as the nodes of the roadmap closest to it at the next time. With this rule we make secure that the generated path will be done as the position in consecutive times without jumps or going back in the time.

Note that the graph search is especially indicated when the number of nodes is huge, and the backtracking is indicated when the number is tiny instead.

The search (the expansion of the graph) has also to be lead in order to get the solution as fast as possible. In order to lead the search a heuristic function is used. This heuristic evaluates the nodes that are going to be expanded. Then, the value provided by the heuristic indicates the order in which the nodes will be expanded. The used heuristic can be only the depth of the node (achieve the exact optimal path) or adjunct with the function distance to the goal or the potential of that node in the configuration space (possibly not the exact optimal path) which provide a greater heuristic power and reduce the computational load.

The starting point of the graph search is the initial configuration and consecutively expansions are produced (saving for each reached node the node from where it comes) until either the stop condition is satisfied or there are no more possibly expansions.

If it a solution is not found within an especified time interval, a new period of time can be appended and then, the goal configuration must be connected to the roadmap and the search can continue.

4. ON-LINE METHOD

The presented algorithm is used when there is no knowledge about what it is going to happend, i.e., when there is no previous information about the positions of the objects in the future. Then, the positions of the objects have to be provided to the algorithm during the search as they change in the time. The method provides then an on-line adaptation to the changing enviroment.

Then, the only information at a time instant is the position of the objects at this time. So, the proposed path planner searchs for a path in the space as if there were no changes in the environment at all during that time. The path shows the way to reach the goal from this time instant and from the actual position. When a new instant is considered and when the objects have changed a new path has to be found. Consecutively the path has to be searched until the goal configuration is reached.

It has been said that a single path planning has to be done at each time. The meaning of this is explained in the following. Here, the starting point will be the roadmap computed at the configuration space. But, now, it is only a set of curves in a bidimensional space. The first step is to connect the initial and goal configurations to the roadmap. The connection is done by following the opposite gradient of the potential (that was calculated with the roadmap and it is repulsive respect the objects). Following the opposite gradient grants that there is no collisions with objects in the connection.

Once the initial and goal configurations have been connected the path can be found in the complete roadmap. A backtracking is used to find the path between the two configurations. Here, the nodes to do the search are not all the points of the roadmap, they are only the cross road points. And the expansion from a node comes by the roads that leave it toward other nodes.

The backtrack is a method that generates a tree. When a node is expanded, one of the reached nodes is chosen to follow the expansion with some criteria (the proximity to the goal, the potential, ...). The information is saved in order to can go back if the chosen node does not carry to the solution. In that case a back step is done, all the information about the path that does not carry to the solution is forgotten, and a new node is chosen.

The output of this single path planner is a path at that instant of time, but the only necessary information is the next configuration that leads to the goal. This configuration will be the initial configuration at the next time and a new path planning will be executed to get the next configuration and so on until the configuration found was the goal.

There is a situation that has not been considered yet. It is when the single path planner does not find a path at a time, so, there are any next points that lead to the goal. In this case, the next point will be the best neighbour. With the best neighbour we mean the neighbour that minimizes some criteria (the proximity to the goal, the potential, ...). By choosing this neighbour we grant that there are no collisions with objects, because a configuration close to C-obstacles have a much higher potential than a configuration that is not so close. As distant configurations to obstacles have very similar potential (the potential approaches to 0) the distance to the goal has much more weight and in this case the mobile robot is lead to the goal because there are no possible collisions.

As a full teplan is made at each time this method seems not to be an efficient aproach. But it is, because it do not take to much time. The roadmap can be computed very fast and the path is found quickly because it jumps from crossroad to crossroad (usually less than 100 in total).

5. RESULTS

In order to test the proposed approaches, an environment with a moving object has been considered in such a way that, in some situations, it allows the connection between two zones of the workspace and in other situations it does not, like a door that separates two rooms. As it has been explained before, the robot is considered as a disc. The starting point and final point of the robot are situated on the separated zones of the workspace.

The bitmaps have been taken with a resolution of 64x64. All the calculation has been done on a SUN Ultra-II (300 Mhz) Workstation. If a Matrox Image Board or a cellular automata is used the time of thinning computation is reduced considerably. It has been realized several animations of the results using the OpenGL libraries from Silicon Graphics.

Figures (2) and (3) show two scenes of the robot environment and the calculated roadmap with both approaches. In figure (2) the robot has not enough space to pass through the door due to the considered dimensions. It has be seen how in Figure (2) there is no connection between the two workspace zones. In figure (3) there are enought

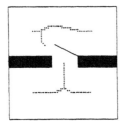

Fig. 2. Workspace and calculated roadmap. The rooms are not connected

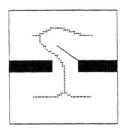

Fig. 3. Workspace and calculated roadmap. The rooms are connected

space and the two rooms are connected. By using both approaches presented in the paper, the result is the same, i.e., a free collision path that connect the initial and final configuration. The differences between them are based on the computational load.

Time optimal path planning

As it has been explained the execution of this task is performed off-line, without any consideration about the workspace changes during its execution. The computational complexity is higher than the associated with the *on-line* method. Nevertheless the computation time was around a few seconds (3 seconds for the algorithm that considers the goal distance in its heuristic and 7 for the one that do not consider it).

On - line Method

At this point the objective is the determination of the Δt that limits the robot speed. This variable will be restricted by the bitmap resolution and the computational load developed at each step. When the bitmap resolution is fixed it is only necessary to define the computational cost. This one consist on the sum of three task times: C-space computation, roadmap generation and search algorithm. With the proposed algorithms the results were: 40 ms to C-space computation, 10 ms to roadmap generation and less than 10 ms to backtrack search. All of them sums less than 60 ms for all the steps, so we can grant a robot speed higher than 16 pixels by second.

6. CONCLUSIONS

In this paper new procedures to perform the path planning task for mobile robots when a

workspace with changing objects is considered. Two aproaches have been presented that solve the problem taking into account two kind of restrictions. They both are based on the use of global techniques but they include innovating facts that make them quite suitable for their real and online application. First, a time optimal application has been developed providing a solution with a reasonably low computational load . Second, if the techniques are implemented as an *on-line* method, then it has been shown that the method allows to manage mobile robots with an acceptable speed. Both results show the power of the considered approach for path planning in real situations

7. REFERENCES

Barraquand, J. and J.C. Latombe (1991). Robot motion planning: a distribuited representation approach. *Int. J. of Robotics Research.*

Blanco, F.J. (1997). New techniques to potential maps evaluation and roadmaps construction at configuration space. Technical report. U. Salamanca.

Brost, R.C. (1989). Computing metric and topological propierties of configuration-obstacles. *Proc. of the IEEE Conf. on Rob. and Autom.* pp. 170–176.

Curto, B. and V. Moreno (1997). Mathematical formalism to perform a fast evaluation of the configuration space. *Proc. of the IEEE International Symposium on Computational Intelligence inRob. and Autom.*

Gonzalez, R.C. and P. Wintz (1987). *Digital Image Processing.* Addison-Wesley Publishing Company, Inc.

Guibas, L., L. Ramshaw and J. Stolfi (1985). A kinetic framework for computational geometry. *Proc. of the IEEE Conf. on Rob. and Autom.*

Janét, J.A., R.C. Luo and M.G. Kay (1997). Autonomous mobile robot global motion planning and geometric beacon collection using traversability vectors. *IEEE Tr. on Rob. and Autom.*

Kavraki, L. (1995). Computation of configuration-space obstacles using the fast fourier transform. *IEEE Tr. on Rob. and Autom.*

Latombe, J.C. (1991). *Robot motion planning.* Kluwer Academic Publishers, Boston, MA,.

Lozano-Pérez, T. (1983). Spatial planning: a configuration space approach. *IEEE Tr. on Comp.*

Tzionas, P.G., A. Thanailakis and P.G. Tsalides (1997). Collision-free path planning for a diamond-shaped robot using two-dimensional cellular automata. *IEEE Tr. on Rob. and Autom.*

MULTI-OBJECTIVE PATH PLANNING FOR AUTONOMOUS SENSOR-BASED NAVIGATION

A. Mandow*, L. Mandow**, V. F. Muñoz*, A. García-Cerezo*

* Dpto. Ingeniería de Sistemas y Automática. Universidad de Málaga, Plaza de El Ejido s/n, 29013, Málaga, SPAIN. Tel: +34 5 213 14 07, Fax: +34 5 213 14 13. E-mail: amandow@ctima.uma.es. Website: http://www.isa.uma.es.

** Dpto. Lenguajes y Ciencias de la Computación Universidad de Málaga. 29071, Málaga (Spain). Fax: +34-5-213.13.97, E-mail: lawrence@lcc.uma.es

Abstract: The paper approaches path planning for autonomous mobile robots based on sensor-based navigation behaviors. The aim is to obtain a plan as a sequence of behaviors that does not only consider minimizing the path's length but also adds robustness by dealing with practical limitations introduced by the sensor and robot systems during task execution. The graph-search algorithm PRIMO-A* has been developed to consider multiple objectives ordered by priority. Moreover, a method is introduced to generate the graph that models the set of possible robot behaviors in the working environment. These methods have been applied to a real robot performing navigation tasks in indoor environments. Copyright © 1998 IFAC

Keywords: mobile robots, path planning, heuristic searches, robot navigation, behaviours.

1. INTRODUCTION

Real world applications for mobile robots rely on planned tasks so that they can be purposefully used in areas such as industry (Byler *et al.*, 1995), agriculture (Mandow *et al.*, 1996) and services (Skewis *et al.*, 1991). Path planning provides these systems with the flexibility they require, allowing task programming as a sequence of goals to be reached.

The specification of a path for a real robot is a relevant problem since it involves coping not only with the topology of the working environment, but also with the kinematics and dynamics that limit the vehicle's manoeuvrability, the uncertainties of sensor data, and restrictions of the sensor system such as switching frequencies, blind areas, and range limits. Furthermore, the plan has to be adapted to the particular path tracking techniques used by the robot (Ollero *et al.*, 1997). Consequently, efficient path planning involves taking into account a number of down-to-earth criteria besides the optimization of the path length.

The paper deals with multi-objective path planning for behavior based navigation in indoor environments. Behaviors are independent pattern activities, such as following walls or turning round

corners, that can be used as building blocks to make up a complete task. In order to build a planner, it is first necessary to stipulate all possible behavior patterns for the robot in a given environment, and then a set of criteria that allows choosing the path or sequence of patterns that better adapts to the particular characteristics of both the mission and the robot. In this sense, mobile robotics can profit from recent research in multi-criteria planning.

Graph search with multiple criteria has been considered for the last fifteen years (Loui, 1983), and has been successfully addressed by several algorithms (Stewart and White, 1991; Navinchandra, 1991; Mandow and Millán, 1997). In fact, it has been demonstrated that the basis of well-known search algorithms such as A* can be easily adapted to multi-criteria optimization problems (Carraway *et al.*, 1990). In particular, the work presented here combines Navinchandra's strategy of reformulating constraints as objectives and a multi-objective search algorithm specially designed to incorporate priorities.

These techniques have been applied to the AURORA robot (see Fig. 1), a mobile platform 1.4 meters long and 0.8 meters wide, which is equipped with a simple sensor system of just four ultrasonic sensors (one at each side and two at the front) which range only up to

1.3 meters. Odometric estimations are based on the readings of angular encoders placed on each of its motors (one for steering and two for differential traction).

Fig. 1. The AURORA mobile robot.

2. BEHAVIORS FOR INDOOR NAVIGATION

The control architecture used by the robot is sketched on Fig. 2. Behaviors are defined as independent modules that use sensor information (internal and/or external) to produce a special pattern activity. A plan is defined as an ordered sequence of pairs $[B_i, G_i]$, where B_i is a behavior descriptor, chosen amongst the available set of *Navigation Behaviors*, and G_i is a goal that implies the ending condition of the associated behavior. The plan is managed during execution by a *Supervisor* module, which activates and deactivates

behaviors according to the perception of goals. Table 1 lists the set of behaviors that have been developed for indoor navigation, together with the goals that can be assigned to them as an ending condition.

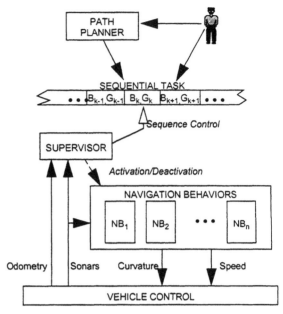

Fig. 2. Behavior based control architecture.

The modularity of this architecture allows for independent design and development of each behavior, even by means of different specific techniques appropriate for each one. Thus, these behaviors have been built by using several possible strategies that have been developed and applied in

Table 1: Behaviors for indoor navigation

Behavior	Pattern	Description	Associated goals
Wall Following		Follows a contour detected by the *side sensor*.	· Reaching some particular distance (*odometry*) · End of the wall (corner) detected with the *side sensor*. · End of the wall (inside corner) detected with the *front sensors*.
Advance		Moves straight ahead with no external sensor feedback.	· Detection of a front wall with the *front sensors*.
Outside Turn		Turns towards a free space detected by the *side sensor*.	· Turning angle completed (*odometry*) OR a wall is found (*side sensor*).
Turn to Align		Turns to align to a perpendicular wall, with both *front sensors* and the corresponding *side sensor*.	· *Odometric* estimation of the distance needed to complete the activity.
Inside Corner		Turns to align to a new wall, with both *front sensors* and the corresponding *side sensor*.	· *Odometric* estimation of the distance needed to complete the activity.

previous works (Mandow *et al.*, 1996; Ollero *et al.*, 1997; García-Cerezo *et al.*, 1997).

3. THE BLOWN FUSE METAPHOR

Planning a navigation task for the robot system described above can be compared to a situation in which a person has to walk through a familiar environment in the dark, such as when a fuse is blown. The usual behavior is to advance by feeling one's way along walls and obstacles, groping for corners and trying to avoid treading away from known landmarks, where perception of direction and travelled distance can be misleading.

In this sense, the *wall following* behavior presents good robustness for navigation since its position uncertainty is limited to a longitudinal odometric error, which is not very important for short stretches. Arriving to a corner allows correcting that uncertainty. On the other hand, the *advance* behavior allows jumping between unconnected contours, but involves longitudinal as well as directional uncertainty that can be represented by an uncertainty cone. This is illustrated in Fig. 3, where uncertainty in localization is greater in the A-B segment corresponding to the *advance* behavior, and can be represented by an uncertainty cone. Thus, going back to the blown fuse metaphor, in order to reduce uncertainty it is desirable to reach a known spot (i.e. a corner) between two consecutive blind advances.

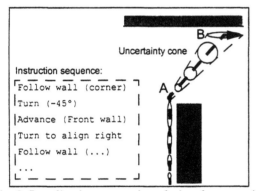

Fig. 3. Localization uncertainty during the execution of a sequence of behaviors.

Therefore, the planner must minimize the number of advances when the real localization of the vehicle is imprecise, as well as minimize the path's length as a secondary objective.

4. CONSTRUCTION OF THE MANOEUVERABILITY GRAPH

The *manoeuvrability graph* models the possible behaviors of the robot for a map of the environment. Taking into account the available behavior repertoire as well as uncertainty considerations, two major rules are defined for the construction of the graph: The first one corresponds to arcs for the wall following activity, while the second refers to advance arcs. Intersections between arcs produce nodes, which coincide with the different turning behaviors. This approach is concerned with the execution of the plan based on tracking contours, which imposes characteristics different from those of classic visibility graphs (Nilsson, 1969). These construction rules are defined as follows:

1) Segments parallel to object contours and to the working environment limits, which are assigned to *wall following* behaviors. Nodes correspond either to *outside turn* or *inside corner*, depending on the graph topology, as can be seen in Fig. 4. These arcs are not directed, so they can be travelled by the robot in both directions.

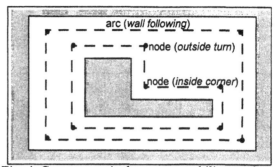

Fig. 4. Contour arcs in the manoeuvrability graph.

2) Segments assigned to the *advance* behavior. These are drawn if the uncertainty cone is fully projected over the target contour. These are directed arcs, in which the vehicle moves straight from a particular node until it finds the edge of a new contour. The arrival to the new arc has the *turn to align* behavior associated to it. In this case two different possibilities are considered:

2.1) Advance from intersection nodes. Arcs starting from the nodes generated in rule 1 are considered for angles of -90°, -45°, 0°, 45° and 90° for each possible orientation with which the node can be approached by the robot. Figure 5 shows such arcs for node n_1 when approached with orientation o_1; angles -90° and -45° have to be discarded, since the projection of their uncertainty cone arrives at more than one edge.

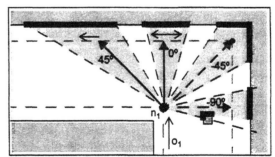

Fig. 5. Advance from a corner in the manoeuvrability graph.

2.2) Advance from one contour to another. For each edge, segments are computed such that all their points are facing another edge by means of

379

perpendicular vectors. The middle point of the segment is considered as a new node, and the perpendicular vector starting from it is the arc joining the two contours if the uncertainty cone projects entirely over the target edge. This rule is illustrated in Fig. 6, where the middle point of the AB segment is used as a node n_2 for a perpendicular *advance* arc.

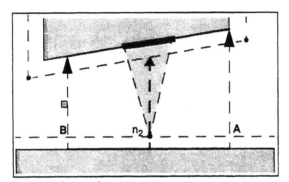

Fig. 6. Advance between two contours.

It must be noted that the method presented to draw arcs for the advance behavior with rule 2 somehow simplifies the real manoeuvrability of the robot. In fact, when reaching a corner, it is possible for the robot to start an advance path with any arbitrary orientation, and two facing edges can be joined by any other point different than the centre of the segment. However, simplification is necessary in order to use the graph as a basis for the search method, since it must be locally finite.

5. MULTI-OBJECTIVE PATH PLANNING

A new algorithm for multi-objective heuristic search has been devised to perform path planning according to the "blown fuse" criteria presented in section 3. This planner represents the state of the robot by means of the following three values:

· *Position*, corresponding to a node of the manoeuvrability graph.

· *Direction*, corresponding to that of the latest traversed arc.

· *Positional imprecision*. A boolean variable, which is true if an advance behavior has been performed but has not yet been followed by a node that represents a known point of the environment (i.e., a corner).

A *state space* is defined allowing transition between two states whenever their positions are connected by an arc in the manoeuvrability graph and the difference between the orientation of the robot and that of the arc is not greater than 90°.

To find a solution to a particular path planning task, a search in the state space is conducted according to two objectives ordered by priority: first, minimizing the number of advances in the state of positional imprecision, and second, minimizing the travelled distance. This leads to obtaining the shortest path among all those that have the minimum number of

advances with positional imprecision. In general, the number of solutions satisfying the first objective can be high, so the second one allows refining the preferences while it guides the search. This kind of multiobjective search is accomplished with a new graph search algorithm, PRIMO-A* (PRIority-ordered Multiple Objective-A*), shown on Fig. 7, which is a simple multicriteria variation of the well-known A* algorithm (Hart *et al.*, 1968; Pearl 1984) and improves upon previous work presented in (Mandow *et al.*, 1997). PRIMO-A* searches a graph for a solution path taking into account a set of several objectives totally ordered by priority.

Algorithm PRIMO-A*

1) Create an empty list of CLOSED nodes, as well as a list of OPEN nodes that contains the initial node. Set the initial node as the root of a SEARCH-TREE.

2) If OPEN is empty, exit with 'FAILURE'.

3) Select a node n' from those in OPEN with the best value of $f_1(n)$. Break ties preferring nodes with best value of $f_2(n),...f_q(n)$ successively. If, in the end, several nodes have equal best cost vectors $\mathbf{F}(n) = (f_1(n), f_2(n), ... f_q(n))$ break the tie among them arbitrarily.

4) If n' is a goal node, then exit with the path in the SEARCH-TREE that goes from the initial node to node n'.

5) If n' is not a goal node, then move it from OPEN to CLOSED, and generate all its successors. For each successor n" of n' do the following,

5.a.) If n" is not in the SEARCH-TREE, add it to the SEARCH-TREE with a pointer to n', and add n" to OPEN.

5.b.) If n" is in the SEARCH-TREE, direct its pointer along the path with best $f_1(n)$. Break ties with the best values of $f_2(n),...f_q(n)$ successively. If both paths yield the same costs, keep the pointer along the old one. If the pointer is changed and n" is in CLOSED, move it back to OPEN.

6) Go back to step 2.

Fig. 7. Algorithm PRIMO-A*.

In multi-criteria search algorithms each path P is evaluated using a cost vector $\mathbf{F}(P) = (f_1(P), f_2(P), ... f_q(P))$ instead of a single cost function f(P). Therefore, the traditional concept of optimal solution is replaced by that of Pareto-optimal or non-dominated solution. A solution is Pareto-optimal if there is no other solution that improves upon it respect to at least one objective and equals respect to the others. PRIMO-A* guarantees that the solution path found is Pareto-optimal and that no other solution path can improve upon one objective without worsening some of the higher priority ones if the following conditions are satisfied: all arc costs are nonnegative and additive and the $f_i(n)$ are of the form,

$$f_i(n) = g_i(n) + h_i(n)$$

where $g_i(n)$ is the i-th cost of the path recorded in the

SEARCH-TREE from the initial node to n, and $h_i(n)$ is an optimistic estimate of the i-th cost of a path from n to a goal node.

6. EXPERIMENTAL RESULTS

This section illustrates with a real example some of the results obtained after applying the presented path planner to the robot AURORA. Fig. 8 shows the map of the working environment where the experiments have taken place. The doors that access to this environment are usually closed, but they can occasionally be opened for the robot to reach adjacent facilities. The manoeuvrability graph, represented by black straight lines, has been drawn over the map according to the method described in section 4. Arcs that can only be traversed in one sense are represented by arrows, and a node exists for every intersection of arcs. Note that some of the generated arcs have had to be suppressed since they posed a hazard for the robot; such is the case of those in the proximity of stairs, which can not actually be detected with the utilized sensor system.

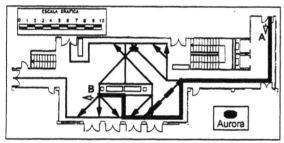

Fig. 8. Manoeuvrability graph over a map of th environment (scale in meters).

Algorithm PRIMO-A* has been applied to this problem to generate a path that takes the robot from state A to state B (represented by the symbol ◁ on Fig 8), while considering the two objectives discussed above. The result is the path represented by a continuous bold line on Fig. 8. Note that there is a shorter path, but it accumulates greater positional imprecision by taking two consecutive advances without reaching a known spot. The path provided by the planner avoids the fist advance (shown in the figure by a bold dotted line) by following the wall and reaching an inside corner.

The resulting path consists of a sequence of 6 nodes of the manoeuvrability graph, which is translated to 12 behavior instructions (alternating straight arcs and turns in nodes). The corresponding sequence of behavior-goal pairs is listed in Fig. 9.

Fig. 10 shows the results obtained by an actual execution of the path. This graph has been obtained by assigning the measured ranges to the estimated odometric position of the robot. The recorded map is clearly deformed with respect of the original one, due to the accumulation of odometric errors, which gives an idea of the difficulty of following long paths with

B_1: Wall Following (Left);	G_1: Front sensors.
B_2: Inside Corner (Right);	G_2: Odometry 1.8m.
B_3: Wall Following (Left);	G_3: Side Sensor (Left).
B_4: Outside Turn (Left);	G_4: Side Sensor (Left) OR Odometry 90°.
B_5: Wall Following (Left);	G_5: Front sensors.
B_6: Inside Corner (Right);	G_6: Odometry 1.8m.
B_7: Wall Following (Left);	G_7: Odometry 5.37m.
B_8: Outside Turn (Right);	G_8: Side Sensor (Right) OR Odometry 90°.
B_9: Advance;	G_9: Front sensors.
B_{10}: Turn to Align (Left);	G_{10}: Odometry 1.8m.
B_{11}: Wall Following (Right);	G_{11}: Side Sensor (Right)
B_{12}: End;	G_{12}: -.

Fig. 9. Sequence of robot instructions corresponding to the planned path.

traditional odometry-based path tracking techniques. Sequencing behaviors according to the plan has allowed the execution of a 32 meter path in a continuous way, reaching the final localization with considerable accuracy. The consideration of multiple objectives has resulted in a plan that combines efficiency with robustness in execution.

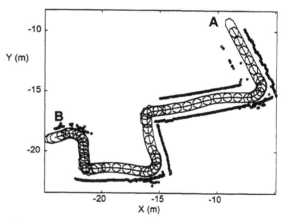

Fig. 10. Real data obtained during the execution of the planned path with the AURORA mobile robot.

7. CONCLUSIONS

The paper has introduced the algorithm PRIMO-A* for graph-search with multiple objectives ordered by priorities. The use of a total order of priority in multiobjective problems induces also a total order of preference between paths. This allows for an algorithm much simpler, more efficient and closer to A* than general multiobjective algorithms like the one presented in (Stewart & White, 1991).

PRIMO-A* has been applied to a path planner that produces robust plans for behavior-based autonomous mobile robots. The search is performed in a space state defined largely over the manoeuvrability graph, which models the possible robot activities for the environment.

It has been shown how the use of multiple objectives allows the search of paths without limiting the manoeuvrability of the robot, by adapting to the limitations posed by real odometric and external sensors. In particular, the planner avoids situations of imprecision in robot localization while it pursues the efficiency of the travelled distance. Results of the application of the method to the AURORA mobile robot have been presented.

The planner produces a path between two robot states specified by the user. The list of robot states resulting from the search conducted by the planner are directly translated to a sequence of behavior-goal pairs that can be directly used as instructions for the AURORA robot.

The method allows increasing the number of considered objectives, so future work involves introducing new practical criteria, such as considering a minimum stabilizing time for a wall following behavior before a new behavior is activated.

ACKNOWLEDGEMENTS

The work presented in the paper has been done within the framework of the TAP96-1184-C04-02 project of the C.I.C.Y.T, Spain.

REFERENCES

Byler, E., W. Chun, W., Hoff, W., Layne, D., (1995) "Autonomous Hazardous Waste Drum inspection Vehicle." IEEE Robotics and Automation Magazine, March 1995.

Carraway, R. L., Morin, T. L. and Moskowitz, H. (1990): "Generalized Dynamic programming for multicriteria optimization". *European Journal of Operational Research* 44 95-104.

García-Cerezo, A., López-Baldán, M. J., Mándow, A., (1997) "Fuzzy Modelling Operator Navigation Behaviors". *Proceedings of the 6th IEEE International Conference on Fuzzy Systems FUZZ IEEE'97*. pp. 1339-1345.

Hart, P.E., Nilsson, N. J. and Raphael, B. (1968): "A fromal basis for the heuristic determination of minimum cost paths". *IEEE Transactions on System Science and Cybernetics* SSC-4 pp 100-107.

Loui, R. P. (1983): "Optimal paths in graphs with stochastic or multidimensional weights". *Communications of the ACM* 26 670-676.

Mandow, A., Gómez-de-Gabriel, J. M., Martínez, J. L., Muñoz, V. F., Ollero, A., García-Cerezo, A., (1996) "The Autonomous Mobile Robot Aurora for Greenhouse Operation". IEEE Robotics and Automation Magazine, vol. 3, no. 4, Diciembre 1996.

Mandow, A., Mandow, L., Muñoz, V.F., García Cerezo, A. (1997): "Planificación Multi-objetivo de Caminos para Navegación Mediante Comportamientos". (in Spanish) CAEPIA'97. Actas de la VII Conferencia de la Asociación Española para la Inteligencia Artificial. V. Botti (Ed.). pp. 55-64.

Mandow, L. and Millán, E. (1997): "Goal programming and heuristic search", in *Advances in Multiple Objective and Goal Programming*. Caballero, R., Fuiz, F. and Steuer, R.E. (Eds.). Lecture Notes in Economics and Mathematical Systems 455, pp 48-56. Springer-Verlag, Berlin Heidelberg.

Navinchandra, D. (1991): *Exploration and Innovation in Design: Towards a computational model*. Springer-Verlag, New York.

Nilsson, N. J., (1969) "A Mobile Automaton: An Application of Artificial Intelligence Techniques". *Proceedings of the 1st International Joint Conference on Artificial Intelligence*, pp. 509-520.

Ollero, A., García-Cerezo, A., Martínez, J. L., Mandow, A. (1997) "Fuzzy Tracking methods for Mobile Robots", Applications of Fuzzy Logic: Towards High Machine Intelligence Quotient Systems, M. Jamshidi, L. Zadeh, A. Titli, S. Boverie, Eds., Cap. 17. Prentice Hall series on Environmental and Intelligent Manufacturing Systems.

Pearl, J. (1984): *Heuristics*. Addison-Wesley, Reading, Mass.

Skewis, T., Evans, J., Lumelsky, V., Krishnamurthy, B., Barrows, B., (1991) "Motion Planning for a Hospital Transport Robot". Proceedings of the 1991 International Conference on Robotics and Automation, Sacramento. pp. 58-63.

Stewart, B. S. and White III, C. C. (1991): "Multiobjective A*". *Journal of the ACM* 38 775-814.

FAST PROCEDURE TO OBSTACLE REPRESENTATION IN THE CONFIGURATION SPACE FOR MOBILE ROBOTS

B. Curto P. Vega V. Moreno

Dept. Informatica y Automatica, Fac. Ciencias, U. of Salamanca
Plaza de la Merced s/n, Salamanca, 37008, Spain
E-mail: control@abedul.usal.es

Abstract: In order to create an autonomous mobile robot it is necessary to develop powerful methods for path planning, avoiding collisions with low computational costs. In this paper, a mathematical formalism for the general evaluation of the representation in the configuration space of any set of obstacles is presented. The use of such representation reduces drastically the complexity of the problem. As it is shown in the paper, with the proposed method, the obstacle representation can be seen as a convolution of two functions that describe the robot and the obstacles respectively. Additionally, the computational load is independent of the shape and number of obstacles and of the robot shape. In the final implementation the Fast Fourier Transform is used to take advantage of the intrinsic parallelable nature of this tool. The resulting algorithm can also be very easily implemented in a parallel way. All the considerations above allow to reduce drastically the computational time making the algorithm suitable for real applications. The method has been applied to robots moving on a plane in a 2D and a 3D dimensional workspace.
Copyright © 1998 IFAC

Keywords: Obstacle avoidance, Configuration Space, Fast Fourier Transform

1. INTRODUCTION

In order to achieve autonomous actuation of robots, algorithms for path planning are needed, together with their execution by the robot control system. The solution to this problem is a set of robot configurations (position and orientation) in such a way that it never collides with the objects in its environment. The collision avoidance approaches can roughly be divided into two categories: global and local. The global techniques have a main advantage that they generate a complete path from the starting point to the target point. The main disadvantages arise from the inherent complexity of these methods that make the

computational load too high to execute them in real time. In constrast, although local approaches use only a small fraction of the world model, reducing the computational complexity very much, in case that an optimal solution is needed, a global approach should be necessarily used.

The purpose of this paper is to provide a general approach to allow the use of using global techniques for path planning, with a reasonable execution time. As it is well known, the solution to the path planning problem is simpler in the configuration space (C-space) than in the robot workspace. Note, for instance, that the robot position and orientation are characterised by just one point in the former(Lozano-Perez, 1983). Consequently, the use of the C-space to solve the problem was the first point taken into account in this work. Additionally, the path planning was

[1] The authors gratefully acknowledge the support of the Castilla y Leon Council and CICYT through the projects C02/197 and TAP96-2410

considered a task to be execute in two steps (see Lozano-Prez for futher details): *findspace*, where the collision free robot configurations are found and *findpath* where a sequence of configurations can be found to move the robot from a given location to another one. This paper is focused on the first step (collision check), when this one is applied to robots moving on a plane in a R^2 and R^3 workspace. The main result is a structure that represents the geometrical constraints, in the C-space, imposed on the robot movements by the obstacles. This structure can be used either for path planning or for the movement robot control (Newman and Branicky, 1991).

Most of the methods (Lozano-Perez, 1983) (Guibas *et al.*, 1985) (Latombe, 1991) (Brost, 1989) that explicitly construct the representation the C-space obstacles model the robot and the obstacles as polygonal or polyhedral objects (convex or non-convex). These methods build their representation by means of a analytic parsing of the contacts between vertices, edges and/or faces of the robot and the obstacles. In any case, the computational time associated to the obstacle projection from the workspace to the C-space, depends on the number of vertices of each object (both robot and obstacles) and also on the robot shape.

In (Brost, 1989) the obstacle representation in the C-space was carried out by means of an algebraic convolution of an obstacle and a mobile robot. This idea is implicitly considered in (Lozano-Perez, 1983), where for a robot A, the mapping of a workspace obstacle B into the C-space obstacle (C-obstacle) is defined like $CO_A(B) \equiv \{x \in \mathrm{Cspace}_A / (A)_x \cap B \neq 0\}$. In (Guibas *et al.*, 1985) is stated that this operation can be seen as the convolution of the set A with the set B. This idea is the basis of the work presented in this paper, that proposes a mathematical formalism to evaluate the C-obstacles by means of the convolution of two different functions: one describing the robot and the other describing the obstacles, if the proper coordinates are chosen. In general, this formalism can be established for mobile and articulated robots (Curto and Moreno, 1997). In this paper specifically its application for robots moving on a plane in 2D or 3D workspaces is presented.

The algorithms for path planning that have been developed in the last years, propose to work in discrete C-spaces (Barraquand and Latombe, 1991). More concisely, a bitmap is used to represent the obstacles in the C-space, reducing drastically the computation time to detect collisions. Kavraki (Kavraki, 1995) proposes the derivation of the C-space bitmap, for a polygonal platform, as the convolution of the obstacle bitmap and the robot bitmap in the workspace. In order to carry out this convolution efficiently, the Fast Fourier Transform is used. This technique provides bitmap with higher accuracy and it is independent of the complexity and shape of the robot and the obstacles.

In this paper the general method that we propose is applied, firstly, to a robot moving in R^2, leading to the same conclusions as in (Kavraki, 1995). Secondly, we are concerned with a more realistic situation by considering that although the robot movement is carried out on a plane, robot and obstacles are defined in a three dimensional workspace. So the path planning task for real mobile robots is solved with a quite fast procedure that includes the advantages of global methods for collisions avoidance. In both cases it is necessary to choose the proper coordinates to work with, considering the particular robot mechanical structure and relating it with the particular degrees of freedom. In this way it is possible to guarantee the existence of the convolution between the robot and the obstacles over one or more coordinates. More precisely, the workspace and robot bitmaps are used and as mathematical tool the FFT was chosen. The intrinsic parallelable nature of this tool at the same time as the very easy parallelation of the resulting algorithm reduce the computational time. Consistently, it is possible the application of this algorithm to changing environments.

The remainder of this paper is organised as follows. In section 2, the mentioned mathematical formalism is proposed, where a function describing the C-obstacles is defined. This is applied to a 2-D and 3-D platform moving on a plane. Next, it shows that by choosing the adequate coordinates, the calculation for the mentioned function is equivalent to the convolution of one function defining the robot and an other function defining the workspace obstacles over certain variables. In section 3, the discretization for the workspace and the C-space is proposed, and a general expression for the discrete domain calculation is given, that is later specified for the two example robots. In section 4, from these discrete expressions, new algorithms for the C-space calculation, by using the FFT, are proposed and implemented for the two considered cases. Finally, in section 5 the main conclusions are presented.

2. PROPOSED MATHEMATICAL FORMALISM

In this section a new expression to calculate the C-obstacles is going to be proposed. This one is based on the evaluation of the integral of two functions product: one describing the robot and other one describing the workspace obstacles.

Let W and C be the workspace and configuration space of a robot, respectively.

Definition 1. Let $\mathbf{A}_{(q)}$ (or $\mathbf{A}(q)$) be the subset of W that represents the robot at configuration q. The function $A : C \times W \to R$ is defined by

$$A(q, x) = \begin{cases} 1 \text{ if } x \in \mathbf{A}_{(q)} \\ 0 \text{ if } x \notin \mathbf{A}_{(q)} \end{cases} \quad (1)$$

Definition 2. Let $\mathbf{B}$ be the subset of W constituted by the obstacles. The function $B : W \to R$ is defined by

$$B(x) = \begin{cases} 1 \text{ if } x \in \mathbf{B} \\ 0 \text{ if } x \notin \mathbf{B} \end{cases} \quad (2)$$

Definition 3. Let $CB : C \to R$ be the function defined by

$$CB(q) = \int A(q, x)B(x)dx \quad \forall q \in C \quad (3)$$

The C-obstacle region $\mathbf{CB_f}$ is defined as

$$\mathbf{CB_f} = \{q \in C / CB(q) > 0\}$$

It can be proved (Curto and Moreno, 1997) that $\mathbf{CB} = \mathbf{CB_f}$, where $\mathbf{CB} = \{q \in C / \mathbf{A}(q) \cap \mathbf{B} \neq 0\}$ is the definition of C-obstacles region that it can be found in (Lozano-Perez, 1983). From the previous definitions, it is straightforward that $q \in C_{free} \Leftrightarrow CB(q) = 0$. So, in order to know whether the robot at a given configuration collides with the obstacles or not it is necessary to evaluate $CB(q)$ (3) and, furthermore, the value of $A(q, x)$. In the following, two cases are considered to illustrate how by choosing the adequate coordinates and frames in the workspace as well as in the C-space, (3) can be calculated in a simpler way.

2.1 Mobile Polygonal Platform

In this section, the robot $\mathbf{A}$ is considered to be a platform with any geometric shape, that can translate and rotate in free way within its workspace $W \subset R^2$. Let F_W be a fixed Cartesian system with the coordinates (x, y) and defined in W. Let F_A be a moving Cartesian frame attached to the robot, whose origin is placed at the center of the robot. In this case, the representation of C may be described by parameterizing each configuration q by $(x_r, y_r, \theta_r) \in R^2 \times [0, 2\pi]$, where (x_r, y_r) and θ_r are, respectively, the origin and orientation of F_A with respect to F_W. So, the previously defined B (2) and A (1) functions, considering that $C \subset R^2 \times [0, 2\pi]$, would be given by

$$B(x, y) = \begin{cases} 1 \text{ if } (x, y) \in \mathbf{B} \\ 0 \text{ if } (x, y) \notin \mathbf{B} \end{cases} \quad (4)$$

$$A(x_r, y_r, \theta_r, x, y) = \begin{cases} 1 \text{ if } (x, y) \in \mathbf{A}_{(x_r, y_r, \theta_r)} \\ 0 \text{ if } (x, y) \notin \mathbf{A}_{(x_r, y_r, \theta_r)} \end{cases} \quad (5)$$

where $\mathbf{B}$ is the subset of W constituted by the obstacles and $\mathbf{A}_{(x_r, y_r, \theta_r)}$ is the set that represents to the robot at configuration (x_r, y_r, θ_r). Then $CB(x_r, y_r, \theta_r)$ (3) would be given by

$$\int A(x_r, y_r, \theta_r, x, y)B(x, y)dxdy \quad (6)$$

Our objective is to simplify the evaluation of this integral. Then, it is necessary to find a new function A' defined over $C' \times W'$ into R, in such a way that it is independent of some configuration parameters. Then, new frames $F_{W'}$ and $F_{A'}$ have to be defined. Particulary, in this case, it is chosen $F_{W'} \equiv F_A$ and $F_{A'} \equiv F_A$. The space C' from the frame $F_{W'}$ contains a configuration set given by $(0, 0, \theta_r')$, where $\theta_r' = \theta_r$. Then, the new function A'

$$C' \times W' \longrightarrow R$$
$$0, 0, \theta_r', x', y' \longrightarrow A'(0, 0, \theta_r', x', y')$$

would be defined as

$$A'(0, 0, \theta_r, x', y') = \begin{cases} 1 \text{ si } (x', y') \in \mathbf{A}_{(0,0,\theta_r)} \\ 0 \text{ si } (x', y') \notin \mathbf{A}_{(0,0,\theta_r)} \end{cases} \quad (7)$$

where $\mathbf{A}_{(0,0,\theta_r)}$ is the subset of points that represents to the robot at configuration $q = (0, 0, \theta_r)$ from $F_{W'}$.

Since $C \times W \overset{A}{\to} R$ and $C' \times W' \overset{A'}{\to} R$, it is only necessary to define a morphism $C \times W \overset{\phi}{\to} C' \times W'$ that makes the coordinate traslation. The morphism definition would be

$$C \times W \longrightarrow C' \times W'$$
$$x_r, y_r, \theta_r, x, y \longrightarrow x_r', y_r', \theta_r', x', y'$$

where

$$\begin{array}{lll} x' = x_r - x & y' = y_r - y \\ x_r' = 0 & y_r' = 0 & \theta_r' = \theta_r \end{array}$$

Then

$$\begin{aligned} A(x_r, y_r, \theta_r, x, y) &= (A' \circ \phi)(x_r, y_r, \theta_r, x, y) \\ &= A'(0, 0, \theta_r, x_r - x, y_r - y) \end{aligned}$$

By the definition of the functions A (5) and A' (7), the following equality holds

$$A'(0, 0, \theta_r, x_r - x, y_r - y) = A(0, 0, \theta_r, x_r - x, y_r - y)$$

Therefore

$$\begin{aligned} A(x_r, y_r, \theta_r, x, y) &= A(0, 0, \theta_r, x_r - x, y_r - y) \\ &= A_{(0,0,\theta_r)}(x_r - x, y_r - y) \quad (8) \end{aligned}$$

where the last equality is due to a change in notation. Clearly, $CB(x_r, y_r, \theta_r)$ can be calculated with (8) and (6) as

$$\int A_{(0,0,\theta_r)}(x_r - x, y_r - y)B(x, y)dxdy \quad (9)$$

For each fixed orientation θ_r of A, a $CB(x_r, y_r, \theta_r)$ can be found, the same as in a situation in which the robot only can translate in the plane. By modifying θ_r in a particular angular range, the C-obstacle can be obtained. Each perpendicular plane to θ_r in the C-space, is the two dimensional C-obstacle corresponding to a fixed orientation.

By considering the convolution product of two functions defined at R^2 over two variables, the expression $CB(q)$ (9) would be given by

$$CB(x_r, y_r, \theta_r) = (A_{(0,0,\theta_r)} * B)(x_r, y_r)$$

By applying the convolution theorem over the previous expression, $\mathcal{F}[CB(x_r, y_r, \theta_r)]$ is given by

$$\mathcal{F}[A_{(0,0,\theta_r)}(x_r, y_r)] \, \mathcal{F}[B(x_r, y_r)]$$

The C-obstacles representation can be obtained by the Inverse Fourier Transform of the previous expression, where the pointwise product of two dimensional transforms is done.

In this way, it has be shown that a set of coordinates defined in the workspace, directly related with the robot mechanical structure, and a representation (x_r, y_r, θ_r) of C, can be found in such a way that CB calculation is simplified.

2.2 *Mobile Polyhedral Platform*

In this section, a polyhedral platform with any geometric shape moving on a plane is considered. First, it is necessary to define a fixed cartesian coordinate system in W given by (x, y, z), and another moving cartesian coordinate system attached to the robot, whose origin is placed at one of the robot vertex on the plane $z = 0$. The problem is quite similar to the previous one. The representation of the C-space is the same: (x_r, y_r, θ_r).

So, the functions B (2) and A (1) would be defined by

$$B(x, y, z) = \begin{cases} 1 \text{ if } (x, y, z) \in \mathbf{B} \\ 0 \text{ if } (x, y, z) \notin \mathbf{B} \end{cases} \quad (10)$$

$$A(x_r, y_r, \theta_r, x, y, z) = \qquad\qquad (11)$$
$$= \begin{cases} 1 \text{ if } (x, y, z) \in \mathbf{A}_{(x_r, y_r, \theta_r)} \\ 0 \text{ if } (x, y, z) \notin \mathbf{A}_{(x_r, y_r, \theta_r)} \end{cases}$$

The function $CB(x_r, y_r, \theta_r)$ would be given by

$$\int A(x_r, y_r, \theta_r, x, y, z) B(x, y, z) dx dy dz \quad (12)$$

The function A' is similar to the one defined in (7) by introducing a new variable z. So, the morphism ϕ, such as $(A = A' \circ \phi)$, will be

$$C \times W \longrightarrow C' \times W'$$
$$x_r, y_r, \theta_r, x, y, z \longrightarrow x_r', y_r', \theta_r', x', y', z'$$

where

$$x' = x_r - x \quad y' = y_r - y \quad z' = z$$
$$x_r' = 0 \qquad y_r' = 0 \qquad \theta_r' = \theta_r$$

It can be obtained that

$$A(x_r, y_r, \theta_r, x, y, z) = A(0, 0, \theta_r, x_r - x, y_r - y, z)$$

So the function A used to define the robot remains invariant by a traslation with a fixed orientation θ_r, placing the robot from the configuration (x_r, y_r, θ_r) to the configuration $(0, 0, \theta_r)$. By applying the previous procedure $CB(x_r, y_r, \theta_r)$ is

$$\int A_{(0,0,\theta_r)}(x_r - x, y_r - y, z) B(x, y, z) dx dy dz (13)$$

The convolution product of two functions defined in R^3 is obtained, although over two variables x and y. It is only necessary to make the integral of this product over z. So, $CB(x_r, y_r, \theta_r)$ will be

$$\int (A_{(0,0,\theta_r)} * B)_{(x,y)}(x_r, y_r, z) dz$$

For each orientation θ_r of the polyhedral robot, a $CB(x_r, y_r, \theta_r)$ is obtained. Each one is calculated by adding the convolution products of two functions $A_{(0,0,\theta_r)}$ and B for each plane $z = constant$, from $z = 0$ to $z = polyhedral height$. If the convolution theorem is applied to obtain $\mathcal{F}[CB(x_r, y_r, \theta_r)]$ then

$$\int \mathcal{F}[A_{(0,0,\theta_r)}(x_r, y_r, z)] \, \mathcal{F}[B(x_r, y_r, z)] \, dz$$

The C-obstacles are obtained by the inverse Fourier transform of the previous expression, where the product of the 2-D FT functions appears.

3. DISCRETIZATION

In order to perform its implementation, a discretization of the workspace W as well as of C-space C is realised, and its resolution N has to be fixed in such a way that the representation of both spaces is correct. When the domains of two functions, A and B, are discretized, the bitmaps that represent both the robot and the obstacles are obtained. Then, the discrete calculation of $\mathbf{CB_f}$ can be given by $\mathbf{CB_f} = \{q_j \in C / CB(q_j) > 0\}$ where

$$CB(q_j) = \sum_{i=0}^{N-1} A(q_j, x_i) B(x_i) \qquad (14)$$

Consequently, a point of C is collision free iff $CB(q_j) = 0$. Therefore, with this discretization, the $CB(q_j)$ calculation, (considering the two analysed cases), becomes a discrete convolution of A

and B over the variables x and y. Applying the convolution theorem to the discrete domain, the DFT of $CB(q_j)$ can be calculated as the pointwise multiplication of the DFT of A and DFT of B by performing the transformation over two variables.

Particularly, for a polygonal platform, a workspace W is limited to $[a, b] \times [c, d]$ and an orientation $\theta_r \in [0, 2\pi]$ is assumed. So, by the expression (9), $CB(x_r, y_r, \theta_r)$ is given by

$$\sum_{i,j=0}^{N-1} A_{(0,0,\theta_r)}(x_r - x_i, y_r - y_j)B(x_i, y_j)$$

where it must be noticed that the variables x_r, y_r and θ_r are discretized too. As it can be observed, this is the discrete convolution product, in both variables, between the workspace and robot bitmap at fixed orientation θ_r. Based on the discrete version of the Convolution theorem, and by using two dimensional discrete transforms, $DF[CB(x_r, y_r, \theta_r)]$ is given by

$$DF\left[A_{(0,0,\theta_r)}(x_r, y_r)\right] DF\left[B(x_r, y_r)\right] \quad (15)$$

For a polyhedral platform moving on a plane $z = 0$ in a workspace limited to $[a, b] \times [c, d] \times [0, height_{robot}]$, $CB(x_r, y_r, \theta_r)$, by the expression (13), can be obtained

$$\sum_{i,j,k=0}^{N-1} A_{(0,0,\theta_r)}(x_r - x_i, y_r - y_j, z_k)B(x_i, y_j, z_k)$$

There is only convolution at the variables x_r and y_r. If discrete convolution theorem is applied, the expression for $DF[CB(x_r, y_r, \theta_r)]$ becomes

$$\sum_{k=0}^{N-1} DF\left[(A_{(0,0,\theta_r)}(x_r, y_r, z_k)\right]_{(x,y)}$$
$$DF\left[B(x_r, y_r, z_k)\right]_{(x,y)} \quad (16)$$

where the DFT are calculated over plane coordinates.

4. ALGORITHM FOR C-SPACE COMPUTATION

In this section, the application of the algorithms to compute the C-obstacle are presented. In both cases, the convolution of bitmaps for robot and workspace could be used directly. If this choice is made, the running time will be $O(N^{d^2})$, being N the discretization resolution and d the dimension. Another choice is to apply the convolution theorem and the FFT algorithm, so the running time will be $O(N^d \log N)$. In order to implement this algorithm, the functions have to be periodical but, at both instances, the involved bitmaps are no cyclic. Kavraki (Kavraki, 1995) shows how to solve the problem, placing a bit "1" at every

workspace contour elements. So, it is ensured that robot can not "wrap around" the W limits.

4.1 Algorithm for a Polygonal Platform

Taking into account the expression (15), next, it is presented the algorithm to compute its C-space and it is in agreement with the proposed one by Kavraki (Kavraki, 1995).

Construct $W(x_r, y_r)$, placing '1' at its limits.
Compute two-dimensional $\mathcal{F}[W(x_r, y_r)]$.
Forevery values of orientation θ_r
    Construct $A_{(0,0,\theta_r)}(x_r, y_r)$.
    Compute two-dimensional $\mathcal{F}\left[A_{(0,0,\theta_r)}\right]$.
    Let $P = \mathcal{F}[W] \cdot \mathcal{F}\left[A_{(0,0,\theta_r)}\right]$.
    Let $IP(x_r, y_r) = \mathcal{F}^{-1}[P]$ (the inverse FFT).
    Let $CB(x_r, y_r, \theta_r) = 1$ iff $|IP(x_r, y_r)| > 0$.

Over the freedom-degree that is not convoluted a "slicing" technique is used. The resolution of the workspace W dicretization, as well as the C-space C, is supposed to be N. In this way, a bitmap W with $N \times N$ dimension and a bitmap A with $N \times N$ dimension, for each robot orientation θ_r, are given. So, as algorithm final result, a bitmap CB, of $N \times N \times N$ dimension is obtained. Because it works with two-dimensional FFT, the algorithm running time is $O(N^3 \log N)$.

4.2 Algorithm for a Polyhedral Platform

From the expression (16), the proposed algorithm can be stated as:

For each coordinate z
    Construct $W(x_r, y_r, z)$, placing '1' at its limits.
    Compute two-dimensional $\mathcal{F}[W(x_r, y_r, z)]$.
For every values of orientation θ_r
    $PT = 0$
    For each coordinate z
        Construct $A_{(0,0,\theta_r)}(x_r, y_r, z)$.
        Compute two-dimensional $\mathcal{F}\left[A_{(0,0,\theta_r)}\right]$.
        Let $P = \mathcal{F}[W] \cdot \mathcal{F}\left[A_{(0,0,\theta_r)}\right]$.
        Accumulate at PT
    Let $IP(x_r, y_r) = \mathcal{F}^{-1}[PT]$ (the inverse FFT).
    Let $CB(x_r, y_r, \theta_r) = 1$ iff $|IP(x_r, y_r)| > 0$.

As in the previous case, the discretization procedure is performed with the same resolution N, at both spaces W and C. So a $N \times N \times N$ dimensional bitmap W and a $N \times N \times N$ dimensional bitmap A for each robot orientation θ_r are constructed. Then, the algorithm final result is a $N \times N \times N$ dimensional bitmap CB. Note that two-dimensional

FFT is used, so this algorithm running time is $O(N^4 \log N)$.

A Silicon Graphics Power Challenge XL platform, with four processors MIPS R8000, has been used to implement both algorithms. As well as, it makes use of a mathematical paralleled library. By considering bitmaps with a resolution of 128, the running times are around a fraction of a second. In order to validate the method, and since they are well-known situations in the literature, at Figure (1), the algorithms results are shown, when the polygonal platform in R^2 is considered. In the same way, the Figure (2) shows the results when it is applied to a polyhedral platform in workspace R^3.

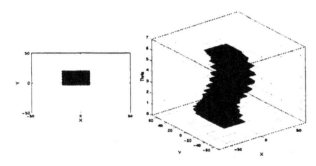

Fig. 1. Polygonal Platform

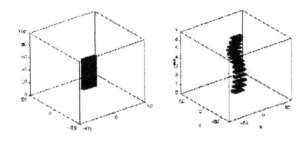

Fig. 2. Polyhedral Platform

5. CONCLUSIONS

In this paper, a general mathematical formalism for C-space computation has been proposed, that is applicable to many kinds of robots. By the choosing of a coordinate system, directly related with the robot mechanical structure, this method derives that C-obstacle is the convolution of two functions: one of them represents the robot and the other one the workspace. This approach has general character and it can be applied to any kind of robot in two dimensional and three dimensional workspaces. To show this characteristic, a object moving on a plane in R^2 and R^3 has been considered. As, it has been described, the convolution is made over two of the robot degrees of freedom. Over the degree of freedom that is not convoluted

a "slicing" technique is used to compute the C-space. Proposed algorithms work with bitmaps that represent the robot and its workspace, and they obtain the C-space bitmap. In both cases, the algorithm running time depends on the used resolution, but it is independent of the exact robot shape, the number of vertices of the obstacles and their geometric shape. To reduce the computation time a parallel implementation of the FFT routines has been applied and, consequently, the running times have been optimized.

6. REFERENCES

Barraquand, J. and J. C. Latombe (1991). Robot motion planning: a distribuited representation approach. *Int. J. of Robotics Research*.

Brost, R. C. (1989). Computing metric and topological propierties of configuration-obstacles. In: *Proc. of the IEEE Conf. on Rob. and Autom*. pp. 170–176.

Curto, B. and V. Moreno (1997). A mathematical formalism for the fast evaluation of the configuration space. In: *Proc. of the IEEE Sym. on Comp. Int. in Rob. and Autom*.

Guibas, L., L. Ramshaw and J. Stolfi (1985). A kinetic framework for computational geometry. In: *Proc. of the IEEE Conf. on Rob. and Autom*.

Kavraki, L. (1995). Computation of configuration-space obstacles using the fast fourier transform. *IEEE Tr. on Rob. and Autom*.

Latombe, J. C. (1991). *Robot motion planning*. Kluwer Academic Publishers. Boston, MA.

Lozano-Perez, T. (1983). Spatial planning: a configuration space approach. *IEEE Tr. on Comp*.

Newman, W. and M. Branicky (1991). Real-time configuration space transforms for obstacle avoidance. *The Int. J. of Robotics Res*.

MULTIPLE-OBSERVER SCHEME FOR SAFE NAVIGATION

Roland Schultze

FH Konstanz, P.O. Box 100543, D – 78405 Konstanz
schultze@fh-konstanz.de

Abstract: The article presents a method for obtaining a fault-free estimate of the state of motion, which is especially suited for coping with temporary faults like ramps or jumps as for instance encountered in position data from GPS receivers. The navigation filter is based upon a bank of observers whose structure is dynamically adapted to the actually available set of sensors. The benefits of this observer scheme are the uninterrupted availability of a filtered state of motion and the large set of innovations which is the basis for fault detection and isolation. *Copyright © 1998 IFAC*

Keywords: Fault detection, observers, redundancy, fault tolerance, safety, position estimation.

1. INTRODUCTION

1.1 General problem setting

One problem occurring in vehicles of all fields of application is the estimation of the current state of motion, which generally consists of position, velocity, angles (heading, roll, pitch) and corresponding turnrates. Depending on the special application the state vector may look different, some elements may be neglected (e.g. the height for ships) or some may be supplemented, e.g. accelerations, actuator angles, or some may be cast into another form, e.g. a curvature instead of a turnrate. Nevertheless, the general problem remains the same for all types of vehicles: How to estimate the state of motion from the available sensor data, including appropriate filtering, automatic fault-detection and automatic recovery after temporary faults. Figure 1 gives an overview over the signals involved. The safety aspect of the navigation filter is to keep faults and dropouts away from the controllers, which would otherwise lead to dangerous and undesired compensating actions (Schultze, 1996). The dashed line indicates closed control loop via actuators. The fourth type of consumer, the transmission of own data to other vehicles and to traffic supervision systems currently gains a lot of interest in maritime community (Hoßfeld, 1996; Holmström, 1996), because it could greatly increase safety at sea. For this purpose, too, one needs consistent, fault-free data.

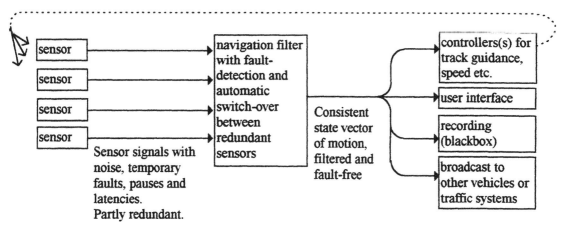

Fig. 1. Typical signal flow. The navigation filter distributes consistent fault-free estimates to several consumers.

1.2 Special features of marine application

This article presents a general strategy with a bank of multiple observers, which can be adapted to any special type of vehicle. The strategy has successfully been realised for the purpose of navigation data filtering on ships. All practical examples in this article come from this field of application. The navigation data filtering problem is characterised by following features:

1. The heading angle measurement by gyro compasses is highly reliable and nearly noise-free so that the heading and turnrate signals coming from the compass electronic unit do not require any further filtering or supervision. The sensors which have to be supervised because they often fail are position and velocity sensors.

2. The roll and pitch motion is neglected for this general-purpose filter because roll sensors with electronic output are normally not available and roll models are complicated to tune.

3. The height component of position vector p is neglected.

4. Two velocity vectors have to be distinguished, the velocity over ground g and the velocity through the water v differ by the current of the waters, called drift d.

$$\dot{p} = g = v + d \qquad (1)$$

5. Beside the North-East coordinates, ship-fixed coordinates f (forward) and c (cross, to starboard) are commonly used for convenience.

6. Typical sensors are position receivers measuring p, water speed logs measuring only the forward component v_f, and *dologs* (Doppler logs) measuring g.

7. At least the forward component of the acceleration $a_f = \dot{v}_f$ is of interest for navigational purposes like speed control, and should also be estimated.

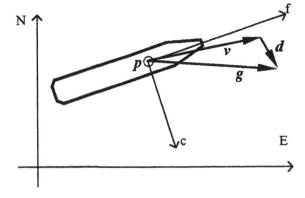

Fig. 2. Some elements of the state vector: position p, velocities v (through water) and g (over ground). The difference d (drift) is caused by currents in the sea. The ship-fixed coordinates are abbreviated f (forward) and c (cross).

The state vector x, used in the filter and fault-detection algorithm, has seven elements, all in ship-fixed coordinates, shown in eq.(2). However, since the ship-fixed coordinate system turns and moves with the ship, the state vector is saved in an absolute format x_{abs} (eq.(3)) between the filtering steps.

$$x = \begin{pmatrix} v_f \\ a_f \\ v_c \\ d_f \\ d_c \\ p_f \\ p_c \end{pmatrix} \qquad (2) \qquad x_{abs} = \begin{pmatrix} v_f \\ a_f \\ v_c \\ d_N \\ d_E \\ p_N \\ p_E \end{pmatrix} \qquad (3)$$

2. MULTIPLE-OBSERVER STRATEGY

A simple strategy to cope with changing availability of partly redundant sensors is to reconfigure the observer every time the set of sensors delivering useful data changes. This simple strategy however produces unreliable data or no data at all in the phases of re-convergence of the reconfigured observer. And it leaves the question of fault detection open, especially for those cases where a sensor continues sending data which can only be judged incorrect if compared to an analytic model and/or other sensors.

To solve the problems mentioned above, this article recommends to use multiple observers in parallel for two reasons:

1. For being prepared for switch-over with the backup observer already converged (hot standby).

2. For having a large number of innovations as residuals for fault detection, which allow to discriminate between sensor faults.

The second reason has long been emphasised in fault detection literature. The arrangement of multiple observers using different subsets of sensors for measurement update is called *bank of observers* (Isermann, 1994b) or *observer scheme* (Patton, Frank and Clark, 1989). This article also emphasises the first reason.

Normally, when the user's favourite sensors are OK, the bank of observers consists of a master and some backup observers. The master observer uses the combination of sensors (e.g. a position, a log, a dolog) favoured by the user, for measurement update. The estimated state of this master observer is used as filtered output of the whole navigation filter, as long as the sensors used are diagnosed OK. For being prepared for a sensor fault, a set of backup observers runs in parallel, each of them with one of the master sensors replaced (Fig.3). So the navigation filter can respond immediately to every single sensor failure by replacing the master

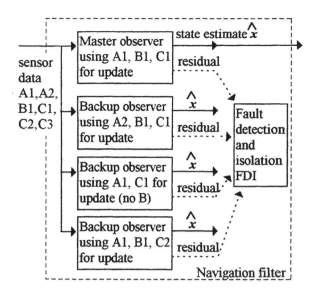

Fig. 3. The bank of observers normally consists of one master and several backups, one for each sensor failure which would affect the master estimate. Here, three sensor types A, B, C are normally used for measurement update in an observer. The numbers behind the letters reflect the user's preference. If a redundant sensor is lacking, it is also possible to produce an estimate with a reduced number of sensors, like in the second backup observer.

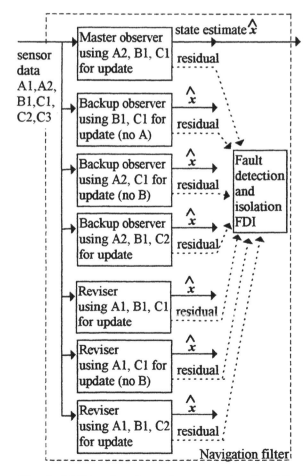

Fig. 4. Additional *revisers*, observers using the failed sensor A1 before it is trusted again.

observer with the appropriate backup observer. An extra algorithm taking care for jump-free switchover may be advantageous, depending on the application (see practical example below). After such an observer replacement, the set of backup observers has to be reconfigured and restarted.

So far, the concept is relatively easy to understand and at least principally known from FDI textbooks (Patton, Frank and Clark, 1989; Isermann 1994a). The only new aspect is the choice of backup observers not only for fault detection but also for replacing the master. The much more important issue of practical application is how to gain confidence again into a sensor after it has once failed. Somehow it must be given a chance of rehabilitation. This can be realised by running another set of observers in parallel, the so-called *revisers* (Fig.4). They use those sensors for updating their estimates which would be used by the master and backup observers if there had not been any failure. These observers are used for replacing the master and backup if the failed sensor has proven fault-free for a certain penalty-time, which can be configured by the user.

3. REALISATION OF THE NAVIGATION FILTER FOR MARINE APPLICATION

3.1 Observer Design

All observers use structurally the same model for extrapolation, shown in equation (4).

$$\frac{d}{dt}\begin{pmatrix} v_f \\ a_f \\ v_c \\ d_f \\ d_c \\ p_f \\ p_c \end{pmatrix} = \begin{pmatrix} 0 & 1 & 0 & 0 & 0 & 0 & 0 \\ 0 & -T_a^{-1} & 0 & 0 & 0 & 0 & 0 \\ 0 & 0 & 0 & 0 & 0 & 0 & 0 \\ 0 & 0 & 0 & -T_d^{-1} & 0 & 0 & 0 \\ 0 & 0 & 0 & 0 & -T_d^{-1} & 0 & 0 \\ 1 & 0 & 0 & 1 & 0 & 0 & 0 \\ 0 & 0 & 1 & 0 & 1 & 0 & 0 \end{pmatrix}\begin{pmatrix} v_f \\ a_f \\ v_c \\ d_f \\ d_c \\ p_f \\ p_c \end{pmatrix} + \begin{pmatrix} 0 \\ 0 \\ f_{sw} \\ 0 \\ 0 \\ 0 \\ 0 \end{pmatrix} \quad (4)$$

T_a, T_d are time constants, which let the acceleration and drift estimate converge to zero in case no sensor data is available. Like the update gains they are design parameters, used to shape the special observers' frequency response in a pole-placement approach. For navigation purposes it is quite important to know how the estimates converge in time. Therefore pole-placement design is more suitable than optimal filtering approaches.

Depending on the seven sensor combinations (table 1), a different structure of the update gain matrix is used. For each one a separate set of parameters (update gains, and T_d, T_a) is designed according to the user's desired time constants. The details of the pole-placement cannot be shown within the 6-page-limit, they will possibly be presented in a longer journal version.

Table 1. Observer structures, differing by combination of sensors used for measurement update.

Sensors used	States updated by measurement	Order of transfer function from sensor to state	Remarks
pos	p	1 (pos → p)	Full state is observed
log	a_f (and v_f via a_f)	2 (log → v_f)	
dolog	d	1 (dolog → g)	
pos	$d\ p$	2 (pos → p)	Full state is observed
log	a_f (and v_f via a_f)	2 (log → v_f)	
pos	p	1 (pos → p)	d_f converges to zero
dolog	$d_c\ a_f$ (v_f via a_f)	2 (dolog$_f$ → v_f) 1 (dolog$_c$ → g_c)	
pos	$d_c\ a_f$(v_f via a_f) p	3 (pos$_f$ → p_f) 2 (pos$_c$ → p_c)	d_f converges to zero
log	a_f (and v_f via a_f)	2 (log → v_f)	p extrapolated, so-called *dead reckoning*
dolog	d	1 (dolog → g)	
dolog	$d_c\ a_f$ (v_f via a_f)	2 (dolog$_f$ → v_f) 1 (dolog$_c$ → g_c)	d_f → 0, *dead reckoning*
log	a_f (and v_f via a_f)	2 (log → v_f)	d → 0, *dead reckoning*

The sway v_c depends on the turnrate: zero for course-keeping, but directed outward in curves. For sway-modelling via the term $f_{sw}(r)$ and for sway-update the non-linear *pivot-range method* is applied (Schultze and Holzhüter, 1997). For describing the filtering and fault-detection strategy, it is sufficient to focus on the normal situation of course-keeping, in which sway can be neglected.

The filters are implemented in real-time environment. A special requirement is event-triggered filter activation. The events of new sensor data arriving occur in an uncoordinated way. So the filter must work with varying time intervals. Design is done in continuous time, and the discrete-but-variant-time implementation is achieved simply by Euler-integration.

The time-varying latency of the incoming raw data is a further aspect of real-world application. At least for those sensors which supply a time-of-measurement with the sensor data, like GPS receivers do, the navigation filter performs a simple but worthwhile compensation, supplied the latency does not exceed a configurable threshold (which would lead to fault detection).

From table 1 it is clear that only the first two observer structures secure a convergence of the state towards the measured values. The five *reduced* observer structures below are only appropriate as short-term backup. However, it is usually better to extrapolate the unobservable states by equation (4) for some minutes than to do no filtering at all.

3.2 Fault Detection

A large number of innovations (number of sensors times number of observers) is passed to the FDI module (Figs.2 and 3). The estimated drift and accelerations are also used as residuals. The residual composition is time-variant with sensors working or not, and with observer configuration. Therefore the fault table 2 is only an example to highlight the procedure of finding residual sets which are *strongly isolating* (Gertler, 1991).

One for instance recognises that is it not sufficient for position fault detection to simply trigger on one position residual, because this will also respond to log faults. Only if all the P1 residuals from all observers have significantly walked into the same direction, one can be sure. A log fault on the other hand can also be concluded from position residuals, if only those observers which use the log for update show the symptom, and the others **not**.

Table 2. Fault table for the simple situation with two position receivers P1 and P2, one log V1 and no dolog.

	residuals of	after P1-jump	after P2-jump	after V1-jump
Master observer using P1, V1 for update	P1		0	
	P2			
	V1	0	0	
Backup observer using P2, V1 for update	P1			
	P2	0		
	V1	0	0	
Backup observer using only P1 for update	P1		0	0
	P2			0
	V1		0	

Regarding the position sensor supervision, the fault detection takes the corresponding position residuals of all observers into account. The residuals are two-dimensional, having one forward (f) and one cross (c) component. In Fig.5 the forward component is dotted. Before deciding about jumps, a *most likely jump direction* is constructed by averaging the residuals of all observers, and then the projection onto this direction serves as one-dimensional residual in the following steps of the FDI algorithm. This ensures that a fault is only detected if all residuals support the same jump direction hypothesis.

Noise and single outliers in the projected residuals make threshold detection impossible (too sensitive to noise, too insensitive to faults). The solution is the Hinkley-detector (Hinkley, 1971; Basseville and Benveniste, 1986; Schultze, 1992, Draber, 1994) leading to fast detection with very few false alarms:

A test value h is built from the one-dimensional residuals by integrating what exceeds a so-called *border* (12 m in the example below). h is restricted to positive values, which is the main reason for the alertness of the Hinkley-detector. If all test values h belonging to one position sensor exceed a threshold λ (72 m·sec in the example below), the sensor is diagnosed to be faulty.

The tests for position sensors and dologs are constructed in a similar manner. For water speed logs, two types of test are applied, one based on the position residuals (cmp. table 2), one independently based only on the one-dimensional forward speed residual and the estimated acceleration a_f.

3.3 Practical Example

Having only one position receiver and one water speed log is a suboptimal situation regarding redundancy, but nevertheless, the navigation filter demonstrates its capability.

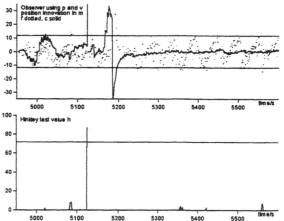

Fig. 5. Position residual (two components) and test value h associated with the observer using position and speed for update.

Until the fault detection at t=5123sec, the observer of Fig. 5 is the master. Afterwards the plotted data comes from the reviser, until at 5334 sec the original observer arrangement is restored again. The horizontal lines show the border for the projected residual and the threshold for the Hinkley test value. (residuals from the other observers not shown due to space limitation.)

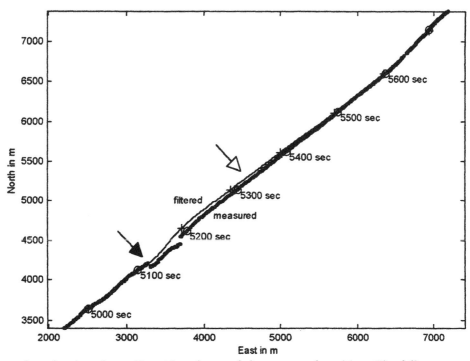

Fig. 6. Position plot, showing the sudden 40 m jump of the measured position. The full arrow marks the moment of fault detection and immediate rearrangement of observers. From that time on the filtered position is only based on dead-reckoning, until the penalty time for the faulty position is over at the open arrow. The observers are rearranged back, but the filtered output is smoothly sliding towards the master observer's state, due to an extra algorithm for jump-free switching.

4. CONCLUSION AND OUTLOOK

This article offers a general strategy to get a consistent fault-free estimate of the state of motion, if one has to cope with the typical problems encountered at many sensors: transient failures like pauses, jumps or ramps and asynchronous data transmission with variable latency. The navigation filter applies a bank of observers which is dynamically rearranged so that the master observer always works with the most favourable set of sensor data which is currently available. The availability and safety improvement of the approach is the automatic fault-detection with immediate switchover to a backup (standby) observer, which has already converged before, when running in parallel to the master observer. Another new aspect of the approach is the additional use of *revisers* after a sensor has failed. The revisers are observers which produce residuals for checking if the sensor can be trusted again, and which prepare a converged estimate before restoring the sensor to the master observer again.

The approach has been realised in a commercial navigation filter for ships. However, the general strategy is assumed to find useful application in other types of vehicles, for instance in mobile robots working indoors and outdoors, which brings about changing availability and accuracy of sensors: GPS only outside, wheel sensors better inside on flat floor.

Some of the implementation details are also expected to be worthwhile in other realisations: Observer design in continuous time with pole-placement, real-time implementation with incoming data triggering the filtering steps, projection of 2- or 3-dimensional residuals on a 'most likely' jump direction, fault detection by means of the Hinkley-detector.

ACKNOWLEDGEMENTS

The basis for this article was laid during my time at Raytheon Anschütz GmbH, Kiel, Germany. The permission to publish this contribution is gratefully acknowledged. I want to express my thanks to all former colleagues for cooperation, innovative spirit and helpful discussions. Especially I would like to thank Arne Carstens who programmed large parts of the position filter software, including the extension to an arbitrary number of sensors and the *reviser*-strategy.

REFERENCES

Basseville, M. and A. Benveniste, Eds. (1986). *Detection of Abrupt Changes in Signals and Systems*. Springer.

Draber, S. (1994). *Methods for Analyzing Fast Single-Channel Data and Studies on the K^+ Channel under Cs^+ Blockade and under Control Conditions*. VDI-Fortschritt-Bericht, Series 17: Number 115, 152 pages. VDI-Verlag, Düsseldorf, ISBN 3 18 311517 4.

Gertler, J. (1991). Analytical redundancy methods in fault detection and isolation. *SAFEPROCESS 91, 1st IFAC/IMACS-Symposium on Fault Detection, Supervision and Safety for Technical Processes*, Baden-Baden, Germany, 10-13 September, 1991. Preprints, Vol. 1, pp. 9-21.

Hinkley, D.V. (1971). Inference about the change-point from cumulative-sum tests. *Biometrika* 57: 1-17.

Holmström, L. (1996). The 4S-transponder - Experience with current development and international implementation. *International Symposium Information on Ships ISIS'96*, Hamburg, 29-30 October, 1996, Proceedings pp. 65-71, DGON (German Institute of Navigation), Düsseldorf.

Holzhüter, T. and R. Schultze (1996). Operating experience with a high-precision track controller for commercial ships. *Control Engineering Practice*, 4: 343-350.

Hoßfeld, B. (1996). Modern transponder technology. *International Symposium Information on Ships ISIS'96*. Hamburg, 29-30 October, 1996, Proceedings pp. 55-59, DGON (German Institute of Navigation), Düsseldorf.

Isermann, R., Ed. (1994a). *Überwachung und Fehler-diagnose* (German textbook) VDI-Verlag, Düsseldorf.

Isermann, R. (1994b). Integration of fault detection and diagnosis methods. *SAFEPROCESS 94, IFAC Symposium on Fault Detection, Supervision and Safety for Technical Processes*, Espoo, Finland, 13-16 June 1994, Postprints Vol. 2, pp. 575-590, Elsevier Science, Oxford, ISBN 0 08 042222 5

Patton, R., P.M. Frank, and R.N. Clark, Eds. (1989). *Fault Diagnosis in Dynamic Systems – Theory and Applications*. Prentice Hall, New York, London. ISBN 0 13 308263 6.

Schultze, R. (1992). Robust identification for adaptive control: The dynamic Hinkley-detector. *4th IFAC Symposium on Adaptive Systems in Control and Signal Processing*. Grenoble, 1-3 July 1992. Preprints pp. 247-252. Postprints pp. 23-28, Pergamon Press.

Schultze, R. (1996). Sensor data supervision with fault-detection applied to satellite navigation and track control. *International Symposium Information on Ships ISIS'96*. Hamburg, 29-30 October, 1996. Proceedings pp. 131-138, DGON (German Institute of Navigation), Düsseldorf.

Schultze, R. and T. Holzhüter (1997). Sway estimation for the track control of ships. *Eleventh Ship Control System Symposium*, Southampton, 14-18 April, 1997. Proceedings (P.A. Wilson, Ed.) Vol.2, pp.291-303, CMP, Southampton and Boston, ISBN 1 85312 503 2.

H∞-CONTROLER DESIGN FOR A VIBRATIONS ISOLATING PLATFORM

Jan Anthonis [1], Jan Swevers, Herman Ramon

Department of Agro-Engineering and Economics, Laboratory of Mechanical Engineering,
K.U. Leuven, Kardinaal Mercierlaan 92, 3001 Heverlee, BELGIUM
[1] Tel: +32-16-321478, fax: +32-16-321994, email: jan.anthonis@agr.kuleuven.ac.be

Abstract: In the paper, an electro-hydraulic control system for a vibrations isolating platform which must behave like a bandpass filter, is designed. A good estimation of the frequency response is obtained by doing an identification on the low and the high frequency band. The ill conditionedness of the H∞ algorithm caused by poles and zeros at the origin is solved by adding uncertainty. Accelerometer drift and vibrations of the excitation table put conflicting constraints on the controller design. Practical implementation issues which arise from numerical problems are tackled. Finally the performance and the robustness are evaluated on the set-up. *Copyright © 1998 IFAC*

Keywords: vibration, frequency domain identification, H∞ control, bandpass filter, validation

1. INTRODUCTION

In agriculture, chemical crop protection by means of liquids, which are commonly distributed by spraying machines, is among the most important field operations. Spray boom vibrations cause a non-uniform spray pattern. Vertical vibrations give rise to variations in spray deposit between 0 % and 1000 % and horizontal vibrations between 20 % and 600 % (100 % is ideal) (Langenakens et al, 1995). However, horizontal movements are much more difficult to reduce than vertical vibrations because, contrary to the latter, there is no absolute reference level, as the soil, and there is no gravity force to stabilize the boom. It has also been proved that very small horizontal vibrations with an amplitude of 30 cm can cause large fluctuations in spray distribution (Ramon, 1993). The fluctuations in spray pattern are less sensitive to vertical vibrations as long as there is overlap between the spray-cones. Well designed passive suspensions suffice to attenuate unwanted vertical vibrations of the boom. To reduce the horizontal movements, more performant active suspensions are required.

A first step in an active horizontal suspension design is the construction of an active vibrations isolating platform. The set-up to be controlled is described in a first section. Black box frequency domain identification is performed in a next section. A controller is designed based on H∞ strategy. Special attention is paid to the problem formulation, to poles and zeros on the imaginary axis and to the design of the shape functions. Finally the controller is validated on a laboratory experiment. Conclusions are drawn with respect to performance and robustness of the feedback control system.

2. DESCRIPTION OF THE SET-UP

Figure 1 shows an outline of the set-up to be identified and controlled. A table which can move in horizontal direction serves as a disturbance signal for the platform. The aim is to control the platform which can slide on the table with the aid of an electrohydraulic cylinder so that it stands absolutely still in a certain bandwidth. An LVDT type position sensor is placed on the cylinder to measure the relative displacements of the platform to the table.

Accelerations are measured with an accelerometer which is mounted at the end of the hydraulic cylinder.

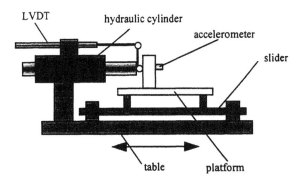

Fig. 1. Sketch of the test set-up.

3. BLACK-BOX IDENTIFICATION

3.1 Measurement set-up and excitation signals.

To avoid drifting of the hydraulic cylinder a very weak position feedback is applied. To reduce the measurement time, special compressed signals can be used as swept sines and multi sines (Schoukens et al, 1991). Another interesting feature of these signals is that no leakage will occur when the measurement period is properly chosen. However for practical reasons, swept sines are not suitable because they induce drift. In a swept sine, the frequency is increased linearly so that at one particular time instant, the cylinder is excited by one frequency. At higher frequencies the position feedback can't follow which results in drift of the actuator piston. In case of a multi sine, the phases of the individual sines are calculated by an optimisation algorithm so that the crest factor (1) is minimised (Guillaume et al, 1991).

$$\text{crest factor} = \frac{\text{peak value}}{\text{effective RMS value}} \quad (1)$$

The effective RMS value is defined as follows:

$$V_{RMSe} = V_{RMS}\sqrt{\frac{\text{energy at the frequencies of interest}}{\text{total energy of the excitation}}} \quad (2)$$

This means that at every time instant, a mixture of frequencies excites the system. The weak position feedback is now able to follow the low frequency signals so that there is no drifting.

The input voltage to the servovalve and the output accelerometer signal are sampled at 200 Hz. To avoid aliasing, hardware 8-order low pass Butterworth filters are used with a cut off frequency of 40 Hz. Input and output are filtered. In this way, the filtering has no effect on the frequency response of the system (3).

$$FRF(j\omega) = \frac{F(j\omega)Y(j\omega)}{F(j\omega)U(j\omega)} \quad (3)$$

ω frequency (rad/s)

As excitation signal, a multi sine consisting of 4096 points with a frequency range between 0.5 Hz and 40 Hz which must be send out at a frequency of 200 Hz is used. The phases are optimised with respect to the time domain i.e. the crest-factor is minimised. No optimisation is performed in the frequency domain to concentrate the energy at those frequencies where it contributes most to the knowledge of the parameters (Schoukens et al, 1991). Sixteen measurement periods are taken

3.2 Frequency domain identification.

Because time domain black-box identification methods like ARX, ARMAX and box Jenkins models didn't give good results, frequency domain identification is performed. Frequency domain identification methods try to fit the FRF as close as possible. Here the non-linear least squares estimator is used which tries to minimise the squared error between the measured FRF $G_m(k)$ and the estimate of the FRF represented by a parametric transfer function $G(\Omega_k,\theta)$ over N frequency lines (4).

$$V_{NLS} = \frac{1}{2}\sum_{k=1}^{N}\left|G_m(k) - G(\Omega_k,\theta)\right|^2 \quad (4)$$

with :

$$G(\Omega_k,\theta) = \frac{B(\Omega_k,\theta)}{A(\Omega_k,\theta)} \quad (5)$$

Ω_k frequency in continuous or discrete time

θ parameter vector

$A(\Omega_k,\theta)$, $B(\Omega_k,\theta)$ complex value at one particular frequency of the denominator respectively numerator for a given parameter vector

An iterative scheme based on Gauss-Newton and Levenberg-Marquardt searches the least squares solution. The linear least squares (6) estimate serves as an initial guess for the iteration. The inverse of the calculated FRF is chosen as a weighting to obtain better estimation results.

$$V_{LS} = \frac{1}{2}\sum_{k=1}^{N}\left|G_m(k)A(\Omega_k,\theta) - B(\Omega_k,\theta)\right|^2 \quad (6)$$

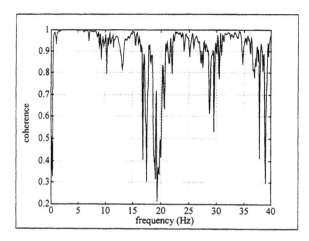

Fig. 2. Coherence of the measured data.

Contrary to time domain identification methods, frequency domain identification methods can be performed in the Z-domain (discrete time) as well as in the Laplace domain (continuous time). Here identification is performed in the Laplace domain although identification in the Z-domain is numerically favourable because the absolute value of the z-variable equals 1 so that all results from the calculation are in the same range. However, ω can take small and large values, so that small and very large figures appear in the calculation which results in bad conditioning of the iterative procedure.

First an estimation of a transfer function is performed on the FRF data up to 40 Hz. Whatever method is used, there is always a considerable mismatch between 17.5 and 21.5 Hz and this also influences the estimation outside this frequency band. Fig. 2. shows that the coherence in this region is bad. To get better results, an identification was done on the low frequency part (0.5 Hz to 17.5 Hz) and on the high frequency part (21.5 Hz to 40 Hz). The 2 identifications are combined by picking the poles and zeros which are in the frequency bands in which tried the FRF is fit. The gain is adjusted until a good fit is obtained.

Trying to estimate the structure or at least the order of the black box model is not straight forward. Fig. 1. shows only a simple sketch of the layout, in practice, the system is a complex structure. The platform itself and the attachments of the cylinder to the table and the platform are not entirely stiff. Experiments showed that the shaking table was not well designed as important eigenfrequencies of the table lie within the bandwidth of interest. Because of the weak position feed back, the behaviour of the cylinder can still be approximated by an integrator from voltage V to position p. In practise, accelerations are measured which result in a double differentiator in the numerator. Combining this previous knowledge results in the following model structure (7).

$$G(s) = \frac{s \cdot B(s)}{A(s)} \qquad (7)$$

Investigations of the FRF show that this supposition is reasonable (fig.3 , fig. 4). For the low frequency part a fifth order model with the structure in (7) seemed to give the best results. For the high frequency part, a third order model was sufficient. Fig. 3. and fig. 4. show the results of the identification procedure.

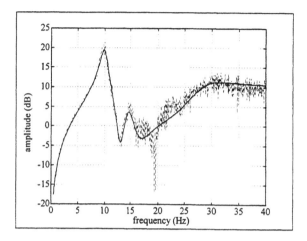

Fig. 3. Comparison between the amplitude of the transfer function estimation (solid line) and the measured FRF (dashed line)

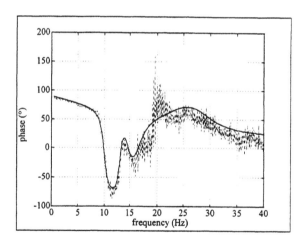

Fig. 4. Comparison between the phase of the transfer function estimation (solid line) and the measured FRF (dashed line)

4. H∞-CONTROLLER DESIGN.

4.1 Design specifications.

As stated in the introduction, the design of a controller for a vibrations isolating platform is a first step in the development process of a horizontal active suspension for a spray boom. Therefore, identical performance specifications will be taken for the platform as for the spray boom. Experiments have shown that vibrations between 0 Hz and 5 Hz give rise to the largest boom movements which must be reduced with the aid of the controller. On the other hand, the spray boom has to follow the tractor when the driver executes accelerations or decelerations or when a bend is taken. By these constraints low frequencies, e.g. 0.5 Hz must be filtered.

4.2 Control lay-out.

The low frequency part of the identified transfer function is used as a design model. Later on, the controller is evaluated on the large model. Remark that in this small design model the weak internal position loop is already incorporated.

Fig. 5. shows a block diagram of the structure to be controlled. The controller of the master loop computes the position of the piston rod of the actuator which is send to the inner loop. This forces the platform to move relatively with regard to the shaking table which has an acceleration w. The absolute acceleration y is the sum of the relative acceleration of the platform o to the table and the acceleration of the table itself. The absolute acceleration is measured by an accelerometer which is able to measure DC values. Accelerometer dynamics can be neglected since the bandwidth is 200 Hz which is far beyond the frequency range of interest. A disadvantage of DC-type accelerometers is drifting which must be taken into account during controller design. By considering accelerations instead of positions as output, an accelerometer model is avoided which results in a simpler controller. Fig. 5. shows that the transfer function from w to y is the sensitivity of the system and from the drift d to y the complementary sensitivity so that a simple sensitivity-complementary sensitivity design can be used. Remark that the block diagram simplifies the problem. The design goal is to reduce indirectly boom movements by attenuating boom accelerations. However large boom vibrations have dominant low frequencies and cause small accelerations which makes it more difficult to reduce actively.

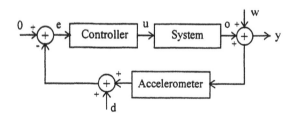

Fig. 5. Block diagram of the total feedback structure.

4.3 Controller design.

Equation (7) shows that the system contains a $j\omega$-axis zero, which in this case renders the state-space H^∞ algorithm ill conditioned. Literature proposes a bilinear pole-shifting transform (Safonov, 1987) to solve that problem. This method is not used here. A special feature of H^∞ is that it takes into account uncertainty which allows deviation of certain physical parameters. By this, the differentiator, giving rise to a $j\omega$-axis zero, can be replaced by a zero in the left-half plane. Applying the previous reasoning solves the ill conditionedness and avoids transformations which makes shaping more easy.

To meet the design specifications, the global system must react as a bandpass filter which is obtained by assigning a weight to the sensitivity. This weight (8) is designed by cascading two second order systems (Herata et al, 1995).

$$\frac{s^2 + 2\zeta_1\omega_n s + \omega_n^2}{s^2 + 2\zeta_2\omega_n s + \omega_n^2} \qquad (8)$$

Numerical values for the parameters of the two transfer functions are listed in table 1. Construction of the weights in this way has the advantage that the D-matrix is not zero which is necessary for the H^∞ algorithm.

Table 1 Numerical values for the parameters of the sensitivity weighting functions

	transfer function 1	transfer function 2
ζ_1	10	10
ζ_2	0.1	0.1
ω_n	$0.6 \times 2 \times \pi$	$4 \times 2 \times \pi$

However this weight is not sufficient because vibrations of 0.1 Hz are suppressed which is not allowable. The feedback system must be insensitive to accelerometer drift. Since the sum of the sensitivity and the complementary sensitivity equals one at every frequency, the sensitivity can be influenced by the complementary sensitivity. Therefore a weight function should be designed so that the complementary sensitivity is suppressed at low frequencies. By this the sensitivity will pass low frequencies and accelerometer drift is removed. A simple first order lead filter is sufficient as a weight on the complementary sensitivity according to the control lay-out as shown in section 4.2 because it will appear as a lag in the final complementary sensitivity $C(s)$ (9).

$$\frac{y}{d} = C(s) = \frac{s}{s + a} C_{p1}(s) \qquad (9)$$

with :

s	Laplace operator
a	zero
$C_{p1}(s)$	other part of the complementary sensitivity

However this lay-out is a simplification of reality because instead of accelerations, boom movements must be controlled which represents the double integral of accelerations so that finally the drift is integrated one time (10).

$$\frac{y}{s^2.d} = \frac{C(s)}{s^2} = \frac{1}{s(s+a)}C_{p1}(s) \qquad (10)$$

A second order lead filter on the complementary sensitivity results in a very poor performance. Therefor, only a first order lead filter is used as a design weight (11).

$$\frac{s + 2\pi 0.6}{s} \qquad (11) \qquad\qquad \frac{s+1}{s} \qquad (12)$$

In a practical implementation, accelerometer signals will pass through a first order high pass filter (12) which is equivalent by using a piezo accelerometer which is unable to measure DC accelerations. It can be theoretically verified that the high pass filter has no influence on the stability of the system.

The weighting functions are plotted in fig. 6..

The controller is designed with the Robust Control Toolbox of Matlab by using the command hinfopt (Chiang et al, 1992) and evaluated on the large identified model described in section 3. The best results were obtained by reducing the weight of the sensitivity with a factor 0.1. Inspection of the eigenvalues and verification of fig. 7. and fig. 8. show that the system is stable and that the performance specifications are met.

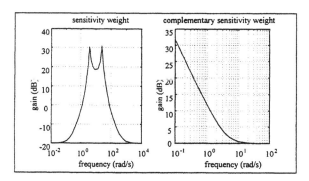

Fig. 6. Weights put on sensitivity and complementary sensitivity..

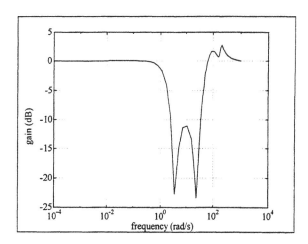

Fig. 7. Sensitivity of the evaluation model with the controller.

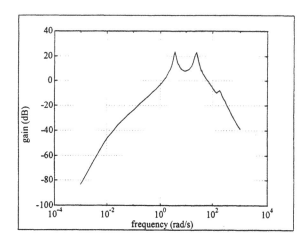

Fig. 8. Complementary sensitivity of the evaluation model with the controller.

5. CONTROLLER IMPLEMENTATION

A 10 Th order controller is obtained. No order reduction is performed, because the speed of the transputer is sufficiently high. Implementation is done by first discretising with the Tustin algorithm after prewarping the poles and the zeros. The sampling frequency is chosen at 1000 Hz so that anti-aliasing filters are avoided (bandwidth accelerometers 200 Hz).

The large dimensions of the controller lead to numerical problems. Combination of small and large coefficients make the controller ill-conditioned. Splitting the controller in small first and second order sub-systems solves the problem. Complex conjugate poles and zeros give rise to a second order sub-system. The first order sub systems are built with real poles and zeros. Well conditionedness of the controller is obtained by combining poles and zeros in the same range of magnitude. A second advantage is that errors on the coefficients have only an influence on the sub-system and are not propagated through the whole filter.

6. TESTING OF THE CONTROLLER ON THE SET-UP

Figure 9 shows the performance of the controller. A random noise acceleration with a bandwidth between 0.3 Hz and 10 Hz is applied. Even with two persons (± 150 kg) on the platform (± 30 kg), the system remains stable which means that the controller is robust to large mass variations.

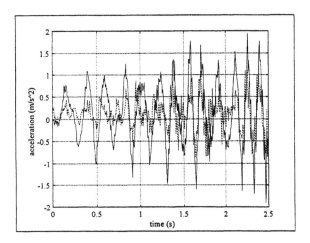

Fig. 9. Performance of the controller (solid line : applied acceleration, dashed line : acceleration measured on the platform).

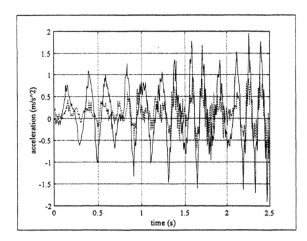

Fig. 10. Performance of the controller with 2 people on the platform (solid line : applied acceleration, dashed line : acceleration measured on the platform).

7. CONCLUSIONS

Optimized excitation signals used together with the non-linear least squares black box frequency domain identification method result in a nice estimation of the frequency responsee. By leaving out low-

coherency parts of the FRF, even a better model is obtained.

The ill conditionedness of the H^∞ algorithm is solved by putting uncertainty on the poles and zeros on the imaginary axis.

A high pass filter, which filters the accelerometer signals, is used after controller design in order to get reasonable performance while solving the conflicting constraints put by accelerometer drift and the shaking table vibrations.

Numerical problems encountered by the implementation of the controller are avoided by splitting it in smaller sub-systems.

Practical experiments show the robustness of the controller against mass variations.

ACKNOWLEDGEMENT

This paper was elaborated within the European Craft Project FA-S2-9009.

REFERENCES

Chiang, R., Y., Safonov, M., G. (1992), *Robust control toolbox user 's guide*, The MathWorks, Inc.

Guillaume, P., Schoukens J., Pintelon R., Kollár I. (1991), *Crest-factor minimization using nonlinear Chebyshev approximation methods*, IEEE Transactions of instrumentation and measurement, Vol. 40, No 6, pp. 982-989

Hirata, T., Koizumi, S., Takahashi, R. (1995), *H^∞ control of a railroad vehicle active suspension*, Automatica Vol. 31, No. 1, pp. 13-14

Langenakens, J. J., Ramon, H., De Baerdemaeker J. (1995), *A model for measuring the effect of tire pressure and driving speed on the horizontal sprayer boom movements and spray patterns*, Transactions of the ASAE, pp. 65-72

Ramon, H. (1993), *A design procedure for modern control algorithms on agricultural machinery applied to active vibration control of a spray boom*, Ph.D. thesis, KULeuven

Safonov, M. G. (1987), '*Imaginary-axis zeros in multivariable H^∞ optimal control*', modeling robustness and sensitivity reduction in control systems, Springer Verlag, New York

Schoukens, J., Pintelon, R. (1991), *Identification of linear systems, A practical guideline to accurate modeling*, Pergamon Press

A HIGHLY PRECISE TRANSLATION SYSTEM FOR ASTRONOMICAL APPLICATIONS

Roland Strietzel, Dietmar Ratzsch

Daimler-Benz Aerospace
Jena-Optronik GmbH
Prüssingstr. 41
07745 Jena, Germany
Tel.: +49 3641 200 177
Fax.: +49 3641 200 220
E-Mail: Roland.Strietzel@djo-jena.de

Abstract: For the processing of star light, highly precise translation and positioning systems are used. They are special robots which carry mirrors, telescopes or retroreflectors. Over a length of more than 50 meters it shall be guaranteed a positioning accuracy of the retroreflector of about 100 nm. A necessary compensation of air vibrations in the light propagation needs a relatively high positioning velocity, expressed by a transient time of 0.25 ms for a point-to-point positioning over 5 µm. The solution of this problem is possible with a three-level translation system with coordinated drives. The three motion levels differ in the operational range, resolution and dynamics. On the base of the model of the mechanical system a suitable transfer function of the open-loop system is derived, which yields a smooth transient behaviour of the closed loop system and a vanishing steady-state control deviation for step command inputs. By means of state feedback the drives are so coordinated, that the desired dynamic behaviour of the whole system under existing conditions of the operational ranges in the three levels can be reached. Simulation results show the solvability of this positioning problem and the influences of control inputs and disturbances. The obtained results are discussed. *Copyright © 1998 IFAC*

Key words: translation system, co-ordination, state feedback, positioning system, crossover frequency, multilevel control

1. INTRODUCTION

For some optical investigations concerning with processing of star light very high accuracy demands of positioning tasks exist (Holm, 1986). Here a special robot is described which carries a retroreflector. Over a length of more than 50 meters, it shall be guaranteed a positioning accuracy of the retroreflector of up to about 100 nm in translatory motion. The retroreflector can be a relatively simple equipment (cat's eye) or a heavy telescope with a positionable mirror in the focal plane. The various masses of different solutions should be considered.

To ensure small lateral deviations, the robot moves on rails. A possible solution is sketched in Fig. 1.

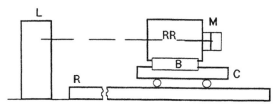

Fig. 1. Translation system with bearings B, carriage C, laser range finder L, positionable mirror M, rails R and retroreflector system RR.

Fig. 1 shows, that the resulting position of the mirror being measurable with the laser metrology system L depends on the position of the carriage C, the position of the retroreflector RR in relation to the carriage C and the position of the mirror M in relation to RR. The problem is, to perform the extremely high positioning accuracy of about 100 nm over the translation distance of about 50 m at existent dynamic demands by means of a suitable coordination of three drives with their own available dynamic properties. That means, that the level 3 has the highest velocity (bandwidth) and the smallest positioning range and the level 1 can move relatively slowly but with a high positioning range.

2. MODELLING

The design of the control algorithms requires a mathematical model of the system to be controlled. Fig. 2 demonstrates the action of drives to the masses as a base for formulating a mathematical model.

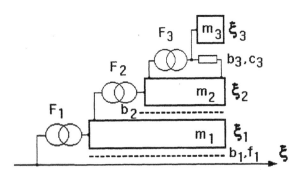

Fig. 2. Schematic representation of masses and drives b_1, b_2, b_3 damping coefficients in the different levels, c_3 spring coefficient, f_1 friction force in the first level, F_1, F_2, F_3 driving forces, m_1, m_2, m_3 masses of the carriage, the cat's eye support (or telescope) and the retroreflector mirror, respectively, ξ_1, ξ_2, ξ_3 the positions of masses m_1, m_2, and m_3, respectively on the common coordinate axis ξ.

Manipulating variables are the forces F_1, F_2, F_3, control variable is the variable ξ_3.
Balancing the forces of inertia, damping, suspension and drives, the equations

$$m_3 \xi_3'' + b_3 (\xi_3' - \xi_2') + c (\xi_3 - \xi_2) = F_3, \quad (2.1)$$

$$m_2 \xi_2'' + b_3 (\xi_2' - \xi_3') + b_2 (\xi_2' - \xi_1') + c (\xi_2 - \xi_3) = F_2 - F_3, \quad (2.2)$$

$$m_1 \xi_1'' + b_2 (\xi_1' - \xi_2') + b_1 \xi_1' = F_1 - F_2, \quad (2.3)$$

with the damping coefficients b_1, b_2, b_3, the spring constant c and the driving forces F_1, F_2, F_3 can be formulated. Dry friction is neglected because of the application of air bearings.

With the substitution

$$\begin{aligned} x_3 &= \xi_3 - \xi_2, \\ x_2 &= \xi_2 - \xi_1, \\ x_1 &= \xi_1, \end{aligned} \quad (2.4)$$

or

$$\begin{aligned} \xi_3 &= x_1 + x_2 + x_3 = x \\ \xi_2 &= x_1 + x_2, \\ \xi_1 &= x_1, \end{aligned} \quad (2.5)$$

one obtains

$$m_3 (x_1'' + x_2'' + x_3'') + b_3 x_3' + c x_3 = F_3, \quad (2.6)$$

$$m_2 (x_1'' + x_2'') - b_3 x_3' + b_2 x_2' - c x_3 = F_2 - F_3 \quad (2.7)$$

$$m_1 x_1'' - b_2 x_2' + b_1 x_1' = F_1 - F_2. \quad (2.8)$$

The mass m_3 is very small in comparison with the masses m_1 and m_2, therefore $m_3 = 0$ and b_3 will only describe the relatively high damping of the spring c. The equations (2.6) - (2.8) are simplified to

$$b_3 x_3' + c x_3 = F_3, \quad (2.9)$$

$$m_2 (x_1'' + x_2'') + b_2 x_2' = F_2, \quad (2.10)$$

$$m_1 x_1'' - b_2 x_2' + b_1 x_1' = F_1 - F_2. \quad (2.11)$$

The introduction of state variables consisting of positions and velocities, x_1, x_2, x_3, $x_4 = x_1'$, $x_5 = x_2'$, gives the equations

$$\begin{aligned} x_1' &= x_4 \\ x_2' &= x_5 \\ b_3 x_3' + c x_3 &= F_3, \\ m_1 x_4' - b_2 x_5 + b_1 x_4 &= F_1 - F_2, \\ m_2 (x_4' + x_5') + b_2 x_5 &= F_2, \end{aligned} \quad (2.12)$$

which are transformed into the usual form

$$\begin{aligned} x_1' &= x_4 \\ x_2' &= x_5 \\ x_3' &= - c/b_3 x_3 + F_3 / b_3, \\ x_4' &= - b_1/m_1 x_4 + b_2/m_1 x_5 + F_1/m_1 - F_2/m_1, \\ x_5' &= b_1/m_1 x_4 - b_2 (1/m_1 + 1/m_2) x_5 - F_1/m_1 + \\ &\quad + F_2 (1/m_1 + 1/m_2). \end{aligned} \quad (2.13)$$

or abbreviated into

$$\begin{aligned} \underline{x}' &= A \underline{x} + B \underline{u} \\ x &= \underline{c}^T \underline{x} \end{aligned} \quad (2.14)$$

$$\underline{x} = (x_1, x_2, x_3, x_4, x_5)^T, \quad \underline{u} = (F_1, F_2, F_3)^T,$$

$$A = \begin{bmatrix} 0 & 0 & 0 & 1 & 0 \\ 0 & 0 & 0 & 0 & 1 \\ 0 & 0 & -c/b_3 & 0 & 0 \\ 0 & 0 & 0 & -b_1/m_1 & b_2/m_1 \\ 0 & 0 & 0 & b_1/m_1 & -b_2 (1/m_1 + 1/m_2) \end{bmatrix}$$

$$(2.15)$$

$$B = \begin{bmatrix} 0 & 0 & 0 \\ 0 & 0 & 0 \\ 0 & 0 & 1/b_3 \\ 1/m_1 & -1/m_1 & 0 \\ -1/m_1 & 1/m_1 + 1/m_2 & 0 \end{bmatrix}$$

(2.16)

$$\underline{c}^T = (1 \ 1 \ 1 \ 0 \ 0).$$ (2.17)

A characteristic property of the model is the interaction of the forces F_1 and F_2 to the positions x_1, x_2, shown in Fig. 3. This figure follows from the model (2.14 ... 2.17).
The values of parameters are

$g_{d1} = 1/m_1$,
$g_{d2} = 1/m_1$,
$g_{d3} = 1/m_1$,
$g_{d4} = 1/m_1 + 1/m_2$,
$g_{d5} = g_{d6} = g_{d7} = g_{d8} = 0.$ (2.18)

Vanishing damping parameters is supposed because of the foreseen application of air bearings. Small damping values can be compensated by the state feedback of the controller.

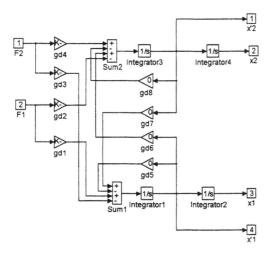

Fig. 3. Simulink representation of the mechanical drives

3. CONTROL ALGORITHM

The control algorithm has to perform the following requirements:

- The driving forces shall be calculated from the control error e so, that a fast and smooth positioning the controlled variable x results.
- The steady-state control error shall be zero.
- By prefiltering of the reference variable w, the driving forces shall be minimised.
- The time lag of the drives can not be neglected. The drives for F_1, F_2 have the same time behaviour.

On the base of this requirements one obtains:

- The transfer function between e and x shall be of the form $G_o(s) = \omega_o/s$ with the crossover frequency ω_o determining the closed loop behaviour.
- By means of a decoupling network the influences of the drive F_1 to the position x_2 and the drive F_2 to the position x_1 are suppressed.
- Thus the internal control variables v_1, v_2, v_3 determine the positions x_1, x_2, x_3, respectively to obtain the desired position x.
- The transfer functions $X_i(s)/V_i(s)$, $i = 1, ... 3$ between the control variables and the belonging positions can be designed under considering the electric dynamic behaviour of the drives, if necessary.
- The state feedback approach is used for the controller design.

The transfer function

$$G_o(s) = \omega_o/s$$ (3.1)

determines the closed-loop behaviour. This open-loop transfer function yields a reference transfer function

$$F(s) = G_o(s) / [1 + G_o(s)] = 1/(1 + s/\omega_o) = 1/(1 + sT)$$ (3.2)

with the bandwidth

$$f = 1/(2\pi T) = \omega_o /(2\pi).$$ (3.3)

Example: f = 300 Hz, T = 0.53 ms, ω_o = 1884 s⁻¹.

The transfer function (3.1) must be realised in three frequency ranges agreeing with the three control inputs.
The compensation of the frequency response is realised according to the relation

$$T_a/(1+sT_a) + T_b/(1+sT_a)(1+sT_b) + 1/[s(1+sT_a)(1+sT_b)] = 1/s.$$ (3.4)

This means, that for $T_b > T_a$ the transfer function $G_o(s)$ is split up into three channels for the higher, the medium and the lower frequencies, respectively.
A realisation of this idea is given by

$$G_o(s) = G_1(s) + G_2(s) + G_3(s) = \omega_o/s$$ (3.5)

with

$$G_1(s) = k/[sT_3 (1+ sT_2)(1+ sT_3)],$$ (3.6)
$$G_2(s) = kT_2 /[T_3 (1+ sT_2)(1+ sT_3)],$$ (3.7)
$$G_3(s) = k/(1+ sT_3)$$ (3.8)

and the value $k = \omega_o T_3$ representing the distance between the cut-off frequency of the drive x_3 and the cross-over frequency being responsible for the closed-loop dynamics. The transfer functions $G_2(s)$

and $G_3(s)$ have the cut-off frequencies $1/T_2$ and $1/T_3$, $T_2 \gg T_3$, respectively. $G_1(s)$ is an integrating transfer function with an integration constant $T_3/k = 1/\omega_o$.

Using the known transfer functions of the drives, the controller shall be designed to perform the desired transfer functions $G_1(s)$, $G_2(s)$, $G_3(s)$.

3.1 Controller 1

The controller 1 is a P-type controller,

$$X_3(s) = E(s) R /(c + sb_3) = G_3(s) E(s),$$
$$T_3 = b_3/c, \; R = c\, k. \tag{3.9}$$

Its transfer constant depends on the amplification k and the spring constant c.

3.2 Controller 2

Controller 2 calculates the motor drives. The control matrix B (2.16) describes the interaction of the driving forces to the positions. It yields the matrix D of the decoupling network for the mechanical drives (Föllinger, 1990),

$$\underline{w} = \begin{bmatrix} 1/m_1 & -1/m_1 \\ -1/m_1 & 1/m_1 + 1/m_2 \end{bmatrix} D\, \underline{v} = \omega_o^2 I \underline{v}, \tag{3.10}$$

$$D = \omega_o^2 \begin{bmatrix} m_1 + m_2 & m_2 \\ m_2 & m_2 \end{bmatrix} \tag{3.11}$$

with the vector $\underline{v} = (v_1 \; v_2)^T$ of the new inputs for the motor drives, the normalising constant ω_o^2, the unity matrix I and the vector $\underline{w}$ of internal values.

In the next design step the translation system is supposed to be without damping, $b_1 = b_2 = 0$. The resulting transfer functions

$$X_1(s)/V_1(s) = \omega_o^2 /s^2 \tag{3.12}$$
$$X_2(s)/V_2(s) = \omega_o^2 /s^2 \tag{3.13}$$

can be transformed into the desired transfer function $G_1(s)$ and $G_2(s)$ by state feedback according to Fig. 4. With this results one obtains the controller 2 according to Fig. 5 having the following parameters:

$$g_{21} = \omega_o^2 (m_1 + m_2),$$
$$g_{22} = \omega_o^2 m_2,$$
$$g_{23} = \omega_o^2 m_2,$$
$$g_{24} = \omega_o^2 m_2,$$
$$g_{25} = 1/(\omega_o^2 T_2 T_3),$$
$$g_{26} = 1/(\omega_o^2 T_2),$$
$$g_{27} = 1/(\omega_o^2 T_2 T_3),$$
$$g_{28} = (T_2 + T_3)/(\omega_o^2 T_2 T_3),$$
$$g_{20} = k/(\omega_o^2 T_2^2). \tag{3.14}$$

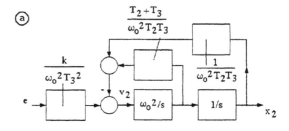

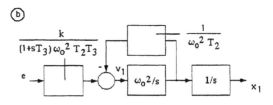

Fig. 4. Resulting transfer functions $G_1(s)$ (b) and $G_2(s)$ (a) realised by state feedback of the decoupled driver transfer functions.

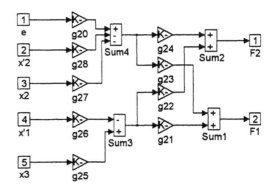

Fig. 5. Structure of the Controller 2

This linear controller consists of the decoupling part ($g_{21} \ldots g_{24}$) and the state feedback part ($g_{25} \ldots g_{28}$). Both parts can be combined, if necessary. Under the condition of Fig. 4 the controller parameters depend only on the crossover frequency ω_o and the time constants T_2, T_3.

With the P-type controller (1) and the state feedback controller (2) the control system of Fig. 6 results.

Fig. 6 is extended by an input filter and a disturbance input, describing the output disturbance of the three output channels. The input filter allows to influence the time behaviour of the driving forces F_1, F_2. With a more extensive filter than in Fig. 6, an approximately time-optimal point-to-point positioning can be realised, when a step input w is supposed.

4. SIMULATION RESULTS

The following simulations are performed with the practical data $m_1 = 150$ kg, $m_2 = 120$ kg, $k = 5$, $\omega_o = 2000$ s^{-1}, $T_3 = 0.0025$ s.

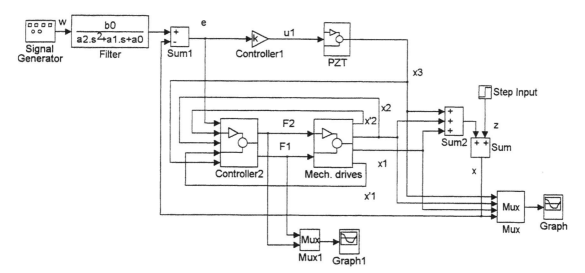

Fig. 6. Complete control system of the translation robot (e control deviation, w reference input, z output disturbance, PZT piezoelectric transducer.

The figures show the time functions of the single translations x_1, x_2, x_3 and their superposition x. In one case the time behaviour of the forces F_1 and F_2 is also represented. The values of x and F are normalised.

To obtain a smooth and precise motion for a wide field of input signals, the controller parameters must be selected so, that a linear operation of the system can be achieved.

The relation between the time constants T_2, T_3 determines the actuating range of the PZT, the support and the carriage for given control loop crossover frequency ω_o.

Fig. 7 shows, that

- the relatively high crossover frequency ω_o gives fast transient behaviours of x(t)

- the obtained aperiodic transient function x(t) follows from the used function of G_o(s) according to Equ. (3.5)

- the coordinated superposition of the motions x_1, x_2, x_3 of different time behaviour guarantees the desired course of x(t).

- the actions of x_2, x_3 go back to zero, the steady-state control deviation is compensated by the integrating part of x_1(t).

Resuming the results of the simulation experiment of Fig. 8, one obtains

- input filtering (a_2 = 0.000025, a_1 = 0.0085, a_o = 1, b_o = 1) reduces the operation range and the velocity of the x_3 drive

- a reduction of the velocity of the point-to-point positioning helps to reduce the accelerating and braking forces of the x_1 and the x_2 drive

- by a suitable design of the input filter, the forces can approach the time-optimal control

- the courses of forces are similar, because of their supporting interaction.

From Fig. 9 results

- the time constant T_2 determines the division of the frequency range of the drives

- the behaviour of the x_3 drive is not affected by time constant T_2

- smaller values of T_2 increase the bandwidth of the x_1 channel.

The possibilities of MATLAB/Simulink create excellent conditions for investigating the dynamic behaviour, if mathematical models are available.

5. CONCLUSIONS

Cascaded drives allow a fast, precise and large scale translation. Therefore this method is suitable for optical and astronomical application. This translation system is designed for precise point-to-point positioning. For obtaining a smooth and shockless motion of the usually heavy equipment, a well designed linear control system is necessary. This requires a suitable selection of the control parameters of the three motion levels, PZT (x_3), drive x_2, drive x_1 for the given reference inputs and the possible disturbances, regarding the available actuating ranges of the three drives. The main task of the input filter is to limit the driving forces F_1, F_2, determining the maximum acceleration.

By means of the MATLAB/Simulink tool, suitable controller parameter can be elaborated.

The authors thank Dr. J. Reiche (Jena-Optronik) for interesting and stimulating discussions.

405

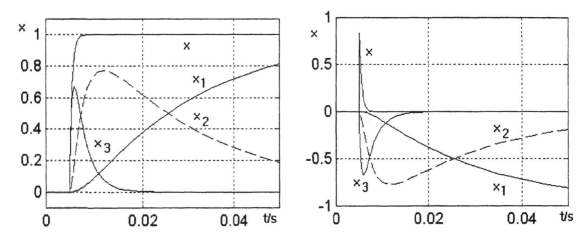

Fig. 7. Transient behaviour caused by a reference input step and an output disturbance step, no input filter, $T_2 = 0.025$ s.

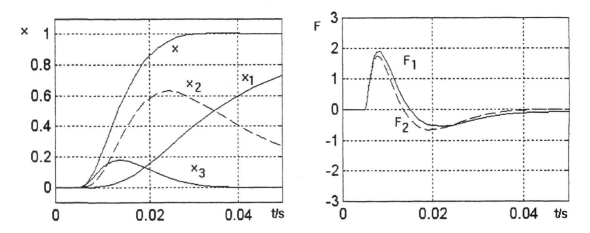

Fig . 8. Transient behaviour with filter at the reference input, representation of x(t) and of the driving forces $F_1(t)$, $F_2(t)$, $T_2 = 0.025$ s.

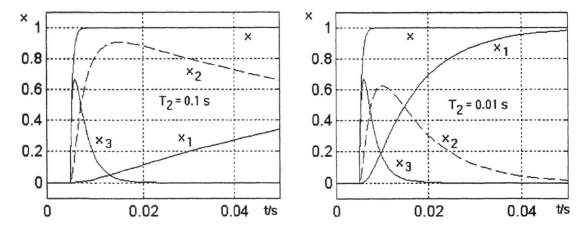

Fig . 9. Transient behaviour without filter at different time constants $T_2 = 0.1$ s, $T_2 = 0.01$ s, respectively.

REFERENCES

Föllinger, O. (1990). *Regelungstechnik*. Hüthing, Heidelberg.

Holm, R. W. (1984). *A high-speed, high-resulution optical delay line for stellar interferometry applications*. Massachusetts Institute of Technology.

USING THE RS LANGUAGE
TO CONTROL AUTONOMOUS VEHICLES

Gustavo Vasconcelos Arnold [Ψ]
Simão Sirineo Toscani [Ω]

[Ψ] *Computer Science Department*
Caxias do Sul University
Caixa Postal 1352, CEP 95001-970. Caxias do Sul, RS, Brazil
(gvarnold@ucs.tche.br)
[Ω] *Institute of Computer Science*
Federal University of Rio Grande do Sul
Caixa Postal 15064, CEP 91501-970. Porto Alegre, RS, Brazil
(simao@inf.ufrgs.br)

Abstract: This paper presents the RS language, and its application to the control of vehicles. The RS language is a useful tool for the development of reactive and real time systems. As a modeling technique, it was used the subsumption architecture, that was initially created to control mobile robots, that has autonomy, robustness and extensibility as its main features. The purpose of this paper is to present the RS language, and to show the use of this language as an implementation tool, to control autonomous vehicles. *Copyright © 1998 IFAC*

Keywords: Computer systems, Deterministic systems, Finite automata, Intelligent control, Programming languages, Real-time systems, Robot programming.

1. INTRODUCTION

One of the main problems of the traditional approach to reactive systems is the need to use a complete representation of the environment where the system will be contained. This need can cause some adversities, since the input data arriving from the environment can be unpredictable. When an unexpected signal arrives, the system can enter into an inconsistent state, since it was not "prepared" to receive this information. The adoption of the subsumption architecture modeling technique avoids the need to represent the environment. Instead of this, the system is decomposed into several simple activities, which are based only in "perception → action", using Finite State Machines. This large set of activities is able to cope with any external event, resulting in an excellent model for a reactive system.

A previous study has been done concerning the use of the subsumption architecture together with the reactive synchronous RS language (Arnold, 1996). The RS language (Toscani, 1993) is a data flow language based on Petri Nets, that uses Finite State Machines at execution time. The subsumption architecture showed its good characteristics to model a reactive system, mainly autonomy, robustness and extensibility. On the other hand, the RS language showed good potentiality for the development of reactive programs and real time controllers. As the subsumption architecture properties are important for many kind of applications, it was initially proposed to adapt the RS language in order to incorporate the control mechanisms found in this architecture. During the planning of the extensions, however, it was realized that the intended modifications were not really necessary. The same effect could be obtained through external mechanisms used at execution time.

A system to control vehicles can be classified as reactive because it has the basic feature of this kind of system. Reactive programs must be able to continuously answer to events originated from an external environment, in an unknown order. Others examples of reactive systems are: robot controllers, industrial process controllers, video-games, communication protocols and user-computer interfaces.

Usually, a reactive program is divided in three layers (Berry and Gonthier, 1992):

• an interface with the environment, that is in charge of input reception and output production. It transforms external physical events into internal logical ones and conversely;

• a reactive kernel, which contains the logic of the system. It handles logical inputs and generates logical outputs;

• a data manipulation layer that executes the trivial computations requested by the reactive kernel.

2. RS LANGUAGE

The RS language (Toscani, 1993) is a synchronous language intended to the programming of reactive kernels that must react immediately to stimulus coming from an external environment. This kernel corresponds to the central and more difficult part of a reactive system. However, the RS language is not self-sufficient because the interface and data manipulation layers must be specified in some host language.

The RS language provides a suitable notation to represent reactive behaviors because a program is an almost direct specification of the internal transformations and signal emissions that happen for each external stimulus.

As a result of the synchrony hypothesis, an RS program can be seen as if its internal operations were executed by an infinitely fast processor. This means that each reaction is accomplished in an infinitesimal time, making the output signals synchronous with the input signals. The synchrony hypothesis corresponds to consider ideal systems, that react instantaneously to each external stimulus, with a transformation of the internal state and with an emission of output signals, like a Mealy machine (Hopcroft, 1979). As the operations are executed in an infinitesimal time, the computer satisfies automatically the external timing constraints and it is not necessary to worry about the

occurrence of other external signals during the execution of a reaction. A consequence of this is the assumption of simple external stimulus, constituted by only one signal by reaction. This assumption, is the second basic hypothesis of RS.

2.1 Structure and functioning

An RS program works with classical variables that are shared in a module level, and with signals that are used for external communication and for internal synchronization.

The signals are partitioned in: *input signals*, the only that originate reactions; *output signals*, used to communicate results to the environment; and *internal signals*, used for synchronization or internal communication. Besides these, there are special input signals, called *sensors*, whose values are always accessible in any reaction.

An RS program is formed by a set of *modules*, which are composed by a set of *boxes*, which are constituted by a set of *reaction rules*. The modules and boxes allow the structuration of the program but they are not really necessary; in fact, any program can be specified by only a single set of rules. A reaction rule has the form: "trigger condition" → "action", where the trigger condition is a not empty list of internal and/or input signal, having at most one input signal. A given set of open (turned on) signals triggers simultaneously off all rules whose trigger conditions are contained in this set, closing (turning off) the signals specified in these conditions. In spite of having this parallel execution of rules, there is no indeterminism in the reaction; this happen because the RS language forces the correct sharing of variables and signals.

The trigger conditions allow the programmer to map the states of the external environment into internal states of the program. Each one of these internal states defines a set of rules which can be triggered by an external stimulus. The internal variables allow the system to keep information from a reaction to another. There are conditional rules which allow the system to react according to conditions that are evaluated at run time.

One interesting construct of the language is the mechanism to handle exceptions, that is, conditions that cause abrupt changes in the state of the reactive system. These conditions can be raised internally or can be signaled from the external environment. Whenever an exception condition occurs, the program suffers an internal change characterized as follows:

• For each module, all internal signals are turned off and all rule boxes are deactivated;

• Each module involved in the exception goes to a new state which is defined by the exception rule corresponding to the signaled condition.

3. SUBSUMPTION ARCHITECTURE

The subsumption architecture was proposed, initially, to build autonomous mobile robots which operate over long periods of time, in dynamic worlds. This architecture does not use any internal representation of the real world. The robots that were built by the traditional Artificial Intelligence approach used to operate in a static environment, or in a specially constructed environment; the perception of this "real world" was used to construct a 2D or 3D internal model which was then used to produce an action plan to achieve the robot goal. Long processing times were lost in the perception and construction of the world models, while just a little computation time was used in planning and acting. As a result, the operation of these robots was very slow.

By the year 1984, some researchers worried about this intelligence organizing problem, proposed that the intelligence should be reactive to the dynamic aspects of the environment, and that it should be able to generate a robust behavior to control the sensors uncertainty, like in an imprevisible environment (Brooks, 1991). Then, a new approach to build autonomous robots was proposed, called subsumption architecture (Brooks, 1986). This architecture was largely used in academic laboratories and also by NASA, in the development of robots to explore Mars (Brooks, 1991).

The subsumption architecture can be used in systems that have the following features (Bryson, 1992): that are required to operate in real time; that needs to pursue multiple goals simultaneously; that exploits a parallel architecture; that needs to be aware of real time events external to it; that needs to process sensor data; when there are conditions involving details not important to the system's central goals that need to be met and maintained in order to pursue those goals.

The main features of the subsumption architecture are (Brooks, 1986, 1991):

• *Vertical decomposition by activities,* which constitute an incremental path from very simple systems to complex autonomous intelligent systems. At each step, it is only necessary to build one module, and to interface it to the existing ones. In this way, an incremental description of the intelligence is achieved (see figure 1). It is important to say that each activity contains the same traditional architecture: perception / planning and decision / action.

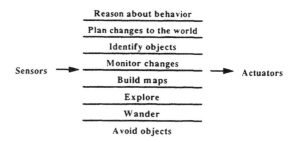

Fig. 1. Vertical decomposition.

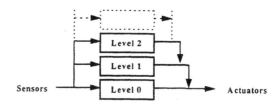

Fig. 2. High level subsumption.

• *Behavior based paradigm,* which explains intelligence in terms of small, simple and coherent behaviors (group of rules) or levels of competence. There are no shared data structures across behaviors, all the communication is by explicit message passing. It is assumed that the behaviors run in parallel and asynchronously. The system could be like the one seen in the figure 2.

• *The world should be its own best model.* Storing models of the world is dangerous in dynamic, unpredictable environments because representations might be incorrect or outdated.

• *Environment.* The subsumption architecture was designed for real and dynamic environments, where they face real-world sensor and actuator problems. Agent modules may attend various aspects of their environment with different levels of urgency because higher-level modules may suppress and inhibit lower level ones from doing their default actions.

• *Reactivity.* The control system could both (1) be naturally opportunistic if fortuitous circumstances are presented, and (2) easily respond to changed circumstances, such as some other object approaching it on a collision course (Brooks, 1991).

• *Representation.* There can be representations which are partial models of the world, but individual layers extract only those aspects of the world which they find relevant.

• *Multiple goals and multiple sensors.* Individual layers can be working on individual goals concurrently. The suppression mechanism then mediates the actions that are taken.

• *Robustness.* This feature can be achieved by developing layers from the lower, more reactive, to the higher ones, incrementally and by debugging each one thoroughly before proceeding with the next one in the hierarchy.

- *Extensibility.* Once a behavior has been built, new ones may be added by guaranteeing that the appropriate type of subsumption occurs, without impair the existing system.

- *Independence in engineering.* Each layer should be engineered separately, then tested and debugged until flawless before proceeding.

- *Test environments.* There should be no simplified test environments. Subsumption applications should cope for themselves with all the noise of inaccurate sensors and all the unpredictability of the real world.

4. THE RS LANGUAGE AND THE SUBSUMPTION ARCHITECTURE

The RS language was extended, and as a result, it is possible to use the subsumption architecture as a modeling technique for its programs. This approach will allow the development of reliable multilevel autonomous systems using its programming facilities.

The main components of the subsumption architecture are implemented in the following way: RS modules represents the levels of competence; the rule boxes represents the behaviors (group of rules). In both the RS language and the subsumption architecture, the communication is only implemented by explicit message passing, and there is no data sharing between different modules. An interesting remark is that the subsumption architecture assumes that the behaviors run in parallel and asynchronously. On the other hand, we know that the components of an RS program are synchronous; this means that a rule is executed instantaneously, and that the output signals are synchronous with the input signals. However, there is no conflict here, because in the RS language, the rules will be synchronous, while the execution of the behaviors will be asynchronous.

4.1 Extension to the RS language

In order to facilitate the use of the subsumption architecture concepts in the RS language, an extension to this language was proposed (Arnold, 1996). In this extension, only the control mechanisms of the subsumption architecture, the inhibitors and the suppressors, were included, each of them corresponding to different modules:

- an inhibitor has one input, that contains two signals: one is the dominant one, and the other, the one that can be inhibited. In this module, if the

dominant signal arrives, the other signal is inhibited, which means that there will be no output from this module. If no dominant signal arrives, the other signal is emitted.

```
module inhibitor:
[ input: [ input_signals(Dominant, Other) ],
  output: [only(X) ],
  t_signal: [ ], p_signal: [ ], var: [ ],
  initially: [ activate(rules) ],
  on exception: [ ],
  input_signals(Dominant, Other) ===> case
        [Dominant != 0 ──> [],
        else          ──> [emit(only(Other))] ]
].
```

- similarly, a suppressor has one input, that contains two signals: one is the dominant one, and the other, the one that can be suppressed. In this module, if the dominant signal arrives, the other signal is inhibited and its value is substituted for the dominant one, producing a new output. If not, the other signal is emitted.

```
module supressor:
[ input : [input_signals (Dominant, Other) ],
  output: [ only(X) ],
  t_signal: [ ], p_signal: [ ], var: [ ],
  initially: [ activate(rules) ],
  on exception: [ ],
  input_signals(Dominant, Other) ===> case
        [Dominant != 0 ──> [emit(only(Dominant))],
        else          ──> [emit(only(Other))] ]
].
```

With this extension, the programmer does not need to consider (in the implementation of a behavior) which signals will be inhibited or suppressed by other ones, making the behaviors more modular. This is one of the advantages of using the subsumption architecture, where the control mechanisms are implemented during the connections between the behaviors, and not at programming time.

4.2 An RS program structured according the subsumption architecture

The following example is an RS program to control a vehicle, which must wander avoiding obstacles. The vehicle has 3 infra-red sensors, two frontals and one at the back, and two motors, one to control the direction (left, right, straight), and another to control the movement (front, behind, stopped).

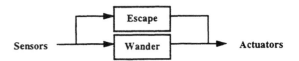

Fig. 3. The controller structure.

According to the subsumption architecture, the vehicle controller is decomposed as shown in figure 3, where each of the modules is developed separately. After the individual tests have been finished, the modules would be linked with the inhibitors and suppressors mechanisms, to compose a final system. The modules are developed separately, without thinking of how they will be connected.

The RS program parts are presented in the following.

```
rs_prog vehicle_controller:
[ input : [sIR(L, R, B, S, D1, D2), tick],
  output: [walk(X), turn(Y)],
```

This is the beginning of the program. In this part, the input signals (coming from the environment) and the output signals (going to the environment) are declared. The *sIR* signal corresponds to the infra-red sensor information, and contains the following data: *L* - left sensor; *R* - right sensor; *B* - back sensor; *S* - sum of the frontal sensors; *D1* - shows when the *L* and *R* sensors are identical; *D2* - used when *D1* is zero, to indicate which one of the *L* or *R* sensor is more significant. The *tick* input corresponds to a clock signal. The *walk* signal puts the moving motor to front, back or stopped, while the *turn* signal puts the direction motor to the left, right or straight direction.

```
module escape:
[ input : [ sIR(L, R, B, S, D1, D2) ],
  output: [ cf supressor_order(X, Y, W, Z) ],
  t_signal: [ ], p_signal: [ ], var: [ ],
  initially: [ activate(rules) ],
  on exception: [ ],
  sIR(L, R, B, S, D1, D2) ===> case
    [ L = 1 & R = 0 & B = 0 --->
        [ emit(suppressor_order(back, left, 0, 0)) ],
      L = 0 & R = 1 & B = 0 --->
        [ emit(suppressor_order(back, right, 0, 0))],
      L = 1 & R = 1 & B = 0 --->
        [ emit(suppressor_order(back, off, 0, 0)) ],
      L = 0 & R = 0 & B = 1 --->
        [ emit(suppressor_order(front, off, 0, 0)) ],
      else                  ---> [] ]
],
```

This layer is responsible for avoiding obstacles and has only the input signal *sIR* as its input data. When one of the *L*, *R* or *B* sensors is on (significant signal), the suppressor order is emitted to avoid the obstacle. In this layer, there is a composition of signals, indicated by the *cf* prefix presented in the *output* declaration. This is necessary because the *suppressor_order* signal can be emitted at the same time by the *wander* layer (the compositional functions belong to the data manipulation layer, and is not explained here). The *emit* command is responsible for sending messages to the environment.

```
module wander:
[ input : [ tick ],
  output: [ cf suppressor_order(X, Y, W, Z) ],
  t_signal:[ next1(N) ], p_signal: [ ], var: [ ],
  initially: [ activate(rules) ],
  on exception: [ ],
  tick ===> [ random(N), up(next1(N)) ],
  #[ next1(N) ] ===> case
    [ N <= 25 --->
        [ emit(suppressor_order(0, 0, front, left)) ],
      N <= 50 --->
        [ emit(suppressor_order(0, 0, front, right))],
      else      --->
        [ emit(suppressor_order(0, 0, front, off))] ]
],
```

This layer is responsible for putting the vehicle to wander at random. When a clock signal comes, the *random* function is used to generate an integer number N, between 1 and 100 (this function belongs to the data manipulation layer, and is not explained here), and the temporary signal *next1(N)* is turned on; the signal *next1(N)* is responsible for triggering off the rule that controls the movement and direction to be used by the vehicle. In this layer, there is also a composition of signals specification, because the *suppressor_order* signal can be emitted in parallel by the *escape* layer.

```
module supressor:
[ input : [ suppressor_order(X, Y, W, Z) ],
  output: [ walk(A), turn(B) ],
  t_signal: [ ], p_signal: [ ], var: [ ],
  initially: [ activate(rules) ],
  on exception: [ ],
  suppressor_order(X, Y, W, Z) ===> case
  [X != 0 v Y != 0 ---> [emit(walk(X)), emit(turn(Y))],
    else           ---> [emit(walk(W)), emit(turn(Z))] ]
] ].
```

This layer is responsible for suppressing the *wander* behavior whenever the vehicle is on a collision course, activating the *escape* behavior. In the first case, the *X* or *Y* parameters have a different from zero value, that activates the *escape* behavior; in the second case, it activates the *wander* behavior. Only in this layer, the *walk* and *turn* signals are emitted to the external environment, activating directly the vehicle motors.

This RS program was implemented in a centralized way, assuming its use in a monoprocessor machine. In this case, the programmer was forced to be aware of the composition of the emitted signals, making use of a compositional function to allow the parallelism in the execution and in the emission of these signals. If this example was implemented in a distributed way, each layer would be completely independent. In this case, some external connections, the suppressors, the ones that would determine the correct behavior to the vehicle, ought to be provided. However, even with this centralized programming, it was possible to see the power of the RS language and the subsumption architecture, making the programming task simpler, modular, and more efficient. Besides that, the program code is easy to understand.

It is important to say that this program was not implemented in a real system. Its code was just simulated in a test environment. The RS language generates an automaton and a set of actions that are interpreted in this environment. To run this program in a real system, it must be developed an executor system for the automaton code in this real system.

For future works, various interfaces between the RS output and some robots like *Khepera* (Mondada, *et al.*, 1993) are being built. The goal is to use the RS language to control others vehicles and systems in real applications.

4.3 RS language versus Behavior language

The Behavior language offers a way of grouping augmented finite state machines into more manageable units, with the capability for whole units being selectively activated or deactivated (Maes, 1989; Brooks, 1990). Real time rules are the key to the Behavior language. They can be isolated or grouped into behaviors, where there is no notion of procedure definition - all abstraction must be done in macros, rules or behaviors. In spite of having differences regarding to the RS language, both languages have various similar features, both of them aiming to reach similar goals.

5. FINAL CONSIDERATIONS

A previous study has been made to apply the subsumption architecture concepts to the RS language (Arnold, 1996). This paper presented the use of the RS language to the development of a simple vehicle controller. The intention is to show that the RS language constitutes a simple and efficient tool.

Depending on the kind of the system, the subsumption architecture should be extended. This architecture is intended to systems that have reflexive features, and some systems do not present this feature. Then, it will be necessary to include some behaviors to control the reactive ones, giving more flexibility and coherence to the execution of the behaviors. However, this is not a real problem, because there are various systems that contains a reactive kernel, and a more intelligent module. These systems are called hybrid systems, and to them, the reactive kernel can be implemented using the subsumption architecture.

The RS language can implement all the subsumption architecture concepts, and an RS program can be interfaced to other components to build these hybrid systems. The goal is, in the future, to develop reactive systems, like vehicle controllers for example, using the RS language as a programming tool.

REFERENCES

Arnold, G.V. (1996). *Arquitetura de Subsunção aplicada à Linguagem Reativa Síncrona RS.* TI-604. CPGCC/UFRGS. Porto Alegre. 46p.

Berry, G. and G. Gonthier (1992). The Esterel Synchronous Programming language: Design, Semantics, Implementation. *Science of Computer Programming*, **19(2)**, 87-152.

Brooks, R.A. (1986). A Robust Layered Control System for a Mobile Robot. *IEEE Journal of robotics and Automation.* **RA-2. No. 1**, 14-23.

Brooks, R.A. (1990). The Behavior Language; User's guide. *A.I. MEMO 1227*, MIT, Cambridge, Massachusetts.

Brooks, R.A. (1991). Intelligence Without Reason. *A.I. MEMO 1293*, MIT, Cambridge, Massachusetts.

Bryson, J.J. (1992). *The subsumption Strategy Development of a Music Modelling System.* Departament of Artificial Intelligence, University of Edinburgh.

Hopcroft, J. E. and J. D. Jullman (1979). *Introduction to Automata Theory.* Addison-Wesley, Reading Massachussetts.

Maes, P. (1989). The Dynamics of Action Selection. *AAAI Spring Symposium on AI Limited Rationality, IJCAI*, Detroit, MI, 991-997.

Mondada, F., E. Franzi and P. Ienne (1993). Mobile robot miniaturization: A tool for investigation in control algorithms. *Third International Symposium on Experimental Robotics*, Kyoto, Japan.

Toscani, S.S. (1993). *RS: Uma Linguagem para Programação de Núcleos Reactivos.* Faculdade de Ciências e Tecnologia. Universidade Nova de Lisboa. Portugal.

APPLICATION OF A NATURAL LANGUAGE INTERFACE TO THE TELEOPERATION OF A MOBILE ROBOT

J.M. González Romano*, J. Gómez Ortega** and E.F. Camacho**

**Dpto. Lenguajes y Sistemas Informáticos. Facultad de Informática. Universidad de Sevilla.
Avda. Reina Mercedes s/n. 41012 Sevilla.
**Dpto. Ing. de Sistemas y Automática. Escuela Superior de Ingenieros. Universidad de
Sevilla. Camino de los Descubrimientos s/n. 41092 Sevilla.*

Abstract: This paper describes the application of a natural language interface to the teleoperation of a mobile robot. Natural language communication with robots is a major goal, since it allows for non expert people to communicate with robots in his or her own language. This communication has to be flexible enough to allow the user to control the robot with a minimum knowledge about its details. In order to do this, the user must be able to perform simple operations as well as high level tasks which involve multiple elements of the system. For this ones, an adequate representation of the knowledge about the robot and its environment will allow the creation of a plan of simple actions whose execution will result in the accomplishment of the requested task. *Copyright © 1998 IFAC*

Keywords: Telerobotics, Man-machine interfaces, Natural language.

1. INTRODUCTION

One important goal in the robotic field is the developing of friendly human-robot interfaces that allow inexperienced operators to deal with robots. For instance, a robotized assembly system could be better controlled by an expert in assembly systems instead of an expert in robotics. The design of such an interface is a complex task because a number of different technologies in robot control, computer vision, man-machine communication and learning fields have to be put together.

With respect to the man-machine interface, one interesting approach is the use of natural language interfaces (NLI), which allow the interaction with the robot through commands given in the own robot operator's language. NLI have been successfully used in a wide variety of fields like communication with expert systems, database access systems or operating systems. Their main advantages are that a very short training period is required and that they seem *natural* to the operator.

In the robotics field, several works can be found in the literature where NLI are used for robot control and operation. Two of the first systems, developed in the seventies, at the beginnings of the Artificial Intelligence, were ROBOT (Harris, 1997) and SHRDLU (Winograd, 1972). Both systems worked with a simulated robot. The improvement in computational facilities has lead to the development of

more powerful systems which can interact in real time with real robots. Selfridge (Selfridge and Vannoy, 1986) developed a NLI for a robotized assembly system which allows the operator to hold a *conversation* with the system in order to carry out several manipulation and vision tasks like object recognition and assembly to built more complex objects. SAM (Brown, 1992) combines written and spoken language; the robot has a video camera in order to recognise the objects. Torrance (Torrance, 1994) developed a NLI to a mobile robot through which the user can command the robot to move through an environment and to memorise it.

This paper presents the application of a NLI to the teleoperation of a NOMAD 200 mobile robot. The work is focused on the communication between the robot and the human operator. The operator will be able to issue high level commands to the robot, and the NLI will decompose them in a set of lower level commands which will be executed directly by the robot navigation and control system. The operator does not need any knowledge about this robot-level commands. The information needed for this task is stored in a previously created knowledge database. Section 2 presents the main characteristics of the NLI. Section 3 describes the tasks that can be carried out by the system. Finally, section 4 shows the conclusions.

2. DESCRIPTION OF THE NLI

The NLI is a part of a more complex system aimed at

communicating in natural language with complex interactive systems (González, 1997b). This system uses a NLI to acquire the knowledge of a generic complex system and another NLI to operate it. The system is composed of two different parts: the acquisition module and the operation module. The first one works off-line, while the second one works on-line. The acquisition module allows the user to describe a system by introducing phrases in his own natural language, and stores the acquired knowledge in a knowledge database. By using this knowledge, the operation module allows to operate the system through simple or complex natural language commands. The accomplishment of the latter ones requires a previous planning of the actions to be carried out.

Fig. 1 shows the block diagram of the developed system, from which there exists a running prototype (González, 1991). This prototype connects to the real system through its control system, which receives the commands from the prototype and executes them over the system. It is also possible to connect the prototype to a program which simulates the behaviour of the system when the connection is not possible or for operator training purposes. The prototype has been applied to different complex systems such as an industrial process (González, 1993) or an electric network (González, 1997a, b). This paper describes its application to a mobile robot wich navigates in a structured environment.

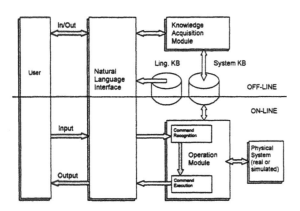

Fig. 1 Block diagram of the developed prototype

2.1 The Knowledge Database

The knowledge database is an important part of the prototype. It consists of two parts: the linguistic knowledge database and the system knowledge database. The first one contains the linguistic knowledge that is necessary to analyze and extract the meaning of the user's phrases, and is composed of a dictionnary and a grammar. The second one contains all the necessary knowledge to describe the system, and can be divided into declarative and procedural knowledge. The first one concerns the different kinds of entities that can be found in the system: **classes**: (different kinds of objects present in the system), **objects** (particular instances of the classes), **connections** (topological relations between objects), **groupings** (groups of objects related to each other by

their topology or their function) and **measurements**: (sensors which allow for some relevant magnitudes of the system to be known).

Procedural knowledge includes a set of functions that represent the means through which the goals of the system can be fulfilled. These functions are related to the different entities found in the system (classes, objects, groupings) or to the whole system. A function has the following components (Fig. 2):

- **Goal** (g): is the goal fulfilled by the function.
- **Prerequisites** (r_i): are conditions that the system must necessary accomplish before applying the means for the fulfillment of the goal of the function. A prerequisite can be a condition on a state or the value of a property.
- **Means** (m_{ij}): are the operations whose execution results in the fulfillment of the goal of the function. They can be simple actions over objects or classes, or other functions. In general there will exist some sequences of means in parallel: some means will execute at the same time, as they are independent, while others will have to do it in sequence, as every means depends on the result of another.
- **Criteria** (c): is the criteria whose accomplishment implies that the goal of the function has been fulfilled. It is a condition over the value of some property, and will always be true provided that the means have been correctly executed.
- **Posterior actions** (p_{ij}): are operations that must be executed once the goal of the function has been fulfilled, in order to leave the system in a specific state. They can be simple actions or functions.

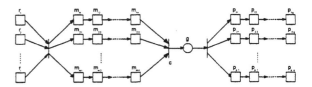

Fig. 2 Representation of a function

The execution of a function may need the previous execution of other functions that act as means of it, and may launch the execution of other functions that act as posterior actions of it. When executing a function, it will be decomposed into its constituents until there are any functions left, thus obtaining a network made of simple actions and conditions that willll be called the **actions network**.

2.2 Knowledge Representation

The elected formalism for the representation of knowledge has been that of frames (Minsky, 1975). This is due to the hierarchical organization of the knowledge, at the declarative (hierarchical structure of clases) and the procedural level (hierachical structure of functions). Thus the knowledge database will be composed of class, object, grouping, measurement and function frames.

3. TELEOPERATION OF THE MOBILE ROBOT

The system to which the prototype has been applied is a Nomad 200 mobile robot (Nomadic, 1997) which navigates in a partially structured environment. It is composed of a base with a turret mounted on it. The base has two driving and one steering wheel, allowing forward and backwards translation movements and left and right turning. The turret is capable of rotating 360° over itself independently of the base. Fig. 3 shows a photograph of the robot. The goal of the application is to teleoperate the robot from a terminal, through which it will be given commands in order to perform certain tasks, such as walk to a named place or walk forward avoiding all the obstacles it can find along its way.

Fig. 3 Photograph of the NOMAD robot

3.1 System description

The robot navigates in a partially structured environment composed of walls, furniture, doors and obstacles. Fig. 4 shows the different classes of objects. The hierarchical structure of the defined functions is shown in Fig. 5. Level 0 corresponds to simple actions over the robot. Functions begin in level 1 and increase its complexity in upper levels. As an example, let it be the level 2 function *walk_watching*. Its goal is to make the robot walk forward avoiding the obstacles it can find along its way. To do this the robot is asked to walk forward. The bumpers are checked during this movement; if a collision is detected, then the robot stops, walks backwards a little bit (*unwalk 100*), walks around the obstacle (*surround_obstacle*), and follows its way (*walk watching*). The frame for this function is shown in Table 1.

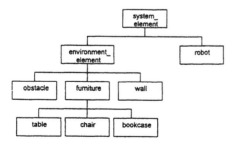

Fig. 4 Class structure of the robotic system

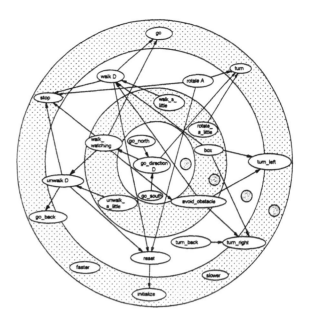

Fig. 5 Structure of the functions of the robotic system

Table 1 Frame for the function walk_watching

NAME	walk_watching
DESCRIPTION	walk forward avoiding obstacles
ASOC_TYPE	class
ASOC_NAME	robot
PRE	()
MEANS	(go)
CRITERIA	bumper = 1
POST	(stop, unwalk 100, surround_obstacle, walk watching)

3.2 Robot operation

The Nomad 200 robot can be operated in two ways. The first one consists in executing the programs which control the robot in the robot itself, as it has its own CPU. Programs are transferred to the robot through the network, and are executed once they are inside it. The second one consists of using the control system located in another machine which is connected to the robot through an ethernet radio link. This second way is more desirable, as it allows to have a friendly development system and a graphical simulation environment which allows to test the programs before executing them on the real robot.

Since the control software of the robot is located in a different machine than the prototype is, it has to be settled a method to communicate both machines in order to send the commands to the robot and to receive its state. For this purpose the sockets have been used. A socket is a way of transferring information between processes which execute under UNIX operating system. Processes can execute in the same machine or in different machines connected through a network, as in this case, which is illustrated in Fig. 6. Communication with sockets is based on the client-server model. There are two server processes which run in the machine where the robot control software is located, and two client processes which are launched by the prototype to send and receive information from the robot. Fig. 7 shows the communications schema.

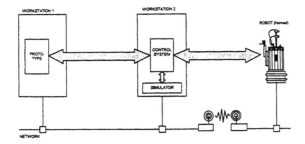

Fig. 6 Connection of the prototype to the Nomad robot

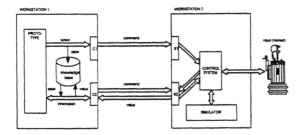

Fig. 7 Communication between prototype and robot

The two possible communication types are:

a) Execution of a command over the robot: the prototype launches a client C1 which connects to the server S1 and sends the command, which is translated and in turn sent to the robot for its execution. The client updates the system knowledge database and ends its execution.

b) Request for information about the robot: the prototype launches a client C2 which connects to the server S2 and receives the actual state of the robot, with which it updates the system knowledge database. Then it shows the user the requested information and ends it execution.

Simple commands. Simple commands are those which apply directly to an element or group of elements, and can be divided into two cathegories: action execution commands and information request commands. The first one includes commands with which the user requests the execution of a specific action over an object or set of objects. The following are examples of this kind of commands:

```
> go the nomad robot.
> accelerate.
> stop nomad.
> turn right.
```

Action execution commands imply the performing of a specific action over a specific object; the recognition process of these commands consists of identifying the action which has to be done and the object to which it has to be applied. Once this has been done, it should be checked whether the action can be applied to the object, and if the actual state of the object allows for the application of that action. If everything is correct, the action will be executed, and the object will be set to its new state.

Information request commands are those commands with which the user aks for a specific information about an object or a set of objects. For instance,

```
> show the position of the nomad robot.
> show its velocity.
```

The requested information corresponds to the state or property value of an object or grouping. The recognition process of these commands consists of identifying the object or grouping whose state or property is to be known and, in this latter case, the corresponding property, which must be a valid property and must have an associated measurement.

Complex commands: actions network. Complex commands are those that imply the execution of a plan of actions in order to fulfill a specific goal. The plan of actions is a sequence of simple actions over some objects in a predetermined order, and is generated from the knowledge stored in the system knowledge database. Given a goal, there will exist in the knowledge database at least a function which will have this goal as its goal. The frame for this function will be the starting point to generate the plan of actions which allows to fulfill the command. Thus, the different means, prerequisites and posterior actions of this frame will be analysed one by one. Each means m_{ij} and posterior action p_{ij} from this function can be a simple or complex command. Every prerequisite r_i can be a complex command or a condition over a state or the value of a property. Simple and complex commands, as well as conditions, have a structure that is represented by a Petri Net (Silva, 1985). Each structure has at least an input place and an output place. The input place will be marked when the net is activated, whereas the marking of the output place will mean the ending of its traversal.

A simple command is represented by two places and one transition (Fig. 8). The input place represents the action to be executed over the object to which the command is applied, and will be marked when the command starts its execution. The transition represents the state to which the object is to be taken, and will be fired when the action over it has been done. At this moment the output place representing the fulfillment of the simple command will be marked.

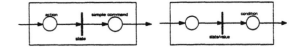

Fig. 8 Structure of a simple command and a condition

For complex commands there will exist in the knowledge database another function whose goal corresponds to that command, and that will have its own means, prerequisites, criteria and posterior actions. The network representing this function will be a subnetwork of the one corresponding to the main function, and should in turn be expanded, should there have another complex command between its means, prerequisites and posterior actions. Its structure is shown in Fig. 9. There is an input place which will be marked when the execution of the complex command begins, and this mark will propagate automatically to all the prerequisites. The output place of the net is the goal place. When the marks reaches this place the goal will be fulfilled, in spite of the end of the posterior actions p_{ij}.

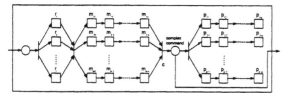

Fig. 9 Structure of a complex command

Finally, a condition type prerequisite is represented by two places and a transition, as it is shown in Fig. 8. The input place will be marked when the evaluation of the prerequisite starts. The transition represents the desired state or property value, and will be fired when it has the adequate value, resulting in the marking of the output place.

If every means, posterior action and prerequisite is successively divided until there only are simple actions and conditions, a network, the **actions network**, will be obtained. Places in this network represent direct actions over objects and transitions represent conditions over the state of the objects or the value of their properties. The initial marking corresponds to the first simple actions to be executed and the first conditions to be checked. The marks will propagate as the transitions are fired, as a result of the accomplishment of the conditions, until the mark reaches the goal place. Fig. 10 shows the generic structure of an actions network.

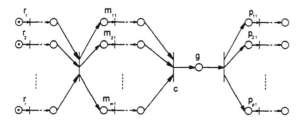

Fig. 10 Structure of an actions network

The goal will be fulfilled when the place g is marked. In order for this to happen, condition c should be true when all of its input places are marked. This will happen once the corresponding means m_{ij} have been accomplished. The initial marking of these means will in turn depend on the marking of the ending places of the prerequisites r_i. On the other hand, once the goal g has been fulfilled its output transition will be fired marking the input places of the posterior actions p_{ij}. The initial marking of the network will correspond to the initial places of the prerequisites r_i, and will propagate towards the goal as transitions are fired.

To summarise, in order for a goal to be fulfilled there have to be accomplished, in the first place, all the requisites. These are, thus, necessary but not sufficient conditions for the fulfillment of the goal. Once the prerequisites are accomplished, the accomplishment of every means results in the fulfillment of the goal, provided that the criteria is true. The accomplishment of the means plus the criteria is, therefore, the necessary and sufficient condition for the main goal to be fulfilled. Finally, once the goal place has been reached, the posterior actions will be executed.

3.3 Operation examples

As an example of simple commands a sequence of actions over the robot is shown.

```
> where is nomad?
     NOMAD IS CURRENTLY AT LABORATORY 1
> go nomad.
     ROBOT NOMAD GOING
> turnright.
     ROBOT NOMAD TURNING
> stop.
     ROBOT NOMAD STOPPED
> turnleft.
     ROBOT NOMAD TURNING
> go.
     ROBOT NOMAD GOING
> turnright.
     ROBOT NOMAD TURNING
> stop.
     ROBOT NOMAD STOPPED
> where is nomad?
     NOMAD IS CURRENTLY AT LABORATORY 2
> what is its velocity?
     THE VALUE OF NOMAD'S VELOCITY IS 0
```

As an example of the execution of complex commands, it is shown the execution of the goal *walk_watching*, which consists in making the robot walk forward avoiding possible obstacles. Fig. 11 shows the graphical representation of the frame representing this function, which appeared in Table 1. The function frames that are necessary to execute this command are those of the function *walk_watching* itself, the functions *unwalk* and *surround_obstacle*, which appear as its posterior actions, the function *walk*, which is a means of *surround_obstacle*, and the function *reset*, which is a means of both *walk* and *unwalk* functions. Fig. 12 shows the relation between all this frames. Fig. 13 shows the simulation environment window during the execution of the complex command, and Fig. 14 represents the generated actions network.

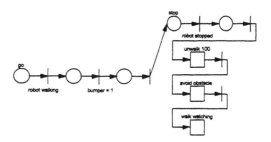

Fig. 11 Structure of the function *walk_watching*

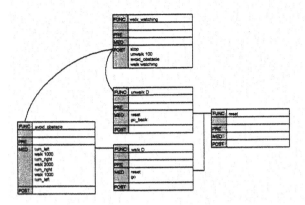

Fig. 12 Frame structure for function *walk_watching*

4. CONCLUSIONS

This paper has shown the application of a natural language interface to the teleoperation of a mobile robot. The adequate description of the system knowledge allows the user to perform high level operations, which are decomposed in other lower-level actions until there is a sequence of simple actions which are executed over the robot, thus having a telescopic vision of the system. The developed prototype could be applied to simulate the execution of goals. In order to do this, it suffices to simulate the behaviour of the system through software and build the plan of actions corresponding to the goal. If the plan is successfully built, the goal can be achieved and could be executed over the real system. In other case, a different plan should be created. This way it is avoided to start executing actions and reach a point in which no more actions can be executed due to a prerequisite not accomplished or an action that cannot be performed, resulting in a half-executed plan and a goal not achieved. This incomplete execution is a problem as it could prevent the goal from being fulfilled with another plan. To summarise, it has been developed a tool which can simplify the operation of complex interactive systems, as the mobile robot to which it has been applied in this paper.

5. REFERENCES

Brown, M.K, B.M. Buntschuh and J.G. Wilpon (1992). SAM: A Perceptive Spoken Language Understanding Robot, *IEEE Transactions on Systems, Man and Cybernetics*, vol. **22**, no. 6, pp. 1390-1402.

Harris, L.R (1977) A High Performance Natural Language Processor for Data Base Query, *ACM SIGART Newsletter*, vol. **61**.

Minsky, M (1975). A Framework for Representing Knowledge, The Psychology of Computer Vision, P.H. Winston, Ed, McGraw-Hill, pp. 211-277.

Nomadic Technologies Inc (1997). NOMAD Language Reference Manual.

González Romano, J.M, J.A. Ternero and E.F. Camacho (1991). Natural Language Interface for Process Control Centers. *Preprints 3rd IFAC International Workshop on Artificial Intelligence in Real Time Control*, Napa (California).

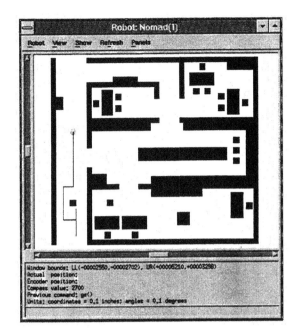

Fig. 13 Execution of the goal *walk_watching*

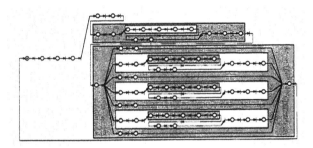

Fig. 14 Actions network for the goal *walk_watching*

González Romano, J.M. and E.F. Camacho (1993). Goal-Oriented Man Machine Interface in Control. Application to a Pilot Plant. *Preprints of the 12th IFAC World Congress*, pp. 455-458, Sydney (Australia).

González Romano, J.M. and E.F. Camacho (1997a). Utilización de un interfaz en lenguaje natural para la realización de operaciones en centros de control de redes eléctricas, *VII Conferencia de la AEPIA (CAEPIA'97)*, Málaga (Spain).

González Romano, J.M (1997b). Aplicación del lenguaje natural a la adquisición de conocimientos y operación de sistemas complejos, Doctoral dissertation, Univ. Sevilla.

Selfridge, M. and W. Vannoy (1986). A Natural Language Interface to a Robot Assembly System, *IEEE Journal of Robotics and Automation*, vol. RA-2, no. 3, pp. 167-171.

Silva, M. (1985). *Las Redes de Petri en la Automática y la Informática*, AC.

Torrance, M.C. (1994). Natural Communication with Robots (doctoral dissertation), Massachusetts Institute of Technology.

Winograd, T. (1972). *Understanding Natural Language*, Academic Press.

A FRAMEWORK FOR INTEGRATING THE SOFTWARE COMPONENTS OF A ROBOTIC VEHICLE*

Juan-Antonio Fernandez, Javier Gonzalez

Departamento de Ingenieria de Sistemas y Automatica (ISA)
E.T.S.I. Informatica, Universidad de Malaga
Campus Teatinos - 29080   Malaga (Spain)
http://www.isa.uma.es
{jafma,jgonzalez}@ctima.uma.es

Abstract: In this paper we present NEXUS, an extension of an operating system designed for integrating the software elements of a mobile robot. It facilitates the integration of programs developed by different people, reducing the cost and complexity of the whole application. It is based on a subscription/production design, and on modular and object-oriented programming techniques. This makes the software implemented on it less sensitive to changes than typical monolithic architectures. NEXUS provides the facilities required in most robot vehicles software architectures, such as inter-process communications, error recovery system, graphical user interfaces, etc. Other important features are its *distributed nature*, its *hierarchical error recovery* system, and the *real-time capabilities* that it inherits from the underlying operating system. NEXUS has been developed for mobile robots, but its design has been made generic enough for programming other robotic system, such as manipulators, teleoperation systems, etc.

Firstly, we comment some reasons that justify the development of this type of framework, then we describe the internal structure and the most important features of NEXUS, and finally we outline some implementation issues and how a set of modules designed for our mobile robot RAM-2 have been integrated using NEXUS. *Copyright © 1998 IFAC*

Keywords: Robotics, Architectures, Distributed Computer Control Systems, Interconnected Systems, Robustness.

* This work has been supported by the Spanish Government under the research project CICYT-TAP96-0763.

1. INTRODUCTION

At the beginning of the research in mobile robotics, many issues still unsolved (intelligent task execution, environment representation, human-robot interfaces, special hardware devices, etc) were tackled through not very flexible solutions, such as hardware implementation of algorithms (Brooks, 1986), or constrained experiments in which software was highly coupled with concrete hardware platforms or devices (Kurz, 1993; Mitchell, 1990). With the growth in the

performance of computers, some of these solutions were improved using the greater flexibility of software, which reduces the cost of development. But actually the design and implementation of modular and reusable software for mobile robots has been an issue often left in the background.

We think that this problem is an important issue in itself due to, the complexity that the current robotic vehicles exhibit, the number of different people that is involved, the unexpected modifications often needed, and the important amount of the work that is nearly the same for different physical platforms.

In order to reduce the influence of these facts in the time and cost of a project, our research has focused on the development of a modular and flexible framework for integrating the software components of a generic robotic vehicle. We have identified some features that such a framework must fulfill:

- It must encourage the use of structured and modular programming techniques in order to maintain a high level of clearness and reusability in the whole design.

- The inclusion of new hardware and software must be a process that does not affect the structure and maintainability of the application.

- It should allow for the optimal exploitation of all the computer boards available for the robotic vehicle (it must manage the robot network).

- It must provide the main features that a robotic vehicle may need (real-time response, error recovery, multiprocessing, robustness, etc.) and a simple and well-defined interface to integrate the software elements.

- Although a highly structured framework may perform slightly worse than one which is more coupled with the low-level software and hardware elements, it is desirable to achieve a reasonable efficiency.

These desirable capabilities have been kept in mind while we have been developing the software framework called NEXUS, which is basically an extension of a real-time operating system. It is a distributed, general and flexible software package that through the use of object-oriented and subscription/production techniques decouples the programs which execute a certain task of a mobile robot[1] from the hardware and software that does not depend on that task (operating system, network, devices). Therefore it reduces the cost of the modifications and the complexity of the whole project, allowing highly structured and reusable designs to be achieved.

NEXUS consists of two main elements: *managers* and *modules*. The *managers* are in charge of dealing with the different requests that the *modules* can send. The *managers* are the fixed part of NEXUS, and do not depend on the particular software that is to be implemented for the robot. On the contrary, the *modules* form the part designed for executing a concrete task and therefore they may vary from one

application to another. The well-defined interface that connects them to NEXUS allows the substitution of any module by other that offers the same functionality without affecting the rest of the architecture. Also, in robotic platforms with several computer boards, the modules can be distributed among them without making changes in other parts of the software. A database resident in every computer and maintained by one of the managers holds all the information needed about each module.

Other approaches to the problem of designing a general framework for integrating robot software can be found in literature. Among them are the TCA (Task Control Architecture) and CODGER, developed at the Carnegie Mellon Robotics Institute (Simmons, *et al.,* 1990; Stentz, 1990), and the complete operating system called CHIMERA (Stewart, *et al.,* 1992). NEXUS is intended to improve the modularity and portability aspects of these and other existing designs.

In the next section the components of NEXUS are described. Section 3 comments the main features that NEXUS provides in order to achieve reusability, flexibility, robustness and efficiency. Section 4 gives some implementation details. Finally some results and future works are outlined.

2. STRUCTURE OF THE SYSTEM

As shown in figure 1, NEXUS is divided into two subsystems: the **Task-Dependent Subsystem** and the **Management Subsystem**. A more detailed description of these components can be found in (Fernandez and Gonzalez, 1998a; Fernandez and Gonzalez, 1998b).

2.1. The Task-dependent Subsystem

It contains the *modules* implemented for a particular application (task). For clearness in the design, they are grouped into sets of modules called *Conceptual Units*.

The task-dependent subsystem can also be seen as a number of *services* suitable for performing different operations[2]. The *services* are the components which define the functionality of the system. Every module offers some services, and it can request any service from another module.

2.2. The Management Subsystem

This contains the *managers* needed for maintaining all the information about the modules and for supporting the communications. They are the fixed part of

---

[1] Although this paper deals mostly with the utility of NEXUS for mobile robots, its design is generic enough for being used in any robotic system.

[2] For example, acquiring distance measurements through a laser range-finder, or controlling manipulators, or decomposing a given user command into simpler actions.

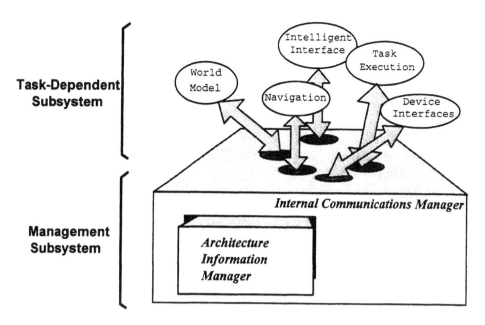

Fig. 1: Components of NEXUS. The Conceptual Units represented in the
task-dependent subsystem are an example (this subsystem depends
on the architecture implemented).

NEXUS. Currently NEXUS includes two managers: the IC manager and the AI manager. All the comunications between different components of the software are managed by the *Internal Communications Manager* (ICM). The structural information about modules and services is stored and managed by the *Architecture Information Manager* (AIM). In every machine of the robot system there is an ICM and an AIM.

The ICM supports the information flow between the different components of the robotic system, sending the communication packages from the source computer to the destination, and, once a package is in the destination machine, sending it to the right module. From the point of view of an application software running on NEXUS the network of computers of the mobile robot appears as a single machine.

The AIM maintains all the information about the distributed architecture in a common database duplicated on every machine. The inconsistency troubles that might exist due to this duplicated database are solved reporting to every AIM the modifications made in each other[3].

3. REDUCING COSTS: INCREASING FLEXIBILITY

The main objective of NEXUS is to reduce the cost of developing and maintaining complex architectures by increasing the flexibility of the software on which these applications are integrated. This is achieved in three important areas: modularity, robustness and efficiency, that are commented in detail in the following paragraphs.

3.1. Modularity

NEXUS has some modular and object-oriented programming characteristics to allow the implementation of applications that are not very sensitive to changes. The most important is the principle of *encapsulation:* the algorithms of a particular application are grouped into modules that hide their implementation details. Only the *services* that the modules provide are public.

Inspired by high-level languages such as ADA, NEXUS incorporates *hierarchical error recovery:* every algorithm that encounters an error must overcome it or propagate it to the higher algorithms in the execution tree. There is no global error recovery agent which knows about all the possible errors that can occur, since it would lead to a lack of modularity.

Modularity is also achieved through some client-server features of NEXUS. On one hand, the managers of NEXUS act like servers when they provide information about the modules/services registered in the database and manage the communications through the network of computers. On the other hand, NEXUS is also based on a client-server paradigm from another perspective: the

---

[3]Although this process can certainly be slow (since each modification is sent through the network to all the computers at the same time), it is limited to occuring only during the module connection phase and consequently it does not affect the performance of the system during its normal operation.

modules can make requests to other modules running on different computers. So every module becomes a *server* for the rest of modules. Although this may be seen as a pure client-server scheme, the existence of an AIM (with all the information that it holds) in every computer makes it more similar to a subscription/production one: the information exists in the network, not in a particular machine. The modules register their services and the transmission of this information to the rest of computers becomes a sort of subscription to these services. The difference with respect a more flexible subscription/production system is that all the machines subscribe to all the services.

3.2. Robustness

There are some special errors called *critical errors* caused by unexpected failures in a process or a machine that prevent the hierarchical error recovery feature from working. For example, when a module[4] is killed, all the routines which are providing services in that module are killed. NEXUS is robust enough for detecting most of the critical errors and for communicating their occurrence to the other parts of the system. Among other things, NEXUS detects and tolerates module killing, out of time service replies, and wrong communications formats. Due to its distributed nature, it also can continue executing despite a machine shutting down. Other critical errors which might lead NEXUS to an unstable status (for example, when a module uses most of the CPU time because of a corrupted code or a wrong algorithm) are handled by the multi-priority and preemptability features that NEXUS inherits from the underlying real-time OS. They limit the system resources that each routine can consume.

3.3. Efficiency

Since NEXUS is implemented as an extension of a real-time operating system, it inherits the real-time characteristics of the OS providing real-time response to the modules. This includes accurate timings, multiple priority levels, and preemptability. In order to maintain these features, we have constrained the implementation of further versions of NEXUS to extend real-time operating systems only.

The managers of NEXUS only execute when a request is made or a message or event is sent from a module, therefore their influence in the computational efficiency of the application is kept little.

4. IMPLEMENTATION ISSUES

NEXUS is currently implemented over LynxOS (Lynx

Fig. 2. The RAM-2 mobile robot equipped with a manipulator.

Real-Time Systems, 1993) and Windows NT. Both of them support real-time processing, concurrent processing, and all the intercommunication facilities of UNIX-like operating systems. NEXUS uses the system resources through several interfaces, so it can be adapted to run on any operating system that provides the same resources, by just reprogramming these interfaces.

NEXUS provides two libraries for linking with the code of the modules. One of them contains all the routines for accessing NEXUS in order to request services, send replies to service requests, etc. The other is an optional interface with the graphical environment OSF-Motif that allows the modules to implement GUIs (Graphical User Interfaces).

The process of implementing a module just consists of defining the services that it will provide and programming the routines that serve requests for these services. The program that contains both the services list and routines shall call a function existing in the first NEXUS library that connects the module and registers the services in the system database automatically.

The design requirements that a routine for providing a service must satisfy to be included in a module are:

1) *Error propagation to the requesters.*
2) *Interfacing with NEXUS and the underlying OS by using the NEXUS library.*
3) *Semaphore protection in the reeentrant portions of the code.*

Notice that the simplified process of designing and connecting modules facilitates the integration of software programmed by different people while offering a simple interface to the common resources needed in any mobile robot.

---

[4] Every module is implemented as a single process which consist of many subprocesses (tasks or *threads*).

5. RESULTS AND FUTURE WORKS

In order to measure broadly the efficiency of our framework to integrate modules implemented by a team of programmers, we have designed an experiment in which our mobile robot RAM-2 (shown in figure 2) navigates in a structured environment using NEXUS (Fernandez and Gonzalez, 1998a). The software attached to NEXUS consist of 10 modules developed by different people. Some of them were available as part of previous applications and others have been implemented for this test. The process of adapting the code of the former to NEXUS has been short: only a few hours for each module. A greater amount of time has been spent in debugging some errors not previously detected that have been discovered due to the error tolerance of NEXUS. The robustness of NEXUS has been demonstrated in these situations, where critical errors made some modules to fall down while the rest of the system was able to continue a limited but correct operation.

Currently we are implementing a more complex software for our mobile robot RAM-2, in order to perform not only navigation, but also manipulation and other high-level tasks.

REFERENCES

Brooks, R.A. (1986). A Robust Layered Control System for a Mobile Robot. *IEEE Journal of Robotics and Automation.* **Vol. RA-2, n° 1**, pp 14-23.

Fernandez, J-A. and Gonzalez, J. (1998a). A Flexible Software System for Implementing Robots Control Architectures. *To appear in the 3rd IFAC Symposium on Intelligent Autonomous Vehicles, Madrid Spain, March 1998.*

Fernandez, J-A. and Gonzalez, J. (1998b). Communicating and Integrating the Modules of a Robotic Software Application. *To appear in the 4th International Symposium on Distributed Autonomous Robotic Systems, Karlsruhe Germany, May 1998.*

Kurz, A. (1993). Building Maps Based on a Learned Classification of Ultrasonic Range Data. *1st IFAC International Workshop on Intelligent Autonomous Vehicles (UK), April 1993.*

Lynx Real-Time Systems, Inc. (1993). LynxOS Application Writer's Guide.

Mitchell, T.M. (1990). Becoming Increasingly Reactive. *Proceedings of the 1990 AAAI Conference, Boston MA, August 1990.*

Simmons, R., Lin, L-J. and Fedor, C. (1990). Autonomous Task Control for Mobile Robots (TCA). *5th IEEE International Symposium on Intelligent Control, Philadelphia PA, Sept 1990.*

Stentz, A. (1990). The CODGER System for Mobile Robot Navigation. In: *Vision and Navigation. The Carnegie Mellon NavLab.* (Charles E. Thorpe Ed.). Kluwer Academic Publishers, 1990.

Stewart, D.B., Schmitz, D.E. and Khosla, P.K. (1992). The Chimera II Real-Time Operating System for Advanced Sensor-Based Robotic Applications. *IEEE Transactions on Systems, Man, and Cybernetics.* **Vol. 22, No. 6**, pp. 1282-1295.

AUTHOR INDEX

426